# STUDENT'S SOLUTIONS MANUAL

*Part II*

to accompany

# Calculus

Elgin H. Johnston ■ Jerold Mathews

Boston San Francisco New York
London Toronto Sydney Tokyo Singapore Madrid
Mexico City Munich Paris Cape Town Hong Kong Montreal

Reproduced by Addison-Wesley from camera-ready copy supplied by Laurel Technical Services.



Printed in the United States of America.

ISBN 0-321-09312-7

3 4 5 6 7 8 9 10 VG 04 03 02 01

# CONTENTS

## Note to the Student

Each chapter of *Calculus, First Edition* by Elgin H. Johnston and Jerold Mathews contains an exercise set for each section and a set of chapter review exercises. The exercise sets provide symbolic exercises as well as applied problems. This manual provides a solution to every odd-numbered exercise from Chapters R, 8–12. The solutions will show detailed calculus steps but will omit algebraic steps that you should have previously mastered. They will contain graphics where appropriate. When an exercise has more than one answer, the solution will indicate that the answers will vary and provide a sample answer. When an exercise asks you to explain your reasoning, the solution gives a brief sketch of what you should provide but your answer should be more complete. The solutions assume that you know how to use a Computer Algebra System (CAS) or else you have someone available to help you. We hope you find this manual to be a useful resource.

## Chapter R Review

### Section R.1 Vectors

**1.** $\overrightarrow{PQ}$ and $\overrightarrow{ST}$ are equivalent because $4-1=10-7=3$ and $-1-1=3-5=-2$;
$\mathbf{r}=\langle 3, -2\rangle$; $\|\mathbf{r}\|=\sqrt{3^2+(-2)^2}=\sqrt{13}$

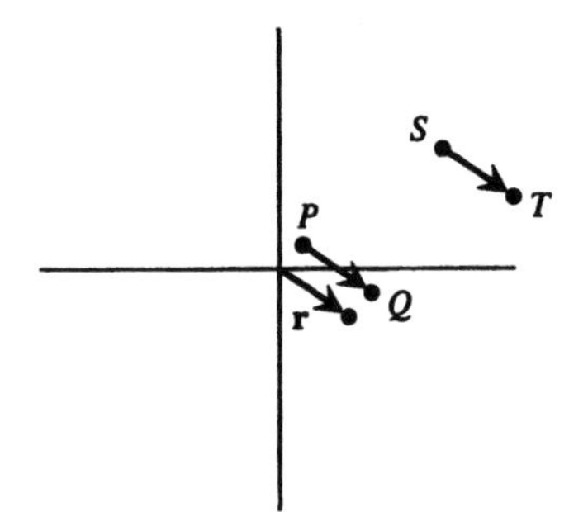


**3.** $\overrightarrow{PQ}$ and $\overrightarrow{ST}$ are equivalent because $5-3=-4-(-6)=2$ and
$-2-\left(-\frac{2}{3}\right)=4-\frac{16}{3}=-\frac{4}{3}$; $\mathbf{r}=\left\langle 2, -\frac{4}{3}\right\rangle$;
$\|\mathbf{r}\|=\sqrt{2^2+\left(-\frac{4}{3}\right)^2}=\sqrt{\frac{52}{9}}=\frac{2}{3}\sqrt{13}$

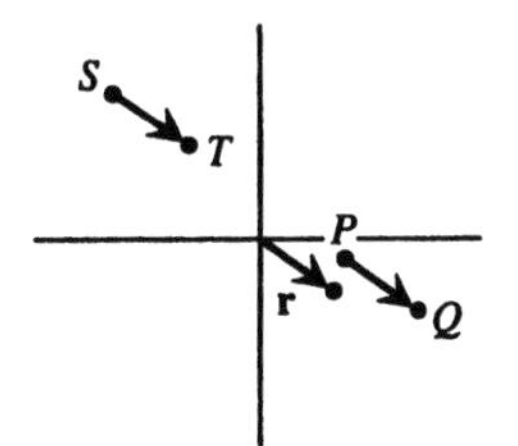


**5.** $\overrightarrow{PQ}$ and $\overrightarrow{ST}$ are equivalent because $(-0.8)-(-2.5)=1.7-0.0=1.7$ and
$1.1-3.0=-3.2-(-1.3)=-1.9$; $\mathbf{r}=\langle 1.7, -1.9\rangle$;
$\|\mathbf{r}\|=\sqrt{1.7^2+(-1.9)^2}=\sqrt{6.5}$

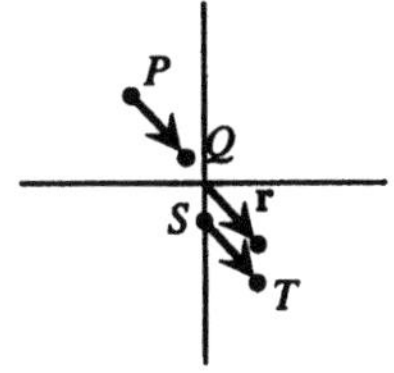


**7.** $\mathbf{a}+\mathbf{b}=\langle 2+5, 5+2\rangle=\langle 7, 7\rangle$

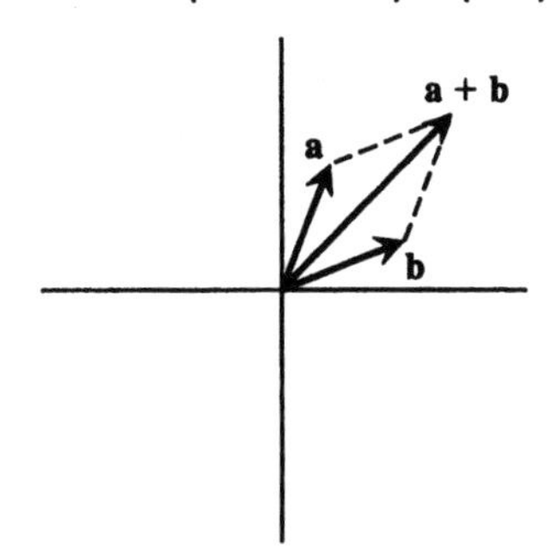


**9.** $\mathbf{a}+\mathbf{b}=\langle -3+1, 3+4\rangle=\langle -2, 7\rangle$

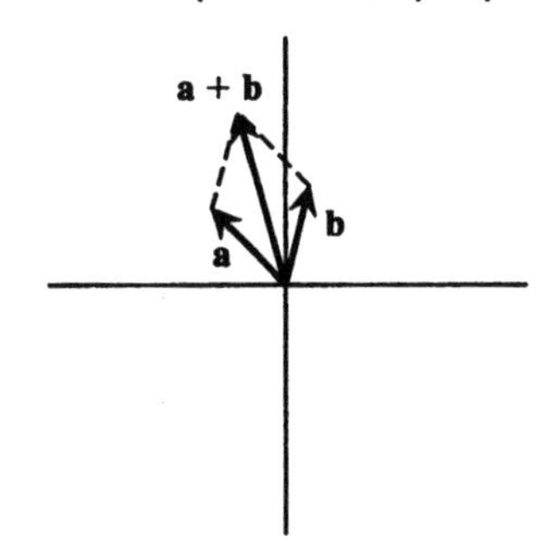


**11.** $\overrightarrow{OR}=\overrightarrow{OP}+\overrightarrow{OQ}$
$=\langle 4-1, 2-1\rangle+\langle 2-1, 7-1\rangle$
$=\langle 3, 1\rangle+\langle 1, 6\rangle$
$=\langle 4, 7\rangle$
$R=(1+4, 1+7)=(5, 8)$

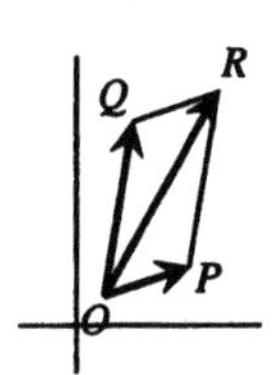


**13.** $\overrightarrow{OR}=\overrightarrow{OP}+\overrightarrow{OQ}$
$=\langle -0.8-(-2.5), 1.1-3.0\rangle$
$+\langle 0.0-(-2.5), 3.3-3.0\rangle$
$=\langle 1.7, -1.9\rangle+\langle 2.5, 0.3\rangle$
$=\langle 4.2, -1.6\rangle$
$R=(-2.5+4.2, 3.0-1.6)=(1.7, 1.4)$

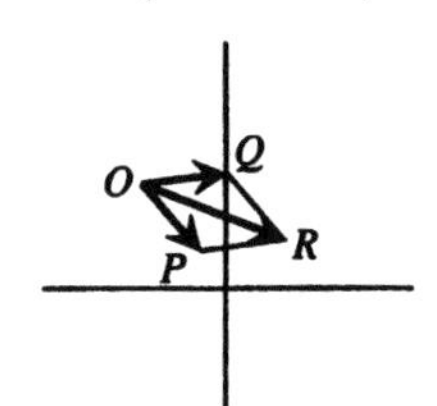

\

**15.** $\|\mathbf{a}\| = \sqrt{5^2 + (-12)^2} = 13$

$\mathbf{a}+\mathbf{b} = \langle 5+1, -12+4\rangle = \langle 6, -8\rangle$

$\mathbf{a}-\mathbf{b} = \langle 5-1, -12-4\rangle = \langle 4, -16\rangle$

$h\mathbf{a}+k\mathbf{b} = 1\cdot\langle 5, -12\rangle + 2\cdot\langle 1, 4\rangle = \langle 7, -4\rangle$

**17.** $\|\mathbf{a}\| = \sqrt{(-4)^2 + (-5)^2} = \sqrt{41}$

$\mathbf{a}+\mathbf{b} = (-4+3)\mathbf{i} + (-5+(-2))\mathbf{j}$
$= -\mathbf{i} - 7\mathbf{j}$

$\mathbf{a}-\mathbf{b} = (-4-3)\mathbf{i} + (-5-(-2))\mathbf{j}$
$= -7\mathbf{i} - 3\mathbf{j}$

$h\mathbf{a}+k\mathbf{b} = -2\cdot(-4\mathbf{i} - 5\mathbf{j}) + 1\cdot(3\mathbf{i} - 2\mathbf{j})$
$= 11\mathbf{i} + 8\mathbf{j}$

**19.** $\|\mathbf{a}\| = \sqrt{1^2 + 6^2} = \sqrt{37}$

$\mathbf{a}+\mathbf{b} = \langle 1+1, 6+(-4)\rangle = \langle 2, 2\rangle$

$\mathbf{a}-\mathbf{b} = \langle 1-1, 6-(-4)\rangle = \langle 0, 10\rangle$

$h\mathbf{a}+k\mathbf{b} = 0.5\cdot\langle 1, 6\rangle + (-0.5)\cdot\langle 1, -4\rangle$
$= \langle 0, 5\rangle$

**21.** $\|\mathbf{a}\| = \sqrt{1.5^2 + (-2.4)^2} \approx 2.83019$

$\mathbf{a}+\mathbf{b} = (1.5+0.3)\mathbf{i} + (-2.4+2.2)\mathbf{j} = 18\mathbf{i} - 0.2\mathbf{j}$

$\mathbf{a}-\mathbf{b} = (1.5-0.3)\mathbf{i} + (-2.4-2.2)\mathbf{j} = 1.2\mathbf{i} - 4.6\mathbf{j}$

$h\mathbf{a}+k\mathbf{b} = 1.6\cdot(1.5\mathbf{i} - 2.4\mathbf{j}) + (-3.4)\cdot(0.3\mathbf{i} + 2.2\mathbf{j})$
$= 1.38\mathbf{i} - 11.32\mathbf{j}$

**23.** $\mathbf{u} = \left\langle \cos\frac{\pi}{6}, \sin\frac{\pi}{6}\right\rangle = \left\langle \frac{\sqrt{3}}{2}, \frac{1}{2}\right\rangle$

**25.** $\mathbf{u} = \langle \cos 210°, \sin 210°\rangle = \left\langle -\frac{\sqrt{3}}{2}, -\frac{1}{2}\right\rangle$

**27.** $\mathbf{u} = \langle \cos 5, \sin 5\rangle \approx \langle 0.283662, -0.958924\rangle$

**29.** Let $\mathbf{F_1}$ be the force on the left; $\mathbf{F_2}$ the force on the right. The sum of their vertical components must equal 44.1 N.

$\|\mathbf{F_1}\|\cos 75° + \|\mathbf{F_2}\|\cos 70° = 44.1$ N.

The sum of their horizontal components must be zero.

$-\|\mathbf{F_1}\|\sin 75° + \|\mathbf{F_2}\|\sin 70° = 0$ or

$\|\mathbf{F_2}\| = \frac{\|\mathbf{F_1}\|\sin 75°}{\sin 70°}$.

Combine the equations and solve for $\|\mathbf{F_1}\|$.

$\|\mathbf{F_1}\|\left(\cos 75° + \frac{\sin 75°}{\sin 70°}\cos 70°\right) = 44.1$ N

$\|\mathbf{F_1}\| \approx 72.2492$ N

Then $\|\mathbf{F_2}\| \approx 74.2662$ N.

**31.** Let $\mathbf{F_1}$ be the force on the left; $\mathbf{F_2}$ the force on the right. The sum of their vertical components must equal 200 N.

$\|\mathbf{F_1}\|\cos 19° + \|\mathbf{F_2}\|\cos 62° = 200$ N.

The sum of their horizontal components must be zero.

$-\|\mathbf{F_1}\|\sin 19° + \|\mathbf{F_2}\|\sin 62° = 0$

$\|\mathbf{F_2}\| = \frac{\|\mathbf{F_1}\|\sin 19°}{\sin 62°}$.

Combine the equations and solve for $\|\mathbf{F_1}\|$.

$\|\mathbf{F_1}\|\left(\cos 19° + \frac{\sin 19°}{\sin 62°}\cos 62°\right) = 200$

$\|\mathbf{F_1}\| \approx 178.7907$ N

Then $\|\mathbf{F_2}\| \approx 65.9253$ N.

**33.** Let $\mathbf{F}$ be the desired vector:

$\mathbf{F} = 8\cdot\frac{\langle 3, 5\rangle}{\|\langle 3, 5\rangle\|} = \left(\frac{8}{\sqrt{34}}\right)\langle 3, 5\rangle.$

**35.** If the direction of $\overrightarrow{AB}$ is $\theta$, than

$\tan\theta = \frac{6-4}{2-1} = 2$. Let

$D = (x, y)$, $\overrightarrow{CD} = \langle x+1, y-3\rangle$. Since the direction of $\overrightarrow{CD}$ is $\theta + 180°$ and $\tan(\theta + 180°) = \tan\theta$,

$\frac{y-3}{x+1} = 2$ or $y - 3 = 2(x+1)$.

$\left\|\overrightarrow{CD}\right\| = \sqrt{(x+1)^2 + (y-3)^2}$
$= \sqrt{(x+1)^2 + 4(x+1)^2}$
$= \sqrt{5}|x+1|$

But $\left\|\overrightarrow{CD}\right\| = 2\left\|\overrightarrow{AB}\right\| = 2\sqrt{5}$, so $\sqrt{5}|x+1| = 2\sqrt{5}$,

$|x+1| = 2$, i.e., $x = 1$ or $x = -3$. Since $\overrightarrow{CD}$ is in the direction opposite that of $\overrightarrow{AB}$, $x < -1$, hence $x = -3$, and $y = -1$. The coordinates of $D$ are $(-3, -1)$.

**37.** Let $T = (x, y)$ be the fourth vertex. Let $\theta$ be the direction of $\overrightarrow{ST}$ and $\overrightarrow{PQ}$. Then

$$\tan\theta = \frac{6-4}{8-3} = \frac{2}{5} = \frac{y-7}{x-5}, \text{ or } y-7 = \frac{2}{5}(x-5).$$

So $\overrightarrow{ST} = \langle x-5, y-7\rangle = \left\langle x-5, \frac{2}{5}(x-5)\right\rangle$ and

$$\left\|\overrightarrow{ST}\right\| = \sqrt{(x-5)^2 + \frac{4}{25}(x-5)^2} = \frac{\sqrt{29}}{5}|x-5|. \text{ But}$$

$$\left\|\overrightarrow{ST}\right\| = \left\|\overrightarrow{PQ}\right\| = \sqrt{29}, \text{ so } \frac{\sqrt{29}}{5}|x-5| = \sqrt{29}, \text{ or}$$

$|x-5| = 5$. Since $T$ lies to the right of $S$, $x = 10$ is the desired solution of the equation.

Thus $y = \frac{2}{5}(x-5) + 7 = 9$.

The coordinates of $T$ are (10, 9).

**39.** The coordinates $(x, y)$ of the final position satisfy: $\langle x, y\rangle = \langle -4, 9\rangle + \langle 4, -5\rangle = \langle 0, 4\rangle$, so $(x, y) = (0, 4)$.

**41.** Let **a** be the displacement vector. Then $\mathbf{a} = 10\langle\cos 30°, \sin 30°\rangle = \langle 5\sqrt{3}, 5\rangle$. So the new coordinates $(x, y)$ satisfy:

$\langle x, y\rangle = \langle 5, 5\rangle + \mathbf{a} = \langle 5 + 5\sqrt{3}, 10\rangle$, so

$(x, y) = \left(5 + 5\sqrt{3}, 10\right)$.

**43.** The single equivalent displacement is $(2\mathbf{i} + 5\mathbf{j}) + (-12\mathbf{i} + 13\mathbf{j}) = -10\mathbf{i} + 18\mathbf{j}$. The new coordinates of $T = (x, y)$ satisfy:

$\langle x, y\rangle = \langle 0, 4\rangle + \langle -10, 18\rangle = \langle -10, 22\rangle$.

So $T = (-10, 22)$.

**45.** $\mathbf{a} = 10\langle\cos 330°, \sin 330°\rangle = \langle 5\sqrt{3}, -5\rangle$

$$\mathbf{b} = 5\langle\cos 60°, \sin 60°\rangle = \left\langle\frac{5}{2}, \frac{5\sqrt{3}}{2}\right\rangle$$

$\mathbf{c} = 7\langle\cos 180°, \sin 180°\rangle = \langle -7, 0\rangle$

The new coordinates $(x, y)$ satisfy:

$\langle x, y\rangle = \langle 1, 1\rangle + \mathbf{a} + \mathbf{b} + \mathbf{c}$

$$= \left\langle -\frac{7}{2} + 5\sqrt{3}, -4 + 5\frac{5\sqrt{3}}{2}\right\rangle \text{ so}$$

$$(x, y) = \left(-\frac{7}{2} + 5\sqrt{3}, -4 + \frac{5\sqrt{3}}{2}\right)$$

**47.** If $\mathbf{r} = \langle x, y\rangle$, then $\frac{y}{x} = m$, or $y = mx$ so $\mathbf{r} = \langle x, mx\rangle$.

For example, if $x = 1$, $r = \langle 1, m\rangle$.

If the direction of $\langle a, b\rangle$ is $\theta$, $\tan\theta = \frac{b}{a}$. So any line parallel to $\langle a, b\rangle$ has slope $\frac{b}{a}$.

**49.** Let $\mathbf{a} = \langle a_1, a_2\rangle$, $\mathbf{b} = \langle b_1, b_2\rangle$ and $\mathbf{c} = \langle c_1, c_2\rangle$

Use the associative law for addition of real numbers.

$$\begin{aligned}
\mathbf{a} + (\mathbf{b} + \mathbf{c}) &= \langle a_1, a_2\rangle + (\langle b_1, b_2\rangle + \langle c_1, c_2\rangle) \\
&= \langle a_1, a_2\rangle + \langle b_1 + c_1, b_2 + c_2\rangle \\
&= \langle a_1 + (b_1 + c_1), a_2 + (b_2 + c_2)\rangle \\
&= \langle (a_1 + b_1) + c_1, (a_2 + b_2) + c_2\rangle \\
&= \langle a_1 + b_1, a_2 + b\rangle + \langle c_1, c_2\rangle \\
&= (\mathbf{a} + \mathbf{b}) + \mathbf{c}
\end{aligned}$$

**51.** For all real numbers $x$ and $y$.

$(x-y)^2 \ge 0$ implies $2xy \le x^2 + y^2$.

Let $\mathbf{a} = \langle a_1, a_2\rangle$, $\mathbf{b} = \langle b_1, b_2\rangle$. Using $x = \frac{|a_1|}{\|a\|}$ and $y = \frac{|b_1|}{\|b\|}$, then

$$2\frac{|a_1|}{\|a\|}\cdot\frac{|b_1|}{\|b\|} \le \frac{|a_1|^2}{\|a\|^2} + \frac{|b_1|^2}{\|b\|^2}$$

Similarly,

$$2\frac{|a_2|}{\|a\|}\cdot\frac{|b_2|}{\|b\|} \le \frac{|a_2|^2}{\|a\|^2} + \frac{|b_2|^2}{\|b\|^2}.$$

Combining the two inequalities, we get

$$2\frac{|a_1b_1| + |a_2b_2|}{\|a\|\|b\|} \le \frac{\|a\|^2}{\|a\|^2} + \frac{\|b\|^2}{\|b\|^2} = 2$$

So $|a_1b_1| + |a_2b_2| \le \|a\|\ \|b\|$.

We use this inequality to prove the triangle inequality.

$$\begin{aligned}
\|a + b\|^2 &= (a_1 + b_1)^2 + (a_2 + b_2)^2 \\
&= a_1^2 + a_2^2 + 2(a_1b_1 + a_2b_2) + b_1^2 + b_2^2 \\
&\le \|a\|^2 + 2|a_1b_1 + a_2b_2| + \|b\|^2 \\
&\le \|a\|^2 + 2(|a_1b_1| + |a_2b_2|) + \|b\|^2 \\
&\le \|a\|^2 + 2\|a\|\|b\| + \|b\|^2 \\
&= (\|a\| + \|b\|)^2.
\end{aligned}$$

So $\|a + b\| \le \|a\| + \|b\|$.

Referring to the figure below, the sum of the distances $\|a\|$ and $\|b\|$ is at least as large as the distance $\|a\|+\|b\|$.

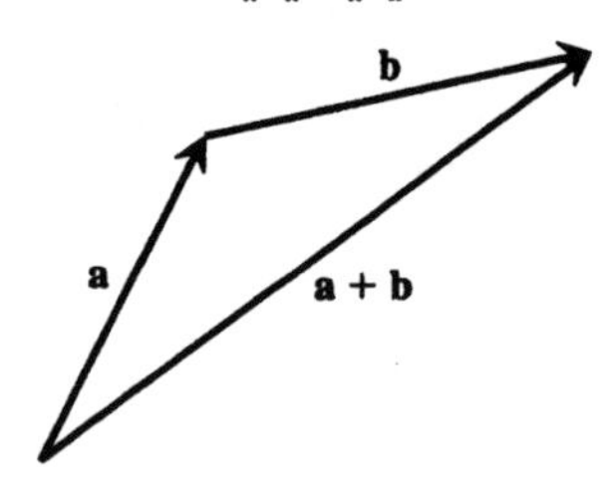


## Section R.2 Parametric Equations

**1.** The direction of motion is 30°, so

$$\mathbf{r}(t) = \langle 0, 1\rangle + 200t\langle \cos 30°, \sin 30°\rangle$$
$$= \langle 200t\cos 30°, 1+200t\sin 30°\rangle$$
$$= \langle 100\sqrt{3}t, 1+100t\rangle$$

Since $\mathbf{r}(t_1) = \mathbf{r}(0.001) \approx \langle 0.173205, 1.1\rangle$, the sled's position at time $t_1$ is approximately (0.173205, 1.1).

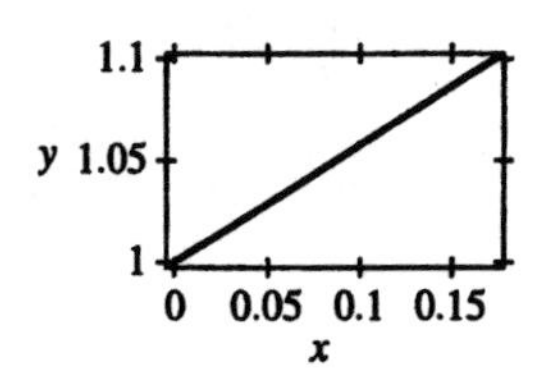


**3.** The direction of motion is 110°, so

$$\mathbf{r}(t) = \langle -5, 3\rangle + 225t\langle \cos 110°, \sin 110°\rangle$$
$$= \langle -5+225t\cos(110°), 3+225t\sin(110°)\rangle$$

Since $\mathbf{r}(t_1) = \mathbf{r}(0.001) \approx \langle -5.07695, 3.21143\rangle$ the sled's position at time $t_1$ is (−5.07695, 3.21143).

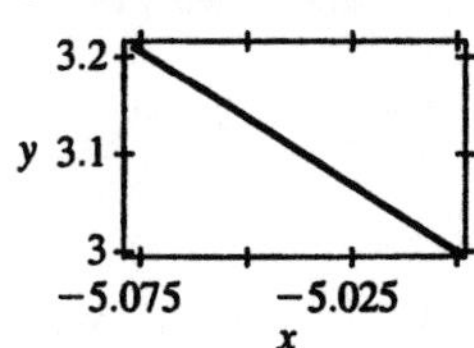


**5.** The direction of motion is 350°, so

$$\mathbf{r}(t) = \langle -4.5, 3.2\rangle + 190t\langle \cos 350°, \sin 350°\rangle$$
$$= \langle -4.5+190t\cos(350°), 3.2+190t\sin(350°)\rangle$$

since $\mathbf{r}(t_1) = \mathbf{r}(0.001) \approx \langle -4.31289, 3.16701\rangle$, the sled's position at time $t_1$ is (−4.31289, 3.16701).

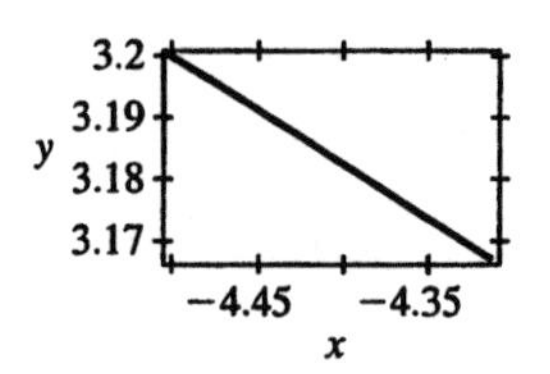


**7.** First we find a vector in the direction of motion:

$$\mathbf{a} = \langle 50-20\sqrt{3}, 20\rangle - \langle 50, 40\rangle = \langle -20\sqrt{3}, -20\rangle.$$

The meteor travels

$$\|\mathbf{a}\| = \sqrt{(20\sqrt{3})^2 + (20)^2} = 40 \text{ m in 5 s}.$$

Its speed is 8 m/s. The unit vector in the direction of motion is

$$\mathbf{u} = \frac{\mathbf{a}}{\|\mathbf{a}\|} = \left\langle -\frac{\sqrt{3}}{2}, -\frac{1}{2}\right\rangle.$$

So the position vector at time $t$ is

$$\mathbf{r}(t) = \langle 50, 40\rangle + 8t\mathbf{u}$$
$$= \langle 50, 40\rangle + \langle -4\sqrt{3}t, -4t\rangle$$
$$= \langle 50-4\sqrt{3}t, 40-4t\rangle.$$

The meteor hits the ground when $40-4t=0$, i.e., when $t$ = 10 s. Since $\mathbf{r}(10) = \langle 50-40\sqrt{3}, 0\rangle$, the position of the meteor when it hits the ground is $\left(50-40\sqrt{3}, 0\right)$.

**9.** First we find a vector in the direction of motion:

$$\mathbf{a} = \langle 35-16\sqrt{2}, 30-16\sqrt{2}\rangle - \langle 35, 30\rangle$$
$$= \langle -16\sqrt{2}, -16\sqrt{2}\rangle.$$

The meteor travels

$$\|\mathbf{a}\| = \sqrt{(16\sqrt{2})^2 + (16\sqrt{2})^2} = 32 \text{ m in 4 s}.$$ Its speed is 8 m/s. The unit vector in the direction of motion is $\mathbf{u} = \frac{\mathbf{a}}{\|\mathbf{a}\|} = \left\langle -\frac{\sqrt{2}}{2}, -\frac{\sqrt{2}}{2}\right\rangle.$

So the position vector at time $t$ is

$$\mathbf{r}(t) = \langle 35, 30\rangle + 8t\left\langle -\frac{\sqrt{2}}{2}, -\frac{\sqrt{2}}{2}\right\rangle$$
$$= \langle 35-4\sqrt{2}t, 30-4\sqrt{2}t\rangle.$$

The meteor hits the ground when $30-4\sqrt{2}t=0$, i.e., $t=\frac{15\sqrt{2}}{4}$ s. Since $\mathbf{r}\left(\frac{15\sqrt{2}}{4}\right) = \langle 5, 0\rangle$, the position of the meteor when it hits the ground is (5, 0).

**11.** First we find a vector in the direction of motion:
$\mathbf{a} = \langle 49.5, 3.3\rangle - \langle 90.2, 52.7\rangle = \langle -40.7, -49.4\rangle$.
The meteor travels
$\|\mathbf{a}\| = \sqrt{(40.7)^2 + (49.4)^2} \approx 64.0066$ m in 8 s. Its speed is about 8 m/s. The unit vector in the direction of motion is
$\mathbf{u} = \frac{\mathbf{a}}{\|\mathbf{a}\|} \approx \langle -0.6359, -0.7718\rangle$. So the position vector at time $t$ is
$\mathbf{r}(t) = \langle 90.2, 52.7\rangle + 8t\mathbf{u}$
$\approx \langle 90.2 - 5.1t, 52.7 - 6.2t\rangle$.
The meteor hits the ground when $52.7 - 6.2t = 0$, i.e., $t \approx 8.5$. Since $\mathbf{r}(8.5) \approx \langle 46.85, 0\rangle$, the position of the meteor when it hits the ground is about (46.85, 0).

**13.** $x = 5t$ and $y = 2 + 3t$. Solving the first equation for $t$ and substituting this result into the second equation yields:
$y = 2 + 3\frac{x}{5} = 2 + \frac{3}{5}x$
$\mathbf{r}(t_1) = \mathbf{r}\left(\frac{1}{2}\right) = \left\langle \frac{5}{2}, \frac{7}{2}\right\rangle$.

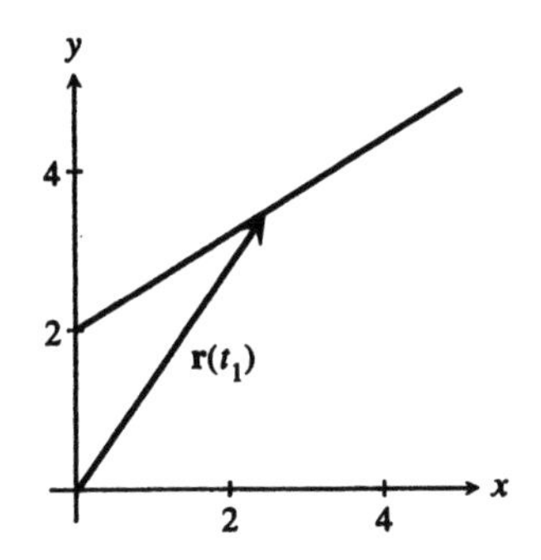


**15.** Since $x = 3\cos t$ and $y = 3\sin t$,
$x^2 + y^2 = 9\cos^2 t + 9\sin^2 t = 9\left(\cos^2 t + \sin^2 t\right) = 9.$
$\mathbf{r}(t_1) = \mathbf{r}\left(\frac{\pi}{2}\right) = \langle 0, 3\rangle$

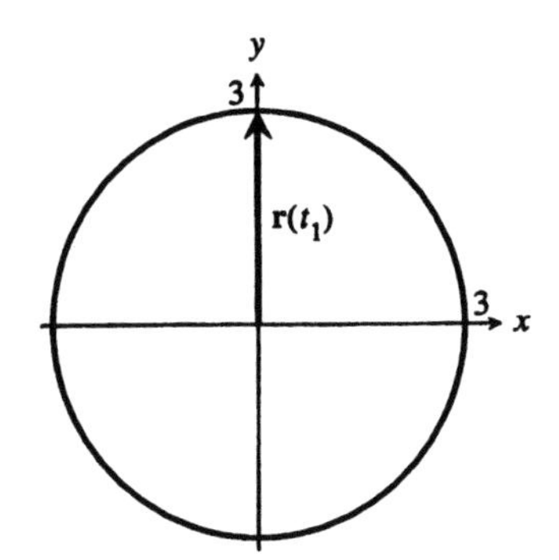


**17.** The position vector of object $P$ after $t$ seconds is
$\mathbf{r}(t) = \langle 3, 3\rangle + 2\langle \cos(2t), \sin(2t)\rangle$
$= \langle 3 + 2\cos(2t), 3 + 2\sin(2t)\rangle$
since object $P$ rotates $2t$ radians in $t$ seconds.

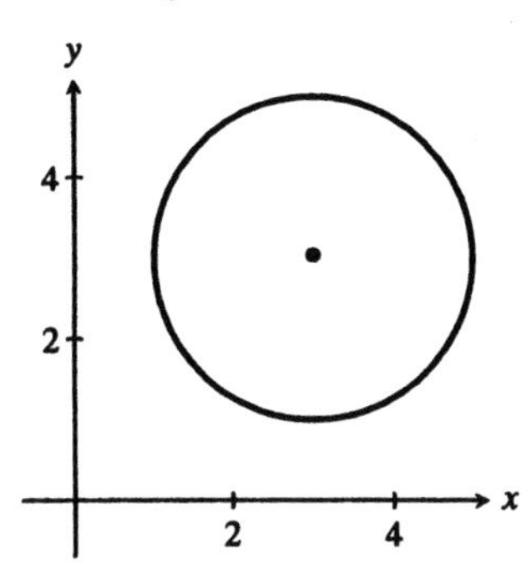


**19.** The position vector of object $P$ after $t$ seconds is
$\mathbf{r}(t) = \langle 0, 2\rangle + 5\left\langle \cos\frac{5\pi t}{9}, \sin\frac{5\pi t}{9}\right\rangle$
$= \left\langle 5\cos\frac{5\pi t}{9}, 2 + 5\sin\frac{5\pi t}{9}\right\rangle$
since object $P$ rotates $\frac{5\pi t}{9}$ radians in $t$ seconds $\left(100° = \frac{5\pi}{9} \text{ radians}\right)$.

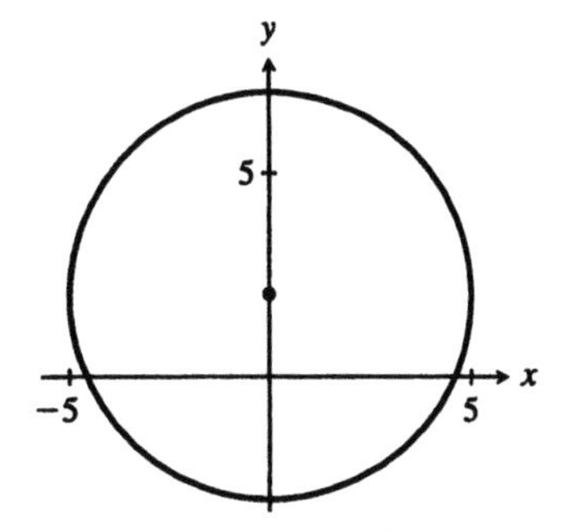


**21.** The object's position with respect to the center of the circle after $t$ seconds is $\pi - 2t$. The initial position is $\pi$ and the object rotates $2t$ radians in $t$ seconds. The negative sign is so that the direction of motion is clockwise. Hence,
$\mathbf{r}(t) = \langle 3, 3\rangle + 2\langle \cos(\pi - 2t), \sin(\pi - 2t)\rangle$
$= \langle 3 + 2\cos(\pi - 2t), 3 + 2\sin(\pi - 2t)\rangle$.

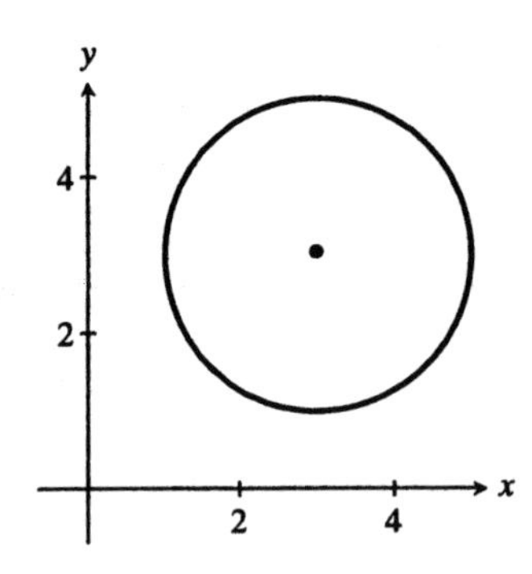


**23.** The passenger boards at the bottom of the ferris wheel. So his or her initial position with respect to the center of the wheel is $\frac{3\pi}{2}$. Hence,

$\mathbf{r}(t)$

$= \langle 0, 62.5\rangle + 50\left\langle \cos\left(\frac{2\pi t}{15} + \frac{3\pi}{2}\right), \sin\left(\frac{2\pi t}{15} + \frac{3\pi}{2}\right)\right\rangle.$

**25.** In one second the proton travels along an arc of length 18,628 miles. The arc is on the boundary of a sector whose central angle measures $\frac{18{,}628}{0.64}$ radians. Hence,

$\mathbf{r}(t) = 0.64\left\langle \cos\frac{18{,}628t}{0.64}, \sin\frac{18{,}628t}{0.64}\right\rangle$

Since $\mathbf{r}(0.0002) \approx \langle 0.572923, -0.285236\rangle$, the particle's position after 0.0002 seconds is about (0.572923, −0.285236). Since the circumference of the Tevatron is $c = 1.28\pi$, the proton completes one revolution in $\frac{c}{18{,}628} = \frac{1.28\pi}{18{,}628}$ seconds. It takes another $\frac{0.5}{18{,}628}$ seconds to reach the target. So the total elapsed time is

$\frac{1.28\pi}{18{,}628} + \frac{0.5}{18{,}628} \approx 0.000242712$ seconds.

**27.** Since $vt$ is the magnitude of the displacement vector $\overrightarrow{PQ}$ at time $t$, $\overrightarrow{PQ} = vt\mathbf{u}$. So $\mathbf{r}(t) = \langle p_1, p_2\rangle + tv\mathbf{u}$.

**29.** The object is moving along a line. It travels a distance of $\|\mathbf{r}(t+1) - \mathbf{r}(t)\| = \|b\mathbf{u}\| = |b|$ in one time unit. Because its speed is constant, the speed is $|b|$ units per time unit. If $\mathbf{u}$ were not a unit vector, it follows in the same way that the speed of the object is $|b| \cdot \|\mathbf{u}\|$ units per time unit. The equation $\mathbf{r}(t) = \langle 2, 3\rangle + 5t\langle 1, 1\rangle$ can be written as $\mathbf{r}(t) = \langle 2, 3\rangle + \left(5\sqrt{2}\right)t\mathbf{u}$, where $\mathbf{u}$ is the unit vector $\left\langle \frac{1}{\sqrt{2}}, \frac{1}{\sqrt{2}}\right\rangle$. It follows that the speed of the object is $5\sqrt{2}$ units per time unit.

**31.** $\mathbf{r}(t) = \begin{cases} \langle 1-t, 2-t\rangle & 0 \le t \le 1 \\ \langle t-1, 2-t\rangle & 1 < t \le 2 \\ \langle t-1, t-2\rangle & 2 < t \le 5 \end{cases}$

If $0 \le t \le 1$, $x = 1 - t$, or $t = 1 - x$, and $y = 2 - t = 2 - (1 - x) = 1 + x$.
If $1 < t \le 2$, $x = t - 1$, or $t = 1 + x$, and $y = 2 - t = 2 - (1 + x) = 1 - x$.
Finally, if $2 < t \le 5$, $y = t - 2 = (1 + x) - 2 = x - 1$.

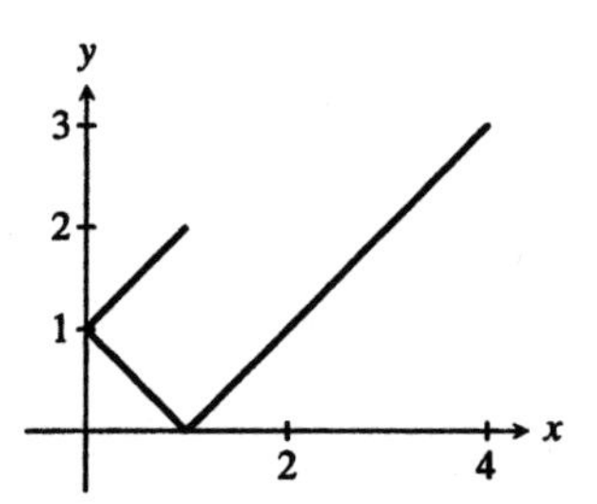


## Section R.3 Velocity and Tangent Vectors

**1.** $\mathbf{r}(1) = \langle 2, 3\rangle$
$\mathbf{v}(t) = \mathbf{r}'(t) = \langle 2, 6t\rangle$
$\mathbf{v}(1) = \langle 2, 6\rangle$

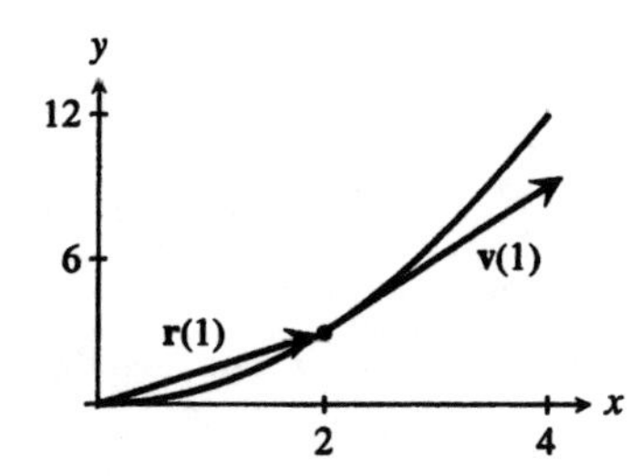


**3.** $\mathbf{r}(1) = 3\mathbf{j}$
$\mathbf{v}(t) = \mathbf{r}'(t) = -\mathbf{i} + 4t\mathbf{j}$
$\mathbf{v}(1) = -\mathbf{i} + 4\mathbf{j}$

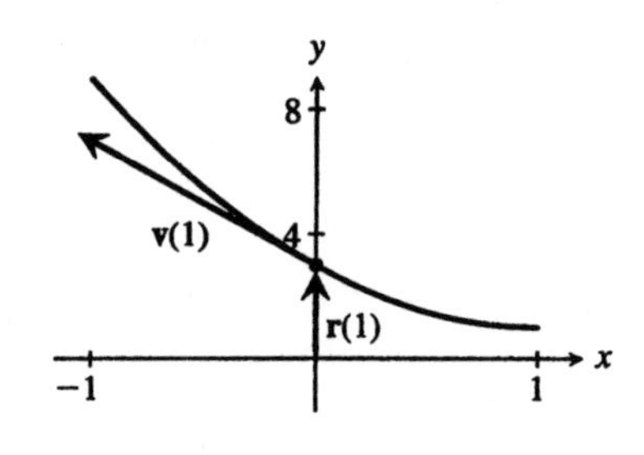

5. $\mathbf{r}\left(\frac{\pi}{6}\right)=5\left\langle\cos\frac{\pi}{3}, \sin\frac{\pi}{3}\right\rangle=\left\langle\frac{5}{2}, \frac{5\sqrt{3}}{2}\right\rangle$

$\mathbf{v}(t)=\mathbf{r}'(t)=5\langle-2\sin 2t, 2\cos 2t\rangle$

$=\langle-10\sin 2t, 10\cos 2t\rangle$

$\mathbf{v}\left(\frac{\pi}{6}\right)=\langle-5\sqrt{3}, 5\rangle$

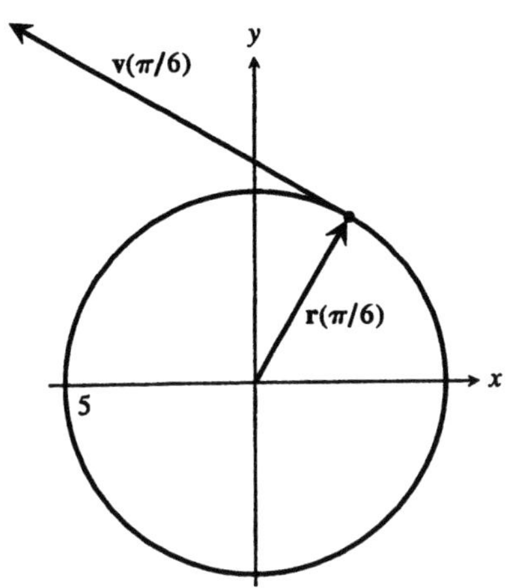


7. $\mathbf{r}(2)=\mathbf{i}+2\mathbf{j}+2\left(\mathbf{i}+\frac{1}{2}\mathbf{j}\right)=3\mathbf{i}+3\mathbf{j}$

$\mathbf{v}(t)=\mathbf{r}'(t)=\mathbf{i}+\frac{1}{2}\mathbf{j}$

$\mathbf{v}(2)=\mathbf{i}+\frac{1}{2}\mathbf{j}$

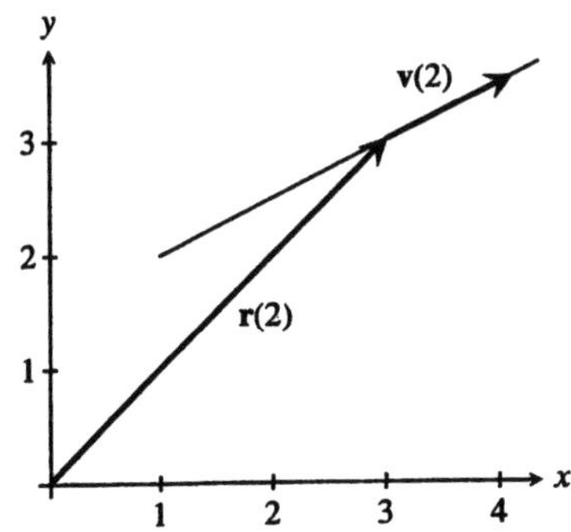


9. $\mathbf{v}\left(\frac{\pi}{6}, \frac{\pi}{6}+0.1\right)$

$=\left\langle\frac{\frac{\pi}{6}+0.1-\frac{\pi}{6}}{\frac{\pi}{6}+0.1-\frac{\pi}{6}}, \frac{\sin\left(\frac{\pi}{6}+0.1\right)-\sin\frac{\pi}{6}}{\frac{\pi}{6}+0.1-\frac{\pi}{6}}\right\rangle$

$\approx(1,0.839604)$

$\mathbf{v}\left(\frac{\pi}{6}, \frac{\pi}{6}+0.01\right)\approx\langle 1, 0.863511\rangle$ m / s

$\mathbf{v}\left(\frac{\pi}{6}, \frac{\pi}{6}+0.001\right)\approx\langle 1, 0.865775\rangle$ m / s

$\mathbf{v}(t)=\mathbf{r}'(t)=\langle 1, \cos t\rangle$

$\mathbf{v}\left(\frac{\pi}{6}\right)=\left\langle 1, \frac{\sqrt{3}}{2}\right\rangle\approx\langle 1, 0.866025\rangle$ m / s.

The differences between the velocity at $t_1$ and the three average velocities are $\langle 0, 0.03\rangle$, $\langle 0, 0.003\rangle$, and $\langle 0, 0.0003\rangle$, approximately.

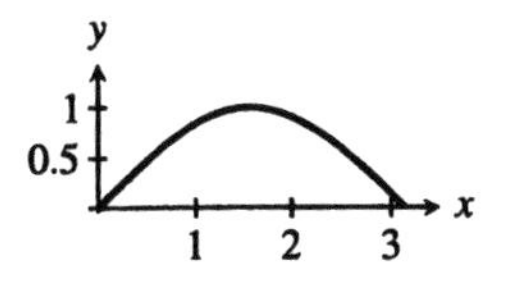


11. $\mathbf{v}(1, 1+0.1)=\left\langle\frac{1+0.1-1}{1+0.1-1}, \frac{e^{1+0.1}-e}{1+0.1-1}\right\rangle$

$\approx\langle 1, 2.85884\rangle$ m / s

$\mathbf{v}(1, 1+0.01)\approx\langle 1, 2.73192\rangle$ m / s

$\mathbf{v}(1, 1+0.001)\approx\langle 1, 2.71964\rangle$ m / s

$\mathbf{v}(t)=\mathbf{r}'(t)=\langle 1, e^t\rangle$

$\mathbf{v}(1)=\langle 1, e^1\rangle\approx\langle 1, 2.71828\rangle$ m / s.

The differences between the velocity at $t_1$ and the three average velocities are $\langle 0, 0.14\rangle$, $\langle 0, 0.014\rangle$, and $\langle 0, 0.0014\rangle$, approximately.

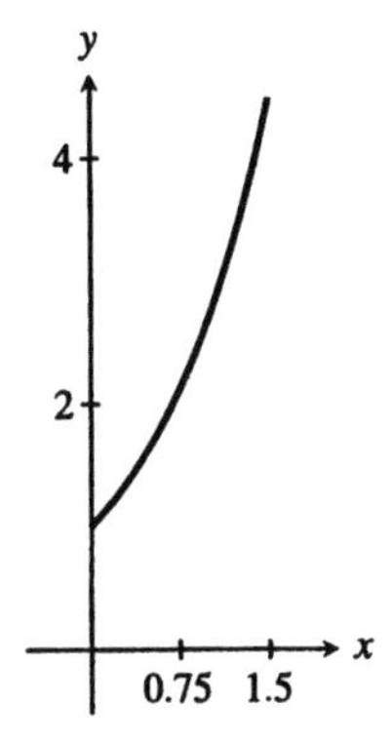


13. $\mathbf{v}(t)=\mathbf{r}'(t)=12.6\langle-\sin(2.1t), \cos(2.1t)\rangle$

$\mathbf{r}(12)=6\langle\cos 25.2, \sin 25.2\rangle$

$\approx\langle 5.98643, 0.403248\rangle$

$\mathbf{v}(12)=12.6\langle-\sin 25.2, \cos 25.2\rangle$

$\approx 12.6\langle-0.067, 0.998\rangle$

So the speed is $\|\mathbf{v}(12)\|=12.6$ m / s and the direction is

$$\theta=\pi-\arctan\left(\frac{\cos(25.2)}{\sin(25.2)}\right)\approx 1.63806.$$

The position at time $t_1+10$ s is given by the vector:

$$\mathbf{r}(12)+10\mathbf{v}(12)=6\langle\cos 25.2, \sin 25.2\rangle + 126\langle-\sin 25.2, \cos 25.2\rangle \approx \langle -2.48178, 126.118\rangle$$

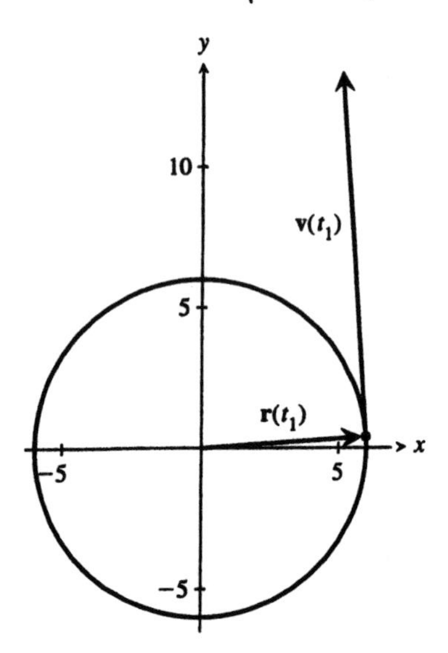


**15.** $\mathbf{v}(t)=\mathbf{r}'(t)=2.5\mathbf{i}+5t\mathbf{j}$
$\mathbf{r}(1) = 2.5\mathbf{i} + 3.5\mathbf{j}$
$\mathbf{v}(1) = 2.5\mathbf{i} + 5\mathbf{j}$
So the speed is
$\|\mathbf{v}(1)\| = \|2.5\mathbf{i}+5\mathbf{j}\| = \sqrt{31.25} \approx 5.59017$ m / s
and the direction is $\theta = \arctan\frac{5}{2.5} \approx 1.10715$.
The position at time $t_1 + 10$ s is given by the vector:
$$\mathbf{r}(1)+10\mathbf{v}(1) = 2.5\mathbf{i}+3.5\mathbf{j}+25\mathbf{i}+50\mathbf{j} = 27.5\mathbf{i}+53.5\mathbf{j}$$

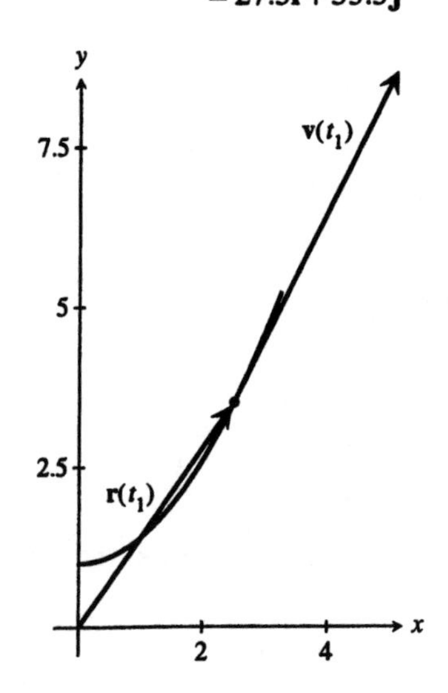


**17.** Since (0, 5) is on the line, let
$\mathbf{a} = \langle 3, 7\rangle - \langle 0, 5\rangle = \langle 3, 2\rangle$
Then $\mathbf{r}(t) = \langle 3, 7\rangle + t\langle 3, 2\rangle, \ -\infty < t < \infty$.

**19.** Since (0, 1) is on the line, let
$\mathbf{a} = \langle 4, 5\rangle - \langle 0, 1\rangle = \langle 4, 4\rangle$.
Then $\mathbf{r}(t) = \langle 4, 5\rangle + t\langle 4, 4\rangle, \ -\infty < t < \infty$.

**21.** Let $\mathbf{a} = \mathbf{q} - \mathbf{p} = \langle 0, 3\rangle - \langle 4, -2\rangle = \langle -4, 5\rangle$.
Then $\mathbf{r}(t) = \langle 4, -2\rangle + t\langle -4, 5\rangle, \ -\infty < t < \infty$.

**23.** Let $\mathbf{a} = \mathbf{q} - \mathbf{p} = \langle -3, 5\rangle - \langle -3, 0\rangle = \langle 0, 5\rangle$.
Then $\mathbf{r}(t) = \langle -3, 0\rangle + t\langle 0, 5\rangle, \ -\infty < t < \infty$.

**25.** $\mathbf{r}(x) = \langle x, x^2\rangle$; the tangent vector is
$\mathbf{r}'(x) = \langle 1, 2x\rangle$, so $\mathbf{r}'(1) = \langle 1, 2\rangle$.
The slope is $\frac{2}{1} = 2$.

**27.** $\mathbf{r}(x) = \langle x, \tan x\rangle$; the tangent vector is
$\mathbf{r}'(x) = \langle 1, \sec^2 x\rangle$, so $\mathbf{r}'\left(\frac{\pi}{4}\right) = \langle 1, 2\rangle$.
The slope is $\frac{2}{1} = 2$.

**29.** $\mathbf{r}(x) = \langle x, x^2 - 6x + 11\rangle$; the tangent vector is
$\mathbf{r}'(x) = \langle 1, 2x - 6\rangle$, so $\mathbf{r}'(4) = \langle 1, 2\rangle$.
The slope is $\frac{2}{1} = 2$.

**31.** $\mathbf{r}(x) = \langle x, e^{2x}\rangle$; the tangent vector is
$\mathbf{r}'(x) = \langle 1, 2e^{2x}\rangle$, so $\mathbf{r}'(1) = \langle 1, 2e^2\rangle$. The slope is $\frac{2e^2}{1} = 2e^2$.

**33.** $\mathbf{r}(x) = x\mathbf{i} + (\ln x)^2\mathbf{j}$; the tangent vector is
$\mathbf{r}'(x) = \mathbf{i} + \frac{2\ln x}{x}\mathbf{j}$, so $\mathbf{r}'(1) = \mathbf{i}$. The slope is
$\frac{0}{1} = 0$.

**35.** If $\mathbf{a} = \langle 6, 7\rangle - \langle -5, 1\rangle = \langle 11, 6\rangle$, then the unit vector in the direction of motion and passing through the two points is
$$\mathbf{u} = \frac{\mathbf{a}}{\|\mathbf{a}\|} = \frac{\langle 11, 6\rangle}{\sqrt{11^2+6^2}} = \frac{\langle 11, 6\rangle}{\sqrt{157}}.$$
So $\mathbf{r}(t) = \langle -5, 1\rangle + vt\mathbf{u} = \langle -5, 1\rangle + \frac{10}{\sqrt{157}}t\langle 11, 6\rangle$.

**37.** $\mathbf{r} = \langle \cos(7t), \sin(7t) \rangle$

**39.** $\mathbf{r} = t\langle 1, 5 \rangle + \langle 0, -2 \rangle$

**41.** $\mathbf{r}(0) = \langle 0, 0 \rangle$; $\mathbf{r}\left(\frac{T}{4}\right) = \mathbf{r}(2) = \langle 2, 28 \rangle$;

$\mathbf{r}\left(\frac{T}{2}\right) = \mathbf{r}(4) = \langle 4, 48 \rangle$; $\mathbf{r}\left(\frac{3T}{4}\right) = \mathbf{r}(6) = \langle 6, 60 \rangle$

$\mathbf{r}(T) = \mathbf{r}(8) = \langle 8, 64 \rangle$

$\mathbf{r}'(t) = \langle 1, -2t + 16 \rangle$; $\mathbf{r}'\left(\frac{T}{2}\right) = \mathbf{r}'(4) = \langle 1, 8 \rangle$. The speed at $\frac{T}{2}$ is $\left\|\mathbf{r}'\left(\frac{T}{2}\right)\right\| = \sqrt{65}$ m / s.

Since $\mathbf{r}(t) = \langle t, -t^2 + 16t \rangle$,

$x = t$ and $y = -t^2 + 16t = -x^2 + 16x$, $0 \le x \le 8$.

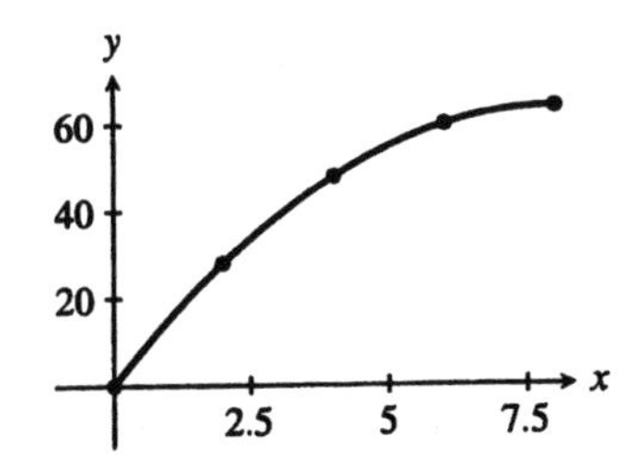


**43.** $\mathbf{r}(0) = \langle 0, 0 \rangle$; $\mathbf{r}\left(\frac{T}{4}\right) = \mathbf{r}(1) = \langle 4, 3 \rangle$;

$\mathbf{r}\left(\frac{T}{2}\right) = \mathbf{r}(2) = \langle 8, 4 \rangle$; $\mathbf{r}\left(\frac{3T}{4}\right) = \mathbf{r}(3) = \langle 12, 3 \rangle$;

$\mathbf{r}(T) = \mathbf{r}(4) = \langle 16, 0 \rangle$

$\mathbf{r}'(t) = \langle 4, -2t + 4 \rangle$; $\mathbf{r}'\left(\frac{T}{2}\right) = \mathbf{r}'(2) = \langle 4, 0 \rangle$.

The speed at $\frac{T}{2}$ is $\left\|\mathbf{r}'\left(\frac{T}{2}\right)\right\| = 4$ m / s

Since $\mathbf{r}(t) = \langle 4t, -t^2 + 4t \rangle$, $x = 4t$ and

$y = -t^2 + 4t = -\left(\frac{x}{4}\right)^2 + 4\left(\frac{x}{4}\right) = -\frac{x^2}{16} + x$,

or $16y = -x^2 + 16x$, $0 \le x \le 16$.

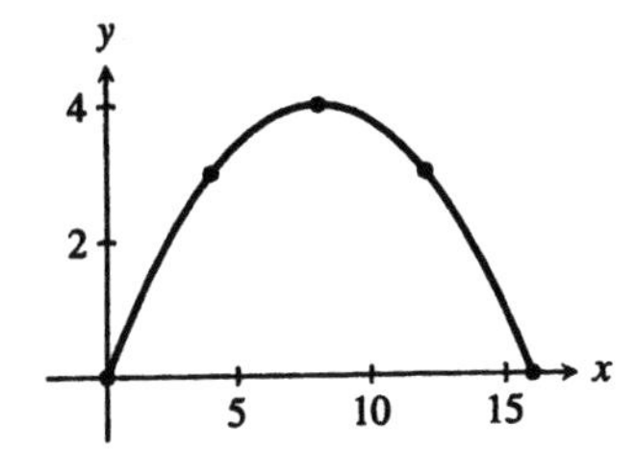


**45.** $\mathbf{r}(-2) = \langle 9, 0 \rangle$, $\mathbf{r}(3) = \langle 19, 5 \rangle$. To eliminate $t$, note that $y = t + 2$, so $t = y - 2$, which gives $x = 1 + 2(y-2)^2 = 2y^2 - 8y + 9$.

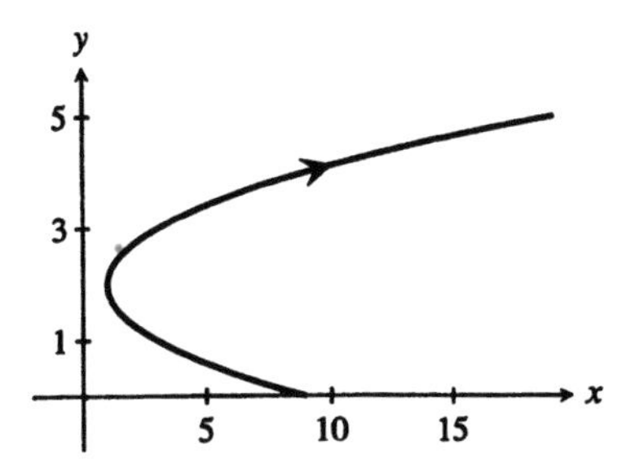


**47.** $\mathbf{r}(-2) = \langle 0, 2 \rangle$, $\mathbf{r}(2) = \langle 4, 6 \rangle$. To eliminate $t$, note the similarity between $x$ and $y$. In fact, $2x - y = t$, so

$$x = \frac{1}{2}(2x - y)^2 + (2x - y)$$
$$= 2x^2 - 2xy + \frac{y^2}{2} + 2x - y$$
$$\Rightarrow 2x^2 + \frac{y^2}{2} - 2xy + x - y = 0, \text{ or}$$
$$4x^2 + y^2 - 4xy + 2x - 2y = 0.$$

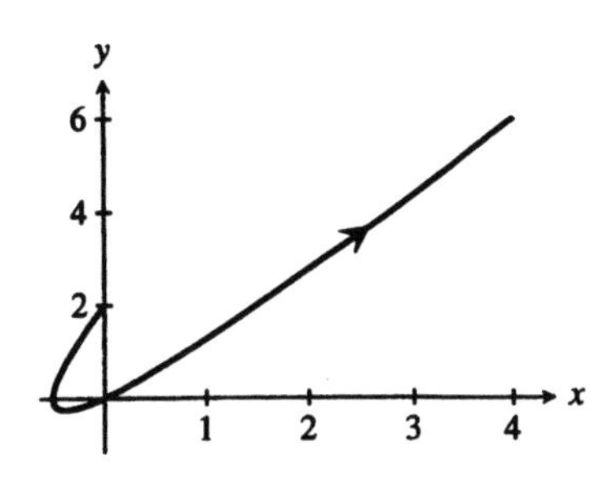


**49.** $\mathbf{r}(-2) = \langle -3, -3 \rangle$, $\mathbf{r}(3) = \langle 7, 2 \rangle$. To eliminate $t$, note that $y = t - 1$, so $t = y + 1$, which gives $x = 2(y+1) + 1 = 2y + 3$.

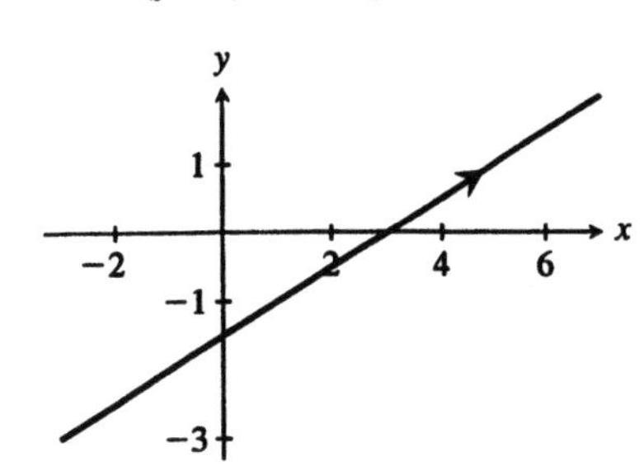

**51.**

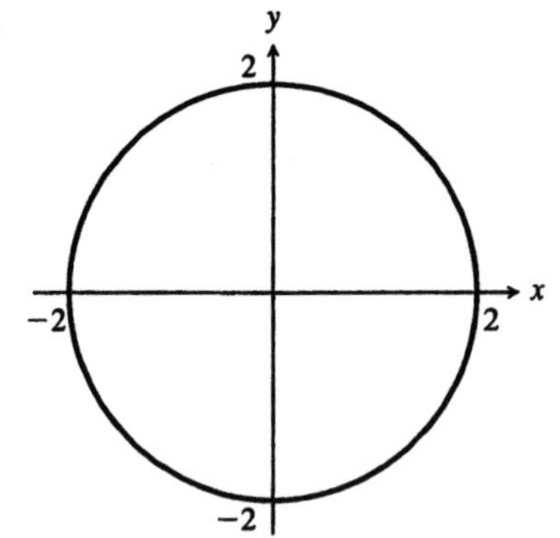


**53.**

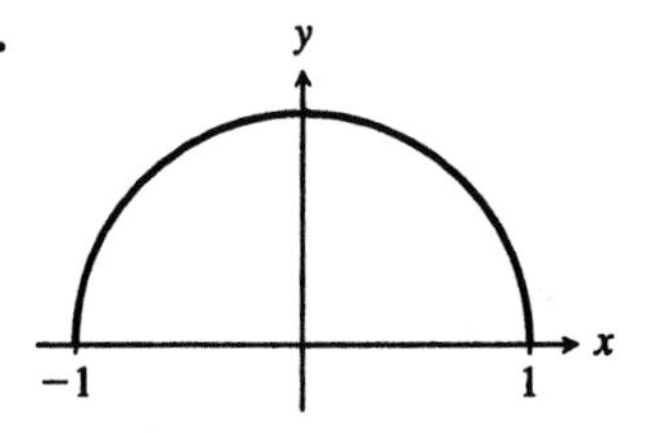


**55.**

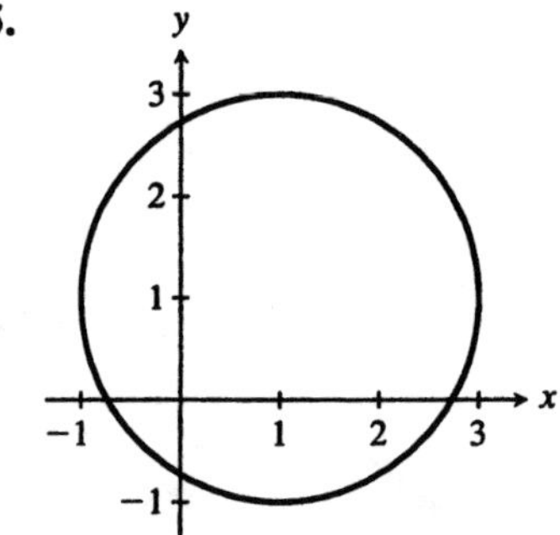


**57.**

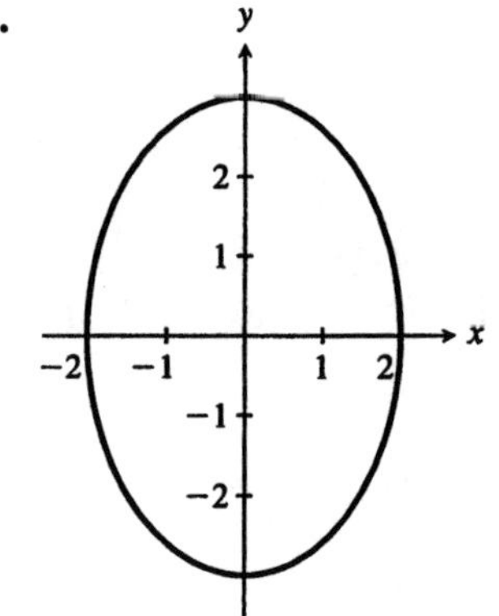


**59.**

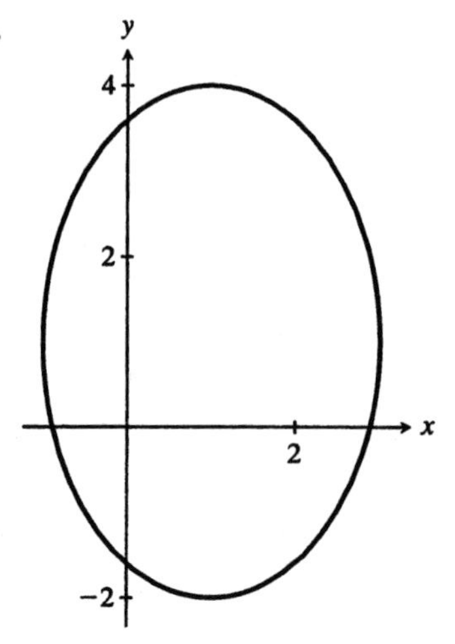


**61.** Because $(x, y)$ satisfies the equation $\frac{x^2}{5^2}+\frac{y^2}{4^2}=1$, the point $\left(\frac{x}{5}, \frac{y}{4}\right)$ satisfies the equation $X^2+Y^2=1$. Hence, there is a number $\theta \in [0, 2\pi)$ for which $\frac{x}{5}=\cos\theta = X$ and $\frac{y}{4}=\sin\theta = Y$.

**63.** Solving $-1 + t = 1$ yields $t = 2$. Since this does not also satisfy $4 + 6t = 15$, the point $(1, 15)$ is not on the line.

**65.** A parametrization is $x+1=2\sin\theta$ and $y-2=2\sin\theta$, $0 \le \theta \le 2\pi$.

**67.** One parametrization is $x=2\cos\theta$ and $y = 3 \sin \theta$.

**69.** By inspection, $\mathbf{r}(0)=\mathbf{r}(\pi)=\langle 0, 0\rangle$. Since $\mathbf{r}'(t)=\langle \cos t, \cos(t+\sin t)(1+\cos t)\rangle$, $\mathbf{r}'(0)=\langle 1, 2\rangle$ and $\mathbf{r}'(\pi)=\langle -1, 0\rangle$. Hence, one tangent line $(t = 0)$ has slope 2 and the other $(t=\pi)$ has slope 0. Hence, the equations of the tangent lines are $y = 0$ and $y = 2x$.

**71.** $\mathbf{r}(t) = \langle \cos^3 t, \sin^3 t \rangle$

$\mathbf{r}'(t) = 3\langle -\cos^2 t \sin t, \sin^2 t \cos t \rangle$

A unit tangent vector at **p** is

$$\mathbf{u} = \frac{\mathbf{r}'\left(\frac{\pi}{4}\right)}{\left\|\mathbf{r}'\left(\frac{\pi}{4}\right)\right\|}$$

$$= \left(\frac{3}{4}\right)\frac{\langle -\sqrt{2}, \sqrt{2} \rangle}{\left\|\left(\frac{3}{4}\right)\langle -\sqrt{2}, \sqrt{2} \rangle\right\|}$$

$$= \frac{\langle -\sqrt{2}, \sqrt{2} \rangle}{2}$$

$$= \left\langle -\frac{1}{\sqrt{2}}, \frac{1}{\sqrt{2}} \right\rangle$$

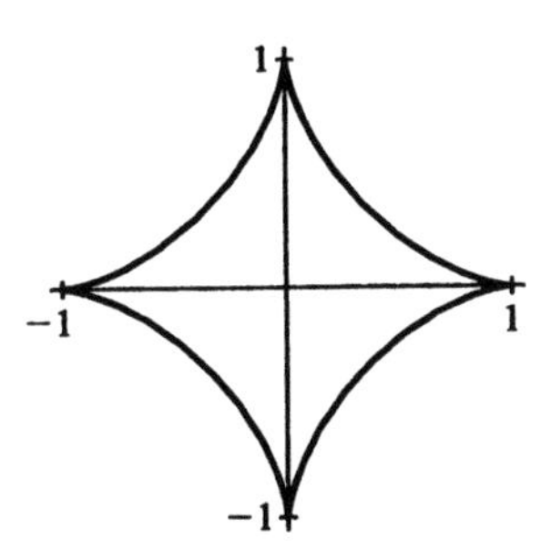


**73.** $\mathbf{r}(t) = \langle t - \sin t, 1 - \cos t \rangle$

$\mathbf{r}'(t) = \langle 1 - \cos t, \sin t \rangle$

A unit tangent vector at **p** is

$$\mathbf{u} = \frac{\mathbf{r}'\left(\frac{\pi}{3}\right)}{\left\|\mathbf{r}'\left(\frac{\pi}{3}\right)\right\|}$$

$$= \frac{\left\langle \frac{1}{2}, \frac{\sqrt{3}}{2} \right\rangle}{\left\|\left\langle \frac{1}{2}, \frac{\sqrt{3}}{2} \right\rangle\right\|}$$

$$= \frac{\frac{1}{2}\langle 1, \sqrt{3} \rangle}{1}$$

$$= \left\langle \frac{1}{2}, \frac{\sqrt{3}}{2} \right\rangle.$$

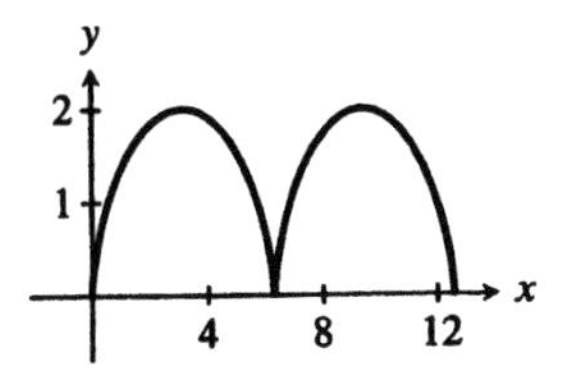


**75.** $\mathbf{r}(t) = \langle 2\tan t, \cos^2 t \rangle$

$\mathbf{r}'(t) = \langle 2\sec^2 t, -2\cos t \sin t \rangle$

$\mathbf{r}'\left(\frac{\pi}{4}\right) = \langle 4, -1 \rangle$

$\left\|\mathbf{r}'\left(\frac{\pi}{4}\right)\right\| = \sqrt{17}$

A unit tangent vector at **p** is

$$\mathbf{u} = \frac{\mathbf{r}'\left(\frac{\pi}{4}\right)}{\left\|\mathbf{r}'\left(\frac{\pi}{4}\right)\right\|} = \left\langle \frac{4}{\sqrt{17}}, \frac{-1}{\sqrt{17}} \right\rangle$$

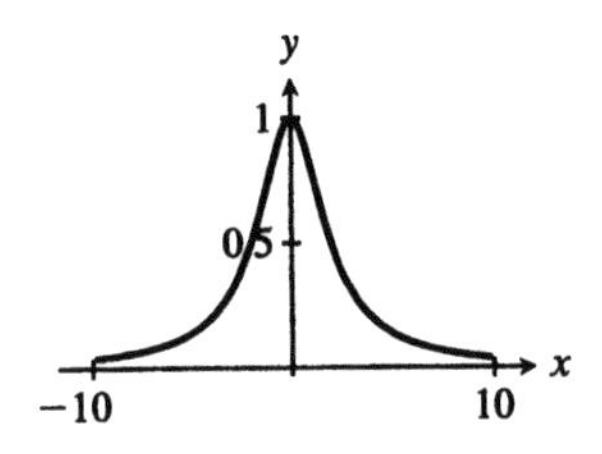


**77.** $\mathbf{r}(t) = \left\langle \frac{t}{1+t^3}, \frac{t^2}{1+t^3} \right\rangle$

$$\mathbf{r}'(t) = \left\langle \frac{1+t^3 - t(3t^2)}{(1+t^3)^2}, \frac{2t(1+t^3) - t^2(3t^2)}{(1+t^3)^2} \right\rangle$$

$$= \left\langle \frac{1-2t^3}{(1+t^3)^2}, \frac{2t-t^4}{(1+t^3)^2} \right\rangle$$

$\mathbf{r}'(2) = \left\langle -\frac{5}{27}, -\frac{4}{27} \right\rangle$

$\|\mathbf{r}'(2)\| = \frac{\sqrt{41}}{27}$

A unit tangent vector at **p** is

$$\mathbf{u} = \frac{\mathbf{r}'(2)}{\|\mathbf{r}'(2)\|} = \left\langle -\frac{5}{\sqrt{41}}, -\frac{4}{\sqrt{41}} \right\rangle.$$

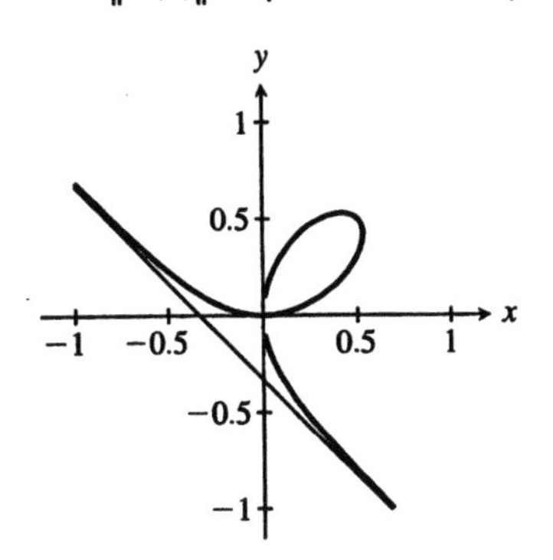

**79.**

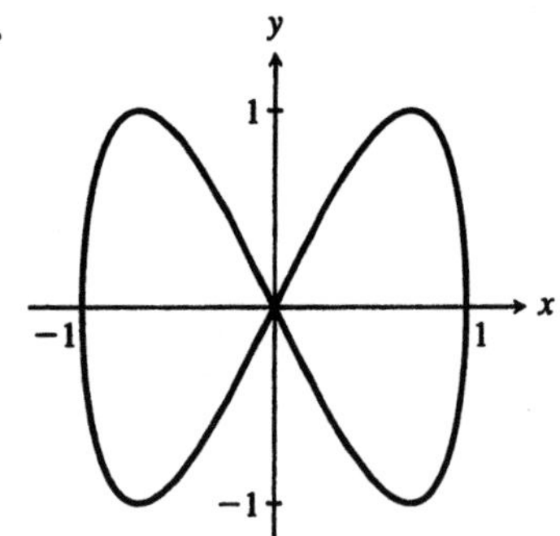


**81.**

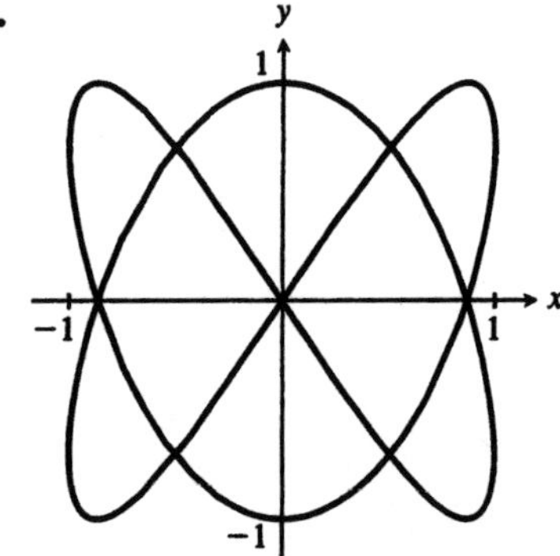


**83.** The positive $x$-intercept is the nonzero root of $t-\frac{1}{2}t^2=t\left(1-\frac{t}{2}\right)=0$, i.e., $t=2$.

Since $\mathbf{r}'(t)=\langle 1+t,\ 1-t\rangle$ and $\mathbf{r}'(2)=\langle 3,\ -1\rangle$, the slope at $t=2$ is $-\frac{1}{3}$.

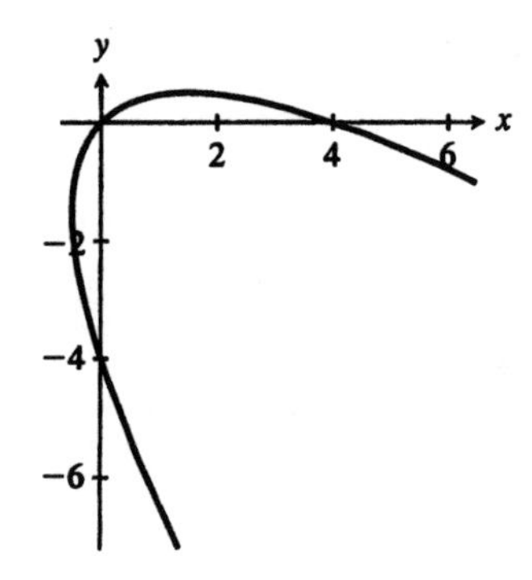


**85.** Let $(x_0, y_0)$ be the point on the line $y=x-\frac{1}{2}$ such that the distance from the point $(x, y)$ to the line is the distance between $(x, y)$ and $(x_0, y_0)$. Then the line through $(x, y)$ and $(x_0, y_0)$ is perpendicular to $y=x-\frac{1}{2}$, so it has slope −1. Hence $y_0-y=-(x_0-x)$ or $y_0=x+y-x_0$. Since $(x_0, y_0)$ is on the line $y=x-\frac{1}{2}$, $y_0=x_0-\frac{1}{2}$. We combine these equations to find $(x_0, y_0)$ in terms of $x$ and $y$.

$$x_0-\frac{1}{2}=x+y-x_0$$

$$x_0=\frac{x}{2}+\frac{y}{2}+\frac{1}{4}$$

$$y_0=x_0-\frac{1}{2}=\frac{x}{2}+\frac{y}{2}-\frac{1}{4}$$

If the point $(x, y)$ is equally distant from $y = x - \frac{1}{2}$ and the point $\left(-\frac{1}{4}, \frac{1}{4}\right)$, then

$$\left(x+\frac{1}{4}\right)^2 + \left(y-\frac{1}{4}\right)^2 = (x-x_0)^2 + (y-y_0)^2.$$

$$\left(x+\frac{1}{4}\right)^2 + \left(y-\frac{1}{4}\right)^2 = \left(\frac{x}{2}-\frac{y}{2}-\frac{1}{4}\right)^2 + \left(-\frac{x}{2}+\frac{y}{2}+\frac{1}{4}\right)^2$$

$$x^2 + \frac{1}{2}x + y^2 - \frac{1}{2}y + \frac{1}{8} = \frac{1}{2}x^2 - xy + \frac{1}{2}y^2 - \frac{1}{2}x + \frac{1}{2}y + \frac{1}{8}$$

$$x - y = -\frac{1}{2}x^2 - xy - \frac{1}{2}y^2$$

$$x - y = -\frac{1}{2}(x+y)^2$$

Thus, the set of all points $(x, y)$ equally distant from the line with equation $y = x - \frac{1}{2}$ and the point $\left(-\frac{1}{4}, \frac{1}{4}\right)$ satisfies (11).

Reversing the algebraic steps shows that any point satisfying (11) is equally distant from the given line and point.

## Section R.4 Dot Product

**1.** $\mathbf{a}\cdot\mathbf{b} = 2\cdot 3 + 9\cdot(-2) = -12;$

$$\theta = \arccos\left(\frac{\mathbf{a}\cdot\mathbf{b}}{\|\mathbf{a}\|\,\|\mathbf{b}\|}\right) = \arccos\left(\frac{-12}{\sqrt{85}\cdot\sqrt{13}}\right) \approx 111.161°$$

**3.** $\mathbf{a}\cdot\mathbf{b} = 1\cdot 3 + 1\cdot(-2) = 1;$

$$\theta = \arccos\left(\frac{\mathbf{a}\cdot\mathbf{b}}{\|\mathbf{a}\|\,\|\mathbf{b}\|}\right) = \arccos\left(\frac{1}{\sqrt{2}\cdot\sqrt{13}}\right) \approx 1.37340$$

**5.** $\mathbf{a}\cdot\mathbf{b} = 1\cdot(-m) + (-m)\cdot 1 = 0;$

$$\theta = \arccos\left(\frac{\mathbf{a}\cdot\mathbf{b}}{\|\mathbf{a}\|\,\|\mathbf{b}\|}\right) = \arccos 0 = \frac{\pi}{2}$$

**7.** $\mathbf{a}\cdot\mathbf{b} = 3\cdot 2.5(\cos 12°\cos 87° + \sin 12°\sin 87°) \approx 1.94114;$
$\theta = 87° - 12° = 75°$

**9.** $\mathbf{a}\cdot\mathbf{b} = 2.3\cdot(-7.05) + 4.7\cdot 3.45 = 0$
$\mathbf{a}\cdot\mathbf{c} = 2.3\cdot 4.7 + 4.7\cdot 2.3 = 21.62$
$\mathbf{b}\cdot\mathbf{c} = -7.05\cdot 4.7 + 3.45\cdot 2.3 = -25.2$
**a** and **b** are perpendicular.
A unit vector perpendicular to **d** is

$$\mathbf{u} = \frac{\langle 2.5, 1.5\rangle}{\|\langle 2.5, 1.5\rangle\|} = \frac{\langle 2.5, 1.5\rangle}{\sqrt{8.5}} \approx \langle 0.857493, 0.514496\rangle.$$

**11.** $\mathbf{a} \cdot \mathbf{b} = 3.08 \cdot 1.88 + 5.17 \cdot (-1.08) = 0.2068$
$\mathbf{a} \cdot \mathbf{c} = 3.08 \cdot (-0.30) + 5.17 \cdot 0.19 = 0.0583$
$\mathbf{b} \cdot \mathbf{c} = 1.88 \cdot (-0.30) + (-1.08) \cdot 0.19$
$= -0.7692$
None of the pairs of vectors is perpendicular.
A unit vector perpendicular to **d** is

$$\mathbf{u} = \frac{-5.7\mathbf{i} + 3.2\mathbf{j}}{\|-5.7\mathbf{i} + 3.2\mathbf{j}\|} = \frac{-5.7\mathbf{i} + 3.2\mathbf{j}}{\sqrt{42.73}} \approx -0.871984\mathbf{i} + 0.489535\mathbf{j}.$$

**13.** $\mathbf{a} \cdot \mathbf{b} = -2 \cdot 3 + 5 \cdot 1 = -1$
$\mathbf{b} \cdot \mathbf{a} = 3 \cdot (-2) + 1 \cdot 5 = -1$

**15.** $(s\mathbf{a}) \cdot \mathbf{b} = (-2(-2\mathbf{i} + 5\mathbf{j})) \cdot (3\mathbf{i} + \mathbf{j})$
$= (4\mathbf{i} - 10\mathbf{j}) \cdot (3\mathbf{i} + \mathbf{j})$
$= 12 - 10$
$= 2$
$s(\mathbf{a} \cdot \mathbf{b}) = -2[(-2\mathbf{i} + 5\mathbf{j}) \cdot (3\mathbf{i} + \mathbf{j})]$
$= -2(-6 + 5)$
$= 2$

For 17–23, note that the length of the projection of **a** on a unit vector in the direction of **b** is

$$\|\mathbf{a}\|\cos\theta| = \|\mathbf{a}\|\frac{|\mathbf{a} \cdot \mathbf{b}|}{\|\mathbf{a}\|\|\mathbf{b}\|} = \frac{|\mathbf{a} \cdot \mathbf{b}|}{\|\mathbf{b}\|}.$$

**17.** $\frac{|\mathbf{a} \cdot \mathbf{b}|}{\|\mathbf{b}\|} = \frac{\frac{3}{\sqrt{2}} + \frac{2}{\sqrt{2}}}{\sqrt{\frac{1}{2} + \frac{1}{2}}} = \frac{5}{\sqrt{2}}$
The length of the projection of **a** on a unit vector in the direction of **b** is $\frac{5}{\sqrt{2}}$.

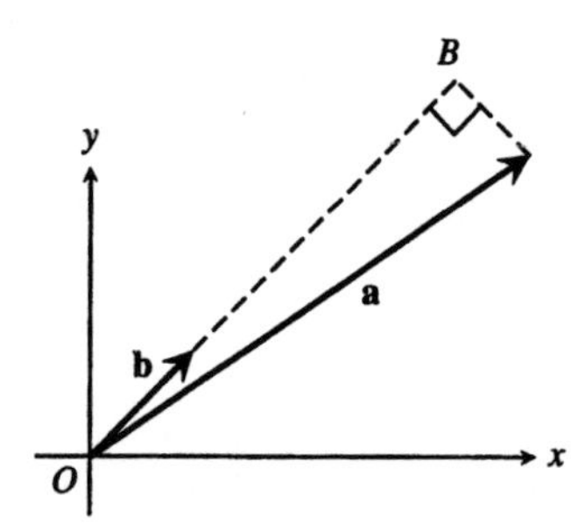


**19.** $\frac{|\mathbf{a} \cdot \mathbf{b}|}{\|\mathbf{b}\|} = \frac{5 + 5}{\sqrt{5^2 + 1}} = \frac{10}{\sqrt{26}}$
The length of the projection of **a** on a unit vector in the direction of **b** is $\frac{10}{\sqrt{26}}$.

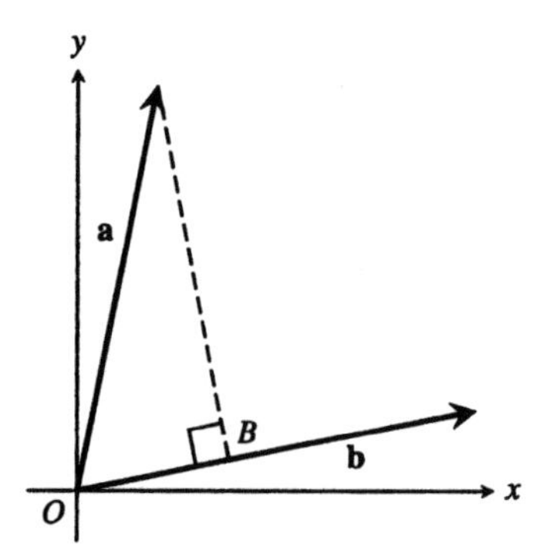


**21.** $\frac{|\mathbf{a} \cdot \mathbf{b}|}{\|\mathbf{b}\|} = \frac{|4.5 - 9.0|}{\sqrt{1 + 1}} = \frac{4.5}{\sqrt{2}} \approx 3.18198$
The length of the projection of **a** on a unit vector in the direction of **b** is $\frac{4.5}{\sqrt{2}} \approx 3.18198.$

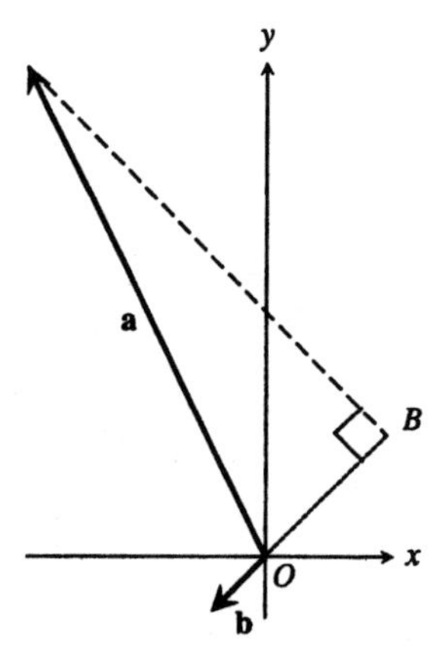


**23.** $\frac{|\mathbf{a} \cdot \mathbf{b}|}{\|\mathbf{b}\|} = \frac{\left|\frac{-2}{\sqrt{2}} + \frac{k}{\sqrt{2}}\right|}{\sqrt{\frac{1}{2} + \frac{1}{2}}} = \frac{|k - 2|}{\sqrt{2}}$
The length of the projection of **a** on a unit vector in the direction of **b** is $\frac{|-2 + k|}{\sqrt{2}}$ with $k = 3$, $\mathbf{a} = \langle 2, 3\rangle$ and the length of the projection is $\frac{1}{\sqrt{2}} \approx 0.707107.$

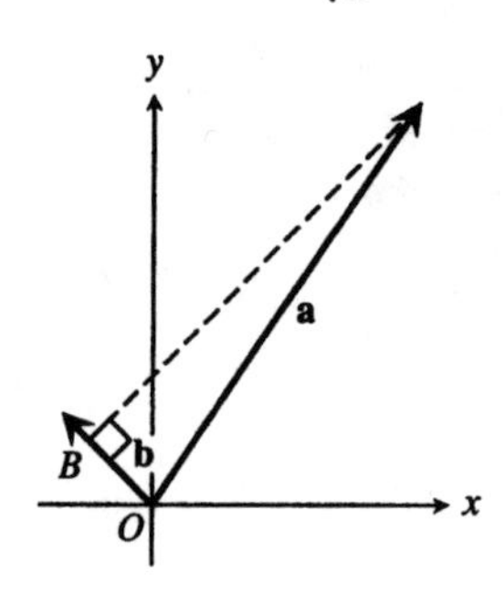

**25.** We follow Example 4:

$\mathbf{w} = 28\langle\cos(-45°), \sin(-45°)\rangle = \langle 14\sqrt{2}, -14\sqrt{2}\rangle$

and $\mathbf{u} = \langle\cos(-90°), \sin(-90°)\rangle = \langle 0, -1\rangle$, so

$\mathbf{t} = (\mathbf{w}\cdot\mathbf{u})\mathbf{u} = 14\sqrt{2}\langle 0, -1\rangle = \langle 0, -14\sqrt{2}\rangle$

$$\begin{aligned}\mathbf{c} = \mathbf{w} - \mathbf{t} &= \langle 14\sqrt{2}, -14\sqrt{2}\rangle - \langle 0, -14\sqrt{2}\rangle \\ &= \langle 14\sqrt{2}, 0\rangle.\end{aligned}$$

**27.** We follow Example 4, using headwind (**h**) instead of tailwind (**t**):

$\mathbf{w} = 20\langle\cos(175°), \sin(175°)\rangle$ and

$\mathbf{u} = \langle\cos(-45°), \sin(-45°)\rangle = \left\langle \frac{1}{\sqrt{2}}, -\frac{1}{\sqrt{2}}\right\rangle$, so

$$\begin{aligned}\mathbf{h} &= (\mathbf{w}\cdot\mathbf{u})\mathbf{u} \\ &= \left(10\sqrt{2}\cos(175°) - 10\sqrt{2}\sin(175°)\right)\mathbf{u} \\ &\approx \langle -10.8335, 10.8335\rangle\end{aligned}$$

and $\mathbf{c} = \mathbf{w} - \mathbf{h} \approx \langle -9.09039, -9.09039\rangle$.

**29.** $\mathbf{F} = \langle 0, -2000\rangle$

$\mathbf{u} = \langle\cos(5°), \sin(5°)\rangle$ is a unit vector in the direction of the hill.

We need the magnitudes of vectors **P** and **N** that are the resolution of **F** in directions parallel and perpendicular to **u**, respectively. Since the angle between **F** and **u** is 95°, $\|\mathbf{P}\| = \|F\|\cos(95°)| = 2000|\cos(95°)| \approx 174$ and $\|\mathbf{N}\| = \|F\|\sin(95°) = 2000\sin(95°) \approx 1992$. The force downhill must exceed $\mu\|\mathbf{N}\| \approx 0.20(1992) \approx 398$. Since $250 + \|\mathbf{P}\| = 250 + 174 = 424 > 398$, the loggers will be able to move the sledge.

**31.** The work is given by $W = (\|\mathbf{F}\|\cos\theta)s$ where **F** is the force, $s$ is the distance, and $\theta$ is the angle between the force and the direction of motion.

$W$ = [500 cos(45°)](15) ≈ 5303 Joules

**33.** The work done by $\mathbf{F}_1$ is

$$\begin{aligned}W_1 = s\|\mathbf{F}_1\|\cos\theta_1 &= 50(380)\cos(24°) \\ &\approx 17{,}357 \text{ J}.\end{aligned}$$

The work done by $\mathbf{F}_2$ is

$$\begin{aligned}W_2 = s\|\mathbf{F}_2\|\cos\theta_2 &= 50(550)\cos(49°) \\ &\approx 18{,}042 \text{ J}.\end{aligned}$$

**35.**

$$\begin{aligned}\langle x, y\rangle &= \frac{\langle -2, 1\rangle}{\|\langle -2, 1\rangle\|} \\ &= \frac{\langle -2, 1\rangle}{\sqrt{5}} \\ &= \left\langle -\frac{2}{\sqrt{5}}, \frac{1}{\sqrt{5}}\right\rangle\end{aligned}$$

**37.** $\langle 1, m\rangle$ has slope $\frac{m}{1} = m$.

**39.** The angle at $A$ is

$$\begin{aligned}\cos^{-1}\left(\frac{\overrightarrow{AB}\cdot\overrightarrow{AC}}{\|\overrightarrow{AB}\|\|\overrightarrow{AC}\|}\right) &= \cos^{-1}\left(\frac{\langle 3, 1\rangle\cdot\langle 1, 3\rangle}{\sqrt{9+1}\sqrt{9+1}}\right) \\ &= \cos^{-1}\left(\frac{6}{10}\right) \approx 0.927295.\end{aligned}$$

The angle at $B$ is

$$\begin{aligned}\cos^{-1}\left(\frac{\overrightarrow{BA}\cdot\overrightarrow{BC}}{\|\overrightarrow{BA}\|\|\overrightarrow{BC}\|}\right) &= \cos^{-1}\left(\frac{\langle -3, -1\rangle\cdot\langle -2, 2\rangle}{\sqrt{9+1}\sqrt{4+4}}\right) \\ &= \cos^{-1}\left(\frac{1}{\sqrt{5}}\right) \approx 1.10715.\end{aligned}$$

The angle at $C$ is

$$\begin{aligned}\cos^{-1}\left(\frac{\overrightarrow{CA}\cdot\overrightarrow{CB}}{\|\overrightarrow{CA}\|\|\overrightarrow{CB}\|}\right) &= \cos^{-1}\left(\frac{\langle -1, -3\rangle\cdot\langle 2, -2\rangle}{\sqrt{1+9}\sqrt{4+4}}\right) \\ &= \cos^{-1}\left(\frac{1}{\sqrt{5}}\right) \approx 1.10715.\end{aligned}$$

**41.** To be perpendicular, their dot product must be zero:

$(9)(2t-3) + (-3)(4) = 0 \Rightarrow t = \frac{13}{6}$.

**43.** Let $Q$ be a point on line **r**, say (3, 4), and call $P$ = (10, 1). See the figure.

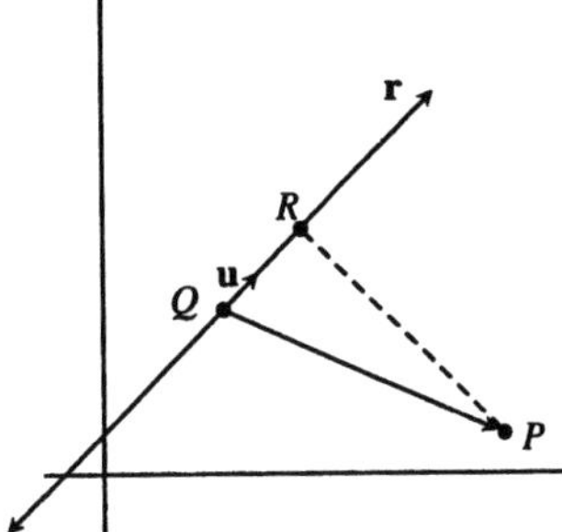


If **u** is a unit vector parallel to **r** and we resolve $\overrightarrow{QP}$ into vectors parallel and perpendicular to **u**, $\overrightarrow{QR}$ and $\overrightarrow{RP}$, respectively, then the distance from $P$ to $\mathbf{r}$ is $\left\|\overrightarrow{RP}\right\|$.

$$\mathbf{u} = \frac{\langle 1, 1\rangle}{\sqrt{1+1}} = \left\langle \frac{1}{\sqrt{2}}, \frac{1}{\sqrt{2}} \right\rangle$$

$$\overrightarrow{QP} = \langle 10-3, 1-4\rangle = \langle 7, -3\rangle$$

$$\overrightarrow{QR} = \left(\overrightarrow{QP}\cdot\mathbf{u}\right)\mathbf{u} = \left(\frac{7}{\sqrt{2}} - \frac{3}{\sqrt{2}}\right)\left\langle \frac{1}{\sqrt{2}}, \frac{1}{\sqrt{2}} \right\rangle = \langle 2, 2\rangle$$

$$\overrightarrow{RP} = \overrightarrow{QP} - \overrightarrow{QR} = \langle 7-2, -3-2\rangle = \langle 5, -5\rangle$$

$$\left\|\overrightarrow{RP}\right\| = \left\|\langle 5, -5\rangle\right\| = 5\sqrt{2}$$

The distance from $P = (10, 1)$ to $\mathbf{r} = \langle 3, 4\rangle + t\langle 1, 1\rangle$ is $5\sqrt{2}$.

**45.** The brakes must produce a force directed up the hill with the same magnitude as $\mathbf{P}$. Since the angle $\theta$ between $\mathbf{F}$ and the hill is 98°,
$\|\mathbf{P}\| = \|\mathbf{F}\||\cos\theta| = 140{,}000|\cos(98°)| \approx 19{,}484$ N.
The force produced by the brakes must be approximately $19{,}484\langle\cos(8°), \sin(80)\rangle$ N.

**47.** The cosine curve intersects the graph of the equation $y = x$ when $x = \cos x$, which is $x \approx 0.739085$. At this point the cosine curve has slope $m = y' = -\sin x \approx -0.673612$, so a vector in this direction is $\langle 1, -0.673612\rangle$. The angle can be found using the dot product.

$$\cos\theta = \frac{\langle 1, -0.673612\rangle \cdot \langle 1, 1\rangle}{\|\langle 1, -0.673612\rangle\| \, \|\langle 1, 1\rangle\|} \approx 0.191414, \text{ so}$$

$\theta \approx 1.37819 \approx 78.96°$.

**49.** Let $\mathbf{a} = \langle a_1, a_2\rangle$.

$$\begin{aligned}\mathbf{a}\cdot\mathbf{a} &= \langle a_1, a_2\rangle \cdot \langle a_1, a_2\rangle \\ &= a_1^2 + a_2^2 \\ &= \left(\sqrt{a_1^2 + a_2^2}\right)^2 \\ &= \|\mathbf{a}\|^2\end{aligned}$$

**51.** Let $\mathbf{a} = \langle a_1, a_2\rangle$, $\mathbf{b} = \langle b_1, b_2\rangle$, and $\mathbf{c} = \langle c_1, c_2\rangle$.

$$\begin{aligned}\mathbf{a}\cdot(\mathbf{b}+\mathbf{c}) &= \langle a_1, a_2\rangle \cdot \langle b_1 + c_1, b_2 + c_2\rangle \\ &= a_1b_1 + a_1c_1 + a_2b_2 + a_2c_2 \\ &= (a_1b_1 + a_2b_2) + (a_1c_1 + a_2c_2) \\ &= \mathbf{a}\cdot\mathbf{b} + \mathbf{a}\cdot\mathbf{c}.\end{aligned}$$

**53.** $\mathbf{a} = \langle\cos\alpha, \sin\alpha\rangle$ and $\mathbf{b} = \langle\cos\beta, \sin\beta\rangle$

$$\begin{aligned}\cos(\alpha-\beta) &= \frac{\mathbf{a}\cdot\mathbf{b}}{\|\mathbf{a}\|\,\|\mathbf{b}\|} = \mathbf{a}\cdot\mathbf{b} \\ &= \cos\alpha\cos\beta + \sin\alpha\sin\beta\end{aligned}$$

Note that since $\cos(-\theta) = \cos\theta$, $\cos(\alpha-\beta) = \cos(\beta-\alpha)$ and the result holds regardless of whether $\alpha$ or $\beta$ is the larger angle.

**55.**
$$\begin{aligned}\mathbf{e}_1\cdot\mathbf{e}_2 &= \cos\theta\cos\left(\theta+\frac{\pi}{2}\right) + \sin\theta\sin\left(\theta+\frac{\pi}{2}\right) \\ &= \cos\left(\theta - \left(\theta+\frac{\pi}{2}\right)\right) \\ &= \cos\left(-\frac{\pi}{2}\right) \\ &= 0.\end{aligned}$$

$x' = \mathbf{w}\cdot\mathbf{e}_1 = x\cos\theta + y\sin\theta;$

$$\begin{aligned}y' = \mathbf{w}\cdot\mathbf{e}_2 &= x\cos\left(\theta+\frac{\pi}{2}\right) + y\sin\left(\theta+\frac{\pi}{2}\right) \\ &= -x\sin\theta + y\cos\theta.\end{aligned}$$

The vector $\mathbf{e}_1 = \langle\cos\theta, \sin\theta\rangle$ makes an angle of $\theta$ with the positive $x$-axis and the vector $\mathbf{e}_2 = \left\langle\cos\left(\theta+\frac{\pi}{2}\right), \sin\left(\theta+\frac{\pi}{2}\right)\right\rangle$ makes an angle of $\theta+\frac{\pi}{2}$ with the positive $x$-axis, hence an angle of $\theta$ with the positive $y$-axis. This means that $x'$- and $y'$-axes in the directions of $\mathbf{e}_1$ and $\mathbf{e}_2$, respectively, are rotated from the standard $x$- and $y$-axes by an angle $\theta$.

**57.** Start with the square of the magnitude of sum.

$$\begin{aligned}\|\mathbf{a}+\mathbf{b}\|^2 &= (\mathbf{a}+\mathbf{b})\cdot(\mathbf{a}+\mathbf{b}) \\ &= \mathbf{a}\cdot(\mathbf{a}+\mathbf{b}) + \mathbf{b}\cdot(\mathbf{a}+\mathbf{b}) \\ &= (\mathbf{a}\cdot\mathbf{a}) + 2(\mathbf{a}\cdot\mathbf{b}) + (\mathbf{b}\cdot\mathbf{b}) \\ &= \|\mathbf{a}\|^2 + 2(\mathbf{a}\cdot\mathbf{b}) + \|\mathbf{b}\|^2 \\ &\le \|\mathbf{a}\|^2 + 2\|\mathbf{a}\|\,\|\mathbf{b}\| + \|\mathbf{b}\|^2 \\ &= (\|\mathbf{a}\| + \|\mathbf{b}\|)^2.\end{aligned}$$

Taking the square root of each side gives us the triangle inequality.

## Section R.5 Newton's Laws

**1.** $\mathbf{v}(t) = -9.8t\mathbf{j} + \mathbf{c}_1$
Since $\mathbf{v}(0) = 26\mathbf{j}$, $\mathbf{c}_1 = 26\mathbf{j}$. Then
$\mathbf{r}(t) = -4.9t^2\mathbf{j} + 26t\mathbf{j} + \mathbf{c}_2$. Since
$\mathbf{r}(0) = 0\mathbf{j}$, $\mathbf{c}_2 = 0\mathbf{j}$, i.e., $\mathbf{r}(t) = \left(26t - 4.9t^2\right)\mathbf{j}$.
The object hits the ground when
$26t_1 - 4.9t_1^2 = 0$, i.e., when
$$t_1 = \frac{26}{4.9} \approx 5.30612 \text{ s.}$$

**3.** $\mathbf{v}(t) = -9.8t\mathbf{j} + \mathbf{c}_1$
Since $\mathbf{v}(0) = 5\mathbf{j}$, $\mathbf{c}_1 = 5\mathbf{j}$. Then
$\mathbf{r}(t) = -4.9t^2\mathbf{j} + 5t\mathbf{j} + \mathbf{c}_2$. Since
$\mathbf{r}(0) = 50\mathbf{j}$, $\mathbf{c}_2 = 50\mathbf{j}$, i.e.,
$\mathbf{r}(t) = \left(-4.9t^2 + 5t + 50\right)\mathbf{j}$.
The object hits the ground when
$-4.9t_1^2 + 5t_1 + 50 = 0$, i.e., when
$$t_1 = \frac{-5 - \sqrt{1005}}{-9.8} \approx 3.74507 \text{ s.}$$

**5.** $\mathbf{v}(t) = -9.8t\mathbf{j} + \mathbf{c}_1$
Since $\mathbf{v}(0) = -5\mathbf{j}$, $\mathbf{c}_1 = -5\mathbf{j}$. Then
$\mathbf{r}(t) = -4.9t^2\mathbf{j} - 5t\mathbf{j} + \mathbf{c}_2$. Since
$\mathbf{r}(0) = 40\mathbf{j}$, $\mathbf{c}_2 = 40\mathbf{j}$, i.e.,
$\mathbf{r}(t) = \left(-4.9t^2 - 5t + 40\right)\mathbf{j}$.
The object hits the ground when
$-4.9t_1^2 - 5t_1 + 40 = 0$, i.e., when
$$t_1 = \frac{5 - \sqrt{809}}{-9.8} \approx 2.39214 \text{ s.}$$

**7.** Using Newton's second law, $\frac{dv}{dt} = k = \frac{F}{m}$, so
$v(t) = kt + c_1$.
Since $v(0) = 4$, $c_1 = 4$, $v(t) = kt + 4$, and
$v(t) = 86$ when $t = \frac{82}{k}$. The position of the plane is $x(t) = \frac{1}{2}kt^2 + 4t + c_2$.

The minimum runway length needed is
$$\begin{aligned} x\left(\frac{82}{k}\right) - x(0) &= \frac{1}{2}k\left(\frac{82}{k}\right)^2 + 4\left(\frac{82}{k}\right) \\ &= \frac{3362}{k} + \frac{328}{k} \\ &= 3690\left(\frac{m}{F}\right) \\ &= 3690\left(\frac{3.600 \times 10^5}{7.700 \times 10^5}\right) \\ &\approx 1725 \text{ m} \end{aligned}$$

**9.** An equivalent method is to keep the muzzle at the origin, and find the time when the $y$-coordinate is $-1.5$ m. The position vector of the bullet is
$$\mathbf{r} = \left\langle 700t\cos 12°, -\frac{1}{2}gt^2 + 700t\sin 12° \right\rangle.$$
Set the $y$-coordinate equal to $-1.5$ m and solve for $t$.
$$-\frac{1}{2}gt^2 + 700t\sin 12° = -1.5$$
$$t = \frac{-700\sin 12° - \sqrt{700^2\sin^2 12° + 3g}}{-g}$$
$$\approx 29.7120$$
The range is the $x$-coordinate at the above time.
$\mathbf{r}(29.7120) = 700(29.7120)\cos 12° \approx 20{,}000$ m.

**11.** $$\frac{2\pi \cdot 0.5 \text{ m}}{\text{rev}} \cdot \frac{2000 \text{ rev}}{\text{min}} \cdot \frac{1 \text{ min}}{60 \text{ sec}} = \frac{100\pi}{3} \text{ m/s.}$$

**13.** From Equation (25), $v^2 = \frac{GM_E}{r}$, where $r$ is the radius of the satellite's orbit in meters, hence $r = R_E + 1.6 \times 10^5$; $G \approx 6.67259 \times 10^{-11}$, $M_E \approx 5.97 \times 10^{24}$, and $R_E \approx 6.378 \times 10^6$.
$$v^2 = \frac{GM_E}{r} \approx \frac{(6.67259 \times 10^{-11})(5.97 \times 10^{24})}{6.378 \times 10^6 + 1.6 \times 10^5}$$
$$v \approx 7806 \text{ m/s} \approx 28{,}100 \text{ km/h}$$
$$\begin{aligned} T = \frac{2\pi r}{v} &\approx \frac{2\pi(6.378 \times 10^6 + 1.6 \times 10^5)}{7806} \\ &\approx 5263 \text{ s} \\ &\approx 87.7 \text{ min} \end{aligned}$$
The orbital speed is approximately 28,100 km/h and the period is about 87.7 minutes.

**15.** $v = 7765 = \sqrt{\dfrac{GM_E}{h + R_E}}$

$$h = \frac{GM_E}{7765^2} - R_E \approx \frac{6.67259 \times 10^{-11} \cdot 5.97 \times 10^{24}}{7765^2} - 6.378 \times 10^6 \approx 228{,}719 \text{ m} \approx 229 \text{ km}$$

**17.** $a = r\omega^2 = \left(\dfrac{5}{12}\text{ ft}\right)\left(3500 \;\dfrac{\text{rev}}{\text{min}}\;\dfrac{2\pi}{\text{rev}}\;\dfrac{1\text{ min}}{60\text{ s}}\right)^2 \approx 56{,}000 \text{ ft / s}^2$

**19.** The force the engines provide is decreased due to the acceleration of gravity, changing the constant $k$. The acceleration vector due to gravity is $\langle 0, -g\rangle$. The magnitude of its projection in the direction of the runway is $-g$ sin 2°. The new value of $k$ is $\dfrac{7.7}{3.6} - 9.8\sin(2°)$. As in Example 1, the new runway length is $x = \dfrac{1}{2}k\left(\dfrac{86}{k}\right)^2 = \dfrac{86^2}{2k} \approx 2058$ m.

**21.** Let $b$ denote the muzzle speed, which gives a flight time of $t = \dfrac{2b\sin 10°}{9.8}$ s. Refer to Example 2 of the text. Substituting for $t$ in the equation $x = bt\cos 10° = 18{,}000$ and solving for the muzzle speed $b$, we get

$$b^2 = \frac{9.8 \cdot 18{,}000}{2\sin 10° \cos 10°}$$

$$b \approx 718 \text{ m / s.}$$

**23.** We can express the range of the bullet in terms of the angle $\theta$ (using an appropriate trig identity).

$$x = \frac{2 \cdot 700^2}{9.8}\sin\theta\cos\theta = \frac{700^2}{9.8}\sin(2\theta).$$

Since sin $(2\theta) = 1$ when $\theta = \dfrac{\pi}{4}$, the angle for maximum range is $\dfrac{\pi}{4}$, or 45°. This gives a range of

$$x = \frac{700^2}{9.8} = 50{,}000 \text{ m.}$$

**25.** 286 orbits in 424 hours and 59 minutes gives a period of $\dfrac{\left(424 + \frac{59}{60}\right)(3600)}{286} \approx 5349.44$ seconds.

From Example 4, $(R_E + h)^3 = \dfrac{GM_E T^2}{4\pi^2}$, where $h$ is the altitude of the satellite.

$$\begin{aligned} h &= \sqrt[3]{\frac{GM_E T^2}{4\pi^2}} - R_E \\ &= \sqrt[3]{\frac{(6.67259 \times 10^{-11})(5.97 \times 10^{24})(5349.44)^2}{4\pi^2}} - (6.378 \times 10^6) \\ &\approx 231{,}601 \text{ m} \\ &\approx 232 \text{ km} \end{aligned}$$

**27.** $\mathbf{r}(3) = 2\langle\cos 2.4, \sin 2.4\rangle \approx \langle -1.475, 1.351\rangle$
$\mathbf{v}(t) = 1.6\langle -\sin(0.8t), \cos(0.8t)\rangle$
$\mathbf{v}(3) \approx \langle -1.081, -1.180\rangle$
$\mathbf{a}(t) = -1.28\langle\cos(0.8t), \sin(0.8t)\rangle$
$\mathbf{a}(3) \approx \langle 0.944, -0.865\rangle$

By Equation (24), the magnitude of the required force is $F = \dfrac{mv^2}{r} = \dfrac{2(1.6)^2}{2} = 2.56$ N.

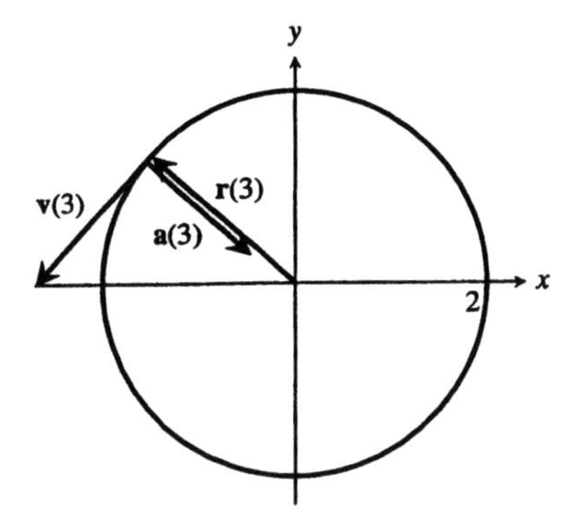


**29.** The satellite is orbiting with the same period as that of the earth, that is, 1 day per revolution, so $T = 86{,}400$ sec.

$$\begin{aligned} h &= \sqrt[3]{\frac{GM_E T^2}{4\pi^2}} - R_E \\ &= \sqrt[3]{\frac{(6.67259\times 10^{-11})(5.97\times 10^{24})(86,400)^2}{4\pi^2}} - (6.378\times 10^6) \\ &\approx 35,854,375 \text{ m} \\ &\approx 36,000 \text{ km} \end{aligned}$$

## Section R.6 Polar Coordinates and Parametric Equations

**1.** $(x, y) = (r\cos\theta, r\sin\theta) = \left(\sqrt{2}\cos\frac{\pi}{4}, \sqrt{2}\sin\frac{\pi}{4}\right) = \left(\sqrt{2}\cdot\frac{1}{\sqrt{2}}, \sqrt{2}\cdot\frac{1}{\sqrt{2}}\right) = (1,1)$

**3.** $(x, y) = (r\cos\theta, r\sin\theta) = (2\cos 1, 2\sin 1) \approx (1.08060, 1.68294)$

**5.** $(x, y) = (r\cos\theta, r\sin\theta) = (\cos 3.5, \sin 3.5) \approx (-0.936457, -0.350783)$

**7.** $(r, \theta) = \left(\sqrt{x^2 + y^2}, \arctan\frac{y}{x}\right)$

$\left(\sqrt{2^2 + 1^2}, \arctan\frac{1}{2}\right) = \left(\sqrt{5}, \arctan\frac{1}{2}\right) \approx (2.23607, 0.463648)$

**9.** $(-2, 0)$ is 2 units from the origin along negative $x$-axis.
$(r, \theta) = (2, \pi)$

**11.** $(r, \theta) = \left(\sqrt{x^2 + y^2},\ \arctan \frac{y}{x}\right)$ before adjustment

$= \left(\sqrt{(-1.5)^2 + 2.5^2},\ \arctan \frac{2.5}{-1.5}\right)$

$= \left(\sqrt{8.5},\ \arctan\left(-\frac{5}{3}\right)\right)$

$\approx (2.91548,\ -1.03038)$

$(r, \theta) = \left(\sqrt{8.5},\ \arctan\left(-\frac{5}{3}\right) + \pi\right)$ $(-1.5, 2.5)$ is in quad. II

$\approx (2.91548,\ 2.11122)$

**13.** The points with polar coordinates $(0, 0)$, $\left(\frac{\pi^2}{16}, \frac{\pi}{4}\right)$, $\left(\frac{\pi^2}{4}, \frac{\pi}{2}\right)$, $\left(\frac{9\pi^2}{16}, \frac{3\pi}{4}\right)$, and $(\pi^2, \pi)$ are part of the graph.

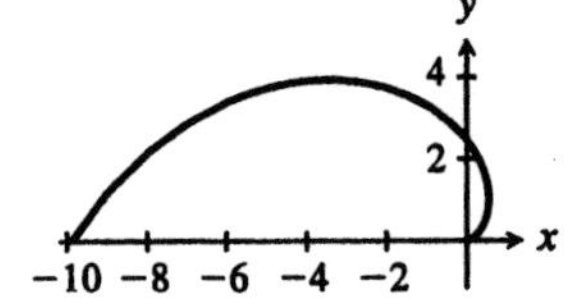


**15.** The points with polar coordinates (0, 0), (1, 1), (0, 2), and (–3, 3) are part of the graph.

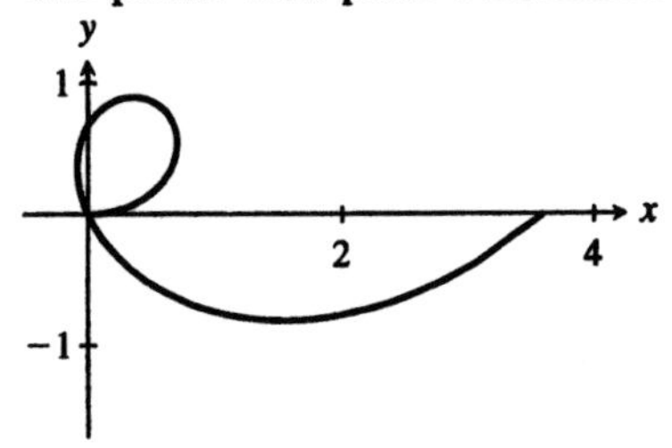


**17.** The points with polar coordinates $(2, 0) = (-2, \pi)$, $\left(\sqrt{2}, \frac{\pi}{4}\right) = \left(-\sqrt{2}, \frac{5\pi}{4}\right)$, $\left(0, \frac{\pi}{2}\right) = \left(0, \frac{3\pi}{2}\right)$, and $\left(-\sqrt{2}, \frac{3\pi}{4}\right) = \left(\sqrt{2}, \frac{7\pi}{4}\right)$ are part of the graph.

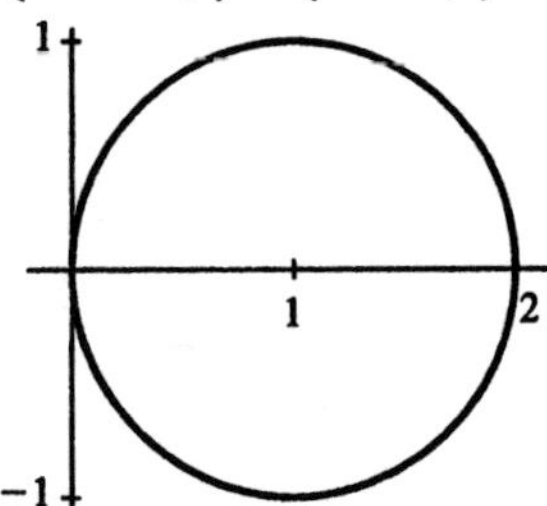

**19.** The points with polar coordinates $(0, 0) = (0, \pi)$, $\left(\sqrt{2}, \frac{\pi}{4}\right) = \left(-\sqrt{2}, \frac{5\pi}{4}\right)$, $\left(2, \frac{\pi}{2}\right) = \left(-2, \frac{3\pi}{2}\right)$, and $\left(\sqrt{2}, \frac{3\pi}{4}\right) = \left(-\sqrt{2}, \frac{7\pi}{4}\right)$ are part of the graph.

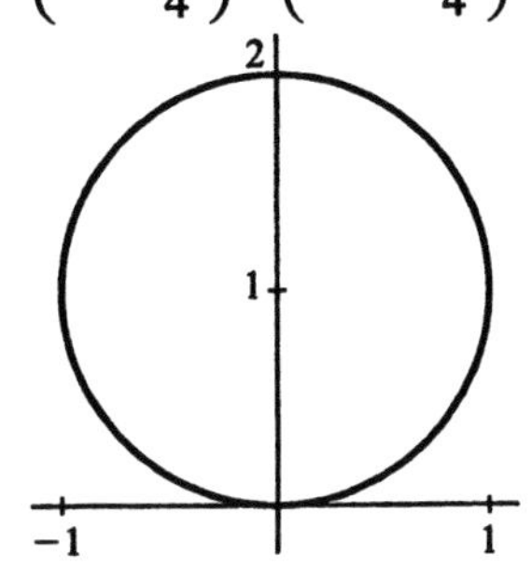


**21.** The points with polar coordinates $(-1, 0)$, $\left(\sqrt{3}-1, \frac{\pi}{2}\right)$, $(1, \pi)$, and $\left(-\sqrt{3}-1, \frac{3\pi}{2}\right)$ are part of the graph.

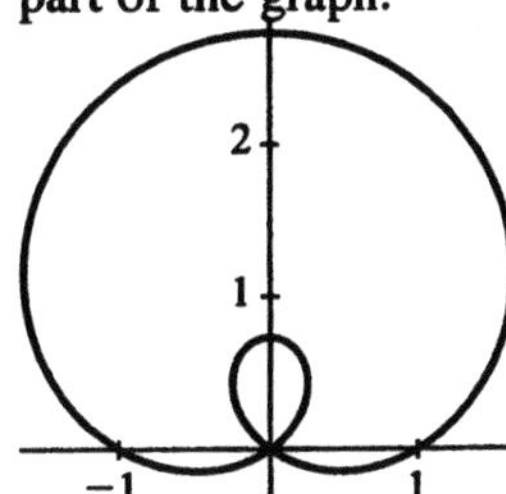


**23.** Cartesian:

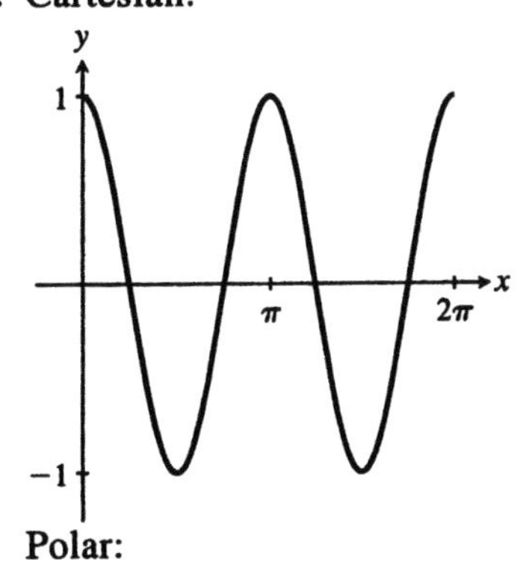


Polar:

**25.**
$$\frac{dy}{dx} = \frac{\frac{dy}{d\theta}}{\frac{dx}{d\theta}} = \frac{\frac{d}{d\theta}(r\sin\theta)}{\frac{d}{d\theta}(r\cos\theta)}$$
$$= \frac{\frac{d}{d\theta}[(1-2\sin\theta)\sin\theta]}{\frac{d}{d\theta}[(1-2\sin\theta)\cos\theta]}$$
$$= \frac{\frac{d}{d\theta}(\sin\theta - 2\sin^2\theta)}{\frac{d}{d\theta}(\cos\theta - 2\sin\theta\cos\theta)}$$
$$= \frac{\frac{d}{d\theta}(\sin\theta - 2\sin^2\theta)}{\frac{d}{d\theta}(\cos\theta - \sin 2\theta)}$$
$$= \frac{\cos\theta - 4\sin\theta\cos\theta}{-\sin\theta - 2\cos 2\theta}$$
$$= \frac{4\sin\left(\frac{5\pi}{4}\right)\cos\left(\frac{5\pi}{4}\right) - \cos\left(\frac{5\pi}{4}\right)}{\sin\left(\frac{5\pi}{4}\right) + 2\cos\left(\frac{5\pi}{2}\right)}$$
$$= -2\sqrt{2} - 1 \approx -3.82843$$

**27.** In the derivation on pp. 468–469, **u** becomes $\langle 2\cos\theta, 2\sin\theta\rangle$, **w** becomes $2\theta\langle\sin\theta, -\cos\theta\rangle$, and $\mathbf{r}(\theta) = \mathbf{u} + \mathbf{w}$ becomes $\langle 2\cos\theta + 2\theta\sin\theta, 2\sin\theta - 2\theta\cos\theta\rangle$.

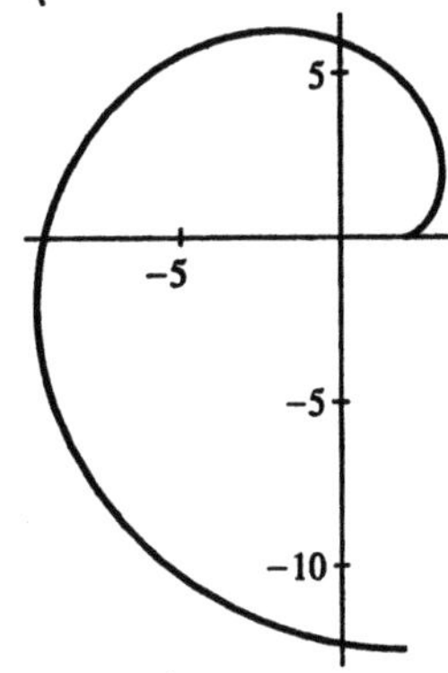


$$\frac{dy}{dx} = \frac{\frac{dy}{d\theta}}{\frac{dx}{d\theta}} = \frac{2\cos\theta + 2\theta\sin\theta - 2\cos\theta}{-2\sin\theta + 2\theta\cos\theta + 2\sin\theta}$$
$$= \tan\theta$$

When 0.6 meters have been unwound, $\theta = \frac{0.6}{2} = 0.3$. At $\theta = 0.3$, $\frac{dy}{dx} = \tan 0.3 \approx 0.309336$.

**29.** In the derivation on pp. 467–468, **u**, **v**, **w**, and **r** all double in magnitude. So $\mathbf{r}(\phi)$ becomes $2\langle\phi - \sin\phi, 1 - \cos\phi\rangle$.

$\mathbf{r}'(\phi) = 2\langle 1 - \cos\phi, \sin\phi\rangle$

When $\mathbf{r}(\phi) = \left\langle \frac{4\pi}{3} - \sqrt{3}, 3 \right\rangle$, $2(1 - \cos\phi) = 3$ and $\phi = \frac{2\pi}{3}$.

$$\left(\text{Check: } 2\left(\frac{2\pi}{3} - \sin\left(\frac{2\pi}{3}\right)\right) = \frac{4\pi}{3} - \sqrt{3}.\right)$$

$$\mathbf{r}'\left(\frac{2\pi}{3}\right) = 2\left\langle 1 - \cos\frac{2\pi}{3}, \sin\frac{2\pi}{3} \right\rangle = \left\langle 3, \sqrt{3} \right\rangle$$

**31.** Haley's comet: the points with polar coordinates (35.300, 0), $\left(1.154, \frac{\pi}{2}\right)$, $(0.587, \pi)$, and $\left(1.154, \frac{3\pi}{2}\right)$ are part of the graph.

Mars: (1.666, 0), $\left(1.510, \frac{\pi}{2}\right)$, $(1.381, \pi)$, and $\left(1.510, \frac{3\pi}{2}\right)$ are on the graph.

Earth: (0, 1.017), $\left(1.000, \frac{\pi}{2}\right)$, $(0.983, \pi)$, and $\left(1.000, \frac{3\pi}{2}\right)$ are part of the graph.

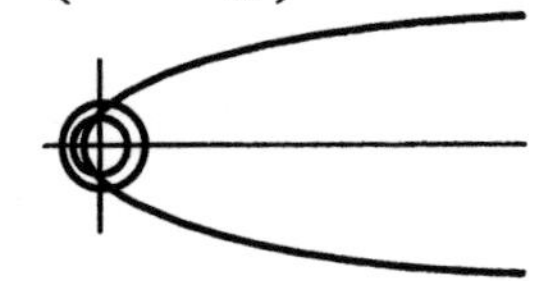

**33.** The vector $\mathbf{u} = \langle \phi, 0 \rangle$ would be $\langle 2t, 0 \rangle$ m after $t$ seconds. Taking $\phi = 2t$, then, the motion is described by $\mathbf{r}(t) = \langle 2t - \sin(2t), 1 - \cos(2t) \rangle$. It follows that the velocity is

$$\mathbf{r}'(t) = \langle 2 - 2\cos(2t), 2\sin(2t) \rangle = 2\langle 1 - \cos(2t), \sin(2t) \rangle \text{ m / s.}$$

For $t = 3$, $\mathbf{r}'(3) = 2\langle 1 - \cos(6), \sin(6) \rangle$ m / s $\approx \langle 0.079659, -0.558831 \rangle$ m / s.

The speed is $\|\mathbf{r}'(3)\| = 2\sqrt{(1-\cos(6))^2 + \sin^2(6)} \approx 0.564480$ m / s.

**35.** $x = \frac{3t}{t^3+1}$, $y = \frac{3t^2}{t^3+1}$

$\frac{y}{x} = t$,

$$x + y = \frac{3t + 3t^2}{t^3+1} = \frac{3t(t+1)}{(t+1)(t^2-t+1)} = \frac{3t}{t^2-t+1}$$

$$x + y = \frac{3\left(\frac{y}{x}\right)}{\left(\frac{y}{x}\right)^2 - \frac{y}{x} + 1}$$

$$x + y = \frac{3xy}{y^2 - xy + x^2}$$

$$xy^2 - x^2y + x^3 + y^3 - xy^2 + x^2y = 3xy$$

$$x^3 - 3xy + y^3 = 0$$

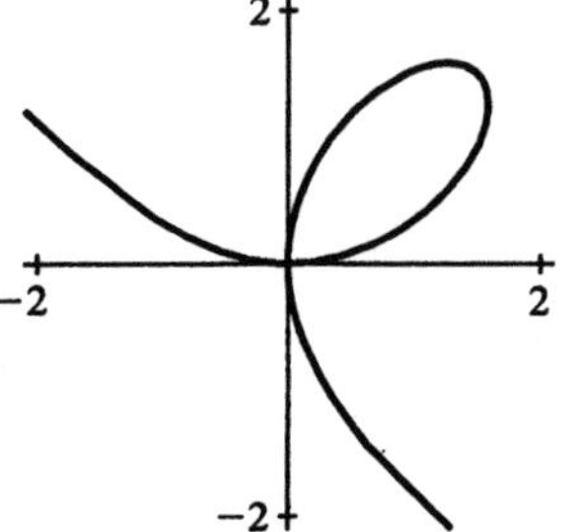


The fourth-quadrant part of the folium ($x > 0$, $y < 0$) is traced as $t$ varies on the interval $(-\infty, -1)$, with the trace moving downward as $t$ increases toward $-1$.
The second-quadrant part ($x \le 0$, $y > 0$) is traced as $t$ varies on $(-1, 0]$, with the trace moving toward the origin as $t$ increases.
The first-quadrant part ($x, y > 0$) is traced as $t$ varies on $[0, \infty)$, with the trace moving counterclockwise.
The tangent line is horizontal when

$$\frac{dy}{dx} = \frac{\frac{dy}{dt}}{\frac{dx}{dt}} = 0$$

$$\Rightarrow \frac{dy}{dt} = \frac{(t^3+1)(6t) - (3t^2)(3t^2)}{(t^3+1)^2} = 0$$

$\Rightarrow (t^3+1)(6t) = (3t^2)(3t^2) \Rightarrow t = 0$ or $t = \sqrt[3]{2}$.

The tangent line is vertical when

$$\frac{dy}{dx} = \frac{\frac{dy}{dt}}{\frac{dx}{dt}} = \text{undef.}$$

$$\Rightarrow \frac{dx}{dt} = \frac{(t^3+1)(3) - (3t^2)(3t)}{(t^3+1)^2} = 0$$

$$\Rightarrow (t^3+1)(3) = (3t^2)(3t) \Rightarrow t = \frac{1}{\sqrt[3]{2}}.$$

So the points on the folium at which the tangent line is horizontal are $\mathbf{r}(0)$ and $\mathbf{r}\left(\sqrt[3]{2}\right)$, and the point on the folium at which the tangent line is vertical is $\mathbf{r}\left(\frac{1}{\sqrt[3]{2}}\right)$.

**37.** We wish to show that
$x = \phi - \sin\phi$, $y = 1 - \cos\phi$
$\Rightarrow \phi = \arccos(1 - y)$ $(0 \le \phi \le \pi)$.
By substitution,
$x = \arccos(1 - y) - \sin(\arccos(1 - y))$.
The range $0 \le \phi \le \pi$ corresponds to a half-revolution of the circle, or the first half of the first period.

**39.** The points with polar coordinates $(-11.5129, 0.0001)$, $(0, 1)$, $(1.1447, \pi)$, and $(1.8379, 2\pi)$ are part of the graph.

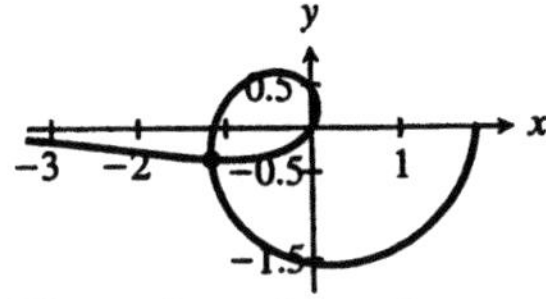


The point where the curve crosses itself corresponds to two values $\theta_1$ and $\theta_2$ such that $\theta_2 = \theta_1 + \pi$ and $r_2 = -r_1$

$$\Rightarrow \ln\theta_1 = -\ln\theta_2 = -\ln(\theta_1 + \pi) = \ln\left(\frac{1}{(\theta_1 + \pi)}\right).$$

Thus, $$\theta_1 = \frac{1}{\theta_1 + \pi}$$

$$\theta_1^2 + \pi\theta_1 - 1 = 0$$

$$\theta_1 = \frac{-\pi + \sqrt{\pi^2 + 4}}{2}$$

$\approx 0.291300$

where we pick the positive root of the quadratic since $\theta > 0$. Then
$(r_1, \theta_1) = (-1.233403, 0.291300)$,
$(r_2, \theta_2) = (1.233403, 3.432892)$.

## Section R.7 Arc Length and Unit Tangent Vectors

**1.**

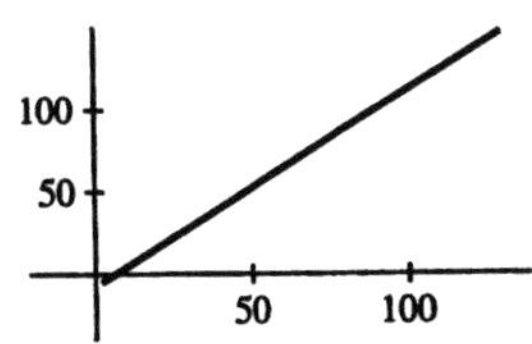


$\mathbf{r}(t) = \langle 3, -4\rangle + t\langle 5, 6\rangle$, $0 \le t \le 25$

$\mathbf{r}'(t) = \langle 5, 6\rangle$

$$s = \int_a^b \sqrt{x'(t)^2 + y'(t)^2}\,dt = \int_0^{25} \sqrt{5^2 + 6^2}\,dt$$
$$= 25\sqrt{61} \text{ meters}$$

**3.**

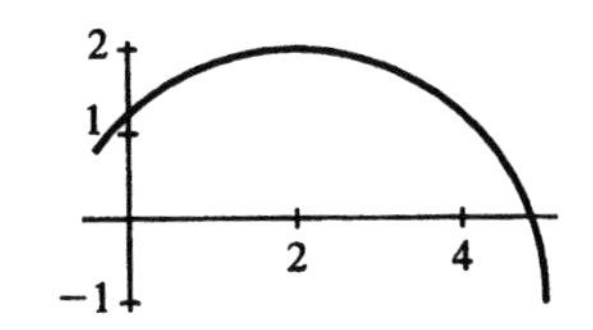


$\mathbf{r}(t) = \langle 2, -1\rangle + 3\langle \cos 0.05t, \sin 0.05t\rangle$, $10 \le t \le 50$

$\mathbf{r}'(t) = 3\langle -0.05\sin 0.05t, 0.05\cos 0.05t\rangle$
$= \langle -0.15\sin 0.05t, 0.15\cos 0.05t\rangle$

$$s = \int_a^b \sqrt{x'(t)^2 + y'(t)^2}\,dt$$
$$= \int_0^{50} \sqrt{(-0.15\sin 0.05t)^2 + (0.15\cos 0.05t)^2}\,dt$$
$$= \int_0^{50} 0.15\sqrt{\sin^2 0.05t + \cos^2 0.05t}\,dt$$
$$= 0.15(50) = 7.5 \text{ meters}$$

**5.**

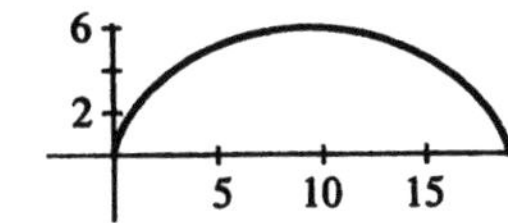


$\mathbf{r}(t) = 3\langle 2t - \sin 2t, 1 - \cos 2t\rangle$, $0 \le t \le \pi$

$\mathbf{r}'(t) = 3\langle 2 - 2\cos 2t, 2\sin 2t\rangle$
$= \langle 6 - 6\cos 2t, 6\sin 2t\rangle$

$$s = \int_a^b \sqrt{x'(t)^2 + y'(t)^2}\,dt$$
$$= \int_0^{\pi} \sqrt{(6 - 6\cos 2t)^2 + (6\sin 2t)^2}\,dt$$
$$= \int_0^{\pi} 6\sqrt{1 - 2\cos 2t + \cos^2 2t + \sin^2 2t}\,dt$$
$$= 6\int_0^{\pi} \sqrt{2 - 2\cos 2t}\,dt$$
$$= 6\int_0^{\pi} \sqrt{4\sin^2 t}\,dt$$
$$= 12\int_0^{\pi} \sin t\,dt$$
$$= -12\cos t\Big|_0^{\pi}$$
$$= 24 \text{ meters}$$

**7.**

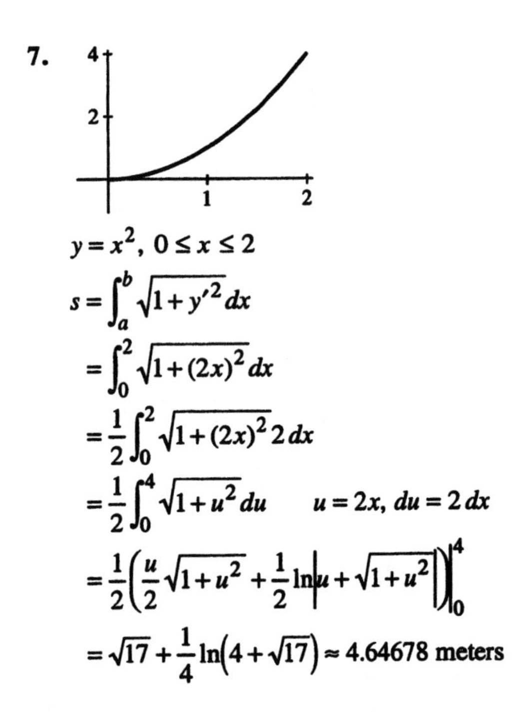


$y = x^2,\ 0 \le x \le 2$

$$s = \int_a^b \sqrt{1 + y'^2}\,dx$$

$$= \int_0^2 \sqrt{1 + (2x)^2}\,dx$$

$$= \frac{1}{2}\int_0^2 \sqrt{1 + (2x)^2}\,2\,dx$$

$$= \frac{1}{2}\int_0^4 \sqrt{1 + u^2}\,du \qquad u = 2x,\ du = 2\,dx$$

$$= \frac{1}{2}\left(\frac{u}{2}\sqrt{1+u^2} + \frac{1}{2}\ln\left|u + \sqrt{1+u^2}\right|\right)\Bigg|_0^4$$

$$= \sqrt{17} + \frac{1}{4}\ln\left(4 + \sqrt{17}\right) \approx 4.64678 \text{ meters}$$

**9.**

$y = x^{3/2},\ 0 \le x \le 5$

$$s = d\int_a^b \sqrt{1 + y'^2}\,dx$$

$$= \int_0^5 \sqrt{1 + \left(\frac{3}{2}x^{1/2}\right)^2}\,dx$$

$$= \int_0^5 \sqrt{1 + \frac{9}{4}x}\,dx$$

$$= \frac{8}{27}\left(1 + \frac{9}{4}x\right)^{3/2}\Bigg|_0^5$$

$$= \frac{335}{27} \approx 12.4074 \text{ meters}$$

**11.**

$r = 1 + \cos\theta,\ 0 \le \theta \le \pi$

$$s = \int_\alpha^\beta \sqrt{r^2 + r'^2}\,d\theta$$

$$= \int_0^\pi \sqrt{(1+\cos\theta)^2 + (-\sin\theta)^2}\,d\theta$$

$$= \int_0^\pi \sqrt{1 + 2\cos\theta + \cos^2\theta + \sin^2\theta}\,d\theta$$

$$= \int_0^\pi \sqrt{2 + 2\cos\theta}\,d\theta$$

$$= \int_0^\pi \sqrt{4\cos^2\frac{\theta}{2}}\,d\theta$$

$$= 2\int_0^\pi \cos\frac{\theta}{2}\,d\theta$$

$$= 2\left(2\sin\frac{\theta}{2}\right)\Bigg|_0^\pi$$

$$= 4 \text{ meters}$$

**13.**

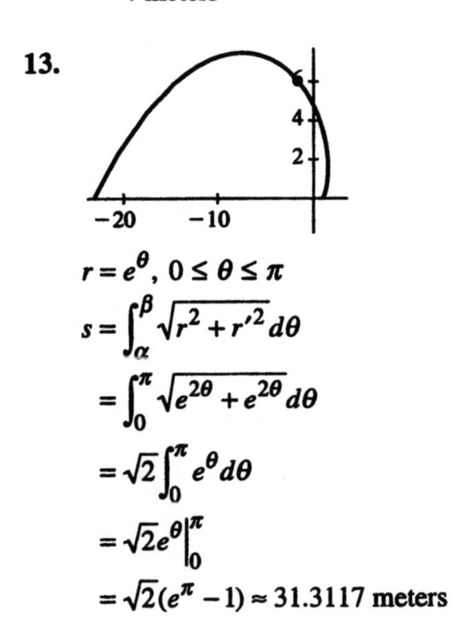


$r = e^\theta,\ 0 \le \theta \le \pi$

$$s = \int_\alpha^\beta \sqrt{r^2 + r'^2}\,d\theta$$

$$= \int_0^\pi \sqrt{e^{2\theta} + e^{2\theta}}\,d\theta$$

$$= \sqrt{2}\int_0^\pi e^\theta\,d\theta$$

$$= \sqrt{2}e^\theta\Big|_0^\pi$$

$$= \sqrt{2}(e^\pi - 1) \approx 31.3117 \text{ meters}$$

**15.**

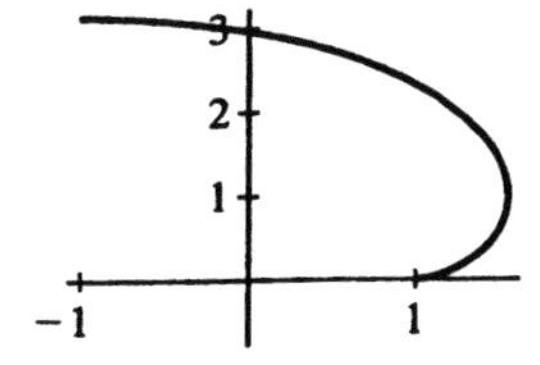


$\mathbf{r}(\theta) = \langle \cos\theta + \theta\sin\theta,\ \sin\theta - \theta\cos\theta \rangle,\ 0 \le \theta \le \pi$

$\mathbf{r}'(\theta) = \langle -\sin\theta + \theta\cos\theta + \sin\theta,\ \cos\theta + \theta\sin\theta - \cos\theta \rangle = \langle \theta\cos\theta,\ \theta\sin\theta \rangle$

$$\begin{aligned} s = \int_\alpha^\beta \sqrt{x'(\theta)^2 + y'(\theta)}\, d\theta &= \int_0^\pi \sqrt{\theta^2\cos^2\theta + \theta^2\sin^2\theta}\, d\theta \\ &= \int_0^\pi \theta\, d\theta \\ &= \frac{1}{2}\theta^2\Big|_0^\pi \\ &= \frac{1}{2}\pi^2 \\ &\approx 4.9348 \text{ meters} \end{aligned}$$

**17.**

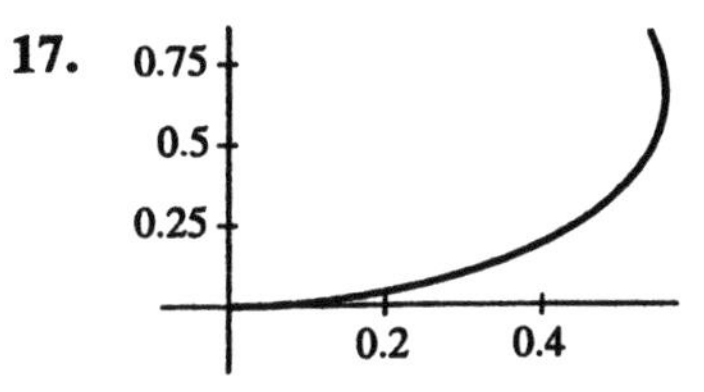


$\mathbf{r}(t) = \langle t\cos t,\ t\sin t \rangle,\ 0 \le t \le 1$

$\mathbf{r}'(t) = \langle -t\sin t + \cos t,\ t\cos t + \sin t \rangle$

$$\begin{aligned} s &= \int_a^b \sqrt{x'(t)^2 + y'(t)^2}\, dt \\ &= \int_0^1 \sqrt{(-t\sin t + \cos t)^2 + (t\cos t + \sin t)^2}\, dt \\ &= \int_0^1 \sqrt{t^2\sin^2 t - 2t\sin t\cos t + \cos^2 t + t^2\cos^2 t + 2t\sin t\cos t + \sin^2 t}\, dt \\ &= \int_0^1 \sqrt{1 + t^2}\, dt \\ &= \left( \frac{t}{2}\sqrt{1+t^2} + \frac{1}{2}\ln\left|t + \sqrt{1+t^2}\right| \right)\Big|_0^1 \\ &= \frac{1}{2}\sqrt{2} + \frac{1}{2}\ln\left|1 + \sqrt{2}\right| \\ &\approx 1.14779 \text{ meters} \end{aligned}$$

**19.**

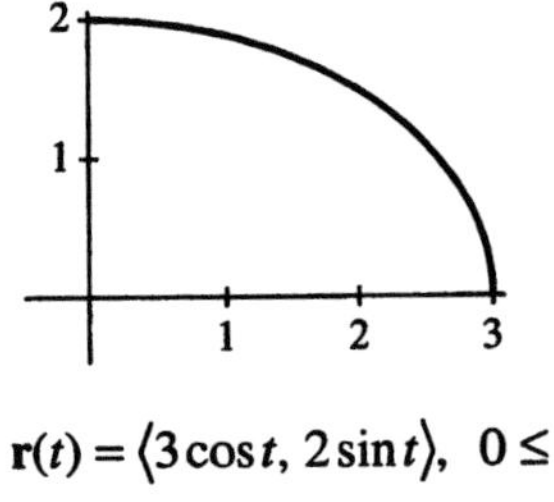


$\mathbf{r}(t) = \langle 3\cos t, 2\sin t\rangle, \ 0 \le t \le \frac{\pi}{2}$

$\mathbf{r}'(t) = \langle -3\sin t, 2\cos t\rangle$

$$s = \int_a^b \sqrt{x'(t)^2 + y'(t)^2}\,dt$$

$$= \int_0^{\pi/2} \sqrt{(-3\sin t)^2 + (2\cos t)^2}\,dt$$

$$= \int_0^{\pi/2} \sqrt{9\sin^2 t + 4\cos^2 t}\,dt$$

$\approx 3.96636$ meters (CAS)

**21.**

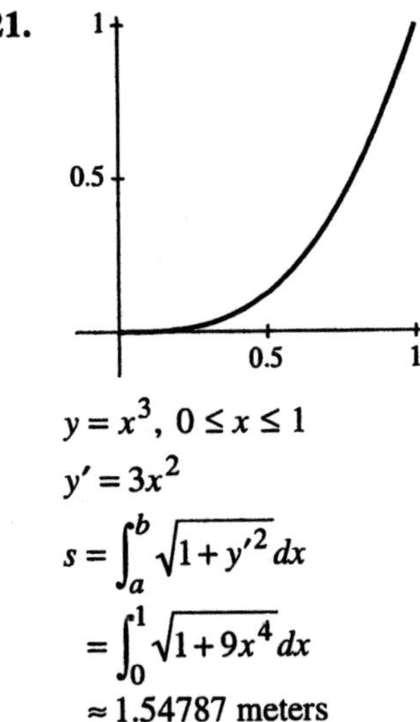


$y = x^3, \ 0 \le x \le 1$

$y' = 3x^2$

$$s = \int_a^b \sqrt{1 + y'^2}\,dx$$

$$= \int_0^1 \sqrt{1 + 9x^4}\,dx$$

$\approx 1.54787$ meters

**23.** $\mathbf{r}(t) = \langle 3\cos t, 2\sin t\rangle, \ 0 \le t \le 2\pi$

$\mathbf{r}'(t) = \langle -3\sin t, 2\cos t\rangle$

$$\mathbf{T}(t) = \frac{1}{\|\mathbf{r}'(t)\|}\mathbf{r}'(t)$$

$$= \frac{1}{\sqrt{9\sin^2 t + 4\cos^2 t}}\langle -3\sin t, 2\cos t\rangle$$

$$\mathbf{T}\left(\frac{\pi}{4}\right) = \frac{1}{\sqrt{9\left(\frac{1}{\sqrt{2}}\right)^2 + 4\left(\frac{1}{\sqrt{2}}\right)^2}}\left\langle -3\left(\frac{1}{\sqrt{2}}\right), 2\left(\frac{1}{\sqrt{2}}\right)\right\rangle$$

$$= \left\langle -\frac{3}{\sqrt{13}}, \frac{2}{\sqrt{13}}\right\rangle$$

**25.** $y = x^3, \ 0 \le x \le 2$

$\mathbf{r} = \langle x, y\rangle = \langle x, x^3\rangle$

$\mathbf{r}' = \langle 1, 3x^2\rangle$

$$\mathbf{T} = \frac{1}{\|\mathbf{r}'\|}\mathbf{r}'$$

$$= \frac{1}{\sqrt{1 + 9x^4}}\langle 1, 3x^2\rangle$$

For $x = 1$,

$$\mathbf{T} = \frac{1}{\sqrt{1+9}}\langle 1, 3\rangle = \left\langle \frac{1}{\sqrt{10}}, \frac{3}{\sqrt{10}}\right\rangle$$

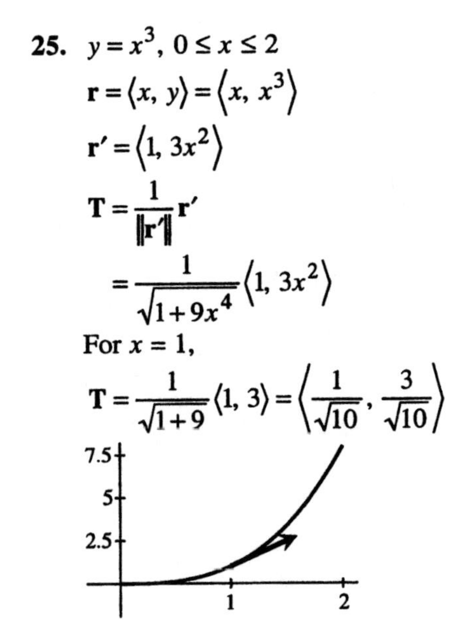

**27.** $\mathbf{r}(\theta) = \langle a\cos\theta,\ b\sin\theta\rangle,\ 0 \le \theta \le 2\pi$

$\mathbf{r}'(\theta) = \langle -a\sin\theta,\ b\cos\theta\rangle$

$$s = \int_\alpha^\beta \sqrt{x'(\theta)^2 + y'(\theta)^2}\, d\theta$$

$$= \int_0^{2\pi} \sqrt{a^2\sin^2\theta + b^2\cos^2\theta}\, d\theta$$

$$= 4\int_0^{\pi/2} \sqrt{a^2\sin^2\theta + b^2\cos^2\theta}\, d\theta \quad \text{two-way symmetry of ellipse}$$

$$= 4\int_0^{\pi/2} \sqrt{a^2\sin^2\theta + b^2(1-\sin^2\theta)}\, d\theta$$

$$= 4\int_0^{\pi/2} \sqrt{c^2\sin^2\theta + b^2}\, d\theta$$

where $c^2 = a^2 - b^2$. Using the trapezoid rule, $s \approx 4(19.27 \pm 0.01)$ AU.

**29.** $\mathbf{r}(t) = \langle t,\ \ln t\rangle,\ \frac{1}{2} \le t \le 2$

$$\mathbf{r}'(t) = \left\langle 1,\ \frac{1}{t}\right\rangle$$

$$s = \int_a^b \sqrt{x'(t)^2 + y'(t)^2}\, dt = \int_{1/2}^2 \sqrt{1 + \frac{1}{t^2}}\, dt \approx \sum_{i=1}^5 \sqrt{1 + \frac{1}{t_{i-1}^2}}\, \Delta t \ \text{ where } t_i = \frac{1}{2} + \frac{3}{10}i \text{ and } \Delta t = \frac{3}{10}.$$

Then $s \approx 2.27322$.

Since $f(t) = \sqrt{1 + \frac{1}{t^2}}$ is a descending function for $t > 0$, the approximation using a Riemann sum based of left endpoints $t_{i-1}$ of intervals $[t_{i-1}, t]$ is too large. But since the function flattens out (slope approaches zero) as $t$ increases, the approximation would improve on the interval $\left[\frac{3}{2}, 3\right]$, other things being equal (i.e., still a left-endpoint approximation with $n = 5$).

**31.** $y = x^2,\ 0 \le x \le 1$

$y' = 2x$

$$s = \int_a^b \sqrt{1 + y'^2}\, dx$$

$$= \int_0^1 \sqrt{1 + (2x)^2}\, dx$$

$$= \frac{1}{2}\int_0^{\arctan 2} \sqrt{1 + \tan^2\theta}\, \sec^2\theta\, d\theta$$

$2x = \tan\theta \Rightarrow dx = \frac{1}{2}\sec^2\theta\, d\theta$

$x = 0 \Rightarrow \theta = 0;\ x = 1 \Rightarrow \theta = \arctan 2$

$$= \frac{1}{2}\int_0^{\arctan 2} \sec^3\theta\, d\theta$$

$$= \frac{1}{2}\left(\frac{1}{2}\sec\theta\tan\theta + \frac{1}{2}\ln|\sec\theta + \tan\theta|\right)\Big|_0^{\arctan 2}$$

$\approx 1.4789$ units

**33.** Since the first equation describes motion at a steady rate, the second equation describes the motion of the hare. At $t = 0.2$ the hare's speed is 6.78584 units per time unit, while the tortoise is going $2\pi$ units per time unit. At $t = 0.5$ the hare is resting and the tortoise is still going at $2\pi$ units per time unit.

**35.** No, because $\mathbf{r}'(t) = \langle 0, 0 \rangle$ at $t = 0, 2\pi$, so that $\|\mathbf{r}'(t)\| = 0$ and $\mathbf{T} = \dfrac{1}{\|\mathbf{r}'(t)\|}\mathbf{r}'(t)$ is undefined.

However, one might argue that unit tangents can be defined at $\mathbf{r}(0)$ and $\mathbf{r}(2\pi)$ by taking a limit as $t \to 0^+$ or $t \to 2\pi^-$.

## Section R.8 Areas of Regions Described by Polar Equations

**1.**

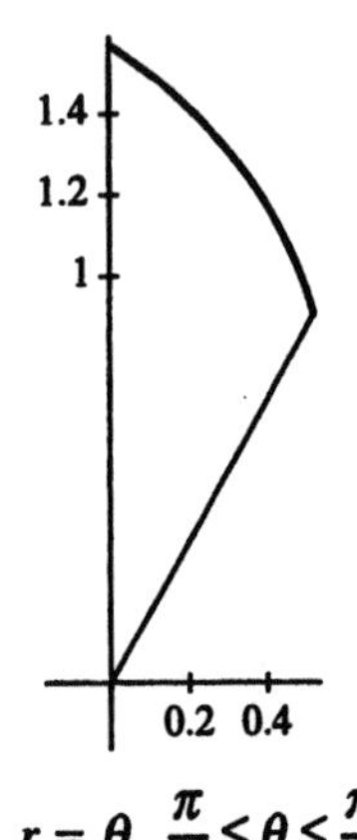


$r = \theta,\ \dfrac{\pi}{3} \le \theta \le \dfrac{\pi}{2}$

$$\begin{aligned} A &= \int_\alpha^\beta \frac{1}{2} r^2 d\theta \\ &= \int_{\pi/3}^{\pi/2} \frac{1}{2}\theta^2 d\theta \\ &= \frac{1}{6}\theta^3 \Big|_{\pi/3}^{\pi/2} \\ &= \frac{19\pi^3}{1296} \text{ square units} \end{aligned}$$

**3.**

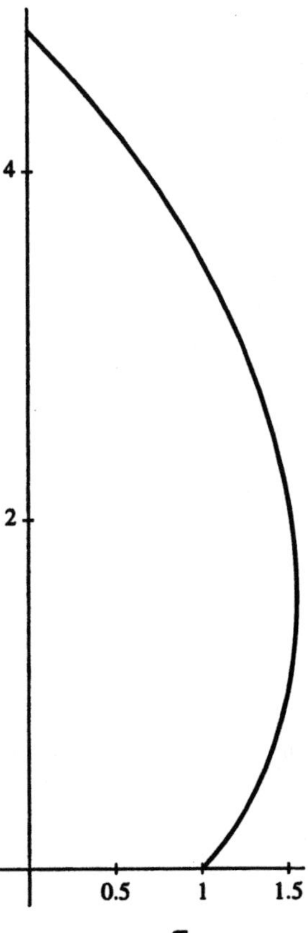


$r = e^\theta,\ 0 \le \theta \le \dfrac{\pi}{2}$

$$\begin{aligned} A &= \int_\alpha^\beta \frac{1}{2} r^2 d\theta \\ &= \int_0^{\pi/2} \frac{1}{2} e^{2\theta} d\theta \\ &= \frac{1}{4} e^{2\theta} \Big|_0^{\pi/2} \\ &= \frac{1}{4}(e^\pi - 1) \text{ square units} \end{aligned}$$

**5.**

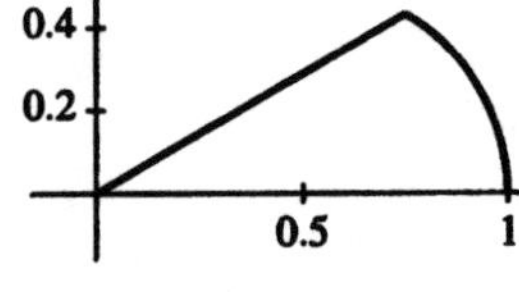


$r = \cos\theta,\ 0 \le \theta \le \dfrac{\pi}{6}$

$$\begin{aligned} A &= \int_\alpha^\beta \frac{1}{2} r^2 d\theta \\ &= \int_0^{\pi/6} \frac{1}{2}\cos^2\theta\, d\theta \\ &= \left(\frac{1}{4}\theta + \frac{1}{8}\sin 2\theta\right)\Big|_0^{\pi/6} \\ &= \frac{1}{48}\left(3\sqrt{3} + 2\pi\right) \text{ square units} \end{aligned}$$

**7.**

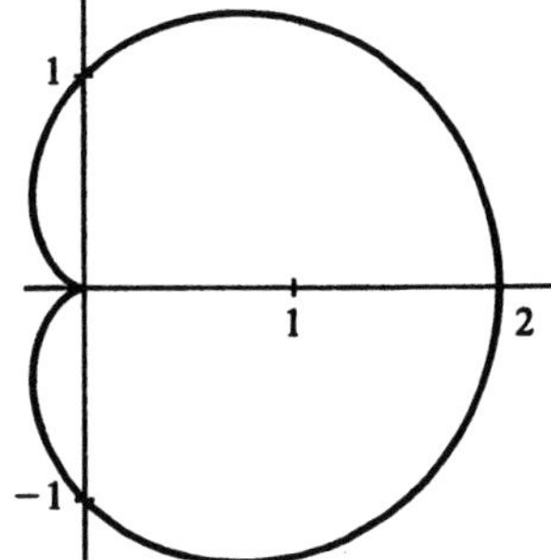


$r = 1 + \cos\theta,\ 0 \le \theta \le 2\pi$

$$A = 2\int_0^{\pi} \frac{1}{2}(1+\cos\theta)^2\,d\theta \quad \text{(by symmetry)}$$

$$= \int_0^{\pi} (1 + 2\cos\theta + \cos^2\theta)\,d\theta$$

$$= \left(\theta + 2\sin\theta + \frac{1}{2}\theta + \frac{1}{4}\sin 2\theta\right)\Big|_0^{\pi}$$

$$= \frac{3\pi}{2} \text{ square units}$$

**9.**

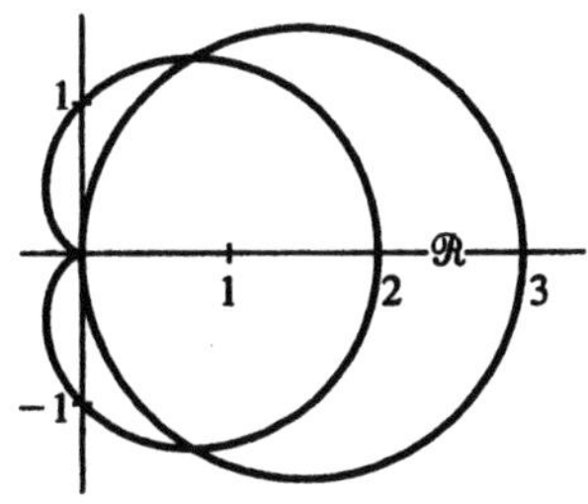


$r = 1 + \cos\theta$ and $r = 3\cos\theta$

Find the intersection points in the first and fourth quadrants.

$1 + \cos\theta = 3\cos\theta \Rightarrow \cos\theta = \frac{1}{2} \Rightarrow \theta = \pm\frac{\pi}{3}$

$$A = 2\int_0^{\pi/3} \frac{1}{2}\left[(3\cos\theta)^2 - (1+\cos\theta)^2\right]d\theta$$

(by symmetry)

$$= \int_0^{\pi/3} (8\cos^2\theta - 2\cos\theta - 1)\,d\theta$$

$$= (4\theta + 2\sin 2\theta - 2\sin\theta - \theta)\Big|_0^{\pi/3}$$

$= \pi$ square units

**11.**

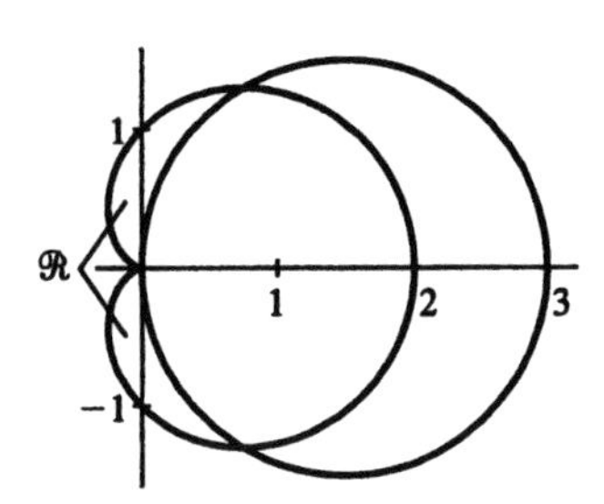


The outer and inner boundaries of the region correspond to different ranges of $\theta$. The outer boundary, part of the cardioid, corresponds to $\frac{\pi}{3} \le \theta \le \frac{5\pi}{3}$. But this is twice through the inner boundary of the region and once through the rest of the circle. So using

$A = \int_{\alpha}^{\beta} \frac{1}{2}\left(r_2^2 - r_1^2\right)d\theta$ will not work. Instead, we take the total area of the circle, subtract the area found in Exercise 9, and subtract the result from the total area of the cardioid, found in Exercise 7. This gives the area of the region in question.

| $\frac{3\pi}{2}$ | $-$ | $\left[\pi\left(\frac{3}{2}\right)^2\right.$ | $-$ | $\left.\pi\right] = \frac{\pi}{4}$ square units. |
|---|---|---|---|---|
| From Exercise 7 | | Area of circle | | From Exercise 9 |

**13.**

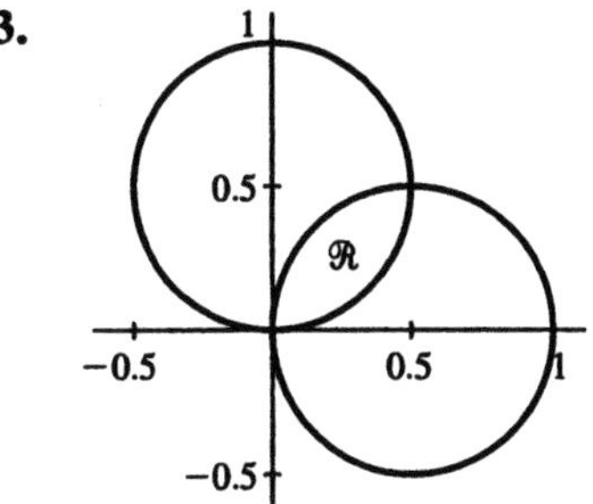


$r = \cos\theta$ and $r = \sin\theta$

$\cos\theta = \sin\theta \Rightarrow \theta = \frac{\pi}{4}\ \left(\text{for } 0 \le \theta \le \frac{\pi}{2}\right)$

$$A = 2\int_0^{\pi/4} \frac{1}{2}\sin^2\theta\,d\theta \quad \text{(by symmetry)}$$

$$= \left(\frac{1}{2}\theta - \frac{1}{4}\sin 2\theta\right)\Big|_0^{\pi/4}$$

$$= \frac{\pi}{8} - \frac{1}{4} \text{ square units}$$

**15.**

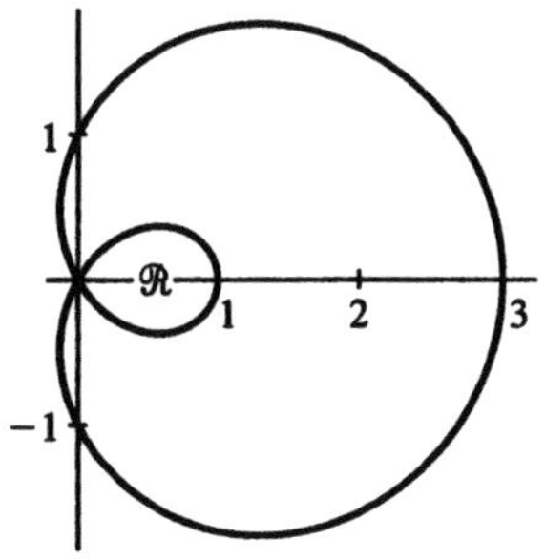

$r = 1 + 2\cos\theta$

$1 + 2\cos\theta = 0 \Rightarrow \cos\theta = -\dfrac{1}{2} \Rightarrow \theta = \dfrac{2\pi}{3}, \dfrac{4\pi}{3}$

$$A = 2\int_{2\pi/3}^{\pi} \frac{1}{2}(1+2\cos\theta)^2\,d\theta \quad \text{(by symmetry)}$$
$$= \int_{2\pi/3}^{\pi} (1+4\cos\theta+4\cos^2\theta)\,d\theta$$
$$= (\theta + 4\sin\theta + 2\theta + \sin 2\theta)\Big|_{2\pi/3}^{\pi}$$
$$= 3\pi - \left(2\pi + \frac{3\sqrt{3}}{2}\right)$$
$$= \pi - \frac{3\sqrt{3}}{2} \text{ square units}$$

**17.**

$r = 2$ and $r = \csc\theta$, $0 \le \theta < \pi$

$2 = \csc\theta \Rightarrow \sin\theta = \dfrac{1}{2} \Rightarrow \theta = \dfrac{\pi}{6}, \dfrac{5\pi}{6}$

$$A = 2\int_{\pi/6}^{\pi/2} \frac{1}{2}(2^2 - \csc^2\theta)\,d\theta \quad \text{(by symmetry)}$$
$$= \int_{\pi/6}^{\pi/2} (4 - \csc^2\theta)\,d\theta$$
$$= (4\theta + \cot\theta)\Big|_{\pi/6}^{\pi/2}$$
$$= 2\pi - \left(\frac{2\pi}{3} + \sqrt{3}\right)$$
$$= \frac{4\pi}{3} - \sqrt{3} \text{ square units}$$

**19.**

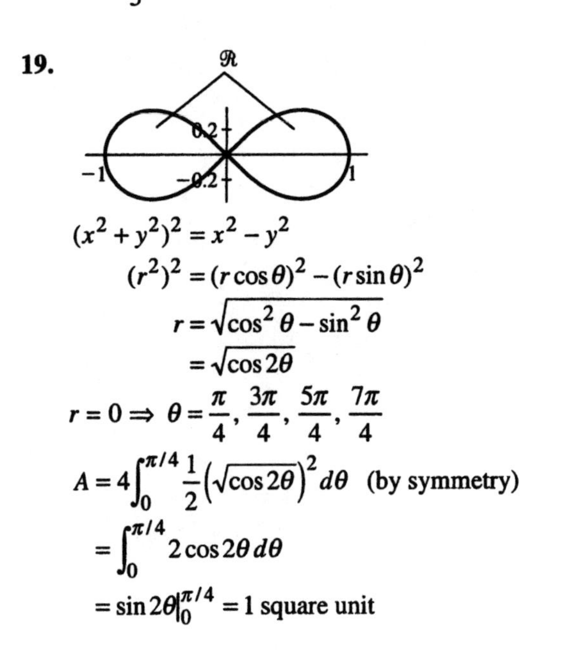

$(x^2+y^2)^2 = x^2 - y^2$

$(r^2)^2 = (r\cos\theta)^2 - (r\sin\theta)^2$

$r = \sqrt{\cos^2\theta - \sin^2\theta}$

$= \sqrt{\cos 2\theta}$

$r = 0 \Rightarrow \theta = \dfrac{\pi}{4}, \dfrac{3\pi}{4}, \dfrac{5\pi}{4}, \dfrac{7\pi}{4}$

$$A = 4\int_0^{\pi/4} \frac{1}{2}\left(\sqrt{\cos 2\theta}\right)^2 d\theta \quad \text{(by symmetry)}$$
$$= \int_0^{\pi/4} 2\cos 2\theta\,d\theta$$
$$= \sin 2\theta\Big|_0^{\pi/4} = 1 \text{ square unit}$$

**21.**

$r = \cos\theta$ and $r = \cos\left(\theta - \dfrac{\pi}{4}\right)$

$\cos\theta = \cos\left(\theta - \dfrac{\pi}{4}\right) \Rightarrow -\theta = \theta - \dfrac{\pi}{4}$

$\Rightarrow \theta = \dfrac{\pi}{8} \quad \left(\text{for } 0 \le \theta \le \dfrac{\pi}{2}\right)$

$$A = 2\int_{\pi/8}^{\pi/2} \frac{1}{2}[\cos(\theta)]^2\,d\theta \quad \text{(by symmetry)}$$
$$= \int_{\pi/8}^{\pi/2} \cos^2\theta\,d\theta$$
$$= \left(\frac{1}{2}\theta + \frac{1}{4}\sin 2\theta\right)\Big|_{\pi/8}^{\pi/2}$$
$$= \frac{\pi}{4} - \left(\frac{\pi}{16} - \frac{1}{4\sqrt{2}}\right) = \frac{3\pi}{16} - \frac{\sqrt{2}}{8} \text{ square units}$$

**23.**

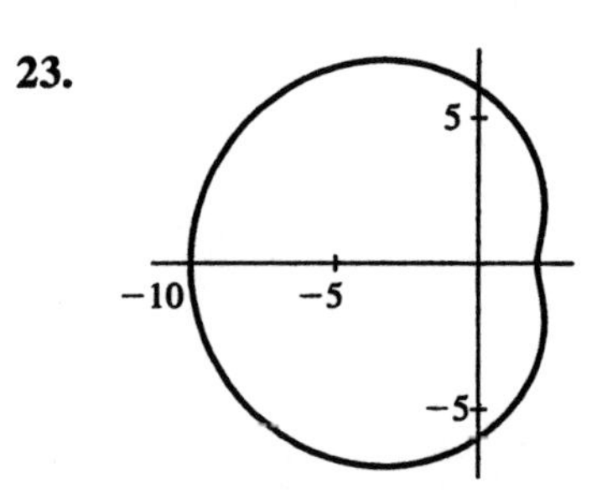


The total cam area is

$$A = \int_0^{2\pi} \frac{1}{2}r^2\,d\theta$$
$$= \int_0^{2\pi} (6.0 - 4.0\cos\theta)^2\,d\theta$$
$$= \int_0^{2\pi} (36 - 48\cos\theta + 16\cos^2\theta)\,d\theta$$
$$= (36\theta - 48\sin\theta + 8\theta + 4\sin 2\theta)\Big|_0^{2\pi}$$
$$= 44\pi \text{ mm}^2.$$

The volume is

$44\pi \times 10 = 440\pi \text{ mm}^3 = 0.44\pi \text{ cm}^3$, and the mass is $0.44\pi \times 7.850 \approx 10.8511$ g.

**25.**

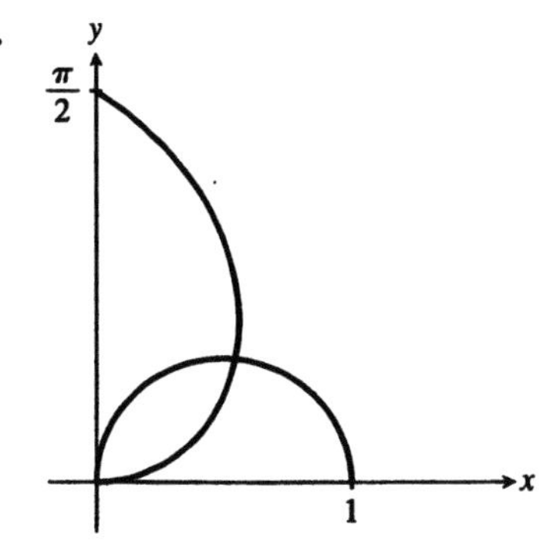


$r = \cos\theta$ and $r = \theta$

$\cos\theta = \theta \Rightarrow \theta \approx 0.73908513$

$$A \approx \int_0^{0.73908513} \frac{1}{2}(\theta)^2\,d\theta + \int_{0.73908513}^{\pi/2} \frac{1}{2}(\cos\theta)^2\,d\theta$$

$$\approx \frac{1}{6}\theta^3\Big|_0^{0.73908513} + \left(\frac{1}{4}\theta + \frac{1}{8}\sin 2\theta\right)\Big|_{0.73908513}^{\pi/2}$$

$$\approx 0.067287 + \left(\frac{\pi}{8} - 0.309235\right)$$

$\approx 0.150751$ square units

**27.**

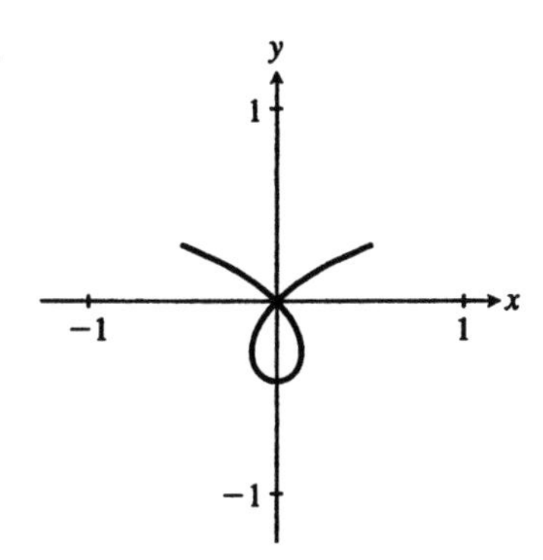


$r = \csc\theta - \sqrt{2}$

$r = 0$ when $\sin\theta = \dfrac{1}{\sqrt{2}} \Rightarrow \theta = \dfrac{\pi}{4}, \dfrac{3\pi}{4}$

$$A = 2\int_{\pi/4}^{\pi/2} \frac{1}{2}\left(\csc\theta - \sqrt{2}\right)^2 d\theta \quad \text{(by symmetry)}$$

$$= \int_{\pi/4}^{\pi/2}\left(\csc^2\theta - 2\sqrt{2}\csc\theta + 2\right)d\theta$$

$$= \left(-\cot\theta - 2\sqrt{2}\ln\left|\csc\theta - \cot\theta\right| + 2\theta\right)\Big|_{\pi/4}^{\pi/2}$$

$$= \pi - \left(-1 - 2\sqrt{2}\ln\left|\sqrt{2} - 1\right| + \frac{\pi}{2}\right)$$

$$= \frac{\pi}{2} + 1 + 2\sqrt{2}\ln\left|\sqrt{2} - 1\right|$$

$\approx 0.0778954$ square units

**29.**

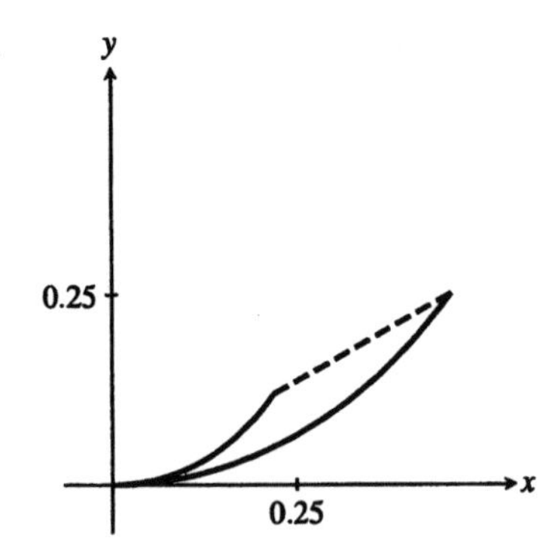


$r = \frac{\theta}{2}$ and $r = \arcsin\theta$

$$A = \int_0^{0.5} \frac{1}{2}\left[(\arcsin\theta)^2 - \left(\frac{\theta}{2}\right)^2\right] d\theta$$

$$= \frac{1}{2}\int_0^{0.5} \arcsin^2\theta\, d\theta - \frac{1}{8}\int_0^{0.5} \theta^2 d\theta$$

To find $\int \arcsin^2\theta\, d\theta$, use integration by parts with

$u = \arcsin\theta$, $dv = \arcsin\theta\, d\theta \Rightarrow du = \frac{1}{\sqrt{1-\theta^2}}d\theta$, $v = \theta\arcsin\theta + \sqrt{1-\theta^2}$.

$$\int u\,dv = uv - \int v\,du$$

$$\int \arcsin^2\theta\, d\theta = \theta\arcsin^2\theta + \sqrt{1-\theta^2}\arcsin\theta - \int\left(\frac{\theta}{\sqrt{1-\theta^2}}\arcsin\theta + 1\right)d\theta$$

$$= \theta\arcsin^2\theta + \sqrt{1-\theta^2}\arcsin\theta - \int\frac{\theta}{\sqrt{1-\theta^2}}\arcsin\theta\, d\theta - \theta$$

To find $\int \frac{\theta}{\sqrt{1-\theta^2}} \cdot \arcsin\theta\, d\theta$, use integration by parts again, with $u = \theta$, $dv = \frac{\arcsin\theta}{\sqrt{1-\theta^2}}d\theta \Rightarrow du = d\theta$,

$v = \frac{1}{2}\arcsin^2\theta$.

$$\int u\,dv = uv - \int v\,du$$

$$\int\frac{\theta}{\sqrt{1-\theta^2}}\arcsin\theta\, d\theta = \frac{1}{2}\theta\arcsin^2\theta - \int\frac{1}{2}\arcsin^2\theta\, d\theta$$

So $\int \arcsin^2\theta\, d\theta = \theta\arcsin^2\theta + \sqrt{1-\theta^2}\arcsin\theta - \theta - \left(\frac{1}{2}\theta\arcsin^2\theta - \frac{1}{2}\int\arcsin^2\theta\, d\theta\right)$

It follows that $\int \arcsin^2\theta\, d\theta = \theta\arcsin^2\theta + 2\sqrt{1-\theta^2}\arcsin\theta - 2\theta$

Then

$$A = \frac{1}{2}\left(\theta\arcsin^2\theta + \sqrt{1-\theta^2}\arcsin\theta - 2\theta\right)\Big|_0^{0.5} - \frac{1}{24}\theta^3\Big|_0^{0.5}$$

$\approx 0.0167804$ square units.

**31.** Letting $u=\tan\frac{\theta}{2}$, we find that

$$\sin\theta = 2\sin\frac{\theta}{2}\cos\frac{\theta}{2} = 2\tan\frac{\theta}{2}\cos^2\frac{\theta}{2} = \frac{2\tan\frac{\theta}{2}}{\sec^2\frac{\theta}{2}} = \frac{2\tan\frac{\theta}{2}}{1+\tan^2\frac{\theta}{2}} = \frac{2u}{1+u^2},$$

$$\cos\theta = \sqrt{1-\sin^2\theta} = \sqrt{\frac{(1+u^2)^2-(2u)^2}{(1+u^2)^2}} = \sqrt{\frac{1+2u^2+u^4-4u^2}{(1+u^2)^2}} = \sqrt{\frac{1-2u^2+u^4}{(1+u^2)^2}} = \frac{1-u^2}{1+u^2}, \text{ and}$$

$$du = \sec^2\frac{\theta}{2}\cdot\frac{1}{2}d\theta$$

$$d\theta = 2\cos^2\frac{\theta}{2}du = (1+\cos\theta)\,du = \left(\frac{1+u^2}{1+u^2}+\frac{1-u^2}{1+u^2}\right)du = \frac{2\,du}{1+u^2}.$$

$$\begin{aligned}
\frac{1}{2}\int\frac{d\theta}{\left(1-\frac{3}{5}\cos\theta\right)^2} &= \frac{1}{2}\int\frac{1}{\left[1-\frac{3}{5}\left(\frac{1-u^2}{1+u^2}\right)\right]^2}\cdot\frac{2}{1+u^2}du \\
&= \int\frac{1+u^2}{\left[1+u^2-\frac{3}{5}(1-u^2)\right]^2}du \\
&= \int\frac{1+u^2}{\left(\frac{2}{5}+\frac{8}{5}u^2\right)^2}du \\
&= \frac{25}{4}\int\frac{(1+u^2)\,du}{(1+4u^2)^2}
\end{aligned}$$

Since $u$ only appears within $u^2$ in the last integral, substitute $x=u^2$ and find $A, B$ for the partial fraction representation.

$$\begin{aligned}
\frac{1+x}{(1+4x)^2} &= \frac{A}{1+4x}+\frac{B}{(1+4x)^2} \\
1+x &= A(1+4x)+B \\
x+1 &= 4Ax+(A+B) \\
A &= \frac{1}{4} \\
B &= \frac{3}{4}
\end{aligned}$$

So

$$\begin{aligned}
\frac{1}{2}\int_0^\beta\frac{d\theta}{\left(1-\frac{3}{5}\cos\theta\right)^2} &= \frac{25}{4}\int_0^\beta\left(\frac{1}{4(1+4n^2)}+\frac{3}{4(1+4n^2)^2}\right)du \\
&= \frac{25}{16}\int_0^\beta\left(\frac{1}{1+4n^2}+\frac{3}{(1+4n^2)^2}\right)du.
\end{aligned}$$

Substitute $2u = \tan\phi$, $du = \frac{1}{2}\sec^2\phi\, d\phi$.

$$= \frac{25}{16}\int_0^\beta \left(\frac{1}{1+\tan^2\phi} + \frac{3}{(1+\tan^2\phi)^2}\right)\frac{1}{2}\sec^2\phi\, d\phi$$

$$= \frac{25}{32}\int_0^\beta \left(\frac{1}{\sec^2\phi} + \frac{3}{\sec^4\phi}\right)\sec^2\phi\, d\phi$$

$$= \frac{25}{32}\int_0^\beta (1+3\cos^2\phi)\, d\phi$$

$$= \frac{25}{32}\left(\phi + \frac{3}{2}\phi + \frac{3}{4}\sin 2\phi\right)\Bigg|_0^\beta$$

$$= \frac{25}{32}\left(\frac{5}{2}\phi + \frac{3}{4}\sin 2\phi\right)\Bigg|_0^\beta$$

$$= \frac{25}{32}\left(\frac{5}{2}\phi + \frac{3}{2}\sin\phi\cos\phi\right)\Bigg|_0^\beta$$

using $\sin\phi = \dfrac{\tan\phi}{\sqrt{1+\tan^2\phi}}$, $\cos\phi = \dfrac{1}{\sqrt{1+\tan^2\phi}}$ where $\tan\phi = 2\tan\left(\dfrac{\theta}{2}\right)$,

$$= \frac{25}{32}\left(\frac{5}{2}\arctan\left(2\tan\frac{\theta}{2}\right) + \frac{3\tan\frac{\theta}{2}}{1+4\tan^2\frac{\theta}{2}}\right)\Bigg|_0^\beta$$

$$= \frac{25}{32}\left(\frac{5}{2}\arctan\left(2\tan\frac{\beta}{2}\right) + \frac{3\tan\frac{\beta}{2}}{1+4\tan^2\frac{\beta}{2}}\right)$$

using $\tan\left(\dfrac{\theta}{2}\right) = \dfrac{\sin\theta}{(1+\cos\theta)}$ and $\tan^2\left(\dfrac{\theta}{2}\right) = \dfrac{(1-\cos\theta)}{(1+\cos\theta)}$,

$$= \frac{25}{32}\left(\frac{5}{2}\arctan\left(2\tan\frac{\beta}{2}\right) + \frac{3\sin\beta}{1+\cos\beta+4(1-\cos\beta)}\right)$$

$$= \frac{125}{64}\arctan\left(2\tan\frac{\beta}{2}\right) - \frac{75\sin\beta}{32(3\cos\beta-5)}$$

**33.** $g(\beta) = \dfrac{125\arctan\left(2\tan\left(\frac{\beta}{2}\right)\right)}{64} - \dfrac{75\sin\beta}{32(3\cos\beta - 5)} - \dfrac{125\pi}{12\cdot 64}$

$$g'(\beta) = \frac{125\sec^2\left(\frac{\beta}{2}\right)}{64\left(1+4\tan^2\left(\frac{\beta}{2}\right)\right)} - \frac{75\left[(3\cos\beta-5)\cos\beta + 3\sin^2\beta\right]}{32(3\cos\beta-5)^2}$$
$$= \frac{125}{64\left(\cos^2\left(\frac{\beta}{2}\right)+4\sin^2\left(\frac{\beta}{2}\right)\right)} - \frac{75(3-5\cos\beta)}{32(3\cos\beta-5)^2}$$
$$= \frac{125}{32(1+\cos\beta+4-4\cos\beta)} - \frac{75(3-5\cos\beta)}{32(3\cos\beta-5)^2}$$
$$= \frac{125}{32(5-3\cos\beta)} - \frac{75(3-5\cos\beta)}{32(3\cos\beta-5)^2}$$
$$= \frac{-125(3\cos\beta-5)-75(3-5\cos\beta)}{32(3\cos\beta-5)^2}$$
$$= \frac{25}{2(3\cos\beta-5)^2}$$

Now use Newton's method.

$\beta_1 = 0.2$

$$\beta_{n+1} = \beta_n - \frac{g(\beta_n)}{g'(\beta_n)}$$

Then $\beta_2 \approx 0.165540$ and $\beta_3 \approx 0.165861$.

**35.**

$$r = \frac{p}{1-e\cos\theta}$$
$$\sqrt{x^2+y^2} = \frac{p}{1-\frac{ex}{\sqrt{x^2+y^2}}}$$
$$\sqrt{x^2+y^2} - ex = p$$
$$x^2+y^2 = (p+ex)^2$$
$$x^2+y^2 = p^2+2epx+e^2x^2$$
$$x^2(1-e^2)-2epx+y^2 = p^2$$
$$x^2(1-e^2)-2epx+\frac{e^2p^2}{1-e^2}+y^2 = p^2+\frac{e^2p^2}{1-e^2}$$
$$\left(x\sqrt{1-e^2}-\frac{ep}{\sqrt{1-e^2}}\right)^2+y^2 = \frac{p^2-p^2e^2+e^2p^2}{1-e^2}$$
$$\frac{\left(x-\frac{ep}{1-e^2}\right)^2}{\frac{1}{1-e^2}}+y^2 = \frac{p^2}{1-e^2}$$
$$\frac{\left(x-\frac{ep}{1-e^2}\right)^2}{\frac{p^2}{(1-e^2)^2}}+\frac{y^2}{\frac{p^2}{1-e^2}} = 1$$

This is of the form $\dfrac{x^2}{a^2}+\dfrac{y^2}{b^2}=1$ with $a = \dfrac{p}{1-e^2}$ and $b = \dfrac{p}{\sqrt{1-e^2}}$.

## Section R.9 Work

In this section, the gravitational acceleration $g$ is assumed equal to 9.81 m/s$^2$.

1. The move from (0, 3) to (2, 0) represents a drop in height of 3 meters.
So $W = mgh = (5)(g)(3) = 15g \approx 150$ J (accurate to two digits).

3. By Hooke's law, $F(x) = kx \Rightarrow 25 = 0.05k \Rightarrow k = 500$ N/m.
$$W = \int_a^b F(x)\,dx = \int_0^{0.30} 500x\,dx = 250x^2\Big|_0^{0.30} = 22.5 \text{ J}$$

5. By Hooke's law, $F(x) = kx \Rightarrow 25 = 0.05k \Rightarrow k = 500$ N/m.
$$W = \int_a^b F(x)\,dx = \int_0^{-0.30} 500x\,dx = 250x^2\Big|_0^{-0.30} = 22.5 \text{ J}$$

7. We parametrize the position vector $\mathbf{r}$ as $\mathbf{r}(t) = \langle t, 1-t^2\rangle$, $-1 \le t \le 1$. Then
$\mathbf{r}'(t) = \langle 1, -2t\rangle$.
With $\mathbf{p} = \langle 1, 0\rangle$, the force points in direction $\mathbf{p}-\mathbf{r}$ and has magnitude $k\|\mathbf{p}-\mathbf{r}\|$, where $k$ is the proportionality constant, so that
$$\begin{aligned}\mathbf{F}(t) &= k\|\mathbf{p}-\mathbf{r}\|\frac{1}{\|\mathbf{p}-\mathbf{r}\|}(\mathbf{p}-\mathbf{r})\\ &= \mathbf{k}(\mathbf{p}-\mathbf{r})\\ &= k\langle 1-t, t^2-1\rangle.\end{aligned}$$
The work done by the force is
$$\begin{aligned}W = \int_a^b \mathbf{F}\cdot\mathbf{r}'\,dt &= \int_{-1}^1 k\langle 1-t, t^2-1\rangle\cdot\langle 1, -2t\rangle\,dt\\ &= k\int_{-1}^1 (-2t^3+t+1)\,dt\\ &= k\left(-\frac{1}{2}t^4+\frac{1}{2}t^2+t\right)\Big|_{-1}^1\\ &= 2k \text{ J.}\end{aligned}$$

9. The semicircular sides can be represented by the bottom half of the circle $x^2+y^2=25^2$. The width of the tank at depth $y$ ($-25 \le y \le 0$) is $2\sqrt{25^2-y^2}$ meters. The work required to lift a thin rectangular sheet to the top of the reservoir is
$$\begin{aligned}dW = hgdm &= (-y)(g)\left(\delta\cdot 200\cdot 2\sqrt{25^2-y^2}\,dy\right)\\ &= -400\delta gy\sqrt{25^2-y^2}\,dy.\end{aligned}$$
The total work is
$$\begin{aligned}W = \int_a^b dW &= \int_{-25}^0 -400\delta gy\sqrt{25^2-y^2}\,dy\\ &= \frac{400}{3}\delta g(25^2-y^2)^{3/2}\Big|_{-25}^0\\ &= \frac{6,250,000}{3}\delta g\end{aligned}$$
which for $\delta = 1000$, $g = 9.81$, equals about $2.04\times 10^{10}$ J.

11. The car moves through a drop in height of 5 meters. So $W = mgh = 5mg$ J.

13. As $x$ varies from 0 to 20, the force varies according to $f(x) = 3250 - 12.5x$. The work is
$$\begin{aligned}W = \int_a^b \mathbf{F}\cdot d\mathbf{r} &= \int_0^{20}\langle 3250-12.5x, 0\rangle\cdot\langle dx, 0\rangle\\ &= \int_0^{20}(3250-12.5x)\,dx\\ &= (3250x-6.25x^2)\Big|_0^{20}\\ &= 62,500 \text{ J.}\end{aligned}$$

15. $x(t) = -1.7e^{-0.5t}\sin(1.2t)$
$x'(t) = -1.7e^{-0.5t}[(1.2\cos(1.2t) - 0.5\sin(1.2t)]$
Time $b$, the moment of maximum compression, is given by $x'(b) = 0$, which implies
$$\begin{aligned}1.2\cos(1.2t) &= 0.5\sin(1.2t)\\ \tan(1.2t) &= 2.4\\ t &= \frac{\arctan 2.4}{1.2}\\ &\approx 0.980004 \text{ seconds}\end{aligned}$$
The position of maximum compression is $x(0.980004) \approx -0.96135$ meters.
The initial (relaxed) position is $x = 0$.
The work during the return stroke is
$$\begin{aligned}W = \int_a^b \mathbf{F}\cdot d\mathbf{r} &= \int_a^b \langle F, 0\rangle\cdot\langle dx, 0\rangle\\ &= \int_a^b F\,dx\\ &= \int_{-0.96135}^0 (-1.6\times 10^4)x\,dx\\ &= (-0.8\times 10^4)x^2\Big|_{-0.96135}^0\\ &\approx 7394 \text{ J.}\end{aligned}$$

**17.** The propulsion force $\mathbf{F}_p$ is equal and opposite to gravity.

$$W = \int_a^b \mathbf{F}_p \cdot \mathbf{dr} = \int_{6.378\times10^6}^{6.538\times10^6} \frac{GM_E m}{x^2}\,dx$$

$$= -\frac{GM_E m}{x}\Bigg|_{6.378\times10^6}^{6.538\times10^6}$$

With the given values of $G$, $M_E$, and $m$,

$W \approx 7.6\times10^{10}$ J.

**19.** When the end of the cable has been lifted a distance $x$, the weight of the hanging portion of the cable is $15gx$. Lifting the cable a small distance $dx$ at this point requires work $dW = 15gx\,dx$. The total work is

$$W = \int_a^b F\,dx = \int_0^{500} 15\,gx\,dx$$

$$= 7.5gx^2\Big|_0^{500}$$

$$\approx 1.8\times10^7 \text{ J.}$$

**21.** When the elevator has been lifted a distance $x$, the combined weight of the elevator and the hanging portion of the cable (the portion not wrapped onto the winch drum) is $[1500 + 15(500 - x)]g$. Lifting the elevator-cable combination through a series of small distances $dx$ that add up to 490 meters requires a total amount of work

$$W = \int_a^b F\,dx = \int_0^{490} [1500 + 15(500 - x)]g\,dx$$

$$= \left[1500x + 15\left(500x - \frac{1}{2}x^2\right)\right]g\Bigg|_0^{490}$$

$$\approx 2.6\times10^7 \text{ J.}$$

**23.** The bucket mass $m$ is 2 kg plus $16 - 0.075t$ kg. From $x(t) = t$ and $m(t) = 2 + (16 - 0.075t)$, it follows that $m(x) = 18 - 0.075x$. The work required to raise the bucket, already hoisted to height $x$, an additional distance $dx$ is $dW = (18 - 0.075x)g\,dx$. The total work is

$$W = \int_0^{20} (18 - 0.075x)g\,dx$$

$$= (18x - 0.0375x^2)g\Big|_0^{20}$$

$$\approx 3400 \text{ J (accurate to two digits).}$$

**25.** With $x$ ranging from $-10$ at the bottom of the tank to 10 at the top, the radius of a thin, circular horizontal slice at $x$ is $\sqrt{10^2 - x^2}$ and its volume is $\pi(10^2 - x^2)dx$. The slice is a distance $10 - x$ from the top of the tank, so $dW = g \cdot h \cdot dm = g \cdot (10 - x) \cdot 680\pi(10^2 - x^2)dx$.

The work required to raise the slice to the top is

$$W = \int_{-10}^{10} 680g\pi(10^2 - x^2)(10 - x)\,dx$$

$$= 680g\pi\int_{-10}^{10} (1000 - 100x - 10x^2 + x^3)\,dx$$

$$= 680g\pi\left(1000x - 50x^2 - \frac{10}{3}x^3 + \frac{1}{4}x^4\right)\Bigg|_{-10}^{10}$$

$$\approx 2.8\times10^8 \text{ J.}$$

**27.** Let $x$ vary from $-3$ at the bottom of the deep end to 0 at the water surface. The length of a thin, rectangular horizontal slice at $x$ is $L(x) = 75 + 25x$ for $-3 \le x \le -1$ (this equals 50 at $x = -1$ and 0 at $x = -3$), and the length is $L(x) = l = 50$ for $-1 \le x \le 0$. The slice's volume is $10L(x)\,dx$. The slice is a distance $-x$ from the surface, so $dW = g \cdot h \cdot dm = g \cdot (-x) \cdot 1000 \cdot 10L(x)\,dx$. The total work required to raise the slice to the top is

$$\begin{aligned}
W &= \int_{-3}^{0} -10{,}000\, gx\, L(x)\, dx \\
&= \int_{-3}^{-1} -10{,}000\, gx\, L(x)\, dx + \int_{-1}^{0} -10{,}000\, gx\, L(x)\, dx \\
&= -10{,}000g\int_{-3}^{-1} (75x + 25x^2)\, dx - 10{,}000g\int_{-1}^{0} 50x\, dx \\
&= -10{,}000g\left(\frac{75}{2}x^2 + \frac{25}{3}x^3\right)\Bigg|_{-3}^{-1} - 10{,}000g(25x^2)\Big|_{-1}^{0} \\
&= \frac{2{,}500{,}000}{3}g + 250{,}000g \\
&\approx 1.1 \times 10^7 \text{ J.}
\end{aligned}$$

**29.** The pile of grain with density $\delta$ has a mass of $m = \delta \times \dfrac{\pi r^2 h}{3} = \dfrac{500\pi\delta}{3}$, and the work to lift that much grain from ground level to 40 m is $40mg = \dfrac{20{,}000\pi g\delta}{3}$.

If the grain is instead taken from the top, calculating the work requires an integral. The thin circular slice at height $x$ has a radius of $10 - 2x$ and a mass $dm$ of $\pi(10-2x)^2\delta\, dx$, it will be lifted through $40 - x$ meters, and so the work to lift it will be $dW = g \cdot h \cdot dm = (10-2x)^2(40-x)\pi g\delta\, dx$. The total work for the whole pile will be

$$\begin{aligned}
W &= \int_{0}^{5} (10-2x)^2(40-x)\pi g\delta\, dx \\
&= \pi g\delta\int_{0}^{5} (4000 - 1700x + 200x^2 - 4x^3)dx \\
&= \pi g\delta\left(4000x - 850x^2 + \frac{200}{3}x^3 - x^4\right)\Bigg|_{0}^{5} \\
&= \frac{19{,}375\pi g\delta}{3}
\end{aligned}$$

The ratio of the two amounts of work is $\dfrac{20{,}000}{19{,}375} = \dfrac{32}{31}$.

**31.** With $x$ ranging from $-3$ at the bottom to 3 at the top, the width of a thin, rectangular horizontal slice is $2\sqrt{3^2 - x^2}$ and its volume is $20\sqrt{3^2 - x^2}\, dx$. The slice is a distance $3 - x$ from the top of the tank, so the work required to raise the slice is $dW = g \cdot h \cdot dm = g \cdot (3-x) \cdot 680 \cdot 20\sqrt{3^2 - x^2}\, dx$. The total work is

$$\begin{aligned}
W &= \int_{-3}^{3} 13{,}600g(3-x)\sqrt{3^2 - x^2}\, dx \\
&= 13{,}600g\int_{-3}^{3}\left(3\sqrt{3^2 - x^2} - x\sqrt{3^2 - x^2}\right)dx \\
&= 13{,}600g\left[3\left(\frac{x}{2}\sqrt{3^2 - x^2} + \frac{3^2}{2}\arcsin\frac{x}{3}\right) + \frac{1}{3}(3^2 - x^2)^{3/2}\right]_{-3}^{3} \\
&\approx 5.7 \times 10^6 \text{ J.}
\end{aligned}$$

**33.** With $x$ ranging from 0 at the bottom to 146.8 at the top, the side of a thin, square horizontal slice is about $230.4 - 1.569482x$ and its volume is about $(230.4 - 1.569482x)^2\,dx$. The work required to raise the slice is $dW \approx g \cdot h \cdot dm = g \cdot x \cdot \delta(230.4 - 1.569482x)^2\,dx$. The total work is

$$\begin{aligned} W &= \int_0^{146.8} g\delta(230.4 - 1.569482x)^2 x\,dx \\ &= g\delta(53{,}084.16x - 723.217306x^2 + 2.463274x^3)\Big|_0^{146.8} \\ &\approx 2.3 \times 10^{12}\text{ J.} \end{aligned}$$

**35.** The speed is about 28 m/s. As shown in the text, the work done by the gravitational force is $\frac{1}{2}mv_b^2$ if we assume that the car started from rest. Because the Earth's gravitational field is conservative, the work done by the gravitational force in moving the car from $A$ to $B$ is $mg \cdot 40$. Hence, $v_b = \sqrt{2g \cdot 40} \approx 28$ m/s.

**37.** **i.** $\|\mathbf{r}(h)\|$ is the length of vector $\mathbf{r}(h)$, or in other words, the distance from the origin to the top of $\mathbf{r}(h)$, where the force is applied. But this is what $h$ is specified to be. So $\|\mathbf{r}(h)\| = h$.

**ii.** The spring is stretched by amount $h - c$, so it exerts a force of magnitude $k(h - c)$ pointing toward the origin. The force pulling on the spring at point $\mathbf{r}(h)$, then, is equal and opposite.

$$\begin{aligned} \mathbf{F}(\mathbf{r}(h)) &= k(h-c)\frac{1}{\|\mathbf{r}(h)\|}\mathbf{r}(h) \\ &= \left(\frac{k(h-c)}{h}\right)\mathbf{r} \end{aligned}$$

**iii.** For a small change $dh$, the end of the spring moves a small step $\mathbf{dr} = \mathbf{r}'\,dh$. The work done by the force is

$$\begin{aligned} dW &= \mathbf{F} \cdot \mathbf{dr} \\ &= \mathbf{F} \cdot \mathbf{r}'\,dh \\ &= \left(\frac{k(h-c)}{h}\right)\mathbf{r} \cdot \mathbf{r}'\,dh. \end{aligned}$$

**iv.** As $h$ increases by amount $dh$, the change in $\|\mathbf{r}\|$ is $\|\mathbf{r}'\|\cos\theta\,dh$, where $\theta$ is the angle between $\mathbf{r}$ and $\mathbf{r}'$. So $\frac{d\|\mathbf{r}\|}{dh} = \|\mathbf{r}'\|\cos\theta$. But $\|\mathbf{r}\| = h$, so $\frac{d\|\mathbf{r}\|}{dh} = 1$, and thus $\|\mathbf{r}'\|\cos\theta = 1$. Then $\mathbf{r} \cdot \mathbf{r}' = \|\mathbf{r}\| \cdot \|\mathbf{r}'\|\cos\theta = \|\mathbf{r}\| = h$. Thus $dW = \left(\frac{k(h-c)}{h}\right) \cdot h\,dh = k(h-c)\,dh$.

The total work, then, is $W = \int_c^d k(h-c)\,dh = \frac{1}{2}k(h-c)^2\Big|_c^d = \frac{1}{2}k(d-c)^2$.

This result is independent of the curve joining $A$ and $B$.

## Section R.10 Center of Mass

**1.** The total mass is $m = 5 + 4 + 6 + 3 = 18$ kg. Using formula (4), the center of mass lies on the $x$-axis at

$$X = \frac{1}{m}\sum_{i=1}^{4} m_i x_i = \frac{1}{18}[(5)(-3) + (4)(-1) + (6)(2) + (3)(7)] = \frac{14}{18} = \frac{7}{9}\text{ m.}$$

**3.** Using formula (2), the center of mass of the system of point masses $\{\langle 5, 1\rangle, \langle 15, 3\rangle, \langle 5, 4\rangle, \langle 20, 8\rangle, \langle 10, 9\rangle\}$ is given by $R = \dfrac{5(1)+15(3)+5(4)+20(8)+10(9)}{5+15+5+20+10} = \dfrac{64}{11}$.

Hence, the equilibrium point is located $\dfrac{64}{11}$ meters from the left end of the board.

**5.** Using formula (2), we have

$$R = \frac{1}{m}\sum_{i=1}^{5} m_i \mathbf{r}_i = \frac{1}{4+6+7+8+1}\left(4\langle -3, 5\rangle + 6\langle -5, -6\rangle + 7\langle 5, -1\rangle + 8\langle 1, 1\rangle + 1\langle 6, 6\rangle\right) = \left\langle \frac{7}{26}, -\frac{9}{26}\right\rangle.$$

The center of mass of the system has position vector $\left\langle \dfrac{7}{26}, -\dfrac{9}{26}\right\rangle$.

**7.** We observe that the density function has the form $\delta(x) = Ax + B$. Since $\delta(0) = 3$ and $\delta(0.5) = 3.7$, it follows that $A = 1.4$ and $B = 3.0$. That is, $\delta(x) = 1.4x + 3.0$. The mass $M$ of the rod is given by

$$\int_0^{0.5} \delta(x)\,dx = \int_0^{0.5} (1.4x + 3.0)\,dx = \frac{67}{40} = 1.675 \text{ kg.}$$

Using formula (5), the center of mass $X$ is given by

$$X = \frac{\int_0^{0.5} x\,\delta(x)\,dx}{\int_0^{0.5} \delta(x)\,dx} = \frac{\int_0^{0.5} (1.4x^2 + 3.0x)\,dx}{\frac{67}{40}} = \frac{\frac{13}{30}}{\frac{67}{40}} = \frac{52}{201}.$$

The center of mass is about 0.258706 meters from the origin.

**9.** The center of mass of the larger square lamina is seen to be located at the point $\langle a, a\rangle$ and the center of mass of the smaller square lamina is located at $\left\langle \dfrac{5a}{2}, \dfrac{a}{2}\right\rangle$. If the mass of the smaller lamina is $m$, then the mass of the larger lamina is $4m$. So, by the Mass Subdivision Theorem, the center of mass of the system is given by $\mathbf{R} = \dfrac{1}{4m+m}\left(4m\langle a, a\rangle + m\left\langle \dfrac{5a}{2}, \dfrac{a}{2}\right\rangle\right) = \left\langle \dfrac{13a}{10}, \dfrac{9a}{10}\right\rangle$.

**11.** Using formula (9), the mass of the lamina is given by $M = \delta\int_0^1 e^x\,dx = \delta e^x\Big|_0^1 = \delta(e-1)$.

Using formulas (10) and (11), the $x$-coordinate of the center of mass is given by

$$X = \frac{\int_0^1 xe^x\,dx}{\int_0^1 e^x\,dx}$$

$$= \frac{e^x(x-1)\Big|_0^1}{e^x\Big|_0^1} \quad \text{(integration by parts)}$$

$$= \frac{1}{e-1}$$

$$\approx 0.581977$$

and the $y$-coordinate of the center of mass is given by

$$Y = \frac{\frac{1}{2}\int_0^1 (e^x)^2\,dx}{\int_0^1 e^x\,dx} = \frac{\frac{e^{2x}}{4}\Big|_0^1}{e^x\Big|_0^1} = \frac{e+1}{4} \approx 0.92957$$

**13.** Using formula (9), the mass of the lamina is given by

$$M = \delta\int_0^b \left[h - \left(h - \frac{hx^2}{b^2}\right)\right]dx$$
$$= \delta\left(\frac{hx^3}{3b^2}\right)\Bigg|_0^b$$
$$= \delta\frac{bh}{3}$$
$$= \frac{bh}{3} \quad (\delta = 1).$$

Using formulas (10) and (11), the $x$-coordinate of the center of mass is given by

$$X = \frac{\int_0^b x\left[h - \left(h - \frac{hx^2}{b^2}\right)\right]dx}{\int_0^b \left[h - \left(h - \frac{hx^2}{b^2}\right)\right]dx} = \frac{\frac{hx^4}{4b^2}\Big|_0^b}{\frac{bh}{3}} = \frac{3b}{4}$$

and the $y$-coordinate of the center of mass is given by

$$Y = \frac{\frac{1}{2}\int_0^b \left[h^2 - \left(h - \frac{hx^2}{b^2}\right)^2\right]dx}{\int_0^b \left[h - \left(h - \frac{hx^2}{b^2}\right)\right]dx}$$
$$= \frac{\frac{1}{2}\left(\frac{2h^2x^3}{3b^2} - \frac{h^2x^5}{5b^4}\right)\Big|_0^b}{\frac{bh}{3}}$$
$$= \frac{\frac{7bh^2}{30}}{\frac{bh}{3}}$$
$$= \frac{7h}{10}.$$

**15.** Find the $y$-coordinate of the center of mass.

$$Y = \frac{\int_a^b y(f(y) - g(y))\,dy}{\int_a^b (f(y) - g(y))\,dy}$$
$$= \frac{\int_0^{a\sin\theta} y\left(\sqrt{a^2 - y^2} - y\cot\theta\right)dy}{\int_0^{a\sin\theta}\left(\sqrt{a^2 - y^2} - y\cot\theta\right)dy}$$
$$= \frac{\left(-\frac{1}{3}(a^2 - y^2)^{3/2} - \frac{1}{3}y^3\cot\theta\right)\Big|_0^{a\sin\theta}}{\left(\frac{y}{2}\sqrt{a^2 - x^2} + \frac{a^2}{2}\arcsin\frac{y}{a} - \frac{1}{2}y^2\cot\theta\right)\Big|_0^{a\sin\theta}}$$
$$= \frac{\frac{a^3}{3}(1 - \cos\theta)}{\frac{a^2}{2}\theta}$$
$$= \frac{2a}{3\theta}(1 - \cos\theta)$$

Since by symmetry, the center of mass must lie on the line $y = \tan\left(\frac{\theta}{2}\right)x = \frac{(1 - \cos\theta)x}{\sin\theta}$, it follows that $X = \frac{2a\sin\theta}{3\theta}$ and that

$$R = \frac{2a}{3\theta}\langle\sin\theta, 1 - \cos\theta\rangle.$$

If the radius is labeled $r$ instead of $a$ and $\theta$ shrinks toward 0, then by l'Hôpital's Rule,

$$\lim_{\theta\to 0} Y = \lim_{\theta\to 0}\frac{2r(1 - \cos\theta)}{3\theta} = \lim_{\theta\to 0}\frac{2r\sin\theta}{3} = 0$$

and $\lim_{\theta\to 0} X = \lim_{\theta\to 0}\frac{2r\sin\theta}{3\theta} = \lim_{\theta\to 0}\frac{2r\cos\theta}{3} = \frac{2r}{3}.$

Thus the centroid approaches $\left\langle 0, \frac{2r}{3}\right\rangle$, which lies a distance $\frac{2r}{3}$ from the vertex.

**Exercises 17–19.** First find the rectangular coordinates, then convert to polar. For each small, wedge-shaped area element, multiply $dA$ by the $x$- or $y$-coordinate, $x_c$ or $y_c$, of the element's centroid, sum all the products by integration, and divide by the cam's total area, also found by integration.

**17.** The cam's area is

$$A = \int_0^{2\pi}\frac{1}{2}r^2\,d\theta$$
$$= \frac{1}{2}\int_0^{2\pi}(6 - 4\cos\theta)^2\,d\theta$$
$$= \frac{1}{2}\int_0^{2\pi}(36 - 48\cos\theta + 16\cos^2\theta)\,d\theta$$
$$= \frac{1}{2}(36\theta - 48\sin\theta + 8\theta + 4\sin 2\theta)\Big|_0^{2\pi}$$
$$= 44\pi.$$

To find $X$ and $Y$, first calculate

$$\int_0^{2\pi} x_c\,dA = \int_0^{2\pi}\left(\frac{2}{3}r\cos\theta\right)\frac{1}{2}r^2\,d\theta$$
$$= \frac{1}{3}\int_0^{2\pi}(6 - 4\cos\theta)^3\cos\theta\,d\theta$$
$$\approx -502.655$$

evaluated numerically. An analytic calculation or a CAS gives the exact result, $-160\pi$.

Then $X = \frac{1}{A}\int_0^{2\pi} x_c\,dA \approx -3.63636.$

By symmetry, $Y = 0$. The polar coordinates corresponding to $(X, Y) = (-3.63636, 0)$ are $(3.63636, \pi)$.

**19.** The part's area is

$$A = \int_0^{\pi} \frac{1}{2} r^2\, d\theta = \int_0^{\pi} \frac{1}{2}\theta^2\, d\theta = \frac{1}{6}\theta^3 \bigg|_0^{\pi} = \frac{\pi^3}{6}$$

To find $X$ and $Y$, first calculate

$$\int_0^{\pi} x_c dA = \int_0^{\pi} \left(\frac{2}{3} r\cos\theta\right)\frac{1}{2} r^2\, d\theta = \frac{1}{3}\int_0^{\pi} \theta^3 \cos\theta\, d\theta \approx -5.869604$$

and

$$\int_0^{\pi} y_c dA = \int_0^{\pi} \left(\frac{2}{3} r\sin\theta\right)\frac{1}{2} r^2\, d\theta = \frac{1}{3}\int_0^{\pi} \theta^3 \sin\theta\, d\theta \approx 4.052240$$

evaluated numerically. The exact values are $4-\pi^2$ and $\frac{\pi^3}{3-2\pi}$. Then

$$(X, Y) = \frac{1}{A}\left(\int_0^{\pi} x_c\, dA, \int_0^{\pi} y_c\, dA\right) \approx (-1.135822, 0.784146).$$

The corresponding polar coordinates are

$$\left(\sqrt{X^2+Y^2}, \pi + \arctan\frac{Y}{X}\right) \approx (1.38021, 2.53735).$$

**21.** By symmetry, $X = 0$. To find $Y$, use formula (11).

$$Y = \frac{\frac{1}{2}\int_0^{18} y^2\, dx}{\int_0^{18} y\, dx} = \frac{\frac{1}{2}\int_0^{18}\left(26-\frac{26}{81}(x-9)^2\right)^2 dx}{\int_0^{18}\left(26-\frac{26}{81}(x-9)^2\right)dx}$$

$$= \frac{\frac{1}{2}\int_0^{18}\left(\frac{676}{6561}x^4 - \frac{2704}{729}x^3 + \frac{2704}{81}x^2\right)dx}{\int_0^{18}\left(\frac{52}{9}x - \frac{26}{81}x^2\right)dx}$$

$$= \frac{\frac{1}{2}\left(\frac{676}{32{,}805}x^5 - \frac{676}{729}x^4 + \frac{2704}{243}x^3\right)\bigg|_0^{18}}{\left(\frac{26}{9}x^2 - \frac{26}{243}x^3\right)\bigg|_0^{18}} = \frac{\frac{16{,}224}{5}}{312} = \frac{52}{5}$$

So $(X, Y) = \left(0, \frac{52}{5}\right)$ meters.

**23.** The lamina lies between $x = 0$ and $x = \frac{\pi}{2}$, with $1 \geq \sin x$ in between. The centroid is given by formulas (10) and (11).

$$X = \frac{\int_0^{\pi/2} x(1-\sin x)\, dx}{\int_0^{\pi/2} (1-\sin x)\, dx} = \frac{\int_0^{\pi/2} (x - x\sin x)\, dx}{\int_0^{\pi/2} (1-\sin x)\, dx}$$

$$= \frac{\left(\frac{x^2}{2} + x\cos x - \sin x\right)\bigg|_0^{\pi/2}}{(x+\cos x)\big|_0^{\pi/2}} = \frac{\frac{\pi^2}{8}-1}{\frac{\pi}{2}-1} \approx 0.409429$$

$$Y = \frac{\frac{1}{2}\int_0^{\pi/2} (1^2 - \sin^2 x) dx}{\int_0^{\pi/2} (1-\sin x) dx} = \frac{\frac{1}{2}\left(x - \frac{2x-\sin 2x}{4}\right)\bigg|_0^{\pi/2}}{(x+\cos x)\big|_0^{\pi/2}} = \frac{\frac{\pi}{4}-\frac{\pi}{8}}{\frac{\pi}{2}-1} \approx 0.687985$$

The centroid is at approximately (0.409429, 0.687985).

**25.** The lamina lies "between" $x = 0$ and $x \to \infty$, with the $y = e^{-x}$ boundary above the $y = 0$ boundary. The coordinates of the centroid are given by formulas (10) and (11).

$$X = \frac{\int_0^{\infty} xe^{-x}\, dx}{\int_0^{\infty} e^{-x}\, dx} = \frac{\lim_{b\to\infty} -(x+1)e^{-x}\Big|_0^b}{\lim_{b\to\infty} -e^{-x}\Big|_0^b} = \frac{\lim_{b\to\infty} (1-(b+1)e^{-b})}{\lim_{b\to\infty} (1-e^{-b})} = 1$$

$\left(\text{Use l'Hôpital's Rule on } \frac{b+1}{e^b}.\right)$

$$Y = \frac{\frac{1}{2}\int_0^{\infty} e^{-2x}\,dx}{\int_0^{\infty} e^{-x}\,dx} = \frac{\lim_{b\to\infty} -\frac{1}{4}e^{-2x}\Big|_0^b}{1} = \lim_{b\to\infty} \frac{1}{4}(1-e^{-2b}) = \frac{1}{4}$$

The centroid is at $\left(1, \frac{1}{4}\right)$.

**27.**

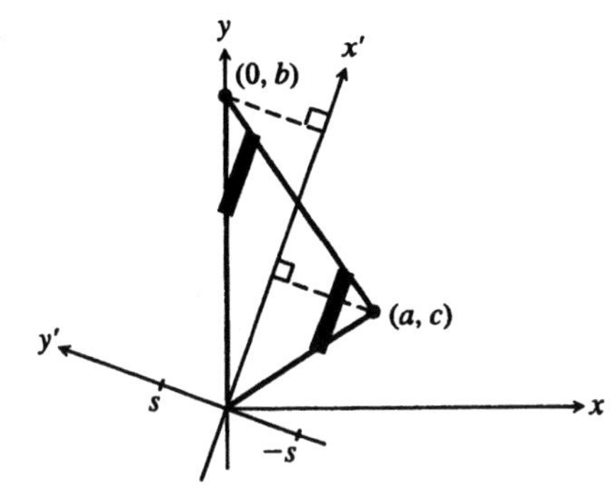


Axes $x'$ and $y'$ are placed so that the $x'$-axis lies on the median from the origin to the side joining $(0, b)$ and $(a, c)$. This median divides the triangle into two smaller triangles. Because a median divides the corresponding side of the triangle into two equal lengths, the altitudes from the $x'$-axis are equal and the two smaller triangles have the same area. Moreover, a thin area element at $y' = s$ has the same length as a thin area element at $y' = -s$, so that the centroids of the two smaller rectangles are the same distance from the $x'$-axis $\left(\frac{1}{3} \text{ of the altitudes}\right)$. It follows by the Mass Subdivision Theorem that the centroid of the whole triangle lies on the
$x'$-axis, i.e., on the median. Similar arguments show that it lies on the other two medians as well, and thus at their intersection. (In fact, this argument establishes that all three medians intersect at a single point, since that is the only way for the centroid to lie on all three.)

**29.** Set up a coordinate system at the center of the circular lamina. Let $m_1 = \delta\pi a^2$ be the mass of the circular region of radius $a$ to be punched and $\mathbf{R}_1 = \langle h, k\rangle$ its center/center of mass; let $m_2 = \delta\pi A^2$ be the mass of the unpunched circular lamina of radius $A$ and $\mathbf{R}_2 = \langle 0, 0\rangle$ its center/center of mass; and let $m_3 = m_2 - m_1$ be the mass of the punched lamina and $\mathbf{R}_3 = \langle X, Y\rangle$ its center of mass. Applying the Mass Subdivision Theorem to the two systems consisting of the punched lamina and the circular region to be punched,

$$\mathbf{R}_2 = \frac{1}{m_1 + m_3}(m_1\mathbf{R}_1 + m_3\mathbf{R}_3).$$

From this equation, we can solve for $X$ and $Y$.

**31.** The linear density is 2702 times the area in square meters of a washer-shaped cross section. The outer radius varies linearly according to $r_o = 0.05 + 0.0025x$, with $x$ representing meters from the thinner end. The inner radius varies according to $r_i = 0.02 + 0.0025x$. So the density is

$$\delta(x) = 2702\pi(r_0^2 - r_i^2) = 2702\pi[(0.05+0.0025x)^2 - 0.02 + 0.0025x)^2] = 2702\pi(0.00015x + 0.0021)$$

Notice that the quadratic term simplifies out.
The mass is

$$\begin{aligned}
M &= \int_0^{10} \delta(x)\,dx \\
&= 2702\pi \int_0^{10} (0.00015x + 0.0021)\,dx \\
&= 2702\pi(0.000075x^2 + 0.0021x)\Big|_0^{10} \\
&= 77.007\pi \\
&\approx 241.925 \text{ kg.}
\end{aligned}$$

The location of the center of mass is given by

$$\begin{aligned}
X &= \frac{1}{M}\int_0^{10} x\,\delta(x)\,dx \\
&= \frac{2702\pi}{M}\int_0^{10} (0.00015x^2 + 0.0021x)\,dx \\
&= \frac{2702\pi}{M}(0.00005x^3 + 0.00105x^2)\Big|_0^{10} \\
&= \frac{418.81}{77.007} \approx 5.4386 \text{ m.}
\end{aligned}$$

The center of mass is about 5.4386 meters from the thinner end.

**33.** $\mathbf{r} = \langle t - \sin t, 1 - \cos t \rangle \Rightarrow \mathbf{r}' = \langle 1 - \cos t, \sin t \rangle$

$$\|\mathbf{r}'\| = \sqrt{(1-\cos t)^2 + \sin^2 t} = \sqrt{2 - 2\cos t} = \sqrt{4\sin^2 \frac{t}{2}} = \left|2\sin\frac{t}{2}\right|$$

$$L = \int_a^b \|\mathbf{r}'\|\,dt = \int_0^{2\pi} \left|2\sin\frac{t}{2}\right| dt = -4\cos\frac{t}{2}\Big|_0^{2\pi} = 8$$

$$\left|2\sin\frac{t}{2}\right| = 2\sin\frac{t}{2} \quad \text{for } 0 \le t \le 2\pi$$

$$\begin{aligned}
X = \frac{1}{L}\int_a^b \|\mathbf{r}'\| x\,dt &= \frac{1}{8}\int_0^{2\pi}\left(2\sin\frac{t}{2}\right)(t - \sin t)\,dt \\
&= \frac{1}{4}\int_0^{2\pi}\left(\sin\frac{t}{2}\right)\left(2\cdot\frac{t}{2} - 2\sin\frac{t}{2}\cos\frac{t}{2}\right)dt \\
&= \frac{1}{2}\int_0^{2\pi}\left(\frac{t}{2}\sin\frac{t}{2} - \sin\frac{t}{2}\cos\frac{t}{2}\right)dt \\
&= \frac{1}{2}\left[2\left(\sin\frac{t}{2} - \frac{t}{2}\cos\frac{t}{2}\right) - \sin^2\frac{t}{2}\right]_0^{2\pi} \\
&= \pi
\end{aligned}$$

$$Y = \frac{1}{L}\int_a^b \|\mathbf{r}'\| y\,dt = \frac{1}{8}\int_0^{2\pi}\left(2\sin\frac{t}{2}\right)(1 - \cos t)\,dt = \frac{1}{8}\int_0^{2\pi} 4\sin^3\frac{t}{2}\,dt = -\frac{1}{3}\left(2 + \sin^2\frac{t}{2}\right)\cos\frac{t}{2}\Big|_0^{2\pi} = \frac{4}{3}$$

So $\mathbf{R} = \left\langle \pi, \frac{4}{3} \right\rangle$.

**35.** The volume of the head, in cubic meters is $V = (0.14)(0.07)(0.07) - \pi(0.0125)^2(0.07) \approx 6.516388 \times 10^{-4}$. The mass is 7800 times that, or 5.082783 kg.
The volume of the handle is $\pi(0.0125)^2(0.35) \approx 1.718058 \times 10^{-4}$, and the mass is 600 times that, or 0.103085 kg.
Let $x = 0$ at the top of the steel head. The centroid of the head is at $x = 0.035$, while the centroid of the handle is at $x = 0.175$. Both centroids lie on the handle's center line.
By the Mass Subdivision Theorem,
$$X = \frac{1}{5.082783 + 0.103085}[5.082783(0.035) + 0.103085(0.175)] \approx 0.037783 \text{ m}.$$
The center of mass of the hammer is on the center line of the handle, down approximately 3.8 cm from the end of the handle flush with a face. (This is near the center of mass of the head alone, at 3.5 cm.)

## Section R.11 Curvature, Acceleration, and Kepler's Second Law

**1.** Let $\mathbf{r}(t) = \langle 2t, t^2 \rangle$. Then $\mathbf{r}'(t) = \langle 2, 2t \rangle$ and $\|\mathbf{r}'(t)\| = 2\sqrt{1+t^2}$.

The unit tangent $\mathbf{T}$ is given by $\mathbf{T} = \dfrac{\mathbf{r}(t)}{\|\mathbf{r}'(t)\|} = \left\langle \dfrac{1}{\sqrt{1+t^2}}, \dfrac{t}{\sqrt{1+t^2}} \right\rangle$ and so $\mathbf{T}(0) = \langle 1, 0 \rangle$.

The curvature vector is given by $\dfrac{d\mathbf{T}}{ds} = \dfrac{\mathbf{T}'(t)}{\|\mathbf{r}'(t)\|} = \left\langle \dfrac{-t}{2(1+t^2)^2}, \dfrac{1}{2(1+t^2)^2} \right\rangle$

It follows that the curvature is $\kappa(t) = \left\| \dfrac{d\mathbf{T}}{ds} \right\| = \dfrac{\sqrt{(1+t^2)^{-3}}}{2}$ and so the curvature at $t = 0$ is $\kappa(0) = \dfrac{1}{2}$.

The unit normal at $t = 0$ is given by $\mathbf{N}_{t=0} = \dfrac{1}{\kappa(0)} \dfrac{\mathbf{T}'(0)}{\|\mathbf{r}'(0)\|} = \langle 0, 1 \rangle$.

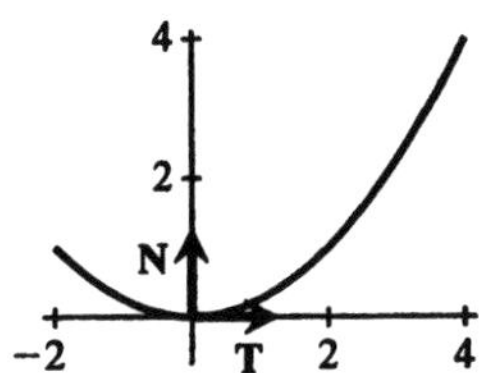


The critical values for $\kappa'(t) = \dfrac{-3t}{2(1+t^2)^{5/2}}$ occur at $t = 0$. Hence, we find the largest curvature is

$\kappa(0) = \dfrac{1}{2}$ and the smallest curvature is $\kappa(2) = \dfrac{1}{10\sqrt{5}}$.

**3.** Let $\mathbf{r}(t) = \langle 3\cos t, 2\sin t \rangle$. Then $\mathbf{r}'(t) = \langle -3\sin t, 2\cos t \rangle$ and $\|\mathbf{r}'(t)\| = \dfrac{\sqrt{13 - 5\cos 2t}}{\sqrt{2}}$.

The unit tangent $\mathbf{T}$ is given by $\mathbf{T} = \dfrac{\mathbf{r}'(t)}{\|\mathbf{r}'(t)\|} = \left\langle \dfrac{-3\sqrt{2}\sin t}{\sqrt{13 - 5\cos 2t}}, \dfrac{2\sqrt{2}\cos t}{\sqrt{13 - 5\cos 2t}} \right\rangle$ and so $\mathbf{T}\left(\dfrac{\pi}{4}\right) = \left\langle \dfrac{-3}{\sqrt{13}}, \dfrac{2}{\sqrt{13}} \right\rangle$.

The curvature vector is given by $\dfrac{d\mathbf{T}}{ds} = \dfrac{\mathbf{T}'(t)}{\|\mathbf{r}'(t)\|} = \left\langle \dfrac{-48\cos t}{(13 - 5\cos 2t)^2}, \dfrac{-72\sin t}{(13 - 5\cos 2t)^2} \right\rangle$.

It follows that the curvature is $\kappa(t) = \left\| \dfrac{d\mathbf{T}}{ds} \right\| = \dfrac{12\sqrt{2}}{(13 - 5\cos 2t)^{3/2}}$ and so the curvature at $t = \dfrac{\pi}{4}$ is

$\kappa\left(\dfrac{\pi}{4}\right) = \dfrac{12}{13}\sqrt{\dfrac{2}{13}}$.

The unit normal at $t = \frac{\pi}{4}$ is given by

$$\mathbf{n}_{t=\pi/4} = \frac{1}{\kappa\left(\frac{\pi}{4}\right)} \frac{\mathbf{T}'\left(\frac{\pi}{4}\right)}{\left\|\mathbf{r}'\left(\frac{\pi}{4}\right)\right\|} = \left\langle \frac{-2}{\sqrt{13}}, \frac{-3}{\sqrt{13}} \right\rangle.$$

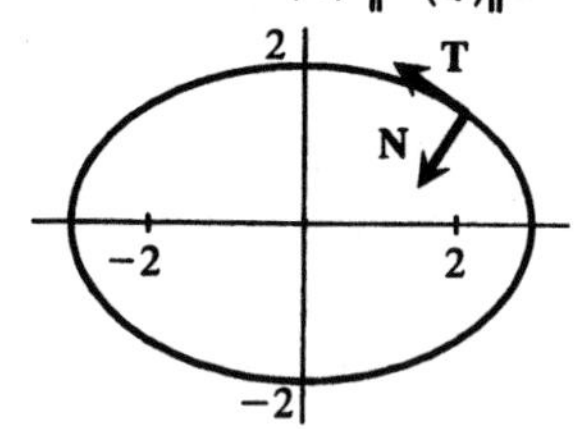


The critical values for $\kappa'(t) = \dfrac{-180\sqrt{2}\sin 2t}{(13 - 5\cos 2t)^{5/2}}$ occur at $t = \frac{k\pi}{2}$, $k = 0, \pm 1, \pm 2, \ldots$. Hence, we find the largest curvature is $\kappa(k\pi) = \frac{3}{4}$ $(k = 0, 1)$ and the smallest curvature is $\kappa\left(\frac{k\pi}{2}\right) = \frac{2}{9}$ $(k = 1, 3)$.

**5.** We consider the vector form of the given curve. Let $\mathbf{r}(t) = \langle t, t^3 \rangle$. Then $\mathbf{r}'(t) = \langle 1, 3t^2 \rangle$ and $\|\mathbf{r}'(t)\| = \sqrt{1 + 9t^4}$.
The unit tangent $\mathbf{T}$ is given by

$$\mathbf{T} = \frac{\mathbf{r}'(t)}{\|\mathbf{r}'(t)\|} = \left\langle \frac{1}{\sqrt{1+9t^4}}, \frac{3t^2}{\sqrt{1+9t^4}} \right\rangle \text{ and so}$$

$$\mathbf{T}(1) = \left\langle \frac{1}{\sqrt{10}}, \frac{3}{\sqrt{10}} \right\rangle.$$

The curvature vector is given by

$$\frac{d\mathbf{T}}{ds} = \frac{\mathbf{T}'(t)}{\|\mathbf{r}'(t)\|} = \left\langle \frac{-18t^3}{(1+9t^4)^2}, \frac{6t}{(1+9t^4)^2} \right\rangle.$$

It follows that the curvature is

$$\kappa(t) = \left\|\frac{d\mathbf{T}}{ds}\right\| = \frac{6t}{(1+9t^4)^{3/2}}.$$

and so the curvature at $t = 1$ is $\kappa(1) = \dfrac{3}{5\sqrt{10}}$.

The unit normal at $t = 1$ is given by

$$\mathbf{N}_{t=1} = \frac{1}{\kappa(1)} \frac{\mathbf{T}'(1)}{\|\mathbf{r}'(1)\|} = \left\langle -\frac{3}{\sqrt{10}}, \frac{1}{\sqrt{10}} \right\rangle.$$

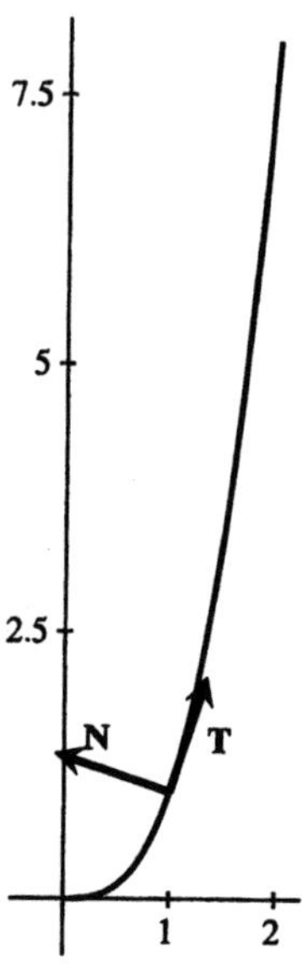


The critical values for $\kappa'(t) = \dfrac{6(1 - 45t^4)}{(1+9t^4)^{5/2}}$ occur at $t = 45^{-1/4}$. Hence, we find the largest curvature is $\kappa(45^{-1/4}) \approx 1.76229$ at $(45^{-1/4}, 45^{-3/4})$ and the smallest curvature is $\kappa(0) = 0$ at $(0, 0)$.

**7.** We consider the vector form of the given half circle. Let $\mathbf{r}(t) = \langle 4\cos^2 t, 4\cos t \sin t \rangle$. Then $\mathbf{r}'(t) = \langle -4\sin 2t, 4\cos 2t \rangle$ and $\|\mathbf{r}'(t)\| = 4$.
The unit tangent $\mathbf{T}$ is given by

$$\mathbf{T} = \frac{\mathbf{r}'(t)}{\|\mathbf{r}'(t)\|} = \langle -\sin 2t, \cos 2t \rangle$$

and so $\mathbf{T}\left(\frac{\pi}{4}\right) = \langle -1, 0 \rangle$.
The curvature vector is given by

$$\frac{d\mathbf{T}}{ds} = \frac{\mathbf{T}'(t)}{\|\mathbf{r}'(t)\|} = \left\langle \frac{-\cos 2t}{2}, \frac{-\sin 2t}{2} \right\rangle.$$

It follows that the curvature is the same everywhere, namely $\kappa(t) = \left\|\frac{d\mathbf{T}}{ds}\right\| = \frac{1}{2}$ and so the curvature at $t = \frac{\pi}{4}$ is $\kappa\left(\frac{\pi}{4}\right) = \frac{1}{2}$.

The unit normal at $t = \frac{\pi}{4}$ is given by

$$\mathbf{N}_{t=\pi/4} = \frac{1}{\kappa\left(\frac{\pi}{4}\right)} \frac{\mathbf{T}'\left(\frac{\pi}{4}\right)}{\left\|\mathbf{r}'\left(\frac{\pi}{4}\right)\right\|} = (0, -1).$$

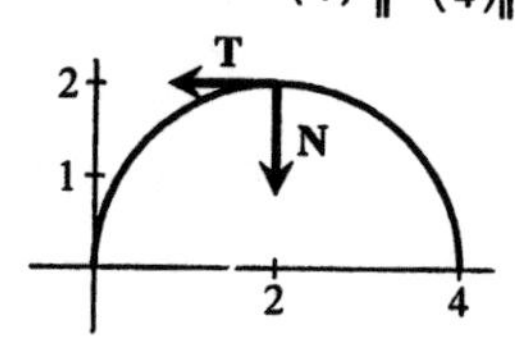

**9.** We consider the vector form of the given curve. Let $\mathbf{r}(t) = \langle t, \sin t\rangle$. Then $\mathbf{r}'(t) = \langle 1, \cos t\rangle$ and $\|\mathbf{r}'(t)\| = \sqrt{1+\cos^2 t}$.
The unit tangent $\mathbf{T}$ is given by

$$\mathbf{T} = \frac{\mathbf{r}'(t)}{\|\mathbf{r}'(t)\|} = \left\langle \frac{1}{\sqrt{1+\cos^2 t}}, \frac{\cos t}{\sqrt{1+\cos^2 t}} \right\rangle$$

and so $\mathbf{T}\left(\frac{\pi}{6}\right) = \left\langle \frac{2}{\sqrt{7}}, \sqrt{\frac{3}{7}} \right\rangle$.

The curvature vector is given by

$$\frac{d\mathbf{T}}{ds} = \frac{\mathbf{T}'(t)}{\|\mathbf{r}'(t)\|} = \left\langle \frac{\sin t\cos t}{(1+\cos^2 t)^2}, -\frac{\sin t}{(1+\cos^2 t)^2} \right\rangle.$$

It follows that the curvature is

$\kappa(t) = \left\|\frac{d\mathbf{T}}{ds}\right\| = \frac{|\sin t|}{(1+\cos^2 t)^{3/2}}$ and so the curvature at $t = \frac{\pi}{6}$ is $\kappa\left(\frac{\pi}{6}\right) = \frac{4}{7\sqrt{7}}$.

The unit normal at $t = \frac{\pi}{6}$ is given by

$$\mathbf{N}_{t=\pi/6} = \frac{1}{\kappa\left(\frac{\pi}{6}\right)} \frac{\mathbf{T}'\left(\frac{\pi}{6}\right)}{\left\|\mathbf{r}'\left(\frac{\pi}{6}\right)\right\|} = \left\langle \sqrt{\frac{3}{7}}, -\frac{2}{\sqrt{7}} \right\rangle.$$

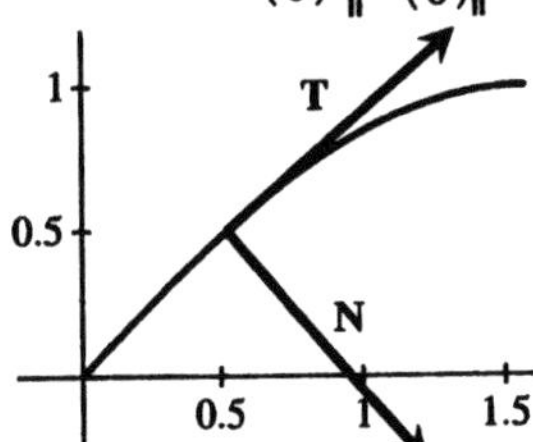


The critical value for

$\kappa'(t) = \frac{2\cos t(1+\sin^2 t)}{(1+\cos^2 t)^{5/2}};\ t \in \left[0, \frac{\pi}{2}\right]$ occurs at $t = \frac{\pi}{2}$. Hence, we find the largest curvature is $\kappa\left(\frac{\pi}{2}\right) = 1$ at $\left(\frac{\pi}{2}, 1\right)$ and the smallest curvature is $\kappa(0) = 0$ at $(0, 0)$.

**11.** We consider the vector form of the given curve. Let $\mathbf{r}(t) = \langle t, e^t\rangle$. Then $\mathbf{r}'(t) = \langle 1, e^t\rangle$ and $\|\mathbf{r}'(t)\| = \sqrt{1+e^{2t}}$.
The unit tangent $\mathbf{T}$ is given by

$$\mathbf{T} = \frac{\mathbf{r}'(t)}{\|\mathbf{r}'(t)\|} = \left\langle \frac{1}{\sqrt{1+e^{2t}}}, \frac{e^t}{\sqrt{1+e^{2t}}} \right\rangle$$

and so $\mathbf{T}(1) = \left\langle \frac{1}{\sqrt{1+e^2}}, \frac{e}{\sqrt{1+e^2}} \right\rangle$.

The curvature vector is given by

$$\frac{d\mathbf{T}}{ds} = \frac{\mathbf{T}'(t)}{\|\mathbf{r}'(t)\|} = \left\langle -\frac{e^{2t}}{(1+e^{2t})^2}, \frac{e^t}{(1+e^{2t})^2} \right\rangle.$$

It follows that the curvature is

$\kappa(t) = \left\|\frac{d\mathbf{T}}{ds}\right\| = \frac{e^t}{(1+e^{2t})^{3/2}}$ and so that the curvature at $t = 1$ is $\kappa(1) = \frac{e}{(1+e^2)^{3/2}}$.

The unit normal at $t = 1$ is given by

$$\mathbf{N}_{t=1} = \frac{1}{\kappa(1)} \frac{\mathbf{T}'(1)}{\|\mathbf{r}'(1)\|} = \left\langle -\frac{e}{\sqrt{1+e^2}}, \frac{1}{\sqrt{1+e^2}} \right\rangle.$$

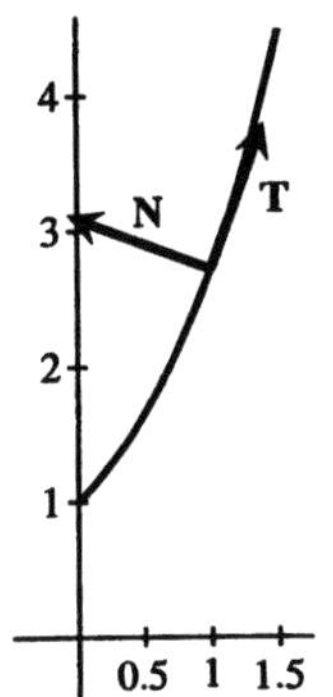


The function $\kappa'(t) = \frac{e^t(2e^{2t}-1)}{(1+e^{2t})^{5/2}}$ has no critical values on $[0, 1.5]$ and so we have endpoint extrema.
We find the largest curvature is $\kappa(0) = 1$ at $(0, 1)$ and the smallest curvature is $\kappa(1.5) \approx 0.046288$ at $(1.5, e^{1.5})$.

**13.** Let $\mathbf{r}(t) = \langle t-1, 2t-t^2 \rangle$. We seek to show that
$$a = \frac{4(t-1)}{\sqrt{4(1-t)^2+1}}\mathbf{T} + \frac{2}{\sqrt{4(1-t)^2+1}}\mathbf{N}.$$
We establish this by using formula (16), and showing that $v' = \frac{4(t-1)}{\sqrt{4(1-t)^2+1}}$ and
$$v^2\kappa = \frac{2}{\sqrt{4(1-t)^2+1}}.$$
Since $\mathbf{r}'(t) = \langle 1, 2-2t \rangle$,
$$v = \|\mathbf{r}'\| = \sqrt{4t^2-8t+5} = \sqrt{4(1-t)^2+1}$$
and so $v^2 = 4(1-t)^2+1$ and
$$v' = \frac{4(t-1)}{\sqrt{4(1-t)^2+1}}.$$
The unit tangent $\mathbf{T}$ is given by
$$\mathbf{T} = \frac{\mathbf{r}'(t)}{\|\mathbf{r}'(t)\|} = \left\langle \frac{1}{\sqrt{4(1-t)^2+1}}, \frac{2-2t}{\sqrt{4(1-t)^2+1}} \right\rangle$$
and so the curvature vector is given by
$$\frac{d\mathbf{T}}{ds} = \frac{\mathbf{T}'(t)}{\|\mathbf{r}'(t)\|}$$
$$= \left\langle \frac{4(1-t)}{(4(1-t)^2+1)^2}, \frac{-2}{(4(1-t)^2+1)^2} \right\rangle.$$
It follows that the curvature is
$$\kappa(t) = \left\|\frac{d\mathbf{T}}{ds}\right\| = \frac{2}{(4(1-t)^2+1)^{3/2}}.$$
The desired result now follows.

**15.** Let $\mathbf{r}(t) = (\sin 3t)\langle \cos 3t, \sin 3t \rangle$. Using the notation of (21), page 514, we have that $r(t) = \sin 3t$ and $\theta(t) = 3t$. The radial acceleration vector is given by
$(r'' - r\theta'^2)e_r = -18\sin 3t\langle \cos 3t, \sin 3t \rangle$.

At $\mathbf{r}\left(\frac{\pi}{18}\right)$ the radial acceleration is
$$\left\langle -\frac{9\sqrt{3}}{2}, -\frac{9}{2} \right\rangle.$$
The transverse acceleration vector is given by
$(2r'\theta' + r\theta'')e_\theta = 18\cos 3t\langle -\sin 3t, \cos 3t \rangle$.

At $\mathbf{r}\left(\frac{\pi}{18}\right)$ the transverse acceleration is
$$\left\langle -\frac{9\sqrt{3}}{2}, \frac{27}{2} \right\rangle.$$
Since the acceleration vector is the sum of the radial acceleration vector and the transverse acceleration vector, the acceleration at $\mathbf{r}\left(\frac{\pi}{6}\right)$ is $\langle -9\sqrt{3}, 9 \rangle$.

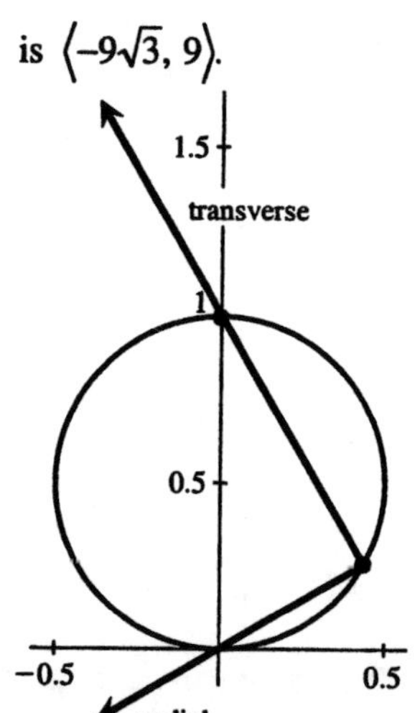


**17.** Let $f(x) = e^x$. Using (13), the curvature for $f(x) = e^x$ is given by $\kappa(x) = \frac{e^x}{(1+e^{2x})^{3/2}}$.
Differentiating $\kappa(x)$ with respect to $x$, we obtain $\kappa'(x) = \frac{e^x - 2e^{3x}}{(1+e^{2x})^{5/2}}$.
The lone critical value of $\kappa(x)$ occurs at
$$x = \ln\frac{1}{\sqrt{2}} \approx -0.346574.$$
The maximum curvature is
$$\kappa\left(-\ln\sqrt{2}\right) = \kappa\left(-\frac{\ln 2}{2}\right) \approx 0.3849.$$

**19.** $(f(r)e_r)\cdot e_\theta = (m(r''-r\theta'^2)e_r + m(2r'\theta'+r\theta'')e_\theta)\cdot e_\theta$

$$f(r)(e_r\cdot e_\theta) = m(r''-r\theta'^2)(e_r\cdot e_\theta) + m(2r'\theta'+r\theta'')(e_\theta\cdot e_\theta)$$

$$f(r)(0) = m(r''-r\theta'^2)(0) + m(2r'\theta'+r\theta'')\|e_\theta\|^2$$

$$0 = m(2r'\theta'+r\theta'')$$

**21.** $r = 2^\theta \Rightarrow r' = (\ln 2)2^\theta \Rightarrow r'' = (\ln 2)^2 2^\theta$

$$\kappa = \frac{\left|2^{2\theta} + 2(\ln 2)^2 2^{2\theta} - (\ln 2)^2 2^{2\theta}\right|}{(2^{2\theta} + (\ln 2)^2 2^{2\theta})^{3/2}} = \frac{1}{2^\theta\sqrt{1+(\ln 2)^2}}$$

**23.** $r = \theta^2 \Rightarrow r' = 2\theta \Rightarrow r'' = 2$

$$\kappa = \frac{\left|\theta^4 + 8\theta^2 - 2\theta^2\right|}{(\theta^4 + 4\theta^2)^{3/2}} = \frac{\theta^2+6}{\theta(\theta^2+4)^{3/2}}$$

**25.** $\mathbf{r} = \langle t, t^2\rangle \Rightarrow \mathbf{r}' = \langle 1, 2t\rangle \Rightarrow \|\mathbf{r}'\| = \sqrt{1+4t^2}$

$$\mathbf{T} = \frac{\mathbf{r}'}{\|\mathbf{r}'\|} = \left\langle \frac{1}{\sqrt{1+4t^2}}, \frac{2t}{\sqrt{1+4t^2}}\right\rangle$$

$$\mathbf{T}' = \left\langle -\frac{4t}{(1+4t^2)^{3/2}}, \frac{2}{(1+4t^2)^{3/2}}\right\rangle$$

$$\kappa = \frac{2}{(1+4t^2)^{3/2}} \quad \text{(from formula (13))}$$

$$\mathbf{N} = \frac{1}{\kappa}\cdot\frac{\mathbf{T}'}{\|\mathbf{r}'\|} = \left\langle -\frac{2t}{\sqrt{1+4t^2}}, \frac{1}{\sqrt{1+4t^2}}\right\rangle$$

At $t = 0$: $\mathbf{r} = \langle 0, 0\rangle$, $\kappa = 2$, and $\mathbf{N} = \langle 0, 1\rangle$.

The osculating circle has a radius of $\frac{1}{\kappa} = \frac{1}{2}$ and is centered at $\mathbf{r} + \frac{\mathbf{N}}{\kappa} = \left\langle 0, \frac{1}{2}\right\rangle$.

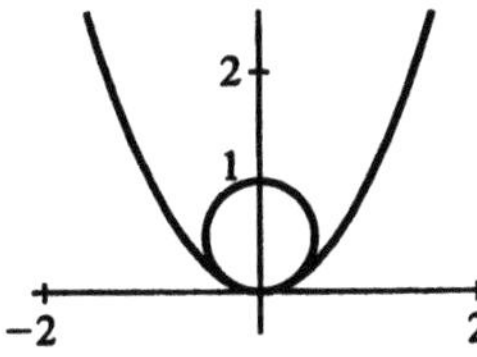


**27.** $\mathbf{r} = \langle t, \sin t\rangle \Rightarrow \mathbf{r}' = \langle 1, \cos t\rangle \Rightarrow \|\mathbf{r}'\| = \sqrt{1+\cos^2 t}$

$$\mathbf{T} = \frac{\mathbf{r}'}{\|\mathbf{r}'\|} = \left\langle \frac{1}{\sqrt{1+\cos^2 t}}, \frac{\cos t}{\sqrt{1+\cos^2 t}}\right\rangle$$

$$\mathbf{T}' = \left\langle \frac{\sin t\cos t}{(1+\cos^2 t)^{3/2}}, -\frac{\sin t}{(1+\cos^2 t)^{3/2}}\right\rangle$$

$$\kappa = \frac{\sin t}{(1+\cos^2 t)^{3/2}} \quad \text{(from formula (13))}$$

$$\mathbf{N} = \frac{1}{\kappa}\cdot\frac{\mathbf{T}'}{\|\mathbf{r}'\|} = \left\langle \frac{\cos t}{\sqrt{1+\cos^2 t}}, -\frac{1}{\sqrt{1+\cos^2 t}}\right\rangle$$

At $t = \frac{\pi}{2}$: $\mathbf{r} = \left\langle \frac{\pi}{2}, 1 \right\rangle$, $\kappa = 1$, $\mathbf{N} = \langle 0, -1 \rangle$.

The osculating circle has a radius of $\frac{1}{\kappa} = 1$

and is centered at $\mathbf{r} + \frac{\mathbf{N}}{\kappa} = \left\langle \frac{\pi}{2}, 0 \right\rangle$.

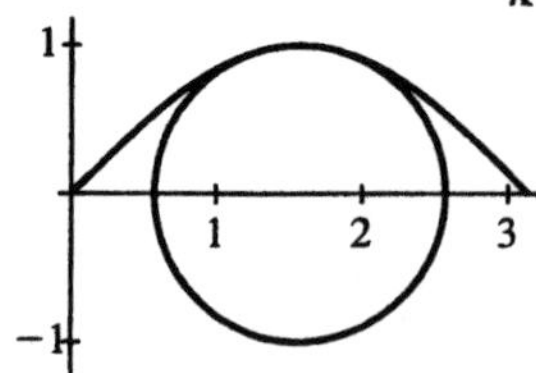


**29.** The curve $y = x^2$ can be described parametrically by $\mathbf{r}(t) = \langle t, t^2 \rangle$. From Exercise 25, $\kappa = \frac{2}{(1+4t^2)^{3/2}}$ and

$$\mathbf{N} = \left\langle -\frac{2t}{\sqrt{1+4t^2}}, \frac{1}{\sqrt{1+4t^2}} \right\rangle.$$

The point of maximum curvature, clearly, is (0, 0), where $\kappa = 2$. The radius of curvature, which is the radius of the drill bit, is $\frac{1}{\kappa} = \frac{1}{2} = 0.5$ units. For any point $\mathbf{r}(t)$, the center of the drill bit should lie at

$$\mathbf{r} + \frac{\mathbf{N}}{\kappa} = \left\langle t - \frac{t}{\sqrt{1+4t^2}}, t^2 + \frac{1}{2\sqrt{1+4t^2}} \right\rangle.$$

This represents curve $A$; substituting $x$ for $t$, we can describe curve $A$ by

$$\mathbf{r}(x) = \left\langle x - \frac{x}{\sqrt{1+4x^2}}, x^2 + \frac{1}{2\sqrt{1+4x^2}} \right\rangle.$$

**31.** The dot product of any vector with itself equals the magnitude squared. So $r^2 = \mathbf{r} \cdot \mathbf{r}$. Differentiating both sides, using formula (15), yields

$$2rr' = \mathbf{r}' \cdot \mathbf{r} + \mathbf{r} \cdot \mathbf{r}'$$
$$rr' = \mathbf{r}' \cdot \mathbf{r}.$$

**33.** The area of the entire ellipse is

$$\int_0^{2\pi} \frac{1}{2} r^2 \, d\theta = \frac{1}{2} \int_0^{2\pi} \frac{d\theta}{(1-0.6\cos\theta)^2} \approx 6.13592.$$

$\left(\text{The exact value is } \frac{125\pi}{64}.\right)$ The point in question corresponds to an angle $\beta$ such that

$$\int_\pi^\beta \frac{1}{2} r^2 \, d\theta = \frac{1}{2} \int_\pi^\beta \frac{d\theta}{(1-0.6\cos\theta)^2} \approx \frac{1}{12}(6.13592) \approx 0.511327.$$

This can be found to be true for $\beta \approx 4.849205$.

Then $r \approx \frac{1}{1-0.6\cos 4.849205} \approx 1.08913$ and $(x, y) = (r\cos\beta, r\sin\beta)$ becomes approximately (0.15, –1.08).

**35.** Let $\mathbf{f}(t) = \langle w(t), x(t) \rangle$, $\mathbf{g}(t) = \langle y(t), z(t) \rangle$.

$$\begin{aligned}(\mathbf{f} \cdot \mathbf{g})' &= \frac{d}{dt}(wy + xz) \\ &= w'y + wy' + x'z + xz' \\ &= w'y + x'z + wy' + xz' \\ &= \langle w', x' \rangle \cdot \langle y, z \rangle + \langle w, x \rangle \cdot \langle y', z' \rangle \\ &= \mathbf{f}' \cdot \mathbf{g} + \mathbf{f} \cdot \mathbf{g}'\end{aligned}$$

## Chapter R Review Exercises

**1.** If at anytime $t$ the position of an object is $x(t)$, then the average velocity of the object on the time interval $[t_1, t_2]$ is $\frac{x(t_2) - x(t_1)}{t_2 - t_1}$. If at any time $t$ the position of an object is $\mathbf{r}(t)$, then the average velocity of the object on the time interval $[t_1, t_2]$ is $\frac{1}{t_2 - t_1}(\mathbf{r}(t_2) - \mathbf{r}(t_1))$.

**3.** 
$$\begin{aligned}v(5.5, 7.2) &= \frac{\langle 1, 4 \rangle - \langle -10, 3 \rangle}{7.2 - 5.5} \\ &= \frac{\langle 11, 1 \rangle}{1.7} \\ &\approx \langle 6.47059, 0.588235 \rangle \text{ m/s}\end{aligned}$$

**5.** Let $\mathbf{u} = \langle u_1, u_2 \rangle$ be a unit vector making an angle of 0.5 radians with $\langle 6, 1 \rangle$. Then

$\|\mathbf{u}\|^2 = u_1^2 + u_2^2 = 1$ and

$$\frac{\mathbf{u} \cdot \langle 6, 1 \rangle}{\|\mathbf{u}\| \, \|\langle 6, 1 \rangle\|} = \frac{6u_1 + u_2}{\sqrt{37}} = \cos 0.5$$

To get the two unit vectors we solve the system of equations:

$$\begin{cases} u_1^2 + u_2^2 = 1 \\ 6u_1 + u_2 = \sqrt{37} \cos 0.5 \end{cases}$$

The two unit vectors are:

$\langle 0.786825, 0.617176 \rangle$ and

$\langle 0.944459, -0.328629 \rangle$

**7.** Let $x = \cosh t$ and $y = \sinh t$. Then

$x^2 - y^2 = 1$. Since $y = \sinh t > 0$, $y = \sqrt{x^2 - 1}$.

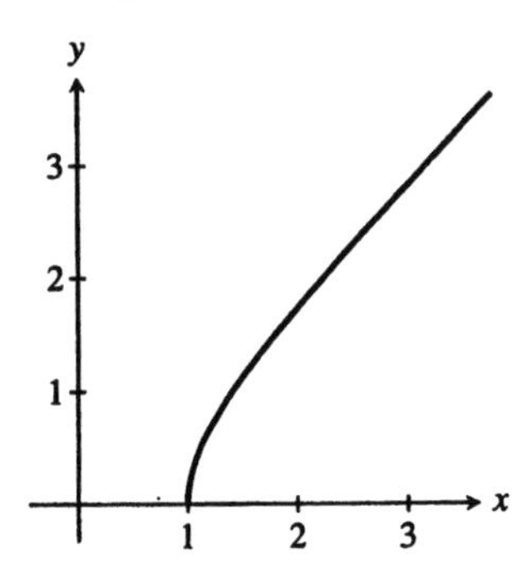


**9.** The paths intersect when $2s = t + t^2$ and $-5 + s = t - t^2$. Thus $2(t - t^2 + 5) = t + t^2$, i.e.,

$3t^2 - t - 10 = (3t + 5)(t - 2) = 0$. Since $t \geq 0$,

$t = 2$. Then $s = \dfrac{t + t^2}{2} = 3$. Since $s \neq t$, the particles do not collide.

**11.** If $\mathbf{a} \neq \pm \mathbf{b}$, then $\mathbf{a} + \mathbf{b} \neq 0$ and $\mathbf{a} - \mathbf{b} \neq 0$.

Since $\|\mathbf{a}\| = \|\mathbf{b}\| = 1$,

$(\mathbf{a} + \mathbf{b}) \cdot (\mathbf{a} - \mathbf{b}) = \mathbf{a} \cdot \mathbf{a} - \mathbf{b} \cdot \mathbf{b} = \|\mathbf{a}\|^2 - \|\mathbf{b}\|^2 = 0$

so $\mathbf{a} + \mathbf{b}$ and $\mathbf{a} - \mathbf{b}$ are perpendicular.

**13.** $\mathbf{v}(t) = \mathbf{r}'(t) = 2\langle -\sin 2t, \cos 2t \rangle$

$\mathbf{a}(t) = \mathbf{v}'(t) = -4\langle \cos 2t, \sin 2t \rangle$

$\mathbf{a}(3.1) = -4\langle \cos 6.2, \sin 6.2 \rangle$

$\approx \langle -3.98617, 0.332358 \rangle \text{ m/s}^2$

**15.** $\mathbf{a} = \dfrac{\mathbf{F}}{m} = \dfrac{\langle 2.4, 1.7 \rangle}{1.2} = \left\langle 2, \dfrac{17}{12} \right\rangle \text{ m/s}^2$;

$\mathbf{v}(t) = \left\langle 2t + c_1, \dfrac{17}{12} t + c_2 \right\rangle$

Since the object was originally at rest, $\mathbf{v}(0) = \langle 0, 0 \rangle$, so $c_1 = c_2 = 0$.

$\mathbf{v}(t) = \left\langle 2t, \dfrac{17}{12} t \right\rangle$

$\mathbf{v}(3.5) \approx \langle 7, 4.95833 \rangle$; $\|\mathbf{v}(3.5)\| \approx 8.57817 \text{ m/s}$;

$\mathbf{r}(t) = \left\langle t^2 + c_3, \dfrac{17}{24} t^2 + c_4 \right\rangle$

Since the object was originally at the origin, $\mathbf{r}(0) = \langle 0, 0 \rangle$, so $c_3 = c_4 = 0$.

$\mathbf{r}(t) = \left\langle t^2, \dfrac{17t^2}{24} \right\rangle$; $\mathbf{r}(3.5) \approx \langle 12.25, 8.67708 \rangle$

**17.** $\mathbf{r}(t) = (-2 + 5\cos t)\mathbf{i} + (3 + 5\sin t)\mathbf{j}$, $0 \leq t \leq 2\pi$

**19.** Observe that $\mathbf{e_1}$ and $\mathbf{e_2}$ are perpendicular since $\mathbf{e_1} \cdot \mathbf{e_2} = 0$. Let $\mathbf{u_1} = \dfrac{\mathbf{e_1}}{\|\mathbf{e_1}\|} = \dfrac{\langle 3, 1 \rangle}{\sqrt{10}}$ and

$\mathbf{u_2} = \dfrac{\mathbf{e_2}}{\|\mathbf{e_2}\|} = \dfrac{\langle -1, 3 \rangle}{\sqrt{10}}$. Then $\mathbf{v} = \langle 5, 7 \rangle$

is the sum of

$$\mathbf{v_1} = (\mathbf{v} \cdot \mathbf{u_1})\mathbf{u_1} = (\mathbf{v} \cdot \mathbf{e_1}) \frac{\mathbf{e_1}}{\|\mathbf{e_1}\|^2}$$

$$= (15 + 7) \frac{\langle 3, 1 \rangle}{10} = \left\langle \frac{33}{5}, \frac{11}{5} \right\rangle$$

and

$$\mathbf{v_2} = (\mathbf{v} \cdot \mathbf{u_2})\mathbf{u_1} = (\mathbf{v} \cdot \mathbf{e_2}) \frac{\mathbf{e_2}}{\|\mathbf{e_2}\|^2}$$

$$= (-5 + 21) \frac{\langle -1, 3 \rangle}{10} = \left\langle -\frac{8}{5}, \frac{24}{5} \right\rangle.$$

**21.** In the diagram shown, angle $A$ is inscribed in a semicircle of radius $a$. We want to show that vectors

$\overrightarrow{AB} = \langle -a - x, -y \rangle$ and $\overrightarrow{AC} = \langle a - x, -y \rangle$ are perpendicular.

$$\overrightarrow{AB} \cdot \overrightarrow{AC} = (-a - x)(a - x) + y^2$$
$$= -a^2 + x^2 + y^2 = 0,$$

since $x^2 + y^2 = a^2$.

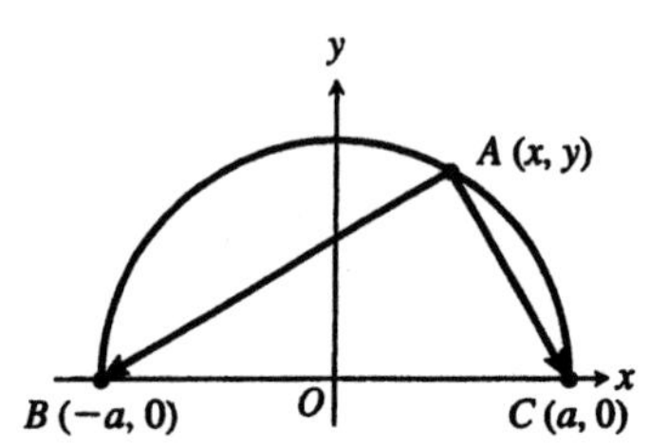


**23.**

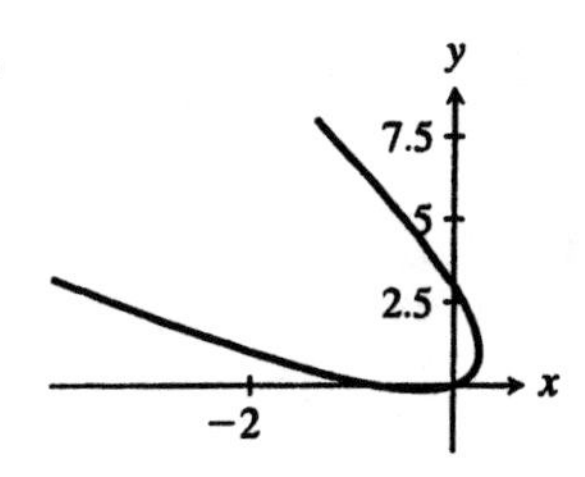


Let $x = t - t^2$ and $y = t + 2t^2$. Then $x - y = -3t^2$ and $2x + y = 3t$. Hence,

$$(2x + y)^2 + 3(x - y) = 0,$$
$$4x^2 + 4xy + y^2 + 3x - 3y = 0.$$

To show that the graph of the equation includes $C$ we need to show that the point $\left(t - t^2, t + 2t^2\right)$ satisfies the equation

$$(2x + y)^2 + 3(x - y) = 0.$$
$$\left(2\left(t - t^2\right) + \left(t + 2t^2\right)\right)^2 + 3\left(t - t^2 - \left(t + 2t^2\right)\right)$$
$$= (3t)^2 + 3\left(-3t^2\right)$$
$$= 9t^2 - 9t^2$$
$$= 0$$

At (–2, 10), $x - y = -12 = -3t^2$ and so $t = 2$ ($t = -2$ corresponds to (–2, 6)).

$\mathbf{r}'(t) = \langle 1 - 2t, 1 + 4t \rangle$, $\mathbf{r}'(2) = \langle -3, 9 \rangle$ and the slope of $C$ at the point (–2, 10) is $\dfrac{9}{-3} = -3$.

**25.** $x^2 - y^2 = 1$ implies $y = \sqrt{x^2 - 1}$ and

$y' = \dfrac{x}{\sqrt{x^2 - 1}}$ (for $y > 0$). The arc length is given by the arc length formula.

$$s = \int_a^b \sqrt{1 + y'^2}\, dx$$
$$= \int_2^3 \sqrt{1 + \left(\frac{x}{\sqrt{x^2 - 1}}\right)^2}\, dx$$
$$= \int_2^3 \sqrt{\frac{x^2 - 1 + x^2}{x^2 - 1}}\, dx$$
$$= \int_2^3 \sqrt{\frac{2x^2 - 1}{x^2 - 1}}\, dx$$

**27.** With $y = a\cosh\left(\dfrac{x}{a}\right) = \left(\dfrac{a}{2}\right)\left(e^{x/a} + e^{-x/a}\right)$,

$y' = \dfrac{1}{2}\left(e^{x/a} - e^{-x/a}\right) = \sinh\left(\dfrac{x}{a}\right)$ and

$$s = \int_{-b}^{b} \sqrt{1 + y'^2}\, dx$$
$$= \int_{-b}^{b} \sqrt{1 + \sinh^2\left(\frac{x}{a}\right)}\, dx$$
$$= \int_{-b}^{b} \cosh\left(\frac{x}{a}\right) dx \qquad 1 + \sinh^2 t = \cosh t^2$$
$$= a\sinh\left(\frac{x}{a}\right)\Bigg|_{-b}^{b}$$
$$= 2a\sinh\left(\frac{b}{a}\right) \qquad \sinh\left(-\frac{b}{a}\right) = -\sinh\left(\frac{b}{a}\right)$$

**29.**

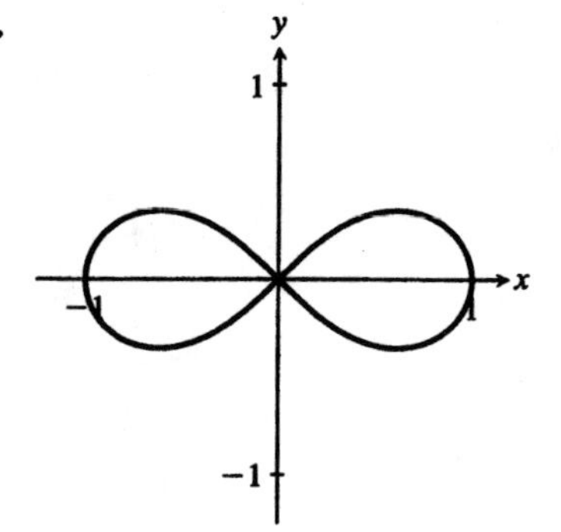

Use $r=\sqrt{\cos 2\theta}$ $-\frac{\pi}{4}\le\theta\le\frac{\pi}{4}$ for one of the loops. The total area is twice the area of one loop.

$$A=2\int_{\alpha}^{\beta}\frac{1}{2}r^2\,d\theta=2\int_{-\pi/4}^{\pi/4}\frac{1}{2}\cos 2\theta\,d\theta=\frac{1}{2}\sin 2\theta\Big|_{-\pi/4}^{\pi/4}=1 \text{ square unit}$$

**31.**

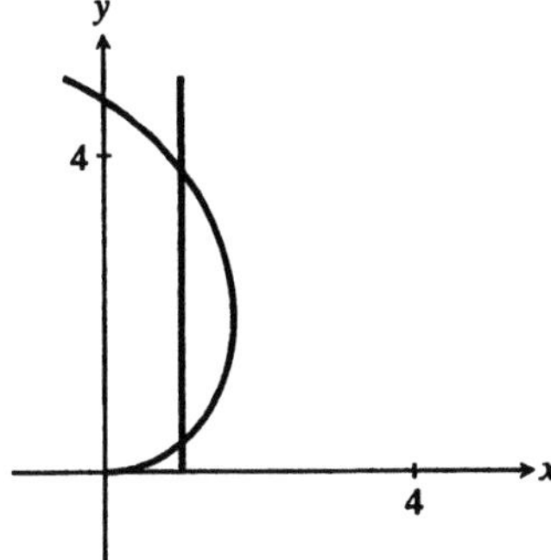


The curves intersect when $3\theta=\sec\theta\Rightarrow\theta\approx 0.355576, 1.314394$. The area is given by

$$A=\int_{\alpha}^{\beta}\frac{1}{2}(r_2^2-r_1^2)\approx\frac{1}{2}\int_{0.355576}^{1.314394}(9\theta^2-\sec^2\theta)\,d\theta=\frac{1}{2}(3\theta^3-\tan\theta)\Big|_{0.355576}^{1.314394}\approx 1.61729.$$

**33.**

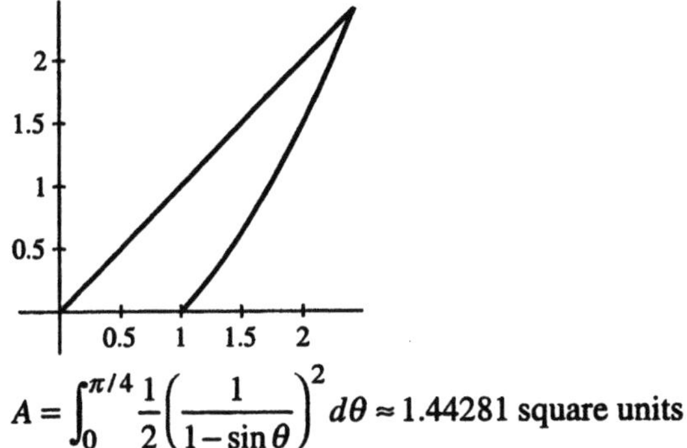


$$A=\int_0^{\pi/4}\frac{1}{2}\left(\frac{1}{1-\sin\theta}\right)^2 d\theta\approx 1.44281 \text{ square units}$$

**35.** The mass of the vehicle is

$$m(x)=4.3\times10^3+2.6\times10^3+\frac{(43.1\times10^3)(160{,}000-x)}{160{,}000}=50{,}000-0.269375x.$$

The force of gravity, against which the engine does its work, is $F(x)=\dfrac{GM_Em(x)}{(x+R_E)^2}$.

So the total work is given by

$$\begin{aligned}W&=\int_0^{160{,}000}\frac{GM_Em(x)}{(x+R_E)^2}\,dx\\&=GM_E\int_0^{160{,}000}\frac{50{,}000-0.269375x}{(x+R_E)^2}\,dx\\&=GM_E\int_0^{160{,}000}\left(\frac{50{,}000+0.269375R_E}{(x+R_E)^2}-\frac{0.269375(x+R_E)}{(x+R_E)^2}\right)dx\\&=GM_E\left(-\frac{50{,}000+0.269375R_E}{x+R_E}-0.269375\ln(x+R_E)\right)\Bigg|_0^{160{,}000}\end{aligned}$$

which, evaluated for $G=6.67259\times10^{-11}$, $M_E=5.97\times10^{24}$, and $R_E=6378145$, equals about $4.38\times10^{10}$ J.

**37.** Using formula (2) from Section 6.7, the center of mass of the system with point masses $\{(50, 0), (40, 4), (55, 9), (45, 12)\}$ is given by $R = \frac{50(0)+40(4)+55(9)+45(12)}{50+40+55+45} = \frac{239}{38}$ m from the zero-meter mark.

**39.** With $x$ ranging from 0 to 2.0, the density is $\delta(x) = 2.8 + 0.2x$. The mass is given by

$$m = \int_a^b \delta(x)\,dx = \int_0^{2.0} (2.8+0.2x)\,dx$$
$$= (2.8x+0.1x^2)\Big|_0^{2.0}$$
$$= 6 \text{ kg}.$$

The distance of the center of mass from the lighter end is

$$X = \frac{1}{m}\int_a^b x\,\delta(x)\,dx = \frac{1}{6}\int_0^{2.0}(2.8x+0.2x^2)\,dx$$
$$= \frac{1}{6}\left(1.4x^2 + \frac{0.2}{3}x^3\right)\Big|_0^{2.0}$$
$$\approx 1.02222 \text{ meters}.$$

**41.** The area of a thin ring of radius $r$ and thickness $dr$ is $dA \approx 2\pi r\,dr$, and so the total sediment is

$$\int_0^{100} \delta(r)\,dA = \int_0^{100} \frac{7}{1+r^2}(2\pi r)\,dr$$
$$= 2\pi\int_0^{100} \frac{7r}{1+r^2}\,dr$$
$$= 7\pi \ln(1+r^2)\Big|_0^{100}$$
$$= 7\pi \ln 10{,}001 \text{ tons}.$$

**43.** For the mass, note that the sector's area is $\frac{\theta}{2\pi}$ times $\pi a^2$, or $\frac{a^2\theta}{2}$, so that the mass must be $\frac{1}{2}a^2\delta\theta$.

With one straight edge of the sector on the $x$-axis, the other straight edge, in the first quadrant, can be represented as $g(y) = y\cot\theta$ and the curved edge can be represented as $f(y) = \sqrt{a^2-y^2}$.

Find the $y$-coordinate of the center of mass.

$$Y = \frac{\int_a^b y(f(y)-g(y))\,dy}{A}$$
$$= \frac{\int_0^{a\sin\theta} y\left(\sqrt{a^2-y^2} - y\cot\theta\right)dy}{\frac{a^2\theta}{2}}$$
$$= \frac{2\left(-\frac{1}{3}(a^2-y^2)^{3/2} - \frac{1}{3}y^3\cot\theta\right)\Big|_0^{a\sin\theta}}{a^2\theta}$$
$$= \frac{\frac{2a^3}{3}(1-\cos\theta)}{a^2\theta} = \frac{2a}{3\theta}(1-\cos\theta)$$

Since by symmetry, the center of mass must lie on the line $y = \tan\left(\frac{\theta}{2}\right)x = \frac{(1-\cos\theta)x}{\sin\theta}$, it follows that $X = \frac{(2a\sin\theta)}{3\theta}$ and that

$$\mathbf{R} = \frac{2a}{3\theta}\langle \sin\theta,\ 1-\cos\theta\rangle.$$

# Chapter 8 Vectors and Linear Functions

## Section 8.1 Vectors in Three Dimensions

**1.** First locate the point (2, 5, 0) on the $(x, y)$-plane. Then move up until you are level with 4 on the $z$-axis.

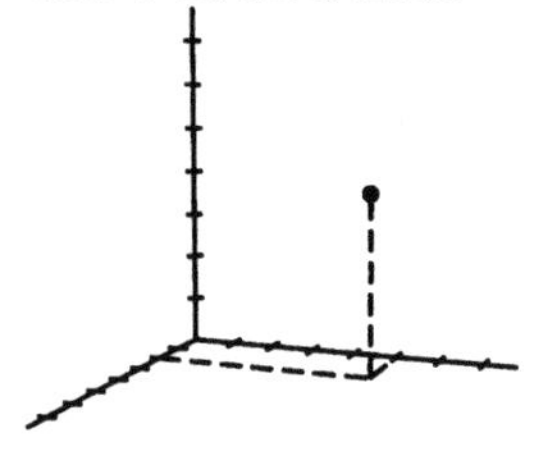

**3.** First locate the point (–1, –2, 0) on the $(x, y)$-plane. Then move down until you are level with –3 on the $z$-axis.

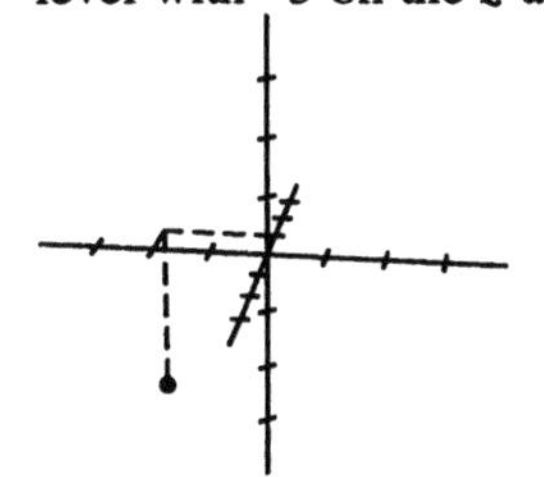

**5.** The vector $\langle 6, 0, 3\rangle$ starts at the origin and ends at (6, 0, 3), in the $(x, z)$-plane.

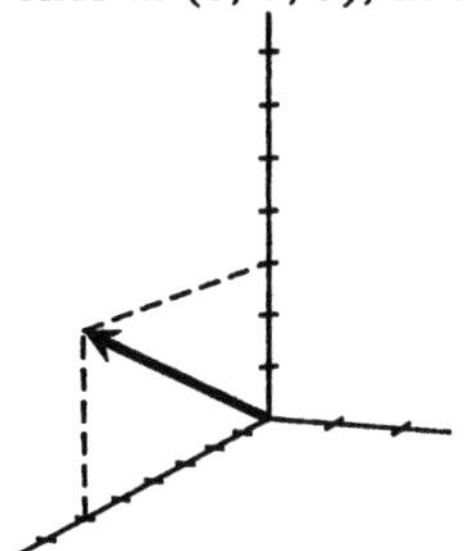

**7.** $\overrightarrow{PQ} = \langle 1, 0, 0\rangle - \langle -4, 3, 3\rangle = \langle 5, -3, -3\rangle$

Use the point at (5, –3, –3) to draw the vector $\langle 5, -3, -3\rangle$.

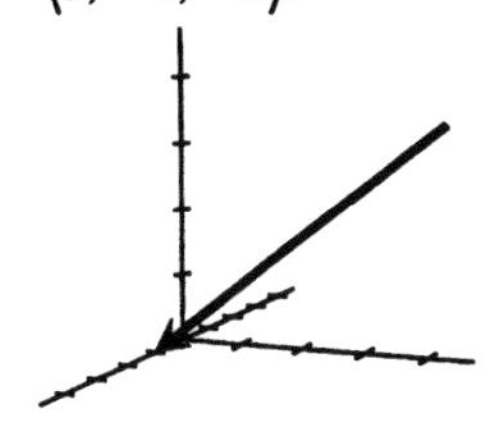

**9.** Plot $P = (2, 0, 5)$ and $Q = (0, 4, 4)$. Then draw the line through the two points.

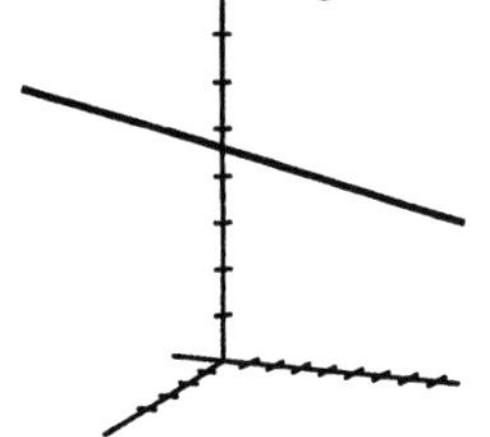

**11.** $\|\mathbf{v}\| = \sqrt{(-5)^2 + 0^2 + 3^2} = \sqrt{34} \approx 5.83095$

**13.** $$\begin{aligned}\mathbf{u} - 2\mathbf{v} &= \langle 2, -1, 1\rangle - 2\langle -5, 0, 3\rangle \\ &= \langle 2, -1, 1\rangle - \langle -10, 0, 6\rangle \\ &= \langle 12, -1, -5\rangle\end{aligned}$$

**15.** $$\begin{aligned}\mathbf{w}\cdot\mathbf{u} &= \langle 4, 3, 2\rangle \cdot \langle 2, -1, 1\rangle \\ &= 4\cdot 2 + 3\cdot(-1) + 2\cdot 1 \\ &= 7\end{aligned}$$

**17.** $\|\mathbf{v}\| = \sqrt{34} \approx 5.83095$ (from Exercise 11)

$$\begin{aligned}\frac{1}{\|\mathbf{v}\|}\mathbf{v} &= \frac{1}{\sqrt{34}}\langle -5, 0, 3\rangle \\ &= \left\langle -\frac{5}{\sqrt{34}}, 0, \frac{3}{\sqrt{34}}\right\rangle \\ &\approx \langle -0.857493, 0, 0.514496\rangle\end{aligned}$$

**19.** $\mathbf{u}\cdot\mathbf{w} = \langle 2, -1, 1\rangle \cdot \langle 4, 3, 2\rangle = 7$, $\|\mathbf{u}\| = \sqrt{6}$, $\|\mathbf{w}\| = \sqrt{29}$

Let $\theta$ be the angle between **u** and **w**.

$$\cos\theta = \frac{\mathbf{u}\cdot\mathbf{w}}{\|\mathbf{u}\|\|\mathbf{w}\|} = \frac{7}{\sqrt{174}}$$

$$\theta = \arccos\left(\frac{7}{\sqrt{174}}\right) \approx 1.011 \approx 57.95^\circ$$

**21.** Let **a** be the unit vector in the direction of **w**.

$$\|\mathbf{w}\| = \sqrt{4^2 + 3^2 + 2^2} = \sqrt{29}$$

$$\mathbf{a} = \frac{1}{\|\mathbf{w}\|}\mathbf{w} = \frac{1}{\sqrt{29}}\langle 4, 3, 2\rangle$$

Let $\ell$ be the length of the projection and let $\theta$ be the angle between **w** and **v**.

$$\ell = \|\mathbf{v}\|\cos\theta = \frac{\|\mathbf{v}\|\|\mathbf{w}\|\cos\theta}{\|\mathbf{w}\|} = \frac{\mathbf{v}\cdot\mathbf{w}}{\|\mathbf{w}\|} = -\frac{14}{\sqrt{29}}$$

The projection of $\mathbf{v}$ onto $\mathbf{a}$ is

$$\ell\mathbf{a} = -\frac{14}{\sqrt{29}} \cdot \frac{1}{\sqrt{29}} \langle 4, 3, 2 \rangle$$
$$= \left\langle -\frac{56}{29}, -\frac{42}{29}, -\frac{28}{29} \right\rangle$$
$$\approx \langle -1.931, -1.448, -0.966 \rangle$$

**23.** $\|\mathbf{v}\| = \left\| -3\mathbf{i} + 8\mathbf{j} - \frac{1}{2}\mathbf{k} \right\|$

$$= \sqrt{(-3)^2 + 8^2 + \left(-\frac{1}{2}\right)^2}$$
$$= \frac{\sqrt{293}}{2}$$
$$\approx 8.5586$$

**25.** $\mathbf{u} - 2\mathbf{v} = (\mathbf{i} - 2\mathbf{j} + 2\mathbf{k}) - 2\left(-3\mathbf{i} + 8\mathbf{j} - \frac{1}{2}\mathbf{k}\right)$

$$= 7\mathbf{i} - 18\mathbf{j} + 3\mathbf{k}$$

**27.** $\mathbf{w} \cdot \mathbf{u} = \mathbf{k} \cdot (\mathbf{i} - 2\mathbf{j} + 2\mathbf{k}) = 2$

**29.** $\|\mathbf{v}\| = \frac{\sqrt{293}}{2}$ (from Problem 23)

A vector of length 1 in the direction of $\mathbf{v}$ is

$$\frac{1}{\|\mathbf{v}\|}\mathbf{v} = \frac{2}{\sqrt{293}}\left(-3\mathbf{i} + 8\mathbf{j} - \frac{1}{2}\mathbf{k}\right)$$
$$= -\frac{6}{\sqrt{293}}\mathbf{i} + \frac{16}{\sqrt{293}}\mathbf{j} - \frac{1}{\sqrt{293}}\mathbf{k}$$

**31.** $\|\mathbf{u}\| = \sqrt{1^2 + (-2)^2 + (2)^2} = 3$; $\|\mathbf{w}\| = \sqrt{1^2} = 1$

$$\cos\theta = \frac{\mathbf{u} \cdot \mathbf{w}}{\|\mathbf{u}\|\|\mathbf{w}\|} = \frac{2}{3};$$
$$\theta = \arccos\frac{2}{3} \approx 0.841 \approx 48.19°$$

**33.** Observe that $\mathbf{w} = \mathbf{k}$ is a unit vector. Let $\ell$ be the length of the projection.

$$\ell = \|\mathbf{v}\|\cos\theta = \frac{\|\mathbf{v}\|\|\mathbf{w}\|\cos\theta}{\|\mathbf{w}\|} = \frac{\mathbf{v} \cdot \mathbf{w}}{\|\mathbf{w}\|} = -\frac{1}{2}$$

The projection of $\mathbf{v}$ onto $\mathbf{w}$ is $\ell\mathbf{w} = -\frac{1}{2}\mathbf{k}$.

**35.** The resultant force is

$$\mathbf{F} = \mathbf{F}_1 + \mathbf{F}_2 + \mathbf{F}_3$$
$$= 2000\mathbf{i} + (400\mathbf{i} - 3000\mathbf{j} + 1000\mathbf{k}) + (-300\mathbf{i} + 2500\mathbf{j} - 2500\mathbf{k})$$
$$= (2000 + 400 - 300)\mathbf{i} + (-3000 + 2500)\mathbf{j} + (1000 - 2500)\mathbf{k}$$
$$= 2100\mathbf{i} - 500\mathbf{j} - 1500\mathbf{k}.$$

**37.** If the three wires are taken to extend, respectively, in directions **i**, **j**, and **k**, with their joining point at the origin, then the hanging light pulls in the direction of $-\mathbf{i}-\mathbf{j}-\mathbf{k}$ with a magnitude of $20g$, which corresponds to a force vector of $-\frac{20g}{\sqrt{3}}(\mathbf{i}+\mathbf{j}+\mathbf{k})$. The component of this force in the direction of the first wire is $-\frac{20g}{\sqrt{3}}(\mathbf{i}+\mathbf{j}+\mathbf{k})\cdot i = -\frac{20g}{\sqrt{3}}$.
So the tension in the first wire, and in the other two also (by symmetry) is $\frac{20g}{\sqrt{3}}$ N.

**39.** $\mathbf{r}(t) = \mathbf{P} + t\mathbf{v} = \langle -3-2t,\ 5+4t,\ 5t\rangle$
For the symmetric equations, $x = -3 - 2t$, $y = 5 + 4t$, and $z = 5t$, so
$t = \frac{-x-3}{2} = \frac{y-5}{4} = \frac{z}{5}$, or $\frac{x+3}{-2} = \frac{y-5}{4} = \frac{z}{5}$.

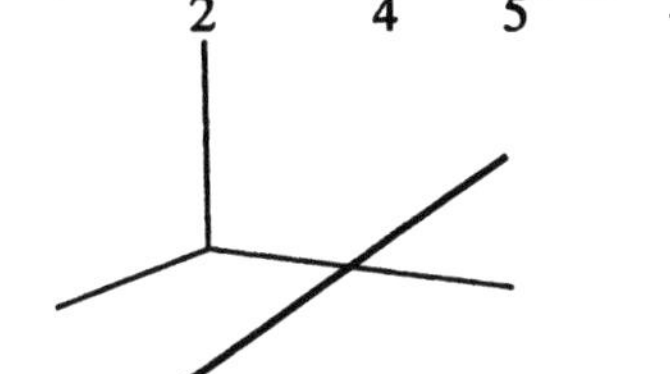

**41.** $\mathbf{r}(t) = \mathbf{P} + t\mathbf{v}$
$= \langle 101+101t,\ 201+201t,\ 301+301t\rangle$,
so $x = 101 + 101t$, $y = 201 + 201t$, and $z = 301 + 301t$.
Therefore, $\frac{x-101}{101} = \frac{y-201}{201} = \frac{z-301}{301}$ are the symmetric equations.

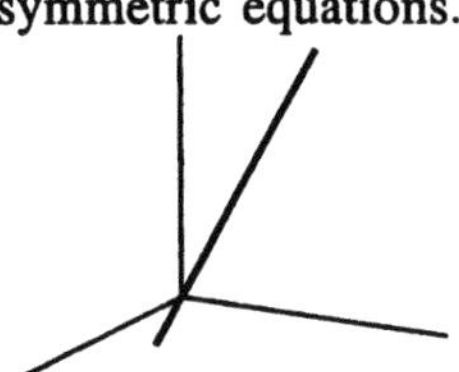

**43.** $\overrightarrow{PQ} = \langle -3, 5, 5\rangle - \langle 1, -1, 2\rangle = \langle -4, 6, 3\rangle$

$\mathbf{r}(t) = \mathbf{P} + t\overrightarrow{PQ} = \langle 1-4t,\ -1+6t,\ 2+3t\rangle$,
so $x = 1 - 4t$, $y = -1 + 6t$, and $z = 2 + 3t$.
Therefore, the symmetric equations are
$\frac{x-1}{-4} = \frac{y+1}{6} = \frac{z-2}{3}$.

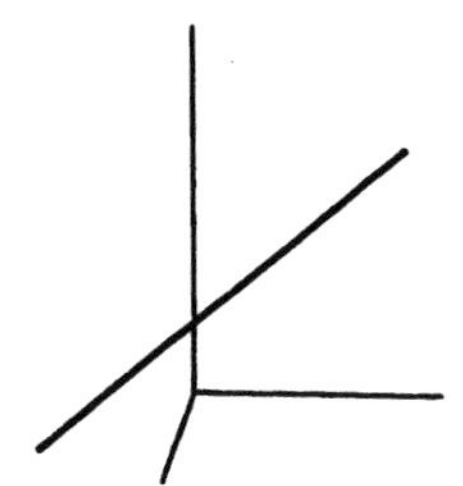

**45.** Check to see if $A = (-2, 3, 1)$ is on the line. If the first coordinate of a point on the line is $-2$, then $-t = -2$, so $t = 2$. However, $\mathbf{r}(2) = (-2, 7, 3) \neq (-2, 3, 1)$. Thus $A$ is not on the line. Now check to see if $B = (-5, 19, 6)$ is on the line. If the first coordinate of a point on the line is $-5$, then $-t = -5$, so $t = 5$. Since $\mathbf{r}(5) = (-5, 19, 6)$, $B$ is on the line.

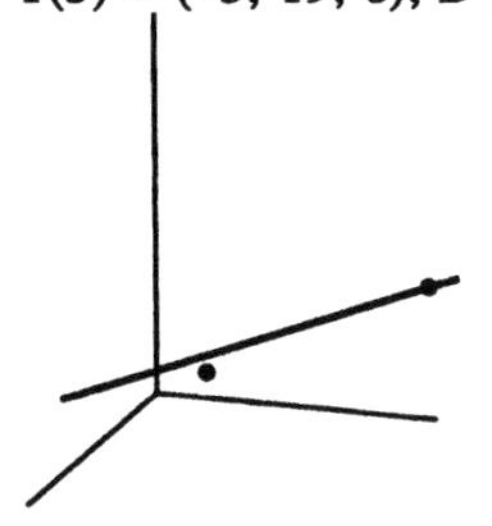

**47.** The midpoint of $\overline{PQ}$ is $\frac{1}{2}(P+Q)$. An equation for the line through $P$ and $Q$ is $\mathbf{r}(t) = P + t(Q-P)$.
$\mathbf{r}\left(\frac{1}{2}\right) = P + \frac{1}{2}(Q-P) = \frac{1}{2}(P+Q)$, so the midpoint is a point on the line through $P$ and $Q$.

**49.** Sample answer (other answers are possible): $\ell_1$ is a line parallel to the vector $\mathbf{v}_1 = \langle -1, 1, 3\rangle$ and $\ell_2$ is a line parallel to the vector $\mathbf{v}_2 = \langle 1, -1, -3\rangle$. Since $\mathbf{v}_1 = -\mathbf{v}_2$, then either $\ell_1$ and $\ell_2$ are parallel lines or they are the same line. If they have one point in common, then they are the same. Since $\mathbf{r}_1(0) = \langle 4, 0, 1\rangle = \mathbf{r}_2(2)$, then $(4, 0, 1)$ is a point on both lines, so $\ell_1$ and $\ell_2$ are the same.

**51.** Sample answer (other answers are possible): $\ell$ is a line parallel to the vector $\mathbf{v}_1 = \langle 2, -5, 0\rangle$ that contains the point (4, 0, 7). Since $\ell$ is parallel to $-\mathbf{v}_1 = \langle -2, 5, 0\rangle$, then $\ell$ can be parametrized by $\mathbf{r}(t) = \langle 4-2t, 5t, 7\rangle$ since $\mathbf{r}(0) = \langle 4, 0, 7\rangle$. Another parametrization can be obtained substituting the linear expression $2s - 2$ in for $t$. Thus, $\langle 4+2t, -5t, 7\rangle$ becomes

$\langle 4+2(2s-2), -5(2s-2), 7\rangle = \langle 4s, 10-10s, 7\rangle$

or letting $s$ be $t$, $\mathbf{r}(t) = \langle 4t, 10-10t, 7\rangle$ is another parametrization. Many other parametrizations are possible.

**53.** Let $\mathbf{r}_1(s) = (-3+s, 2s, -5s+3)$ describe $\ell_1$ and let $\mathbf{r}_2(t) = (2t, -4t+1, t)$ describe $\ell_2$. If $\ell_1$ and $\ell_2$ intersect, then there exists some $s$ and $t$ such that $\mathbf{r}_1(s) = \mathbf{r}_2(t)$. This equality gives

$$-3+s = 2t$$
$$2s = -4t+1$$
$$-5s+3 = t$$

Substitute $t = -5s + 3$ into the first two equations to get

$$-3+s = -10s+6$$
$$2s = 20s-11$$

Solving the equations gets $s = \dfrac{9}{11}$ and $s = \dfrac{11}{18}$. Thus there is no solution for $\mathbf{r}_1(s) = \mathbf{r}_2(t)$, so the lines do not intersect.

**55.** Let $\mathbf{r}_1(s) = (-3s+6, 4s, -s+7)$ describe $\ell_1$ and let $\mathbf{r}_2(t) = (-t-9, t+18, -t-2)$ describe $\ell_2$. If $\ell_1$ and $\ell_2$ intersect, then there exists some $s$ and $t$ such that $\mathbf{r}_1(s) = \mathbf{r}_2(t)$. This equality gives

$$-3s+6 = -t-9$$
$$4s = t+18$$
$$-s+7 = -t-2$$

From the third equation, $t = s - 9$. Substitute $t = s - 9$ into the first two equations to get

$$-3s+6 = -s$$
$$4s = s+9$$

Solving the equations gives $s = 3$ and $s = 3$. When $s = 3$, $t = -6$, so $\mathbf{r}_1(3) = \mathbf{r}_2(-6) = (-3, 12, 4)$. Thus the lines intersect.

**57.** Sample answer (other answers are possible): Write a parameterization for $\ell_1$ and $\ell_2$. $t = \dfrac{x-4}{3} = \dfrac{y+7}{2}$ yields $x = 3t + 4$, $y = 2t - 7$, and $z = 3$ for $\ell_1$, and $s = -x = \dfrac{y+19}{4} = \dfrac{z-7}{-2}$ yields $x = -s$, $y = 4s - 19$, and $z = -2s + 7$ for $\ell_2$. If $\ell_1$ and $\ell_2$ intersect, then $-2s + 7 = 3$, so that $s = 2$. Therefore, $x = -s = -2$ and $y = 4(2) - 19 = -11$. In this case, $-2 = x = 3t + 4$, so that $t = -2$. Since $y = 2(-2) - 7 = -11$, then both lines intersect the point $(-2, -11, 3)$. $\ell_1$ intersects $(-2, -11, 3)$ when $t = -2$, and $\ell_2$ intersects the point when $s = 2$.

**59.** Since the $y$-coordinates of points on $\ell_1$ are always 4, and the $y$-coordinates of points on $\ell_2$ are always $-1$, then $\ell_1$ and $\ell_2$ cannot have any points in common.

**61.** Sample answer (other answers are possible): A line that intersects points $P$ and $Q$ is parallel to the vector $\mathbf{Q} - \mathbf{P}$ and thus can be parametrized by $\mathbf{r}(t) = \mathbf{P} + (\mathbf{Q} - \mathbf{P})t = \mathbf{P}(1 - t) + \mathbf{Q}t$. Since $\mathbf{r}(0) = \mathbf{P}$ and $\mathbf{r}(1) = \mathbf{Q}$, then $\mathbf{r}(t) = \mathbf{P}(1 - t) + \mathbf{Q}t$, $0 \le t \le 1$ is a parameterization for the line segment between the points $P$ and $Q$.

**63.** Use the result of Exercise 61 with $\mathbf{P} = \langle 0, 4, -4\rangle$ and $\mathbf{Q} = \langle 7, 9, -5\rangle$. Thus, $\mathbf{r}(t) = (1-t)\langle 0, 4, -4\rangle + t\langle 7, 9, -5\rangle$, $0 \le t \le 1$ is the desired parameterization.

## Section 8.2 Matrices and Determinants

**1.** $-3A = -3\begin{pmatrix} -2 & 3 \\ 4 & 1 \\ 0 & 8 \end{pmatrix} = \begin{pmatrix} 6 & -9 \\ -12 & -3 \\ 0 & -24 \end{pmatrix}$

**3.** The matrix operation is undefined since $A$ and $-5C$ are not of the same dimensions.

**5.** $ED = (2 \quad -2 \quad 3)\begin{pmatrix} 5 \\ 3 \\ 1 \end{pmatrix} = (7)$

**7.** The matrix operation is undefined since the number of columns of $C$ is not equal to the number of rows of $A$.

**9.** The matrix operation is undefined since the number of columns of $B$ is not equal to the number of rows of $A$.

**11.** Since the number of columns of $C$ is not equal to the number of rows of $B$, $CB$ is undefined. Thus $CBA$ is undefined.

**13.** **a.** $a_{12} = 3$

**b.** $e_{13} = 3$

**c.** $b_{32} = -2$

**d.** $b_{23}$ is undefined.

**e.** $d_{12}$ is undefined.

**15.** $AB = \begin{pmatrix} 2 & 3 \\ -1 & 6 \end{pmatrix}\begin{pmatrix} -1 & 5 \\ 0 & 2 \end{pmatrix} = \begin{pmatrix} -2 & 16 \\ 1 & 7 \end{pmatrix};$

$BA = \begin{pmatrix} -1 & 5 \\ 0 & 2 \end{pmatrix}\begin{pmatrix} 2 & 3 \\ -1 & 6 \end{pmatrix} = \begin{pmatrix} -7 & 27 \\ -2 & 12 \end{pmatrix}$

Thus $AB \neq BA$.

**17.** Sample answer (other answers are possible):

$A = \begin{pmatrix} 2 & 0 \\ 3 & -2 \end{pmatrix}$ and $B = \begin{pmatrix} \frac{1}{2} & 0 \\ \frac{3}{4} & -\frac{1}{2} \end{pmatrix}$. Clearly, $A \neq B$.

$AB = \begin{pmatrix} 2 & 0 \\ 3 & -2 \end{pmatrix}\begin{pmatrix} \frac{1}{2} & 0 \\ \frac{3}{4} & -\frac{1}{2} \end{pmatrix} = \begin{pmatrix} 1 & 0 \\ 0 & 1 \end{pmatrix}$ and

$BA = \begin{pmatrix} \frac{1}{2} & 0 \\ \frac{3}{4} & -\frac{1}{2} \end{pmatrix}\begin{pmatrix} 2 & 0 \\ 3 & -2 \end{pmatrix} = \begin{pmatrix} 1 & 0 \\ 0 & 1 \end{pmatrix}$, so $AB = BA$.

**19.** $\det(B) = \begin{vmatrix} a & b \\ c & d \end{vmatrix} = ad - bc$

**21.** The determinant of $B$ does not exist since $B$ is not a square matrix.

**23.** $\det(A) = \begin{vmatrix} 1 & 0 & 1 \\ 1 & 0 & 1 \\ 1 & 0 & 1 \end{vmatrix}$

$= 1\begin{vmatrix} 0 & 1 \\ 0 & 1 \end{vmatrix} - 0\begin{vmatrix} 1 & 1 \\ 1 & 1 \end{vmatrix} + 1\begin{vmatrix} 1 & 0 \\ 1 & 0 \end{vmatrix}$

$= 1(0\cdot 1 - 1\cdot 0) - 0 + 4(1\cdot 0 - 0\cdot 1)$

$= 0$

**25.** $AB = \begin{pmatrix} 4 & 5 \\ 6 & 7 \end{pmatrix}\begin{pmatrix} 4 & 5 \\ 6 & 7 \end{pmatrix} = \begin{pmatrix} 46 & 55 \\ 66 & 79 \end{pmatrix}$

$\det(AB) = \begin{vmatrix} 46 & 55 \\ 66 & 79 \end{vmatrix} = 4$

$\det(A) = \det(B) = \begin{vmatrix} 4 & 5 \\ 6 & 7 \end{vmatrix} = -2$, so

$\det(A)\det(B) = 4$.

**27.** $AB = \begin{pmatrix} -2 & 3 & 3 \\ 4 & 0 & 7 \\ -4 & 3 & -2 \end{pmatrix}\begin{pmatrix} 1 & 2 & 3 \\ -1 & -2 & -3 \\ 3 & 5 & 7 \end{pmatrix}$

$= \begin{pmatrix} 4 & 5 & 6 \\ 25 & 43 & 61 \\ -13 & -24 & -351 \end{pmatrix}$

$\det(AB) = \begin{vmatrix} \frac{75}{2} & 30 & \frac{23}{2} \\ -19 & -19 & 9 \\ 20 & -10 & 14 \end{vmatrix} = 0$

$\det(A) = \begin{vmatrix} -2 & 3 & 3 \\ 4 & 0 & 7 \\ -4 & 3 & -2 \end{vmatrix} = 18$ and

$\det(B) = \begin{vmatrix} 1 & 2 & 3 \\ -1 & -2 & -3 \\ 3 & 5 & 7 \end{vmatrix} = 0$, so

$\det(A)\det(B) = 0$.

**29.** If the matrices on the left side of the equation are multiplied, this will yield the system of equations

$$\begin{aligned} -3x + 4y &= -10 \\ x + 5y &= -3. \end{aligned}$$

One method for solving the system is to multiply the second equation by 3 before adding it to the first equation.

$$\begin{aligned} -3x + 4y &= -10 \\ 3x + 15y &= -9 \\ 19y &= -19 \\ y &= -1 \end{aligned}$$

Next substitute $y = -1$ into one of the original equations.

$$\begin{aligned} x + 5(-1) &= -3 \\ x &= 2 \end{aligned}$$

The values $x = 2$, $y = -1$ should now be substituted into the two original equations as a check. In this case, these values check out and so $\begin{pmatrix} x \\ y \end{pmatrix} = \begin{pmatrix} 2 \\ -1 \end{pmatrix}$ is the correct solution.

Other methods of solution should yield the same result.

**31.** If the matrices on the left side of the given equation are multiplied, this will yield the system of equations

$$x+y+z=1$$
$$-y+5z=-15$$
$$3x+2y-2z=18.$$

One method of solution is to solve the second equation for $y$ and then substitute this expression in for $y$ in the first and third equations.

$y = 5z + 15$

$$x+(5z+15)+z=1$$
$$3x+2(5z+15)-2z=18.$$

Simplifying the last two equations in $x$ and $z$ yields the following system of two equations in two variables.

$$x+6z=-14$$
$$3x+8z=-12$$

Solving the first equation for $x$ and substituting this expression in for $x$ in the second equation yields

$x = -6z - 14$

$3(-6z - 14) + 8z = -12$

Now simplify the equation in $z$ and solve.

$$-10z=30$$
$$z=-3$$

Now substitute $z = -3$ into $x = -6z - 14$ to solve for $x$.

$x = -6(-3) - 14 = 4$

Substitute $z = -3$ into $y = 5z + 15$.

$y = 5(-3) + 15 = 0$

The values $x = 4$, $y = 0$, and $z = -3$ should be substituted into the original system as a check. In this case, these values satisfy all three equations, so $x = 4$, $y = 0$, and $z = -3$ is the correct solution. Other methods of solution should yield the same result.

**33.** Let $A=\begin{pmatrix} a_{11} & a_{12} \\ a_{21} & a_{22} \\ a_{31} & a_{32} \end{pmatrix}$, $B=\begin{pmatrix} b_{11} & b_{12} \\ b_{21} & b_{22} \\ b_{31} & b_{32} \end{pmatrix}$, and $C=\begin{pmatrix} c_{11} & c_{12} \\ c_{21} & c_{22} \\ c_{31} & c_{32} \end{pmatrix}$.

$$\begin{aligned} A+B &= \begin{pmatrix} a_{11}+b_{11} & a_{12}+b_{12} \\ a_{21}+b_{21} & a_{22}+b_{22} \\ a_{31}+b_{31} & a_{32}+b_{32} \end{pmatrix} \\ &= \begin{pmatrix} b_{11}+a_{11} & b_{12}+a_{12} \\ b_{21}+a_{21} & b_{22}+a_{22} \\ b_{31}+a_{31} & b_{32}+a_{32} \end{pmatrix} \\ &= B+A \end{aligned}$$

$$\begin{aligned} &(A+B)+C \\ &= \begin{pmatrix} a_{11}+b_{11} & a_{12}+b_{12} \\ a_{21}+b_{21} & a_{22}+b_{22} \\ a_{31}+b_{31} & a_{32}+b_{32} \end{pmatrix} + \begin{pmatrix} c_{11} & c_{12} \\ c_{21} & c_{22} \\ c_{31} & c_{32} \end{pmatrix} = \begin{pmatrix} a_{11}+b_{11}+c_{11} & a_{12}+b_{12}+c_{12} \\ a_{21}+b_{21}+c_{21} & a_{22}+b_{22}+c_{22} \\ a_{31}+b_{31}+c_{31} & a_{32}+b_{32}+c_{32} \end{pmatrix} \\ &= \begin{pmatrix} a_{11} & a_{12} \\ a_{21} & a_{22} \\ a_{31} & a_{32} \end{pmatrix} + \begin{pmatrix} b_{11}+c_{11} & b_{12}+c_{12} \\ b_{21}+c_{21} & b_{22}+c_{22} \\ b_{31}+c_{31} & b_{32}+c_{32} \end{pmatrix} \\ &= A+(B+C) \end{aligned}$$

**35.** Use a computer algebra system to define $A=\begin{pmatrix} a & b & c \\ d & e & f \\ g & h & i \end{pmatrix}$ and $B=\begin{pmatrix} j & k & l \\ m & n & o \\ p & q & r \end{pmatrix}$. Then use the computer algebra system to evaluate det($AB$) – det($A$)det($B$). You should get 0. This verifies that det($AB$) = det($A$)det($B$).

**37.** $$\det(A)=\begin{vmatrix} a_1x+b_1y & a_2x+b_2y & a_3x+b_3y \\ a_1 & a_2 & a_3 \\ b_1 & b_2 & b_3 \end{vmatrix}$$
$$=(a_1x+b_1y)\begin{vmatrix} a_2 & a_3 \\ b_2 & b_3 \end{vmatrix}-(a_2x+b_2y)\begin{vmatrix} a_1 & a_3 \\ b_1 & b_3 \end{vmatrix}+(a_3x+b_3y)\begin{vmatrix} a_1 & a_2 \\ b_1 & b_2 \end{vmatrix}$$
$$=(a_1x+b_1y)(a_2b_3-a_3b_2)-(a_2x+b_2y)(a_1b_3-a_3b_1)+(a_3x+b_3y)(a_1b_2-a_2b_1)$$
$$=a_1a_2b_3x-a_1a_3b_2x+a_2b_1b_3y-a_3b_1b_2y-a_1a_2b_3x+a_2a_3b_1x+a_1b_2b_3y$$
$$-a_3b_1b_2y+a_1a_3b_2x-a_2a_3b_1x+a_1b_2b_3y-a_2b_1b_3y=0$$

## Section 8.3 The Cross Product

**1.** $$\mathbf{u}\times\mathbf{v}=\langle 2,-1,1\rangle\times\langle -5,0,3\rangle$$
$$=\langle -1\cdot 3-1\cdot 0,\ 1\cdot(-5)-2\cdot 3,\ 2\cdot 0-(-1)(-5)\rangle$$
$$=\langle -3,-11,-5\rangle$$

**3.** $$\mathbf{w}\times\mathbf{r}=\langle 4,3,2\rangle\times\langle 3,4,-2\rangle$$
$$=\langle -2\cdot 3-4(2),\ 3(2)-(-2)4,\ 4\cdot 4-3\cdot 3\rangle$$
$$=\langle -14,14,7\rangle$$

**5.** Note that $\mathbf{w}\times\mathbf{u}$ is perpendicular to the plane containing $\mathbf{w}$ and $\mathbf{u}$. Hence it is perpendicular to $\mathbf{u}$. Thus $\mathbf{u}\cdot(\mathbf{w}\times\mathbf{u})=0$.

**7.** $$\mathbf{r}\times\mathbf{v}=\langle 3,4,-2\rangle\times\langle -5,0,3\rangle$$
$$=\langle 4\cdot 3-0(-2),\ (-5)(-2)-3\cdot 3,\ 3\cdot 0-(-5)4\rangle$$
$$=\langle 12,1,20\rangle$$

is a vector perpendicular to $\mathbf{r}$ and $\mathbf{v}$ with length $\|\langle 12,1,20\rangle\|=\sqrt{12^2+1^2+20^2}=\sqrt{545}$.

Thus, $\frac{1}{\sqrt{545}}\langle 12,1,20\rangle=\left\langle \frac{12}{\sqrt{545}},\frac{1}{\sqrt{545}},\frac{20}{\sqrt{545}}\right\rangle$ is a unit vector perpendicular to $\mathbf{r}$ and $\mathbf{v}$.

**9.** $$\mathbf{u}\times\mathbf{w}=\langle 2,-1,1\rangle\times\langle 4,3,2\rangle$$
$$=\langle -1\cdot 2-1\cdot 3,\ 1\cdot 4-2\cdot 2,\ 2\cdot 3-(-1)4\rangle$$
$$=\langle -5,0,10\rangle$$

is a vector perpendicular to $\mathbf{u}$ and $\mathbf{w}$ with length $\|-5,0,10\|=\sqrt{125}=5\sqrt{5}$. Thus

$\frac{2}{5\sqrt{5}}\langle -5,0,10\rangle=\left\langle -\frac{2}{\sqrt{5}},0,\frac{4}{\sqrt{5}}\right\rangle$ is a vector of length 2 perpendicular to $\mathbf{u}$ and $\mathbf{w}$.

**11.** $\mathbf{u}\times\mathbf{v}=(\mathbf{i}-2\mathbf{j}+2\mathbf{k})\times\left(-3\mathbf{i}+8\mathbf{j}-\frac{1}{2}\mathbf{k}\right)$

$$=\left((-2)\left(-\frac{1}{2}\right)-2\cdot 8\right)\mathbf{i}+\left(2(-3)-1\left(-\frac{1}{2}\right)\right)\mathbf{j}+(1\cdot 8-(-2)(-3))\mathbf{k}$$

$$=-15\mathbf{i}-\frac{11}{2}\mathbf{j}+2\mathbf{k}$$

**13.** $\mathbf{w}\times\mathbf{r}=\mathbf{k}\times(-2\mathbf{i}-3\mathbf{k})$

$$=(0(-3)-0\cdot 1)\mathbf{i}+(1(-2)-0(-3))\mathbf{j}+(0\cdot 0-(-2)0)\mathbf{k}$$

$$=-2\mathbf{j}$$

**15.** Note that $\mathbf{w}\times\mathbf{u}$ is perpendicular to the plane containing $\mathbf{w}$ and $\mathbf{u}$. Hence it is perpendicular to $\mathbf{u}$. Thus $\mathbf{u}\cdot(\mathbf{w}\times\mathbf{u})=0$.

**17.** $\mathbf{r}\times\mathbf{v}=(-2\mathbf{i}-3\mathbf{k})\times\left(-3\mathbf{i}+8\mathbf{j}-\frac{1}{2}\mathbf{k}\right)$

$$=\left(0\left(-\frac{1}{2}\right)-8(-3)\right)\mathbf{i}+\left((-3)(-3)-(-2)\left(-\frac{1}{2}\right)\right)\mathbf{j}+((-2)8-(-3)0)\mathbf{k}$$

$$=24\mathbf{i}+8\mathbf{j}-16\mathbf{k}$$

is a vector perpendicular to $\mathbf{r}$ and $\mathbf{v}$ of length $\sqrt{24^2+8^2+(-16)^2}=\sqrt{896}=8\sqrt{14}$. Thus $\frac{1}{8\sqrt{14}}(24\mathbf{i}+8\mathbf{j}-16\mathbf{k})=\frac{3}{\sqrt{14}}\mathbf{i}+\frac{1}{\sqrt{14}}\mathbf{j}-\frac{2}{\sqrt{14}}\mathbf{k}$ is a unit vector perpendicular to each of $\mathbf{r}$ and $\mathbf{v}$.

**19.** First find a vector perpendicular to $\mathbf{u}$ and $\mathbf{w}$.

$\mathbf{u}\times\mathbf{w}=(\mathbf{i}-2\mathbf{j}+2\mathbf{k})\times\mathbf{k}=-2\mathbf{i}-\mathbf{j}$

$\|\mathbf{u}\times\mathbf{w}\|=\sqrt{(-2)^2+(-1)^2}=\sqrt{5}$

Thus, $-\frac{2}{\sqrt{5}}\mathbf{i}-\frac{1}{\sqrt{5}}\mathbf{j}$ is a unit vector perpendicular to $\mathbf{u}$ and $\mathbf{w}$, so a vector of length two is $-\frac{4}{\sqrt{5}}\mathbf{i}-\frac{2}{\sqrt{5}}\mathbf{j}$ $\left(\text{or else } \frac{4}{\sqrt{5}}\mathbf{i}+\frac{2}{\sqrt{5}}\mathbf{j}\right)$.

**21.**

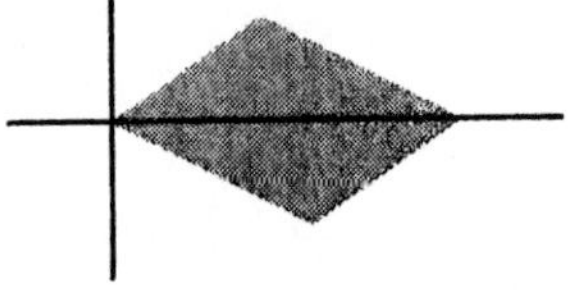

$$\text{Area}=\left|\det\begin{pmatrix}3 & 2\\ 4 & -2\end{pmatrix}\right|=|3(-2)-4\cdot 2|=|-14|=14$$

**23.**

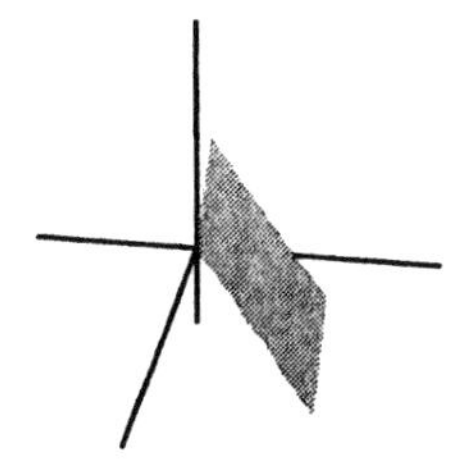

$\langle -1, 0, 1\rangle \times \langle 4, 2, 0\rangle$

$= \det\begin{pmatrix} \mathbf{i} & \mathbf{j} & \mathbf{k} \\ -1 & 0 & 1 \\ 4 & 2 & 0 \end{pmatrix}$

$= \mathbf{i}(0-2) + \mathbf{j}(4-0) + \mathbf{k}(-2-0)$

$= -2\mathbf{i} + 4\mathbf{j} - 2\mathbf{k}$

$\text{Area} = \|-2\mathbf{i} + 4\mathbf{j} - 2\mathbf{k}\|$

$= \sqrt{(-2)^2 + 4^2 + (-2)^2}$

$= \sqrt{24}$

$= 2\sqrt{6}$

**25.**

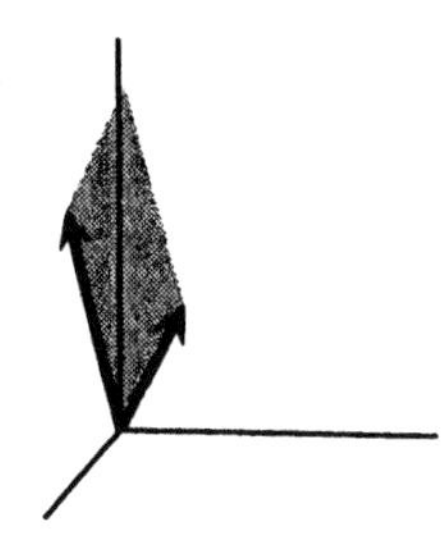

$\langle 1, 2, 3\rangle \times \langle -2, -2, 5\rangle$

$= \det\begin{pmatrix} \mathbf{i} & \mathbf{j} & \mathbf{k} \\ 1 & 2 & 3 \\ -2 & -2 & 5 \end{pmatrix}$

$= (10-(-6))\mathbf{i} + (-6-5)\mathbf{j} + (-2-(-4))\mathbf{k}$

$= 16\mathbf{i} - 11\mathbf{j} + 2\mathbf{k}$

$\text{Area} = \|16\mathbf{i} - 11\mathbf{j} + 2\mathbf{k}\|$

$= \sqrt{16^2 + (-11)^2 + 2^2}$

$= \sqrt{381}$

**27.**

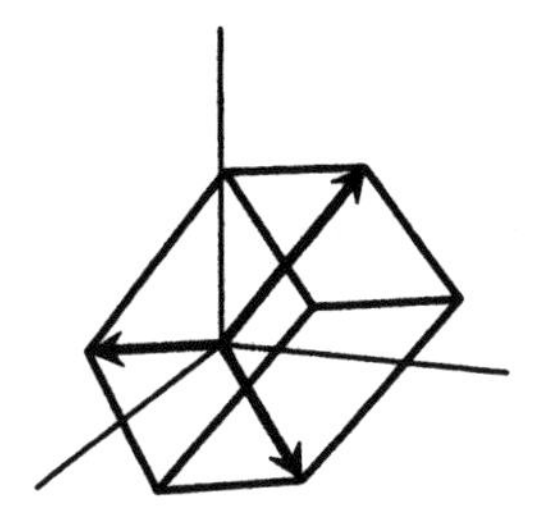

$\mathbf{b} \times \mathbf{c} = \det\begin{pmatrix} \mathbf{i} & \mathbf{j} & \mathbf{k} \\ 2 & -4 & 0 \\ 3 & 4 & -3 \end{pmatrix}$

$= (12-0)\mathbf{i} + (0-(-6))\mathbf{j} + (8-(-12))\mathbf{k}$

$= 12\mathbf{i} + 6\mathbf{j} + 20\mathbf{k}$

$\text{Volume} = |\mathbf{a}\cdot(\mathbf{b}\times\mathbf{c})|$

$= |\langle 0, 5, 6\rangle \cdot \langle 12, 6, 20\rangle|$

$= |0 + 30 + 120|$

$= 150$

**29.** Since $\mathbf{v} \times \mathbf{w}$ is a perpendicular to $\mathbf{v}$ and $\mathbf{w}$, the line should be in the same direction as

$\mathbf{v} \times \mathbf{w} = \det\begin{pmatrix} \mathbf{i} & \mathbf{j} & \mathbf{k} \\ -1 & 3 & 4 \\ -2 & 4 & 6 \end{pmatrix}$

$= (18-16)\mathbf{i} + (-8+6)\mathbf{j} + (-4+6)\mathbf{k}$

$= 2\mathbf{i} - 2\mathbf{j} + 2\mathbf{k}.$

Since $\langle 1, -1, 1\rangle = \frac{1}{2}\langle 2, -2, 2\rangle$, the line can be parametrized by $\mathbf{r}(t) = \langle 4, -3, 1\rangle + t\langle 1, -1, 1\rangle$.

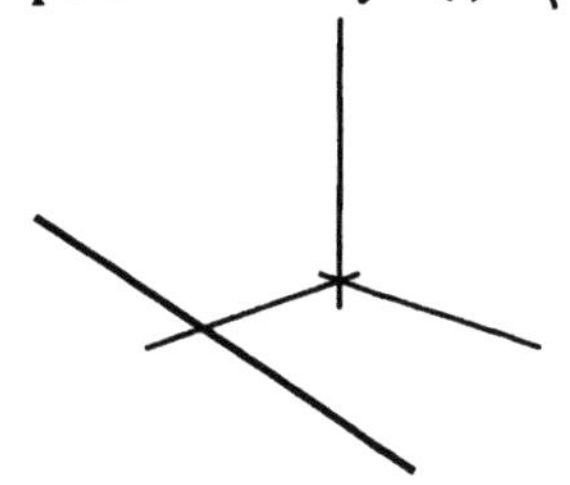

**31.** $\mathbf{i} \times \mathbf{j} = \det\begin{pmatrix} \mathbf{i} & \mathbf{j} & \mathbf{k} \\ 1 & 0 & 0 \\ 0 & 1 & 0 \end{pmatrix} = 0\mathbf{i} + 0\mathbf{j} + 1\mathbf{k} = \mathbf{k}$

$\mathbf{j} \times \mathbf{i} = \det\begin{pmatrix} \mathbf{i} & \mathbf{j} & \mathbf{k} \\ 0 & 1 & 0 \\ 1 & 0 & 0 \end{pmatrix} = -\det\begin{pmatrix} \mathbf{i} & \mathbf{j} & \mathbf{k} \\ 1 & 0 & 0 \\ 0 & 1 & 0 \end{pmatrix} = -\mathbf{k}$

using facts about determinants (see Exercise 36, Sec. 8.2). Next graph the vectors **i**, **k**, **–k**, and the parallelogram defined by **i** and **j**.

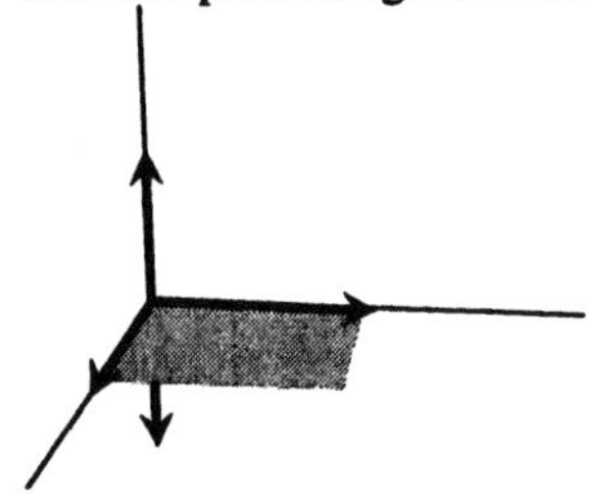

**33.** If we assume the wrench corresponds to a vector lying along the positive $x$-axis and this vector is pointing toward the origin where the nut is attached to a bolt that is aligned with the $z$-axis, then when the wrench turns the nut, it will swing in the $xy$-plane. In this case, the wrench corresponds to the vector $\langle 0.4, 0, 0\rangle$ and from properties of vectors in 2 dimensions, the force vector is given by $\mathbf{F}=\langle 50\cos\theta, 50\sin\theta, 0\rangle$, where $\theta$ is the angle between $\mathbf{F}$ and $\langle 0.4, 0, 0\rangle$.

**a.** $\theta = 90°$, so in this case $\mathbf{F}=\langle 0, 50, 0\rangle$. Since the torque is $\tau=\langle 0.4, 0, 0\rangle\times\langle 0, 50, 0\rangle=20\mathbf{k}$, then the magnitude of the torque is $\|\tau\|=\|20\mathbf{k}\|=20$ joules.

**b.** $\theta = 30°$, so in this case $\mathbf{F}=\langle 25\sqrt{3}, 25, 0\rangle$. The torque is $\tau=\langle 0.4, 0, 0\rangle\times\langle 25\sqrt{3}, 25, 0\rangle=10\mathbf{k}$, so the magnitude of the torque is $\|\tau\|=\|10\mathbf{k}\|=10$ joules.

**c.** $\theta = 135°$, so in this case $\mathbf{F}=\langle -25\sqrt{2}, 25\sqrt{2}, 0\rangle$. The torque is $\tau=\langle 0.4, 0, 0\rangle\times\langle -25\sqrt{2}, 25\sqrt{2}, 0\rangle=10\sqrt{2}\mathbf{k}$, so the magnitude of the torque $\|\tau\|=\|10\sqrt{2}\mathbf{k}\|=10\sqrt{2}$ joules.

**35.** If we assume that the pedal shaft corresponds to a vector that lies along the positive $x$-axis, then the pedal shaft corresponds to the vector $\langle 20, 0, 0\rangle$. To calculate the magnitude of the torque, for which the units are joules = newtons · meters, we must convert the length of the pedal shaft into meters. Since 20 cm = 0.20 m, the pedal shaft will correspond to the vector $\mathbf{v}=\langle 0.20, 0, 0\rangle$. If we consider the force $\mathbf{F}$ applied by the biker to $\mathbf{v}$ as taking place in the $xy$-plane, then from properties of vectors in two dimensions, $\mathbf{F}=\langle 50\cos\theta, 50\sin\theta, 0\rangle$, where $\theta = 135°$ is the angle between $\mathbf{F}$ and $\mathbf{v}$.

Therefore,
$\mathbf{F}=\langle 50\cos 135°, 50\sin 135°, 0\rangle$
$=\langle -25\sqrt{2}, 25\sqrt{2}, 0\rangle$.
The torque then is
$\tau=\langle 0.20, 0, 0\rangle\times\langle -25\sqrt{2}, 25\sqrt{2}, 0\rangle=5\sqrt{2}\mathbf{k}$,
and the magnitude of the torque is
$\|\tau\|=\|5\sqrt{2}\mathbf{k}\|=5\sqrt{2}$ joules.

**37.** The line $\ell$ described by $\mathbf{r}(t)=\langle -2t, -t+4, 2t-1\rangle$ contains the point $A = (0, 4, -1)$ and is parallel to the vector $\mathbf{v}=\langle -2, -1, 2\rangle$. Referring to Equation (8) of Example 6 of the section, we know the distance from $P = (-3, 5, 6)$ to $\ell$ is given by the formula $d=\dfrac{\|\mathbf{v}\times\overrightarrow{AP}\|}{\|\mathbf{v}\|}$. Since

$\overrightarrow{AP}=\langle -3, 5, 6\rangle-\langle 0, 4, -1\rangle=\langle -3, 1, 7\rangle$, then

$$\|\mathbf{v}\times\overrightarrow{AP}\|=\|\langle -2, -1, 2\rangle\times\langle -3, 1, 7\rangle\|$$
$$=\|\langle -9, 8, -5\rangle\|$$
$$=\sqrt{170}.$$

Therefore, since $\|\mathbf{v}\|=\sqrt{4+1+4}=3$, the distance from $P$ to $\ell$ is $d=\dfrac{\sqrt{170}}{3}$.
To find the point $F$ on $\ell$ that is closest to $P$, we must solve the equation $\text{dist}(P, F)=\dfrac{\sqrt{170}}{3}$. Since $F$ is on $\ell$, for some real number $t$, $F = (-2t, -t + 4, 2t - 1)$.
Use the distance formula to get the equation

$$\sqrt{(-3-(-2t))^2+(5-(-t+4))^2+(6-(2t-1))^2}$$
$$=\frac{\sqrt{170}}{3}.$$

Simplifying the radicand, squaring both sides, and then gathering like terms yields the quadratic equation $81t^2-342t+361=0$. Facts concerning perpendicular lines imply that the point $F$ is unique, hence the above quadratic equation has a unique real solution $t$. This means that $81t^2-342t+361$ will factor as $(at+b)^2$, where $a$ and $h$ are real numbers such that $a^2=81$ and $b^2=361$. In this case, $81=9^2$ and $361=19^2$ are perfect squares and the $t$ coefficient $-342 = -2 \cdot 9 \cdot 19$. It can then be deduced (and checked) that the above quadratic equation can be solved by factoring:

$(9t-19)^2=0$. That is, $t=\frac{19}{9}$ is the unique solution. $F$ is then given by the vector

$\mathbf{r}\left(\frac{19}{9}\right)=\left\langle -\frac{38}{9}, \frac{17}{9}, \frac{29}{9}\right\rangle$. So, $F=\left(-\frac{38}{9}, \frac{17}{9}, \frac{29}{9}\right)$ is the closest point to $P$ on $\ell$.

**39.** Let $P = (0, 0, 0)$. The line contains the point $Q=(a_1, a_2, a_3)$ and is parallel to vector $\mathbf{v}=\langle b_1, b_2, b_3\rangle$.

As in Exercise 37, the distance $d$ is $\dfrac{\left\|\mathbf{v}\times\overrightarrow{PQ}\right\|}{\|\mathbf{v}\|}$.

$$\begin{aligned}\mathbf{v}\times\overrightarrow{PQ}&=\langle b_1, b_2, b_3\rangle\times\langle a_1, a_2, a_3\rangle\\&=\langle a_3b_2-a_2b_3, a_1b_3-a_3b_1, a_2b_1-a_1b_2\rangle\end{aligned}$$

$$\left\|\mathbf{v}\times\overrightarrow{PQ}\right\|=\sqrt{(a_3b_2-a_2b_3)^2+(a_1b_3-a_3b_1)^2+(a_2b_1-a_1b_2)^2}$$

$$\|\mathbf{v}\|=\sqrt{b_1{}^2+b_2{}^2+b_3{}^2}$$

$$d=\frac{\sqrt{(a_3b_2-a_2b_3)^2+(a_1b_3-a_3b_1)^2+(a_2b_1-a_1b_2)^2}}{\sqrt{b_1{}^2+b_2{}^2+b_3{}^2}}$$

**41.** The line $\ell$ in question can be parametrized by $\mathbf{r}(t)=\langle -2t-2, 5, 3t-1\rangle$. $\ell$ contains the point $Q = (-2, 5, -1)$ and is parallel to the vector $\mathbf{v}=\langle -2, 0, 3\rangle$. The distance from $P = (-3, 1, 1)$ to $\ell$ is given by $d=\dfrac{\left\|\mathbf{v}\times\overrightarrow{QP}\right\|}{\mathbf{v}}$. Since $\overrightarrow{QP}=\langle -3, 1, 1\rangle-\langle -2, 5, -1\rangle=\langle -1, -4, 2\rangle$, then

$$\begin{aligned}\left\|\mathbf{v}\times\overrightarrow{QP}\right\|&=\|\langle -2, 0, 3\rangle\times\langle -1, -4, 2\rangle\|\\&=\|\langle 12, 1, 8\rangle\|\\&=\sqrt{209}.\end{aligned}$$

Therefore, since $\|\mathbf{v}\|=\sqrt{13}$, then $d=\sqrt{\frac{209}{13}}$.

To find the point $F$ on $\ell$ closest to $P$, we must solve the equation $\text{dist}(P, F)=\sqrt{\frac{209}{13}}$. Since $F$ is on $\ell$, then $F = (-2t - 2, 5, 3t - 1)$ for some real number $t$. Use the distance formula to get the equation $\sqrt{(-3-(-2t-2))^2+(1-5)^2+(1-(3t-1))^2}=\sqrt{\frac{209}{13}}$.

Simplifying the radicand, squaring both sides, and then collecting like terms yields the quadratic equation $169t^2-208+64=0$, which can be solved by factoring. $(13t-8)^2=0$, so $t=\frac{8}{13}$ is the unique solution. $F$ is then given by vector $\mathbf{r}\left(\frac{8}{13}\right)=\left\langle -\frac{42}{13}, 5, \frac{11}{13}\right\rangle$, so $F=\left(-\frac{42}{13}, 5, \frac{11}{13}\right)$ is the point on $\ell$ closest to $P$.

**43. a.** Use the distance formula to find the distance between the points $P = (-2, 4, 1)$ and $(-3t, 5, t + 3)$.

$$\begin{aligned}d(t)&=\sqrt{(-2-(-3t))^2+(4-5)^2+(1-(t+3))^2}\\&=\sqrt{(2-3t)^2+1^2+(t+2)^2}\end{aligned}$$

**b.** Since $d(t) \geq 0$ for all $t$, then if there is a value $t = a$ such that $d(a)^2 \leq d(t)^2$ for all $t$, then $d(a) \leq d(t)$ for all $t$. Therefore, $d(t)$ can be minimized by minimizing $D(t) = d(t)^2$. Simplifying the radicand of $d(t)$ yields $D(t) = 10t^2 - 8t + 9$. Use the second derivative test to minimize $D(t)$. $D'(t) = 20t - 8 = 0$ implies that $t = \frac{8}{20} = \frac{2}{5}$. Since $D''(t) = 20$, then $D''\left(\frac{2}{5}\right) > 0$, so $D(t)$ hence $d(t)$ reaches its minimum value at $t = \frac{2}{5}$. Therefore, $d\left(\frac{2}{5}\right) = \sqrt{10\left(\frac{2}{5}\right)^2 - 8\left(\frac{2}{5}\right) + 9} = \sqrt{\frac{37}{5}}$ is the minimum value of $d(t)$. This means the closest the line $\ell$ with equation $\mathbf{r}(t) = \langle -3t, 5, t+3 \rangle$ gets to the point (–2, 4, 1) is a distance of $\sqrt{\frac{37}{5}}$.

**c.** Since $\mathbf{r}\left(\frac{2}{5}\right) = \left\langle -\frac{6}{5}, 5, \frac{17}{5} \right\rangle$, then the point $\left(-\frac{6}{5}, 5, \frac{17}{5}\right)$ is the point on $\ell$ closest to (–2, 4, 1).

**45.** Let $\mathbf{v} = \langle v_1, v_2, v_3 \rangle$.
Then
$$\mathbf{v} \times \mathbf{v} = \langle v_1, v_2, v_3 \rangle \times \langle v_1, v_2, v_3 \rangle$$
$$= \langle v_2v_3 - v_3v_2, v_3v_1 - v_1v_3, v_1v_2 - v_2v_1 \rangle$$
$$= \langle 0, 0, 0 \rangle.$$
The parallelogram formed by the vectors $\mathbf{v}$ and $\mathbf{v}$ has zero area. Thus $\|\mathbf{v} \times \mathbf{v}\| = 0$. The only vector whose length is zero is the zero vector, so $\mathbf{v} \times \mathbf{v} = \langle 0, 0, 0 \rangle$.

**47.** If $\mathbf{v} \times \mathbf{u} = \mathbf{w}$, then $\mathbf{v}$ must be perpendicular to $\mathbf{w}$. Since $\mathbf{v} \cdot \mathbf{w} = \langle 2, 1, 1 \rangle \cdot \langle -7, 3, 11 \rangle = 0$, $\mathbf{v}$ is indeed perpendicular to $\mathbf{w}$. Suppose $\mathbf{a}$ is some other vector perpendicular to $\mathbf{w}$ that is not a scalar multiple of $\mathbf{v}$. Hence, $\mathbf{v}$ and $\mathbf{a}$ lie on a plane perpendicular to $\mathbf{w}$, so $\mathbf{w}$ is a scalar multiple of $\mathbf{v} \times \mathbf{a}$. Let $k$ be the value of the scalar, so $k\langle \mathbf{v} \times \mathbf{a} \rangle = \mathbf{v} \times k\mathbf{a} = \mathbf{w}$. Thus $\mathbf{u} = k\mathbf{a}$. Since there are an infinite number of possible vectors $\mathbf{a}$ that are not scalar multiples of each other, there are an infinite number of $\mathbf{u}$.

**49.** Let $P$ = (–3, 4, 0), $Q$ = (1, 2, 1), and $R$ = (–5, 2, 1). Let $S$ be the fourth vertex. There are three possibilities for creating a parallelogram. Suppose $PQ$ and $PR$ form two sides of a parallelogram. Then $\overrightarrow{PQ} + \overrightarrow{PR} = \overrightarrow{PS}$.
$\overrightarrow{PQ} = \langle 4, -2, 1 \rangle$ and $\overrightarrow{PR} = \langle -2, -2, 1 \rangle$.

$\overrightarrow{PQ} + \overrightarrow{PR} = \langle 2, -4, 2 \rangle$, so
$S$ = (2, –4, 2) + (–3, 4, 0) = (–1, 0, 2).
Suppose $QP$ and $QR$ form two sides of a parallelogram. Then $\overrightarrow{QP} + \overrightarrow{QR} = \overrightarrow{QS}$.

$\overrightarrow{QP} = \langle -4, 2, -1 \rangle$ and $\overrightarrow{QR} = \langle -6, 0, 0 \rangle$.

$\overrightarrow{QP} + \overrightarrow{QR} = \langle -10, 2, -1 \rangle$, so
$S$ = (–10, 2, –1) + (1, 2, 1) = (–9, 4, 0).
Suppose $RP$ and $RQ$ form two sides of parallelogram. Then $\overrightarrow{RP} + \overrightarrow{RQ} = \overrightarrow{RS}$.

$\overrightarrow{RP} = \langle 2, 2, -1 \rangle$ and $\overrightarrow{QR} = \langle 6, 0, 0 \rangle$.

$\overrightarrow{RP} + \overrightarrow{RQ} = \langle 8, 2, -1 \rangle$, so
$S$ = (8, 2, –1) + (–5, 2, 1) = (3, 4, 0).

**51.** $\|\langle 3, 8, 10 \rangle - \langle -2, 3, 5 \rangle\| = \|\langle 5, 5, 5 \rangle\| = 5\sqrt{3}$
$\|\langle 3, 8, 5 \rangle - \langle -2, 3, 5 \rangle\| = \|\langle 5, 5, 0 \rangle\| = 5\sqrt{2}$
$\|\langle 3, 8, 10 \rangle - \langle 3, 8, 5 \rangle\| = \|\langle 0, 0, 5 \rangle\| = 5$
Since the points are three of the vertices on the cube, it appears that the cube is determined by the vectors
$\mathbf{a} = \langle 5, 0, 0 \rangle$, $\mathbf{b} = \langle 0, 5, 0 \rangle$, and $\mathbf{c} = \langle 0, 0, 5 \rangle$
located at (–2, 3, 5).
The other vertices are
(–2, 3, 5) + (5, 0, 0) = (3, 3, 5),
(–2, 3, 5) + (0, 5, 0) = (–2, 8, 5),
(–2, 3, 5) + (0, 0, 5) = (–2, 3, 10),
(–2, 3, 5) + (5, 0, 0) + (0, 0, 5) = (3, 3, 10),
and
(–2, 3, 5) + (0, 5, 0) + (0, 0, 5) = (3, 3, 10).

**53.** $\mathbf{a} \times \mathbf{b}$ is perpendicular to $\mathbf{a}$ and $\mathbf{b}$, so the parallelepiped is a right cylinder with a base that is a parallelogram determined by $\mathbf{a}$ and $\mathbf{b}$.
volume $= |(\mathbf{a} \times \mathbf{b}) \cdot (\mathbf{a} \times \mathbf{b})| = \|\mathbf{a} \times \mathbf{b}\|^2$

**55.** Since $a\mathbf{u} + b\mathbf{v}$ is in the same plane that contains **u** and **v**, the three vectors form a hexagon. Thus the volume should be zero.

$$\begin{aligned}\text{volume} &= |(\mathbf{u}\times\mathbf{v})\cdot(a\mathbf{u}+b\mathbf{v})| \\ &= |a(\mathbf{u}\times\mathbf{v})\cdot\mathbf{u}+b(\mathbf{u}\times\mathbf{v})\cdot\mathbf{v}| \\ &= 0\end{aligned}$$

($\mathbf{u} \times \mathbf{v}$ is perpendicular to **u** and **v**.)

**57.** Let $\mathbf{a}=\langle a_1, a_2, a_3\rangle$, $\mathbf{b}=\langle b_1, b_2, b_3\rangle$, and $\mathbf{c}=\langle c_1, c_2, c_3\rangle$.

$$\begin{aligned}\mathbf{a}\times(\mathbf{b}+\mathbf{c}) &= \langle a_1, a_2, a_3\rangle\times\langle b_1+c_1, b_2+c_2, b_3+c_3\rangle \\ &= \langle a_2(b_3+c_3)-a_3(b_2+c_2), a_3(b_1+c_1)-a_1(b_3+c_3), a_1(b_2+c_2)-a_2(b_1+c_1)\rangle \\ &= \langle a_2b_3-a_3b_2, a_3b_1-a_1b_3, a_1b_2-a_2b_1\rangle+\langle a_2c_3-a_3c_2, a_3c_1-a_1c_3, a_1c_2-a_2c_1\rangle \\ &= \mathbf{a}\times\mathbf{b}+\mathbf{a}\times\mathbf{c}\end{aligned}$$

**59.**
$$\begin{aligned}\mathbf{a}\times\mathbf{b} &= \langle r_1\cos\alpha, r_1\sin\alpha, 0\rangle\times\langle r_2\cos(\alpha+\theta), r_2\sin(\alpha+\theta), 0\rangle \\ &= \langle 0, 0, r_1r_2\cos\alpha\sin(\alpha+\theta)-r_1r_2\sin\alpha\cos(\alpha+\theta)\rangle \\ &= \langle 0, 0, r_1r_2(\cos\alpha\sin(\alpha+\theta)-\sin\alpha\cos(\alpha+\theta))\rangle \\ &= \langle 0, 0, r_1r_2\sin((\alpha+\theta)-\alpha)\rangle \\ &= \langle 0, 0, r_1r_2\sin\theta\rangle\end{aligned}$$

Since $0 \le \theta < \pi$, $r_1r_2\sin\theta > 0$. Thus $\mathbf{a}\times\mathbf{b}=\langle 0, 0, c\rangle$ for some positive number $c$. This verifies the right-hand rule since your thumb should point in the positive direction.

**61.** $\mathbf{v}=\langle 0, -1, -3\rangle$ is a vector parallel to the given line and $Q = (5, -t + 1, -3t + 5)$ is a point on the given line. $\overrightarrow{PQ}=\langle 3, -t+2, -3t+4\rangle$. We want to find a $t$ such that $\overrightarrow{PQ}\cdot\mathbf{v}=0$.

$$\begin{aligned}\overrightarrow{PQ}\cdot\mathbf{v} &= \langle 3, -t+2, -3t+4\rangle\cdot\langle 0, -1, -3\rangle \\ &= 10t-14,\end{aligned}$$

so $t=\frac{7}{5}$. If $t=\frac{7}{5}$, $Q=\left(5, -\frac{2}{5}, \frac{4}{5}\right)$ and $\overrightarrow{PQ}=\left\langle 3, \frac{3}{5}, -\frac{1}{5}\right\rangle$ is a vector perpendicular to the given line.

The line described by $\mathbf{w}(t)=\langle 2, -1, 1\rangle+t\left\langle 3, \frac{3}{5}, -\frac{1}{5}\right\rangle$ intersects the given line at a right angle and passes through $P$.

**63.** For $\mathbf{a}=\langle a_1, a_2, a_3\rangle$, $\mathbf{b}=\langle b_1, b_2, b_3\rangle$, and $\mathbf{c}=\langle c_1, c_2, c_3\rangle$,

$$\det\begin{pmatrix} a_1 & a_2 & a_3 \\ b_1 & b_2 & b_3 \\ c_1 & c_2 & c_3\end{pmatrix}=a_1(b_2c_3-c_2b_3)+a_2(c_1b_3-b_1c_3)+a_3(b_1c_2-c_1b_2).$$

$\mathbf{b}\times\mathbf{c}=\langle b_2c_3-c_2b_3, c_1b_3-b_1c_3, b_1c_2-c_1b_2\rangle$, so

$$\begin{aligned}\mathbf{a}\cdot(\mathbf{b}\times\mathbf{c}) &= \langle a_1, a_2, a_3\rangle\cdot\langle b_2c_3-c_2b_3, c_1b_3-b_1c_3, b_1c_2-c_1b_2\rangle \\ &= a_1(b_2c_3-c_2b_3)+a_2(c_1b_3-b_1c_3)+a_3(b_1c_2-c_1b_2).\end{aligned}$$

Therefore, $\mathbf{a}\cdot(\mathbf{b}\times\mathbf{c})=\det\begin{pmatrix} a_1 & a_2 & a_3 \\ b_1 & b_2 & b_3 \\ c_1 & c_2 & c_3\end{pmatrix}$.

## Section 8.4 Linear Functions

**1.** $L\begin{pmatrix}2\\-3\end{pmatrix}=\begin{pmatrix}-7\\-10\end{pmatrix}$; $L\begin{pmatrix}0\\1\end{pmatrix}=\begin{pmatrix}1\\4\end{pmatrix}$

**3.** $L(-5)=\left(-\frac{5}{2}\right)$; $L(2002)=(1001)$

**5.** $L\begin{pmatrix}0\\7\\-6\end{pmatrix}=\begin{pmatrix}1&2&3\\1&0&8\end{pmatrix}\begin{pmatrix}0\\7\\-6\end{pmatrix}=\begin{pmatrix}-4\\-48\end{pmatrix}$;

$L\begin{pmatrix}1\\0\\1\end{pmatrix}=\begin{pmatrix}1&2&3\\1&0&8\end{pmatrix}\begin{pmatrix}1\\0\\1\end{pmatrix}=\begin{pmatrix}4\\9\end{pmatrix}$

**7.** $L\begin{pmatrix}x\\y\end{pmatrix}=\begin{pmatrix}-3&1\\4&7\end{pmatrix}\begin{pmatrix}x\\y\end{pmatrix}$

$2\times 2$ matrix

The domain is $R^2$ and the range is in $R^2$.

**9.** $L\begin{pmatrix}x\\y\\z\end{pmatrix}=\begin{pmatrix}1&0&0\\0&1&0\\0&0&1\end{pmatrix}\begin{pmatrix}x\\y\\z\end{pmatrix}$

$3\times 3$ matrix

The domain is $R^3$ and the range is in $R^3$.

**11.** $L\begin{pmatrix}x\\y\\z\end{pmatrix}=\begin{pmatrix}-2&3&4\end{pmatrix}\begin{pmatrix}x\\y\\z\end{pmatrix}$

$1\times 3$ matrix

The domain is $R^3$ and the range is in $R$.

**13.** $L\begin{pmatrix}x\\y\end{pmatrix}=\begin{pmatrix}-2&3\\6&0\end{pmatrix}\begin{pmatrix}x\\y\end{pmatrix}=\begin{pmatrix}-2x+3y\\6x\end{pmatrix}$

The domain is $R^2$ and the range is in $R^2$.

**15.** $L\begin{pmatrix}x\\y\\z\end{pmatrix}=\begin{pmatrix}a&b&c\end{pmatrix}\begin{pmatrix}x\\y\\z\end{pmatrix}=(ax+by+cz)$

The domain is $R^3$ and the range is in $R^1$.

**17.** $L(x) = (14.92)(x) = (14.92x)$

The domain is $R^1$ and the range is in $R^1$.

**19.** $L\begin{pmatrix}x\\y\end{pmatrix}=\begin{pmatrix}-2&1\\3&2\end{pmatrix}\begin{pmatrix}x\\y\end{pmatrix}$; $M\begin{pmatrix}x\\y\end{pmatrix}=\begin{pmatrix}-1&0\\1&4\end{pmatrix}\begin{pmatrix}x\\y\end{pmatrix}$

The matrix associated with $L$ is $A=\begin{pmatrix}-2&1\\3&2\end{pmatrix}$.

The matrix associated with $M$ is $B=\begin{pmatrix}-1&0\\1&4\end{pmatrix}$.

Thus, the matrix associated with $M\circ L$ is

$BA=\begin{pmatrix}-1&0\\1&4\end{pmatrix}\begin{pmatrix}-2&1\\3&2\end{pmatrix}=\begin{pmatrix}2&-1\\10&9\end{pmatrix}$.

The matrix associated with $L\circ M$ is

$AB=\begin{pmatrix}-2&1\\3&2\end{pmatrix}\begin{pmatrix}-1&0\\1&4\end{pmatrix}=\begin{pmatrix}3&4\\-1&8\end{pmatrix}$.

**21.** $L\begin{pmatrix}x\\y\end{pmatrix}=\begin{pmatrix}2&0\\3&-2\end{pmatrix}\begin{pmatrix}x\\y\end{pmatrix}$; $M\begin{pmatrix}x\\y\end{pmatrix}=\begin{pmatrix}\frac{1}{2}&0\\\frac{1}{2}&-\frac{1}{2}\end{pmatrix}\begin{pmatrix}x\\y\end{pmatrix}$

The matrix associated with $L$ is $A=\begin{pmatrix}2&0\\3&-2\end{pmatrix}$.

The matrix associated with $M$ is

$B=\begin{pmatrix}\frac{1}{2}&0\\\frac{3}{4}&-\frac{1}{2}\end{pmatrix}$.

Thus the matrix associated with $M\circ L$ is

$BA=\begin{pmatrix}\frac{1}{2}&0\\\frac{3}{4}&-\frac{1}{2}\end{pmatrix}\begin{pmatrix}2&0\\3&-2\end{pmatrix}=\begin{pmatrix}1&0\\0&1\end{pmatrix}$.

The matrix associated with $L\circ M$ is

$AB=\begin{pmatrix}2&0\\3&-2\end{pmatrix}\begin{pmatrix}\frac{1}{2}&0\\\frac{3}{4}&-\frac{1}{2}\end{pmatrix}=\begin{pmatrix}1&0\\0&1\end{pmatrix}$.

The composite functions are the identity function. Thus for vector **v** in $R^2$ $(L\circ M)(\mathbf{v})=(M\circ L)(\mathbf{v})=\mathbf{v}$.

**23.** $L\begin{pmatrix}1\\0\end{pmatrix}=\begin{pmatrix}a_{11}&a_{12}\\a_{21}&a_{22}\end{pmatrix}\begin{pmatrix}1\\0\end{pmatrix}=\begin{pmatrix}a_{11}\\a_{21}\end{pmatrix}$

$L\begin{pmatrix}0\\1\end{pmatrix}=\begin{pmatrix}a_{11}&a_{12}\\a_{21}&a_{22}\end{pmatrix}\begin{pmatrix}0\\1\end{pmatrix}=\begin{pmatrix}a_{12}\\a_{22}\end{pmatrix}$

**25.** Since for any vector $\mathbf{v}=\begin{pmatrix}x\\y\end{pmatrix}$ in $R^2$, $L$ can be written $L(\mathbf{v}) = A\mathbf{v}$, where $A=\begin{pmatrix}a_{11}&a_{12}\\a_{21}&a_{22}\end{pmatrix}$, then $L$ is a linear function from $R^2$ to $R^2$.

**27.** $\begin{pmatrix} x_1 \\ y_1 \end{pmatrix} = \begin{pmatrix} 0.99 & 0.055 \\ 0.023 & 0.94 \end{pmatrix}\begin{pmatrix} 176,995,000 \\ 71,715,000 \end{pmatrix} = \begin{pmatrix} 179,169,375 \\ 71,482,985 \end{pmatrix}$

Thus in 1991 the 20 and over population was 179,169,375 and the under 20 population was 71,482,985.

To find the populations in 1992 use the equation with $\begin{pmatrix} x \\ y \end{pmatrix} = \begin{pmatrix} 179,169,375 \\ 71,482,985 \end{pmatrix}$. Then the new $\begin{pmatrix} x_1 \\ y_1 \end{pmatrix}$ gives the populations in 1992. To find the populations in 1993 use the equation with the new $\begin{pmatrix} x_1 \\ y_1 \end{pmatrix}$ as $\begin{pmatrix} x \\ y \end{pmatrix}$.

**29.** $\begin{pmatrix} -2 & 3 \\ 1 & 7 \end{pmatrix}\begin{pmatrix} x \\ y \end{pmatrix} = \begin{pmatrix} -2x+3y \\ x+7y \end{pmatrix} = \begin{pmatrix} 0 \\ 2 \end{pmatrix}$, so $-2x + 3y = 0$ and $x + 7y = 2$. Solving this system, we get $y = \frac{4}{17}$ and $x = \frac{6}{17}$.

**31.** $\begin{pmatrix} 1 & 1 & 1 \\ 3 & 0 & -2 \end{pmatrix}\begin{pmatrix} x \\ y \\ z \end{pmatrix} = \begin{pmatrix} x+y+z \\ 3x-2z \end{pmatrix} = \begin{pmatrix} 3 \\ 1 \end{pmatrix}$, so $x + y + z = 3$ and $3x - 2z = 1$. Let $z = t$. Solving this system, we get $x = \frac{2}{3}t + \frac{1}{3}$ and $y = -\frac{5}{3}t + \frac{8}{3}$. Sample answer (other answers are possible): If $t = 1, -2, 4$, three different vectors satisfying the equation are $\begin{pmatrix} 1 \\ 1 \\ 1 \end{pmatrix}$, $\begin{pmatrix} -1 \\ 6 \\ -2 \end{pmatrix}$, and $\begin{pmatrix} 3 \\ -4 \\ 4 \end{pmatrix}$. In general, any vector of the form $\begin{pmatrix} \frac{2}{3}t+\frac{1}{3} \\ -\frac{5}{3}t+\frac{8}{3} \\ t \end{pmatrix}$ is a solution.

**33.** Observe that $L(\mathbf{v}) = \begin{pmatrix} \mathbf{r}_1 \cdot \mathbf{v} \\ \mathbf{r}_2 \cdot \mathbf{v} \end{pmatrix}$. Since $\mathbf{v} = k(\mathbf{r}_1 \times \mathbf{r}_2)$ is a vector perpendicular to both $\mathbf{r}_1$ and $\mathbf{r}_2$, $\mathbf{r}_1 \cdot \mathbf{v} = 0$ and $\mathbf{r}_2 \cdot \mathbf{v} = 0$. Thus $L(\mathbf{v}) = \begin{pmatrix} 0 \\ 0 \end{pmatrix}$.

**35.** $\mathbf{v} = k(\langle 1, 2, -3\rangle \times \langle 4, 1, -6\rangle)$
$= k\langle -9, -6, -7\rangle$
$= -k\langle 9, 6, 7\rangle$

Re-labeling the constant $-k$ as $k$ and writing as a column vector gives the answer $k\begin{pmatrix} 9 \\ 6 \\ 7 \end{pmatrix}$, or $\begin{pmatrix} 9k \\ 6k \\ 7k \end{pmatrix}$.

**37.** $\mathbf{v} = k(\langle 1, -7, 2\rangle \times \langle 4, 3, 0\rangle) = k\langle -6, 8, 31\rangle$

As a column vector, the answer is $k\begin{pmatrix} -6 \\ 8 \\ 31 \end{pmatrix}$, or $\begin{pmatrix} -6k \\ 8k \\ 31k \end{pmatrix}$.

**39.** $L(\mathbf{v}+\mathbf{w}) = L\begin{pmatrix} -1 \\ 3 \end{pmatrix} = \begin{pmatrix} 8 \\ -7 \end{pmatrix}$

$L(\mathbf{v}) = L\begin{pmatrix} 2 \\ 2 \end{pmatrix} = \begin{pmatrix} 2 \\ -2 \end{pmatrix}$; $L(\mathbf{w}) = L\begin{pmatrix} -3 \\ 1 \end{pmatrix} = \begin{pmatrix} -6 \\ -5 \end{pmatrix}$

$L(\mathbf{v}) + L(\mathbf{w}) = \begin{pmatrix} -4 \\ -7 \end{pmatrix}$

Thus $L(\mathbf{v} + \mathbf{w}) \neq L(\mathbf{v}) + L(\mathbf{w})$, so $L$ is not a linear function since (7) does not hold.

**41.** Sample answer (other answers are possible): (8) does not hold.

$2L\begin{pmatrix} 0 \\ 0 \end{pmatrix} = 2\begin{pmatrix} -1 \\ 3 \end{pmatrix} = \begin{pmatrix} -2 \\ 6 \end{pmatrix}$; $L\begin{pmatrix} 0 \\ 0 \end{pmatrix} = \begin{pmatrix} -1 \\ 3 \end{pmatrix}$

$2L\begin{pmatrix} 0 \\ 0 \end{pmatrix} \neq L\begin{pmatrix} 2\cdot 0 \\ 2\cdot 0 \end{pmatrix} = L\begin{pmatrix} 0 \\ 0 \end{pmatrix}$

**43.** Sample answer (other answers are possible): (7) does not hold.

$L\begin{pmatrix} 1 \\ 1 \end{pmatrix} = 3$, $L\begin{pmatrix} 0 \\ 1 \end{pmatrix} = 1$, and $L\begin{pmatrix} 1 \\ 2 \end{pmatrix} = 5$, so

$L\begin{pmatrix} 1 \\ 1 \end{pmatrix} + L\begin{pmatrix} 0 \\ 1 \end{pmatrix} = 4$. Therefore,

$L\begin{pmatrix} 1 \\ 1 \end{pmatrix} + L\begin{pmatrix} 0 \\ 1 \end{pmatrix} \neq L\begin{pmatrix} 1+0 \\ 1+1 \end{pmatrix} = 5.$

## Section 8.5 The Geometry of Linear Functions

**1.** $S$ is the line segment in $R^2$ joining the points designated by $\mathbf{r}(0)=\begin{pmatrix}0\\1\end{pmatrix}$ and $\mathbf{r}(1)=\begin{pmatrix}-4\\2\end{pmatrix}$. Then $L(S)$ is the line segment in $R^2$ described by

$$\mathbf{w}(t)=L(\mathbf{r}(t))=L\left(\begin{pmatrix}0\\1\end{pmatrix}+t\begin{pmatrix}-4\\1\end{pmatrix}\right)$$
$$=L\begin{pmatrix}0\\1\end{pmatrix}+tL\begin{pmatrix}-4\\1\end{pmatrix}$$
$$=\begin{pmatrix}-1\\-3\end{pmatrix}+t\begin{pmatrix}-5\\-11\end{pmatrix},\ 0\le t\le 1.$$

Thus $L(S)$ is the line segment in $R^2$ joining the points designated by $\mathbf{w}(0)=\begin{pmatrix}-1\\-3\end{pmatrix}$ and $\mathbf{w}(1)=\begin{pmatrix}-6\\-14\end{pmatrix}$.

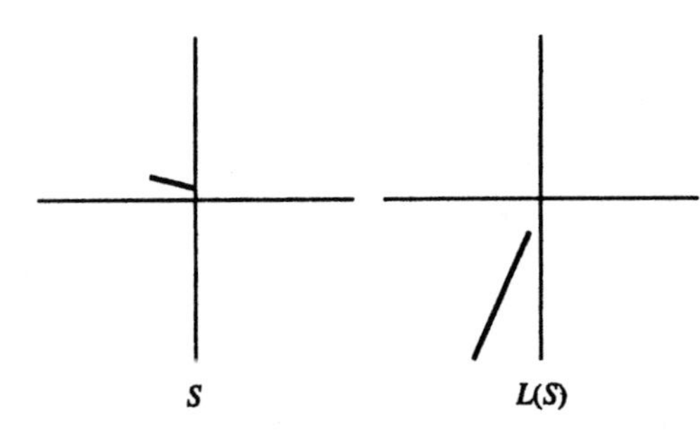


**3.** $S$ is the line segment in $R^2$ joining the points designated by $\mathbf{r}(-2)=\begin{pmatrix}-6\\0\end{pmatrix}$ and $\mathbf{r}(3)=\begin{pmatrix}9\\0\end{pmatrix}$. Then $L(S)$ is the line segment in $R^3$ described by

$$\mathbf{w}(t)=L(\mathbf{r}(t))=L\left(t\begin{pmatrix}3\\0\end{pmatrix}\right)=tL\begin{pmatrix}3\\0\end{pmatrix}=t\begin{pmatrix}-9\\18\\0\end{pmatrix},$$

$-2\le t\le 3$. Thus $L(S)$ is the line segment in $R^3$ joining the points designated by

$$\mathbf{w}(-2)=\begin{pmatrix}18\\-36\\0\end{pmatrix} \text{ and } \mathbf{w}(3)=\begin{pmatrix}-27\\54\\0\end{pmatrix}.$$

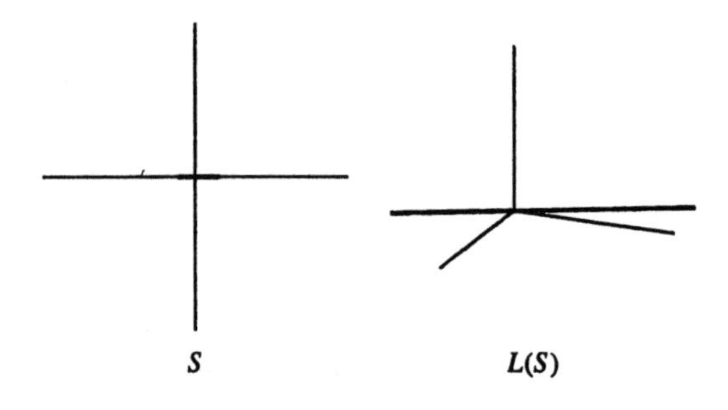


**5.** $S$ is the unit square in $R^2$, so it is determined by the vectors $\mathbf{a}=\begin{pmatrix}1\\0\end{pmatrix}$ and $\mathbf{b}=\begin{pmatrix}0\\1\end{pmatrix}$. Then $L(S)$ is determined by the vectors $L(\mathbf{a})=L\begin{pmatrix}1\\0\end{pmatrix}=\begin{pmatrix}2\\2\end{pmatrix}$ and $L(\mathbf{b})=L\begin{pmatrix}0\\1\end{pmatrix}=\begin{pmatrix}1\\3\end{pmatrix}$. Thus $L(S)$ is the region on and inside of the parallelogram determined by these vectors in $R^2$.

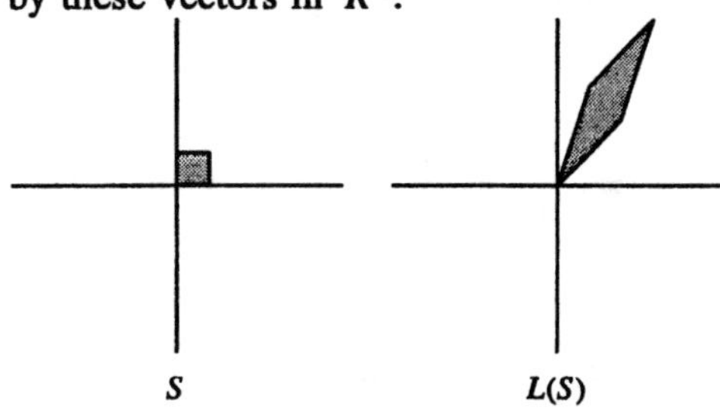


**7.** $S$ is the parallelogram in $R^2$ determined by the vectors $\mathbf{a}=\begin{pmatrix}2\\3\end{pmatrix}$ and $\mathbf{b}=\begin{pmatrix}8\\-6\end{pmatrix}$. Then $L(S)$ is determined by the vectors

$L(\mathbf{a})=L\begin{pmatrix}2\\3\end{pmatrix}=\begin{pmatrix}8\\9\end{pmatrix}$ and

$L(\mathbf{b})=L\begin{pmatrix}8\\-6\end{pmatrix}=\begin{pmatrix}32\\-18\end{pmatrix}$. Thus $L(S)$ is the region on and inside of the parallelogram determined by these vectors in $R^2$.

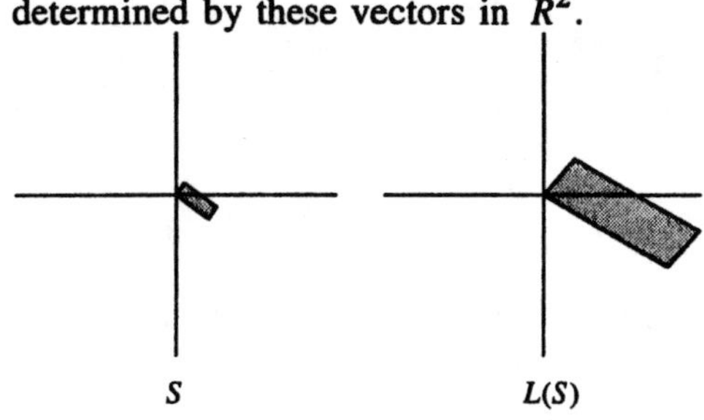

**9.** $S$ is the parallelogram in $R^2$ determined by the vectors $\mathbf{a}=\begin{pmatrix}1\\2\end{pmatrix}$ and $\mathbf{b}=\begin{pmatrix}-1\\2\end{pmatrix}$. Then $L(S)$ is determined by the vectors

$$L(\mathbf{a})=L\begin{pmatrix}1\\2\end{pmatrix}=(12) \text{ and } L(\mathbf{b})=L\begin{pmatrix}-1\\2\end{pmatrix}=(16).$$

Thus $L(S)$ is the line segment in $R^1$ from $0 = 0L(\mathbf{a}) + 0L(\mathbf{b})$ to $28 = L(\mathbf{a}) + L(\mathbf{b})$.

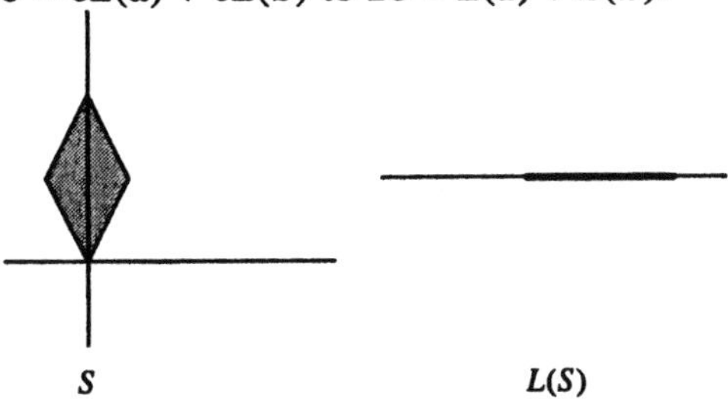

$S$ $\qquad$ $L(S)$

**11.** $S$ is the line segment in $R^3$ joining the points designated by $\mathbf{r}(-1)=\begin{pmatrix}7\\0\\2\end{pmatrix}$ and $\mathbf{r}(1)=\begin{pmatrix}1\\0\\8\end{pmatrix}$. Then $L(S)$ is the line segment in $R^3$ described by

$$\begin{aligned}\mathbf{w}(t)=L(\mathbf{r}(t))&=L\left(\begin{pmatrix}4\\0\\5\end{pmatrix}+t\begin{pmatrix}-3\\0\\3\end{pmatrix}\right)\\&=L\begin{pmatrix}4\\0\\5\end{pmatrix}+tL\begin{pmatrix}-3\\0\\3\end{pmatrix}\\&=\begin{pmatrix}-37\\9\\-5\end{pmatrix}+t\begin{pmatrix}21\\0\\-3\end{pmatrix}, -1\le t\le 1.\end{aligned}$$

Thus $L(S)$ is the line segment in $R^3$ joining the points designated by $\mathbf{w}(-1)=\begin{pmatrix}-58\\9\\-2\end{pmatrix}$ and $\mathbf{w}(1)=\begin{pmatrix}-16\\9\\-8\end{pmatrix}$.

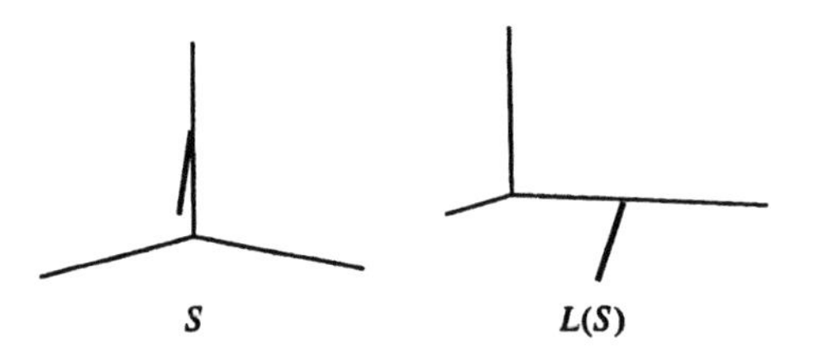

$S$ $\qquad$ $L(S)$

**13.** $S$ is the unit cube in $R^3$, so it is determined by the vectors $\mathbf{a}=\begin{pmatrix}1\\0\\0\end{pmatrix}$, $\mathbf{b}=\begin{pmatrix}0\\1\\0\end{pmatrix}$, and $\mathbf{c}=\begin{pmatrix}0\\0\\1\end{pmatrix}$. Then $L(S)$ is determined by the vectors $L(\mathbf{a})=L\begin{pmatrix}1\\0\\0\end{pmatrix}=\begin{pmatrix}2\\1\\0\end{pmatrix}$,

$L(\mathbf{b})=L\begin{pmatrix}0\\1\\0\end{pmatrix}=\begin{pmatrix}0\\1\\-2\end{pmatrix}$, and

$L(\mathbf{c})=L\begin{pmatrix}0\\0\\1\end{pmatrix}=\begin{pmatrix}-1\\1\\0\end{pmatrix}$. Thus $L(S)$ is the region on and inside of the parallelepiped determined by these vectors in $R^3$.

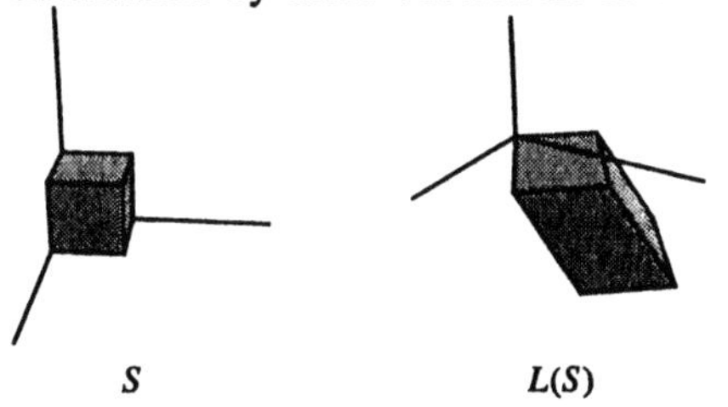

$S$ $\qquad$ $L(S)$

**15.** $S$ is the unit cube in $R^3$, so it is determined by the vectors $\mathbf{a}=\begin{pmatrix}1\\0\\0\end{pmatrix}$, $\mathbf{b}=\begin{pmatrix}0\\1\\0\end{pmatrix}$, and $\mathbf{c}=\begin{pmatrix}0\\0\\1\end{pmatrix}$. Then $L(S)$ is determined by the vectors $L(\mathbf{a})=L\begin{pmatrix}1\\0\\0\end{pmatrix}=\begin{pmatrix}1\\0\end{pmatrix}$,

$L(\mathbf{b}) = L\begin{pmatrix}0\\1\\0\end{pmatrix} = \begin{pmatrix}0\\1\end{pmatrix}$, and $L(\mathbf{c}) = L\begin{pmatrix}0\\0\\1\end{pmatrix} = \begin{pmatrix}2\\-1\end{pmatrix}$.

Thus $L(S)$ is the hexagon on and inside of the region determined by these vectors in $R^2$.

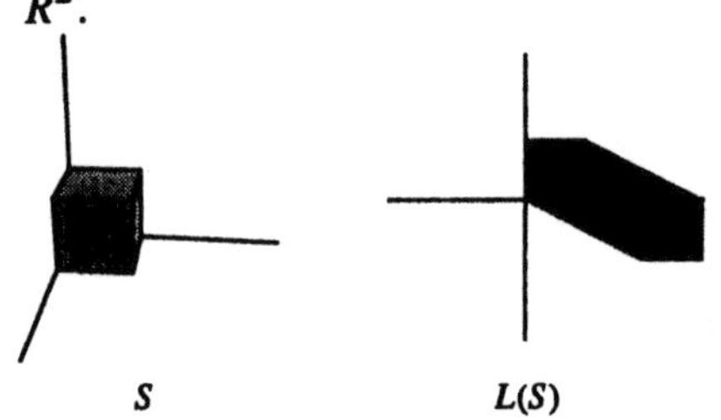

$S$ $\qquad$ $L(S)$

**17.** We can describe $L$ as $L\begin{pmatrix}x\\y\end{pmatrix} = \begin{pmatrix}4 & 0\\0 & -3\end{pmatrix}\begin{pmatrix}x\\y\end{pmatrix}$.

Then $\text{area}(L(S_1)) = 12 \cdot \text{area}(S_1)$. Since $\text{area}(S_1) = 5 \cdot 5 = 25$, $\text{area}(L(S_1)) = 12 \cdot 25 = 300$.

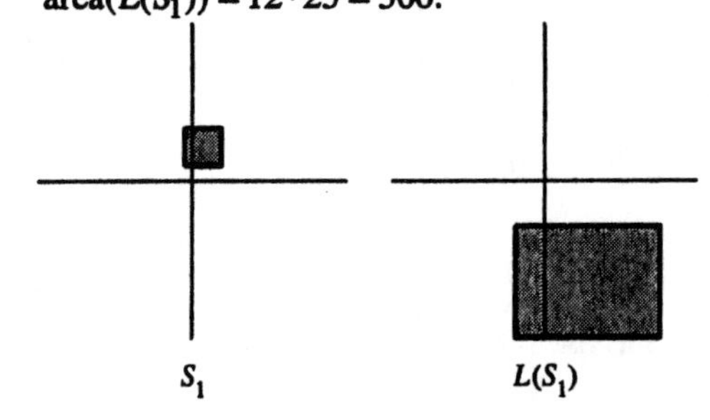

$S_1$ $\qquad$ $L(S_1)$

**19.** Since $L\begin{pmatrix}x\\y\end{pmatrix} = \begin{pmatrix}2 & 5\\-1 & 3\end{pmatrix}\begin{pmatrix}x\\y\end{pmatrix}$. Then $\text{area}(L(S_1)) = 11 \cdot \text{area}(S_1)$. $S_1$ is the parallelogram determined by the vectors $\mathbf{a} = \begin{pmatrix}4\\4\end{pmatrix}$ and $\mathbf{b} = \begin{pmatrix}-1\\3\end{pmatrix}$, so $\text{area}(S_1) = \|\mathbf{a} \times \mathbf{b}\| = 16$. Thus $\text{area}(L(S_1)) = 11 \cdot 16 = 176$.

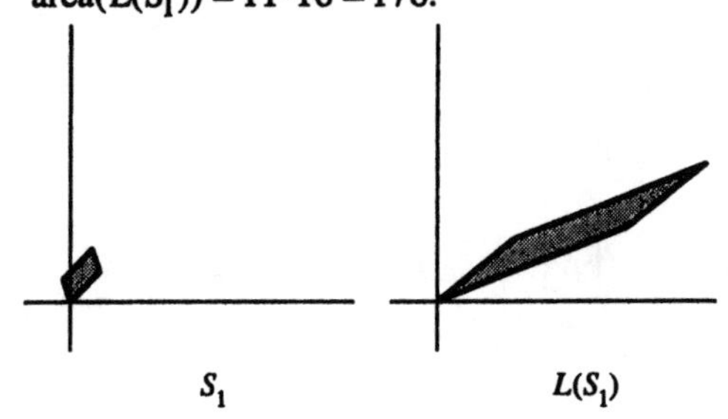

$S_1$ $\qquad$ $L(S_1)$

**21.** $C$ is determined by the vectors $\mathbf{i} = \begin{pmatrix}1\\0\\0\end{pmatrix}$, $\mathbf{j} = \begin{pmatrix}0\\1\\0\end{pmatrix}$, and $\mathbf{k} = \begin{pmatrix}0\\0\\1\end{pmatrix}$, so the parallelepiped $L(C)$ is determined by the vectors $L(\mathbf{i}) = \begin{pmatrix}1\\0\\0\end{pmatrix}$, $L(\mathbf{j}) = \begin{pmatrix}0\\-2\\0\end{pmatrix}$, and $L(\mathbf{k}) = \begin{pmatrix}0\\0\\3\end{pmatrix}$.

This is a rectangular box, so $\text{volume}(L(C)) = 1 \cdot 2 \cdot 3 = 6$. Since $\text{volume}(S) = 3 \cdot 6 \cdot 3 = 54$,

$$\begin{aligned}\text{volume}(L(S)) &= \text{volume}(L(C))\text{volume}(S)\\ &= 6 \cdot 54\\ &= 324.\end{aligned}$$

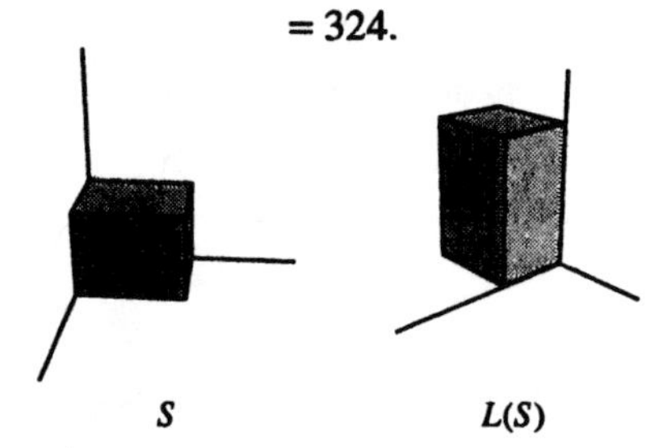

$S$ $\qquad$ $L(S)$

**23.** $C$ is determined by the vectors $\mathbf{i} = \begin{pmatrix}1\\0\\0\end{pmatrix}$, $\mathbf{j} = \begin{pmatrix}0\\1\\0\end{pmatrix}$, and $\mathbf{k} = \begin{pmatrix}0\\0\\1\end{pmatrix}$, so the parallelepiped $L(C)$ is determined by the vectors $L(\mathbf{i}) = \begin{pmatrix}2\\-3\\2\end{pmatrix}$, $L(\mathbf{j}) = \begin{pmatrix}-1\\1\\0\end{pmatrix}$, and $L(\mathbf{k}) = \begin{pmatrix}5\\0\\-5\end{pmatrix}$.

Thus

$$\begin{aligned}&\text{volume}(L(C))\\ &= |L(\mathbf{i}) \cdot (L(\mathbf{j}) \times L(\mathbf{k}))|\\ &= |\langle 2, -3, 2\rangle \cdot (\langle -1, 1, 0\rangle \times \langle 5, 0, -5\rangle)|\\ &= |\langle 2, -3, 2\rangle \cdot \langle -5, -5, -5\rangle|\\ &= 5.\end{aligned}$$

Since $\text{volume}(S) = 5 \cdot 8 \cdot 8 = 320$,

$$\begin{aligned}\text{volume}(L(S)) &= \text{volume}(L(C))\text{volume}(S)\\ &= 5\cdot 320\\ &= 1600.\end{aligned}$$

$S$ $\qquad$ $L(S)$

**25.** $$T\begin{pmatrix}x\\y\end{pmatrix}=\begin{pmatrix}-x+3y-4\\4x+1\end{pmatrix}=\begin{pmatrix}-x+3y\\4x\end{pmatrix}+\begin{pmatrix}-4\\1\end{pmatrix}=\begin{pmatrix}-1&3\\4&0\end{pmatrix}\begin{pmatrix}x\\y\end{pmatrix}+\begin{pmatrix}-4\\1\end{pmatrix}$$

Thus if $\mathbf{v}$ is a vector in $R^2$, then $T(\mathbf{v}) = L(\mathbf{v}) + \mathbf{b}$, where $L$ is the linear function defined by $L\begin{pmatrix}x\\y\end{pmatrix}=\begin{pmatrix}-1&3\\4&0\end{pmatrix}\begin{pmatrix}x\\y\end{pmatrix}$ and $\mathbf{b}=\begin{pmatrix}-4\\1\end{pmatrix}$. $L(S)$ is the parallelogram determined by the vectors $L\begin{pmatrix}1\\0\end{pmatrix}=\begin{pmatrix}-1\\4\end{pmatrix}$ and $L\begin{pmatrix}0\\1\end{pmatrix}=\begin{pmatrix}3\\0\end{pmatrix}$. $T(S)$ is found by translating $L(S)$ by $\mathbf{b}$.

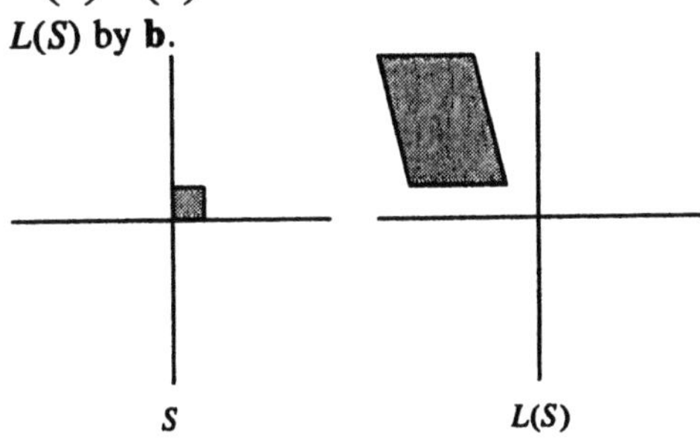

$S$ $\qquad$ $L(S)$

**27.** $$T\begin{pmatrix}x\\y\end{pmatrix}=\begin{pmatrix}x+3y-8\\2x-y\end{pmatrix}=\begin{pmatrix}x+3y\\2x-y\end{pmatrix}+\begin{pmatrix}-8\\0\end{pmatrix}=\begin{pmatrix}1&3\\2&-1\end{pmatrix}\begin{pmatrix}x\\y\end{pmatrix}+\begin{pmatrix}-8\\0\end{pmatrix}$$

Thus if $\mathbf{v}$ is a vector in $R^2$, then $T(\mathbf{v}) = L(\mathbf{v}) + \mathbf{b}$, where $L$ is the linear function defined by $L\begin{pmatrix}x\\y\end{pmatrix}=\begin{pmatrix}1&3\\2&-1\end{pmatrix}\begin{pmatrix}x\\y\end{pmatrix}$ and $\mathbf{b}=\begin{pmatrix}-8\\0\end{pmatrix}$. $L(S)$ is the parallelogram determined by the vectors $L\begin{pmatrix}1\\0\end{pmatrix}=\begin{pmatrix}1\\2\end{pmatrix}$ and $L\begin{pmatrix}0\\1\end{pmatrix}=\begin{pmatrix}3\\-1\end{pmatrix}$. $T(S)$ is found by translating $L(S)$ by $\mathbf{b}$.

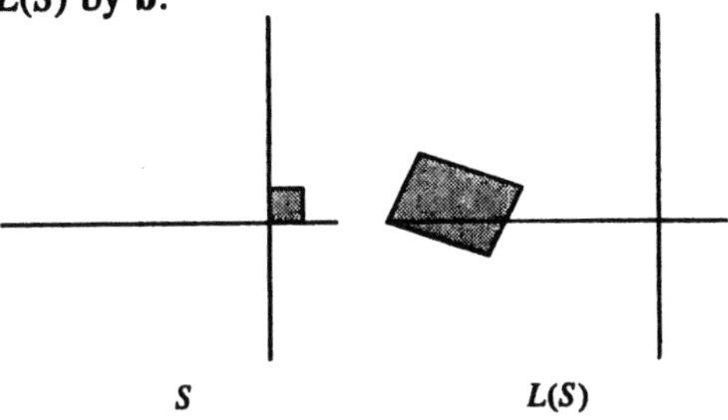

$S$ $\qquad$ $L(S)$

**29. a.** Sample answer (other answers are possible): The unit square is determined by the vectors $\begin{pmatrix}1\\0\end{pmatrix}$ and $\begin{pmatrix}0\\1\end{pmatrix}$, while $S_1$ is determined by the vectors $\begin{pmatrix}-2\\5\end{pmatrix}$ and $\begin{pmatrix}6\\-1\end{pmatrix}$. One linear function can be defined so $L\begin{pmatrix}1\\0\end{pmatrix}=\begin{pmatrix}-2\\5\end{pmatrix}$ and $L\begin{pmatrix}0\\1\end{pmatrix}=\begin{pmatrix}6\\-1\end{pmatrix}$. The matrix associated with $L$ has the form $A=\begin{pmatrix}a_{11}&a_{12}\\a_{21}&a_{22}\end{pmatrix}$, so $L\begin{pmatrix}1\\0\end{pmatrix}=\begin{pmatrix}a_{11}\\a_{21}\end{pmatrix}=\begin{pmatrix}-2\\5\end{pmatrix}$ and $L\begin{pmatrix}0\\1\end{pmatrix}=\begin{pmatrix}a_{12}\\a_{22}\end{pmatrix}=\begin{pmatrix}6\\-1\end{pmatrix}$. Thus $A=\begin{pmatrix}-2&6\\5&-1\end{pmatrix}$ and

$$L\begin{pmatrix}x\\y\end{pmatrix}=\begin{pmatrix}-2&6\\5&-1\end{pmatrix}\begin{pmatrix}x\\y\end{pmatrix}=\begin{pmatrix}-2x+6y\\5x-y\end{pmatrix}.$$

**b.** There are two different linear functions $L$ such that $L(S)=S_1$. The other linear function can be defined so $L\begin{pmatrix}1\\0\end{pmatrix}=\begin{pmatrix}6\\-1\end{pmatrix}$ and $L\begin{pmatrix}0\\1\end{pmatrix}=\begin{pmatrix}-2\\5\end{pmatrix}$.

**c.** One linear function can be defined so $K\begin{pmatrix}-2\\5\end{pmatrix}=\begin{pmatrix}1\\0\end{pmatrix}$ and $K\begin{pmatrix}6\\-1\end{pmatrix}=\begin{pmatrix}0\\1\end{pmatrix}$. The matrix associated with $K$ has the form $B=\begin{pmatrix}b_{11} & b_{12}\\b_{21} & b_{22}\end{pmatrix}$, so

$K\begin{pmatrix}-2\\5\end{pmatrix}=\begin{pmatrix}-2b_{11}+5b_{12}\\-2b_{21}+5b_{22}\end{pmatrix}=\begin{pmatrix}1\\0\end{pmatrix}$ and

$K\begin{pmatrix}6\\-1\end{pmatrix}=\begin{pmatrix}6b_{11}-b_{12}\\6b_{21}-b_{22}\end{pmatrix}=\begin{pmatrix}0\\1\end{pmatrix}$. This results in solving two systems of two equations, so we get $B=\begin{pmatrix}\frac{1}{28} & \frac{3}{14}\\\frac{5}{28} & \frac{1}{14}\end{pmatrix}$ and

$$K\begin{pmatrix}x\\y\end{pmatrix}=\begin{pmatrix}\frac{1}{28} & \frac{3}{14}\\\frac{5}{28} & \frac{1}{14}\end{pmatrix}\begin{pmatrix}x\\y\end{pmatrix}=\begin{pmatrix}\frac{1}{28}x+\frac{3}{14}y\\\frac{5}{28}x+\frac{1}{14}y\end{pmatrix}.$$

**d.** As in the case for $L$, there are two different linear functions $K$ such that $K(S_1)=S$.

**31. a.** Vertices of $S$ must be mapped to vertices of $S_1$, but one of the vertices of $S_1$ is (0, 0) while none of the vertices of $S$ are (0, 0). Since linear functions map (0, 0) to (0, 0), there are no linear functions $L$ with $L(S)=S_1$.

**b.** By similar reasoning, there are no linear functions $K$ such that $K(S_1)=S$.

**33. a.** Sample answer (other answers are possible):
The unit cube is determined by the vectors $\begin{pmatrix}1\\0\\0\end{pmatrix}$, $\begin{pmatrix}0\\1\\0\end{pmatrix}$, and $\begin{pmatrix}0\\0\\1\end{pmatrix}$. One linear function can be defined so $L\begin{pmatrix}1\\0\\0\end{pmatrix}=\begin{pmatrix}1\\2\\1\end{pmatrix}$,

$L\begin{pmatrix}0\\1\\0\end{pmatrix}=\begin{pmatrix}-1\\0\\2\end{pmatrix}$, and $L\begin{pmatrix}0\\0\\1\end{pmatrix}=\begin{pmatrix}3\\-4\\5\end{pmatrix}$. The matrix associated with $L$ has the form

$A=\begin{pmatrix}a_{11} & a_{12} & a_{13}\\a_{21} & a_{22} & a_{23}\\a_{31} & a_{32} & a_{33}\end{pmatrix}$, so

$L\begin{pmatrix}1\\0\\0\end{pmatrix}=\begin{pmatrix}a_{11}\\a_{21}\\a_{31}\end{pmatrix}=\begin{pmatrix}1\\2\\1\end{pmatrix}$,

$L\begin{pmatrix}0\\1\\0\end{pmatrix}=\begin{pmatrix}a_{12}\\a_{22}\\a_{32}\end{pmatrix}=\begin{pmatrix}-1\\0\\2\end{pmatrix}$, and

$L\begin{pmatrix}0\\0\\1\end{pmatrix}=\begin{pmatrix}a_{13}\\a_{23}\\a_{33}\end{pmatrix}=\begin{pmatrix}3\\-4\\5\end{pmatrix}$.

Thus $A=\begin{pmatrix}1 & -1 & 3\\2 & 0 & -4\\1 & 2 & 5\end{pmatrix}$ and

$$L\begin{pmatrix}x\\y\\z\end{pmatrix}=\begin{pmatrix}1 & -1 & 3\\2 & 0 & -4\\1 & 2 & 5\end{pmatrix}\begin{pmatrix}x\\y\\z\end{pmatrix}=\begin{pmatrix}x-y+3z\\2x-4z\\z+2y+5z\end{pmatrix}.$$

**b.** There are six different linear functions $L$ such that $L(C)=S$ since there are six different ways to map the vectors $\begin{pmatrix}1\\0\\0\end{pmatrix}$, $\begin{pmatrix}0\\1\\0\end{pmatrix}$, and $\begin{pmatrix}0\\0\\1\end{pmatrix}$ to the vectors $\begin{pmatrix}1\\2\\1\end{pmatrix}$, $\begin{pmatrix}-1\\0\\2\end{pmatrix}$, and $\begin{pmatrix}3\\-4\\5\end{pmatrix}$.

**c.** One linear function can be defined so $K\begin{pmatrix}1\\2\\1\end{pmatrix}=\begin{pmatrix}1\\0\\0\end{pmatrix}$ and $K\begin{pmatrix}-1\\0\\2\end{pmatrix}=\begin{pmatrix}0\\1\\0\end{pmatrix}$, and $K\begin{pmatrix}3\\-4\\5\end{pmatrix}=\begin{pmatrix}0\\0\\1\end{pmatrix}$. The matrix associated with $K$ has the form $B=\begin{pmatrix}b_{11} & b_{12} & b_{13}\\b_{21} & b_{22} & b_{23}\\b_{31} & b_{23} & b_{33}\end{pmatrix}$, so

$K\begin{pmatrix}1\\2\\1\end{pmatrix}=\begin{pmatrix}b_{11}+2b_{12}+b_{13}\\b_{21}+2b_{22}+b_{23}\\b_{31}+2b_{32}+b_{33}\end{pmatrix}=\begin{pmatrix}1\\0\\0\end{pmatrix}$,

$$K\begin{pmatrix}-1\\0\\2\end{pmatrix}=\begin{pmatrix}-b_{11}+2b_{13}\\-b_{21}+2b_{23}\\-b_{31}+2b_{33}\end{pmatrix}=\begin{pmatrix}0\\1\\0\end{pmatrix}, \text{ and}$$

$$K\begin{pmatrix}3\\-4\\5\end{pmatrix}=\begin{pmatrix}3b_{11}-4b_{12}+5b_{13}\\3b_{21}-4b_{22}+5b_{23}\\3b_{31}-4b_{32}+5b_{33}\end{pmatrix}=\begin{pmatrix}0\\0\\1\end{pmatrix}.$$

This results in solving three systems of three equations, so we get

$$B=\begin{pmatrix}\frac{4}{17}&\frac{11}{34}&\frac{2}{17}\\-\frac{7}{17}&\frac{1}{17}&\frac{5}{17}\\\frac{2}{17}&-\frac{3}{34}&\frac{1}{17}\end{pmatrix} \text{ and}$$

$$K\begin{pmatrix}x\\y\\z\end{pmatrix}=\begin{pmatrix}\frac{4}{17}&\frac{11}{34}&\frac{2}{17}\\-\frac{7}{17}&\frac{1}{17}&\frac{5}{17}\\\frac{2}{17}&-\frac{3}{34}&\frac{1}{17}\end{pmatrix}\begin{pmatrix}x\\y\\z\end{pmatrix}=\begin{pmatrix}\frac{4}{17}x+\frac{11}{34}y+\frac{2}{17}z\\-\frac{7}{17}x+\frac{1}{17}y+\frac{5}{17}z\\\frac{2}{17}x-\frac{3}{34}y+\frac{1}{17}z\end{pmatrix}.$$

**d.** As in the case for $L$, there are six different linear functions $K$ such that $K(S) = C$.

**35.** Using Exercise 34,

$$L\begin{pmatrix}x\\y\end{pmatrix}=\begin{pmatrix}x\cos\frac{\pi}{3}-y\sin\frac{\pi}{3}\\x\sin\frac{\pi}{3}+y\cos\frac{\pi}{3}\end{pmatrix}=\begin{pmatrix}\frac{1}{2}x-\frac{\sqrt{3}}{2}y\\\frac{\sqrt{3}}{2}x+\frac{1}{2}y\end{pmatrix}=\begin{pmatrix}\frac{1}{2}&-\frac{\sqrt{3}}{2}\\\frac{\sqrt{3}}{2}&\frac{1}{2}\end{pmatrix}\begin{pmatrix}x\\y\end{pmatrix}.$$

**37.** Using Problem 34, a linear function $L_1$ that rotates vectors through an angle of $-100°$ and preserves the length of all vectors is

$$L_1\begin{pmatrix}x\\y\end{pmatrix}=\begin{pmatrix}x\cos(-100°)-y\sin(-100°)\\x\sin(-100°)+y\cos(-100°)\end{pmatrix}=\begin{pmatrix}x\cos 100°+y\sin 100°\\-x\sin 100°+y\cos 100°\end{pmatrix}.$$

Thus

$$L\begin{pmatrix}x\\y\end{pmatrix}=2L_1\begin{pmatrix}x\\y\end{pmatrix}=\begin{pmatrix}2x\cos 100°+2y\sin 100°\\-2x\sin 100°+2y\cos 100°\end{pmatrix}=\begin{pmatrix}2\cos 100°&2\sin 100°\\-2\sin 100°&2\cos 100°\end{pmatrix}\begin{pmatrix}x\\y\end{pmatrix}.$$

**39.** Let $\theta$ be the angle between the vectors. Then $\cos\theta=\left\langle\frac{1}{2},\frac{\sqrt{3}}{2}\right\rangle\cdot\langle 1,0\rangle=\frac{1}{2}$, so $\theta=\frac{\pi}{3}$. Thus a rotation of $-\frac{\pi}{3}$ will map $\begin{pmatrix}\frac{1}{2}\\\frac{\sqrt{3}}{2}\end{pmatrix}$ to $\begin{pmatrix}1\\0\end{pmatrix}$, preserve the angle between any two vectors, and preserves the length of all vectors. Using Problem 34, such a linear function $L$ is

$$L\begin{pmatrix}x\\y\end{pmatrix}=\begin{pmatrix}x\cos\left(-\frac{\pi}{3}\right)-y\sin\left(-\frac{\pi}{3}\right)\\x\sin\left(-\frac{\pi}{3}\right)+y\cos\left(-\frac{\pi}{3}\right)\end{pmatrix}=\begin{pmatrix}\frac{1}{2}x+\frac{\sqrt{3}}{2}y\\-\frac{\sqrt{3}}{2}x+\frac{1}{2}y\end{pmatrix}=\begin{pmatrix}\frac{1}{2}&\frac{\sqrt{3}}{2}\\-\frac{\sqrt{3}}{2}&\frac{1}{2}\end{pmatrix}\begin{pmatrix}x\\y\end{pmatrix}.$$

**41.** Let $\mathbf{v}=\langle -2,3\rangle$ and $\mathbf{w}=\langle -1,7\rangle$. We want to find a linear function that rotates $\mathbf{v}$ so that it is parallel to and in the same direction as $\mathbf{w}$. Let $\theta$ be the angle of rotation between the vectors. Then

$$\cos\theta=\frac{1}{\|\mathbf{v}\|\|\mathbf{w}\|}\mathbf{v}\cdot\mathbf{w}=\frac{1}{\sqrt{13}\sqrt{50}}23=\frac{23}{\sqrt{650}}.$$

Plotting the two vectors, we see that $\theta < 0$, so $\sin\theta=-\frac{\sqrt{650-23^2}}{\sqrt{650}}=-\frac{11}{\sqrt{650}}$. Thus the linear function $M$ defined by

$$M\begin{pmatrix}x\\y\end{pmatrix}=\begin{pmatrix}x\cos\theta-y\sin\theta\\x\sin\theta+y\cos\theta\end{pmatrix}=\begin{pmatrix}\frac{23}{\sqrt{650}}x+\frac{11}{\sqrt{650}}y\\-\frac{11}{\sqrt{650}}x+\frac{23}{\sqrt{650}}y\end{pmatrix}$$

preserves the angle between vectors and rotates $\mathbf{v}$ so that it is parallel to $\mathbf{w}$. Since $\|L(\mathbf{v})\|=\|\mathbf{v}\|=\sqrt{13}$ and $\|\mathbf{w}\|=\sqrt{50}$, the linear function $L$ described by

$$L\begin{pmatrix} x \\ y \end{pmatrix} = \frac{\sqrt{50}}{\sqrt{13}} M\begin{pmatrix} x \\ y \end{pmatrix}$$
$$= \frac{\sqrt{50}}{\sqrt{13}} \begin{pmatrix} \frac{23}{\sqrt{650}}x + \frac{11}{\sqrt{650}}y \\ -\frac{11}{\sqrt{650}}x + \frac{23}{\sqrt{650}}y \end{pmatrix}$$
$$= \begin{pmatrix} \frac{23}{13}x + \frac{11}{13}y \\ -\frac{11}{13}x + \frac{23}{13}y \end{pmatrix}$$
$$= \begin{pmatrix} \frac{23}{13} & \frac{11}{13} \\ -\frac{11}{13} & \frac{23}{13} \end{pmatrix} \begin{pmatrix} x \\ y \end{pmatrix}$$

maps **v** to **w** and preserves the angle between any two vectors.

**43.** Observe that $L_1$ keeps the $z$-coordinates fixed. Thus, we look at how $L_1$ acts on the $x$- and $y$-coordinates.

Let $M$ be the function acting on $R^2$ that is the restriction of $L_1$ defined by $M\begin{pmatrix} x \\ y \end{pmatrix} = \begin{pmatrix} \frac{\sqrt{2}}{2}x + \frac{\sqrt{2}}{2}y \\ -\frac{\sqrt{2}}{2}x + \frac{\sqrt{2}}{2}y \end{pmatrix}$.

Observe that $\left\| M\begin{pmatrix} x \\ y \end{pmatrix} \right\| = \left\| \begin{pmatrix} x \\ y \end{pmatrix} \right\| = \sqrt{x^2 + y^2}$. If $\theta$ is the angle between $M\begin{pmatrix} x \\ y \end{pmatrix}$ and $\begin{pmatrix} x \\ y \end{pmatrix}$,

$$\cos\theta = \frac{1}{\left\| M\begin{pmatrix} x \\ y \end{pmatrix} \right\| \left\| \begin{pmatrix} x \\ y \end{pmatrix} \right\|} M\begin{pmatrix} x \\ y \end{pmatrix} \cdot \begin{pmatrix} x \\ y \end{pmatrix}$$
$$= \frac{1}{x^2 + y^2} \left( \frac{\sqrt{2}}{2}x^2 + \frac{\sqrt{2}}{2}y^2 \right)$$
$$= \frac{\sqrt{2}}{2}.$$

Thus $M$ rotates every vector by $\theta = \frac{\pi}{4}$. Notice that $M\begin{pmatrix} 1 \\ 0 \end{pmatrix} = \begin{pmatrix} \frac{\sqrt{2}}{2} \\ -\frac{\sqrt{2}}{2} \end{pmatrix}$ so $M$ rotates every vector clockwise.

Therefore $L_1(\mathbf{v})$ can be obtained by rotating **v** through a clockwise angle of $\frac{\pi}{4}$ about the $z$-axis.

**45.** $L(\mathbf{v})$ can be obtained by first rotating **v** through a clockwise angle $\frac{\pi}{4}$ about the $z$-axis and then by rotating the resulting vector **v** through a clockwise angle of $\frac{\pi}{4}$ about the $x$-axis.

**47.** Let $\mathbf{v} = \langle x, y, z\rangle$ and $\mathbf{w} = \langle x_1, y_1, z_1\rangle$.

$$\|L(\mathbf{v})\|^2 = \left(-\frac{1}{\sqrt{26}}x + \frac{1}{\sqrt{3}}y + \frac{7}{\sqrt{78}}z\right)^2 + \left(\frac{3}{\sqrt{26}}x - \frac{1}{\sqrt{3}}y + \frac{5}{\sqrt{78}}z\right)^2 + \left(\frac{4}{\sqrt{26}}x + \frac{1}{\sqrt{3}}y - \frac{2}{\sqrt{78}}z\right)^2$$

$$= \left(\frac{1}{26}x^2 + \frac{1}{3}y^2 + \frac{49}{78}z^2 - \frac{2}{\sqrt{78}}xy - \frac{7}{13\sqrt{3}}xz + \frac{14}{3\sqrt{26}}yz\right)$$

$$+\left(\frac{9}{26}x^2 + \frac{1}{3}y^2 + \frac{25}{78}z^2 - \frac{6}{\sqrt{78}}xy + \frac{15}{13\sqrt{3}}xz - \frac{10}{3\sqrt{26}}yz\right)$$

$$+\left(\frac{16}{26}x^2 + \frac{1}{3}y^2 + \frac{4}{78}z^2 + \frac{8}{\sqrt{78}}xy - \frac{8}{13\sqrt{3}}xz - \frac{4}{3\sqrt{26}}yz\right)$$

$$= x^2 + y^2 + z^2$$

Thus $\|L(\mathbf{v})\| = \|\mathbf{v}\| = \sqrt{x^2 + y^2 + z^2}$.

$$L(\mathbf{v}) \cdot L(\mathbf{w}) = \left(-\frac{1}{\sqrt{26}}x + \frac{1}{\sqrt{3}}y + \frac{7}{\sqrt{78}}z\right)\left(-\frac{1}{\sqrt{26}}x_1 + \frac{1}{\sqrt{3}}y_1 + \frac{7}{\sqrt{78}}z_1\right)$$

$$+\left(\frac{3}{\sqrt{26}}x - \frac{1}{\sqrt{3}}y + \frac{5}{\sqrt{78}}z\right)\left(\frac{3}{\sqrt{26}}x_1 - \frac{1}{\sqrt{3}}y_1 + \frac{5}{\sqrt{78}}z_1\right)$$

$$+\left(\frac{4}{\sqrt{26}}x + \frac{1}{\sqrt{3}}y - \frac{2}{\sqrt{78}}z\right)\left(\frac{4}{\sqrt{26}}x_1 + \frac{1}{\sqrt{3}}y_1 - \frac{2}{\sqrt{78}}z_1\right)$$

This simplifies to $xx_1 + yy_1 + zz_1$, so $L(\mathbf{v}) \cdot L(\mathbf{w}) = \mathbf{v} \cdot \mathbf{w}$. Let $\theta_1$ be the angle between $L(\mathbf{v})$ and $L(\mathbf{w})$ and let $\theta_2$ be the angle between $\mathbf{v}$ and $\mathbf{w}$. Then $\cos\theta_1 = \frac{1}{\|L(\mathbf{v})\|\|L(\mathbf{w})\|}L(\mathbf{v}) \cdot L(\mathbf{w}) = \frac{1}{\|\mathbf{v}\|\|\mathbf{w}\|}\mathbf{v} \cdot \mathbf{w} = \cos\theta_2$.

So the angle between $L(\mathbf{v})$ and $L(\mathbf{w})$ is equal to the angle between $\mathbf{v}$ and $\mathbf{w}$.

**49.** Let $A = \begin{pmatrix} a_{11} & a_{12} \\ a_{21} & a_{22} \end{pmatrix}$ and $B = \begin{pmatrix} b_{11} & b_{12} \\ b_{21} & b_{22} \end{pmatrix}$ be matrices associated with $L$ and $K$ respectively. Then

$$(L+K)\begin{pmatrix} x \\ y \end{pmatrix} = L\begin{pmatrix} x \\ y \end{pmatrix} + K\begin{pmatrix} x \\ y \end{pmatrix}$$

$$= \begin{pmatrix} a_{11} & a_{12} \\ a_{21} & a_{22} \end{pmatrix}\begin{pmatrix} x \\ y \end{pmatrix} + \begin{pmatrix} b_{11} & b_{12} \\ b_{21} & b_{22} \end{pmatrix}\begin{pmatrix} x \\ y \end{pmatrix}$$

$$= \left(\begin{pmatrix} a_{11} & a_{12} \\ a_{21} & a_{22} \end{pmatrix} + \begin{pmatrix} b_{11} & b_{12} \\ b_{21} & b_{22} \end{pmatrix}\right)\begin{pmatrix} x \\ y \end{pmatrix}$$

$$= \begin{pmatrix} a_{11}+b_{11} & a_{12}+b_{12} \\ a_{21}+b_{21} & a_{22}+b_{22} \end{pmatrix}\begin{pmatrix} x \\ y \end{pmatrix}.$$

Thus $L + K$ is a linear function with the associated matrix $\begin{pmatrix} a_{11}+b_{11} & a_{12}+b_{12} \\ a_{21}+b_{21} & a_{22}+b_{22} \end{pmatrix}$.

**51.** Let $\mathbf{a} = (a, b)$ and $\mathbf{b} = (c, d)$. If $P = (x, y)$ is a point in $S$, we see in the figure that $\mathbf{P}$ can be written as $\mathbf{P} = k\mathbf{a} + m\mathbf{b}$ where $k$ is the length of the projection of $\mathbf{P}$ on $\mathbf{a}$ divided by the length of $\mathbf{a}$ and $m$ is the length of the projection of $\mathbf{P}$ on $\mathbf{b}$ divided by the length of $\mathbf{b}$.

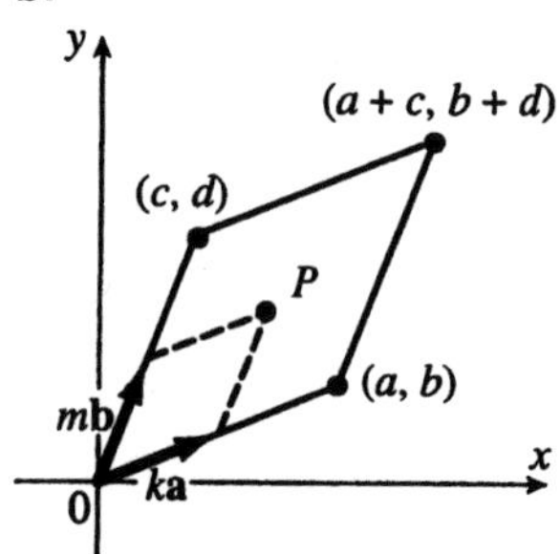


Precisely, $k = \frac{\mathbf{P}\cdot\mathbf{a}}{\|\mathbf{a}\|^2}$ and $m = \frac{\mathbf{P}\cdot\mathbf{b}}{\|\mathbf{b}\|^2}$. Since the length of each projection is less than the length of vector being projected onto, both $k$ and $m$ are between 0 and 1.
$L(\mathbf{P}) = L(k\mathbf{a} + m\mathbf{b}) = kL(\mathbf{a}) + mL(\mathbf{b})$ by the linearity properties for $L$. There are three cases.
Case 1: If $L(\mathbf{a}) = L(\mathbf{b}) = \mathbf{0}$ then $L(\mathbf{P}) = \mathbf{0}$, so $L(S)$ is a single point.
Case 2: If $L(\mathbf{a})$ and $L(\mathbf{b})$ are parallel and not both equal to $\mathbf{0}$, $L(S)$ is a line segment that contains the origin and is parallel to $L(\mathbf{a})$.
Case 3: If $L(\mathbf{a})$ and $L(\mathbf{b})$ are not parallel and neither is equal to $\mathbf{0}$, $L(S)$ is the set of points on or inside the parallelogram determined by $L(\mathbf{a})$ and $L(\mathbf{b})$.

**53.** Suppose $P = (x, y, z)$ is a point in $S$. In a manner similar to Exercise 51, we can write $\mathbf{P}$ as $\mathbf{P} = k\mathbf{a} + l\mathbf{b} + m\mathbf{c}$ where $k$ is the length of the projection of $\mathbf{P}$ on $\mathbf{a}$ divided by the length of $\mathbf{a}$, $l$ is the length of the projection of $\mathbf{P}$ on $\mathbf{b}$ divided by the length of $\mathbf{b}$, and $m$ is the length of the projection of $\mathbf{P}$ on $\mathbf{c}$ divided by the length of $\mathbf{c}$. Precisely,
$k = \frac{\mathbf{P}\cdot\mathbf{a}}{\|\mathbf{a}\|^2}$, $l = \frac{\mathbf{P}\cdot\mathbf{b}}{\|\mathbf{b}\|^2}$, and $m = \frac{\mathbf{P}\cdot\mathbf{c}}{\|\mathbf{c}\|^2}$. Since the length of each projection is less than the length of vector being projected, $k$, $l$, and $m$ are between 0 and 1.

$$\begin{aligned} L(\mathbf{P}) &= L(k\mathbf{a} + l\mathbf{b} + m\mathbf{c}) \\ &= kL(\mathbf{a}) + lL(\mathbf{b}) + mL(\mathbf{c}) \end{aligned}$$

by linearity properties for $L$. There are four cases.
Case 1: If $L(\mathbf{a}) = L(\mathbf{b}) = L(\mathbf{c}) = \mathbf{0}$, then $L(\mathbf{P}) = \mathbf{0}$, so $L(S)$ is a single point, the origin.
Case 2: If $L(\mathbf{a})$, $L(\mathbf{b})$, and $L(\mathbf{c})$ are parallel to each other and not all equal to $\mathbf{0}$, $L(S)$ is a line segment that contains the origin and is parallel to $L(\mathbf{a})$.
Case 3: If $L(\mathbf{a})$, $L(\mathbf{b})$, and $L(\mathbf{c})$ are parallel to a plane in $R^3$ and at least two are not $\mathbf{0}$ and parallel to each other, $L(S)$ is the set of points on or inside the region determined by $L(\mathbf{a})$, $L(\mathbf{b})$, and $L(\mathbf{c})$. This region is a hexagon if no two of the vectors $L(\mathbf{a})$, $L(\mathbf{b})$, and $L(\mathbf{c})$ are parallel to each other and none are equal to $\mathbf{0}$. If two of the vectors $L(\mathbf{a})$, $L(\mathbf{b})$, and $L(\mathbf{c})$ are parallel and not $\mathbf{0}$ or if one is $\mathbf{0}$ and the other two are not parallel to each other, then the region is a parallelogram.
Case 4: If $L(\mathbf{a})$, $L(\mathbf{b})$, and $L(\mathbf{c})$ are not parallel to each other and not equal to $\mathbf{0}$, $L(S)$ is the set of points on or inside the parallelepiped determined by $L(\mathbf{a})$, $L(\mathbf{b})$, and $L(\mathbf{c})$.

**55.** It is not always true that $T(\mathbf{v} + \mathbf{w}) = T(\mathbf{v}) + T(\mathbf{w})$. Since $T(\mathbf{v}) = L(\mathbf{v}) + \mathbf{b}$ for some linear function $L$ and constant vector $\mathbf{b}$,
$T(\mathbf{v} + \mathbf{w}) = L(\mathbf{v} + \mathbf{w}) + \mathbf{b} = L(\mathbf{v}) + L(\mathbf{w}) + \mathbf{b}$.
On the other hand

$$\begin{aligned} T(\mathbf{v}) + T(\mathbf{w}) &= L(\mathbf{v}) + \mathbf{b} + L(\mathbf{w}) + \mathbf{b} \\ &= L(\mathbf{v}) + L(\mathbf{w}) + 2\mathbf{b}, \end{aligned}$$

so $T(\mathbf{v} + \mathbf{w}) = T(\mathbf{v}) + T(\mathbf{w})$ if and only if $\mathbf{b} = \mathbf{0}$.

**57.** $T(S) = L(S) + \mathbf{b}$ for some linear function $L$ and constant vector $\mathbf{b}$. We know that $L(S)$ is a single point, a nondegenerate line segment, or the set of points on and inside a parallelogram. Hence $T(S)$ is a translation by $\mathbf{b}$ of $L(S)$, so the possibilities for $T(S)$ are a single point, a nondegenerate line segment, and the set of points on and inside a parallelogram.

**59.** $T(\mathbf{v}) = M(\mathbf{v}) + \mathbf{b}$ for some linear function $M$ and constant vector $\mathbf{b}$. Since $M$ is a linear function, $M(\mathbf{0}) = \mathbf{0}$.

$$\begin{aligned} T(\mathbf{v}) - T(\mathbf{0}) &= M(\mathbf{v}) + \mathbf{b} - (M(\mathbf{0}) + \mathbf{b}) \\ &= M(\mathbf{v}) \end{aligned}$$

Thus $L$ is the linear function $M$.

## Section 8.6 Planes

1. $\langle -1, 3, 5\rangle \cdot \langle x-2, y, z+5\rangle = 0$

$$-1(x-2)+3(y)+5(z+5)=0$$
$$-x+3y+5z=-27$$

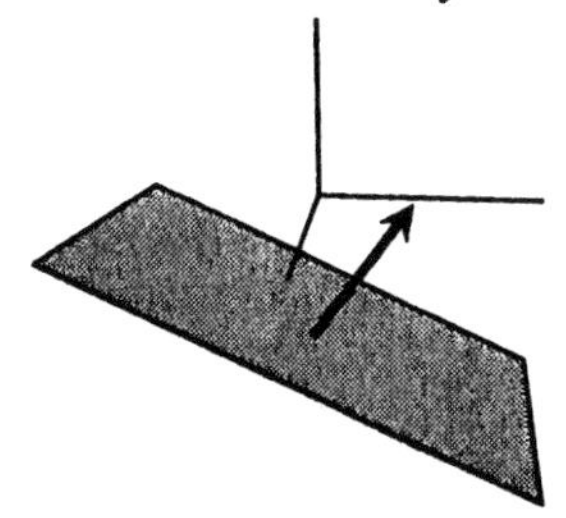

3. $\langle 0, 0, 1\rangle \cdot \langle x, y, z\rangle = 0$

$$z=0$$

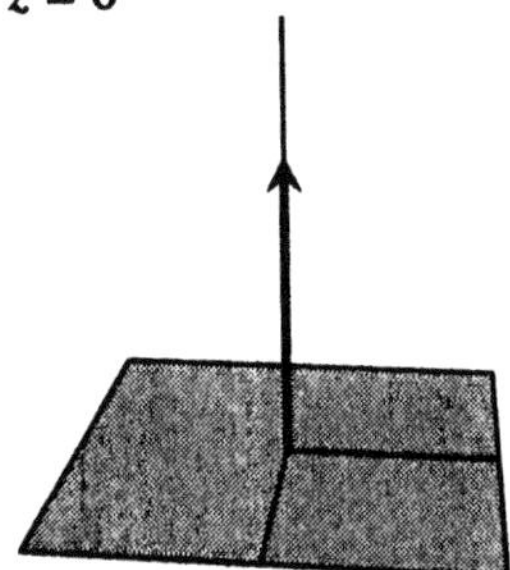

5. $\mathbf{N} = \langle 3, 1, -7\rangle$

$$\langle 3, 1, -7\rangle \cdot \langle x-7, y-8, z-2\rangle = 0$$
$$3(x-7)+1(y-8)-7(z-2)=0$$
$$3x+y-7z=15$$

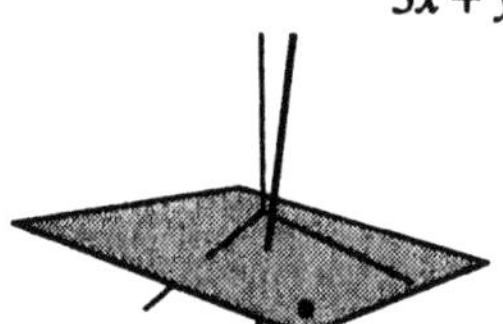

7. Since $\overrightarrow{PQ} = \langle 5, 3, -7\rangle$ and $\overrightarrow{PR} = \langle 1, -1, -2\rangle$ lie in the plane, a vector normal to the plane is

$$\mathbf{N} = \overrightarrow{PQ} \times \overrightarrow{PR} = \langle 5, 3, -7\rangle \times \langle 1, -1, -2\rangle = \langle -13, 3, -8\rangle$$
$$\langle -13, 3, -8\rangle \cdot \langle x+1, y-3, z-4\rangle = 0$$
$$-13(x+1)+3(y-3)-8(z-4)=0$$
$$-13x+3y-8z=-10$$

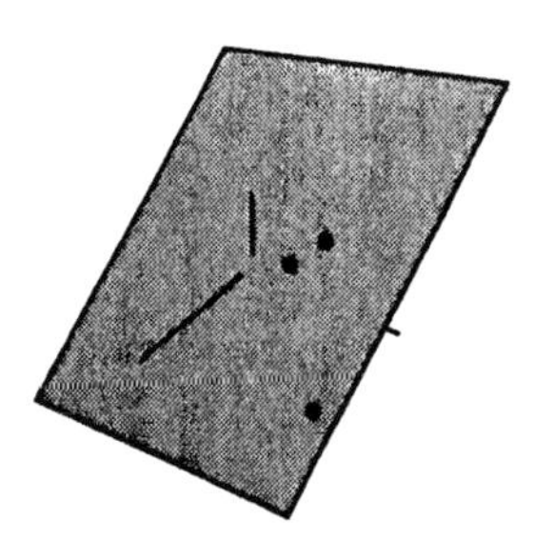

9. $\mathbf{N}_1 = \langle -2, 3, -4\rangle$ and $\mathbf{N}_2 = \langle 4, 0, -3\rangle$ are vectors normal to the given planes. A vector normal to the plane we are looking for is

$$\mathbf{N} = \mathbf{N}_1 \times \mathbf{N}_2 = \langle -2, 3, -4\rangle \times \langle 4, 0, -3\rangle = \langle -9, -22, -12\rangle$$
$$\langle -9, -22, -12\rangle \cdot \langle x-4, y+3, z+1\rangle = 0$$
$$-9(x-4)-22(y+3)-12(z+1)=0$$
$$-9x-22y-12z=42$$

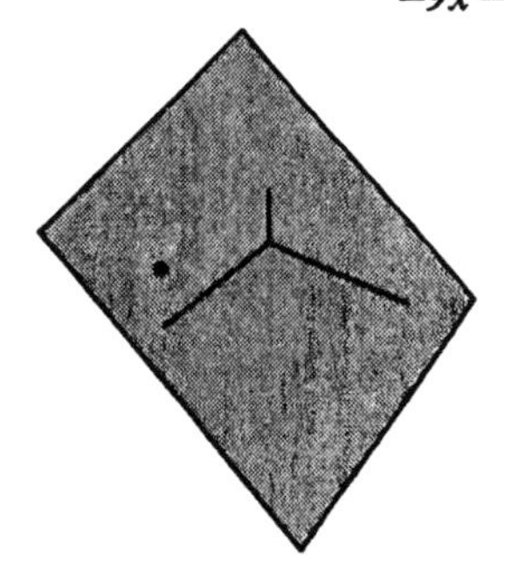

11. $\mathbf{N}_1 = \langle -3, 2, 2\rangle$ and $\mathbf{N}_2 = \langle -1, 0, 3\rangle$; let $\theta$ be the angle between $\mathbf{N}_1$ and $\mathbf{N}_2$.

$$\cos\theta = \frac{\mathbf{N}_1 \cdot \mathbf{N}_2}{\|\mathbf{N}_1\|\|\mathbf{N}_2\|} = \frac{9}{\sqrt{17}\sqrt{10}} = \frac{9}{\sqrt{170}}$$
$$\theta = \arccos\frac{9}{\sqrt{170}} \approx 0.808936$$

13. $\mathbf{N}_1 = \langle -7, 4, -10\rangle$ and $\mathbf{N}_2 = \langle 2, 2, -4\rangle$; let $\theta$ be the angle between $\mathbf{N}_1$ and $\mathbf{N}_2$.

$$\cos\theta = \frac{\mathbf{N}_1 \cdot \mathbf{N}_2}{\|\mathbf{N}_1\|\|\mathbf{N}_2\|} = \frac{34}{\sqrt{165}\sqrt{24}} = \frac{17}{3\sqrt{110}}$$
$$\theta = \arccos\frac{17}{3\sqrt{110}} \approx 1.00001$$

15. Sample answer (other answers are possible): $\mathbf{N}_1 = \langle -3, 1, -1\rangle$ is a vector normal to the given plane. We need to find $\mathbf{N}$ such that the angle $\theta$ between $\mathbf{N}_1$ and $\mathbf{N}$ is $\frac{\pi}{4}$. We can assume that $\mathbf{N}$ is a unit vector and that $\mathbf{N} = \langle a, b, c\rangle$ for real numbers $a$, $b$, and $c$.

$\dfrac{\mathbf{N}\cdot\mathbf{N}_1}{\|\mathbf{N}\|\|\mathbf{N}_1\|} = \cos\theta = \dfrac{\sqrt{2}}{2}$, so

$\langle a, b, c\rangle \cdot \langle -3, 1, -1\rangle = \dfrac{\sqrt{2}}{2}\sqrt{11}$. Therefore,

$-3a + b - c = \dfrac{\sqrt{22}}{2}$, and we know

$a^2 + b^2 + c^2 = 1$.

Since we have two equations and three unknowns, there are many possible solutions. We narrow the possibilities by setting $c = 0$. Then

$-3a + b = \dfrac{\sqrt{22}}{2}$,

$a^2 + b^2 = 1$.

Solving the first equation for $b$ and substituting in the second produces

$$a^2 + \left(\frac{\sqrt{22}}{2} + 3a\right)^2 = 1$$

$$10a^2 + 3\sqrt{22}a + \frac{18}{4} = 0$$

$$a = \frac{-3\sqrt{22} \pm \sqrt{9\cdot 22 - 10\cdot 18}}{20} = \frac{-3\sqrt{22} - 3\sqrt{2}}{20}$$

if we choose the smaller of the two values found for $a$. Then

$$b = \frac{-9\sqrt{22} - 9\sqrt{2}}{20} + \frac{\sqrt{22}}{2} = \frac{\sqrt{22} - 9\sqrt{2}}{20}.$$

Plugging the values obtained into the equation $ax + by + cz = 0$ and multiplying both sides by $-\frac{20}{\sqrt{2}}$ gives the answer

$$\left(3 + 3\sqrt{11}\right)x + \left(9 - \sqrt{11}\right)y = 0.$$

**17.** Sample answer (other answers are possible): $\mathbf{N}_1 = \langle 2, -4, 1\rangle$ is a vector normal to the given plane $\Pi$. We need to find a vector $\mathbf{N} = \langle a, b, c\rangle$ such that the angle $\theta$ between $\mathbf{N}_1$ and $\mathbf{N}$ is 1 radian. For simplicity, we can assume $\|\mathbf{N}\| = 1$. Since

$\cos\theta = \dfrac{1}{\|\mathbf{N}\|\|\mathbf{N}_1\|}\mathbf{N}\cdot\mathbf{N}_1 = \cos 1$, then

$\langle a, b, c\rangle\cdot\langle 2, -4, 1\rangle = \sqrt{21}\cos 1$, and so $a$, $b$, and $c$ must satisfy $2a - 4b + c = \sqrt{21}\cos 1$ and $a^2 + b^2 + c^2 = 1$.

Since we have two equations and three unknowns, there are many possible solutions. We narrow the possibilities by setting $c = 0$. Then $2a - 4b = \sqrt{21}\cos 1$, $a^2 + b^2 = 1$ or $(2a)^2 + (2b)^2 = 4$.

Solving the first equation for $2a$ and substituting in the second produces

$$\left(4b + \sqrt{21}\cos 1\right)^2 + (2b)^2 = 4$$

$20b^2 + \left(8\sqrt{21}\cos 1\right)b + 21\cos^2 1 - 4 = 0$

$b \approx -0.122777$

if we choose the larger of the two values found for $b$ using the quadratic formula. Then

$a = \dfrac{\sqrt{21}\cos 1 + 4b}{2} \approx 0.992434$. Thus an equation for a plane that intersects $\Pi$ at an angle of 1 radian is

$0.992434x - 0.122777y = 0$.

Any constant multiple of this equation will work as well. For instance, multiplying by $\sqrt{3} \approx 1.73205421$ produces

$1.71895x - 0.212656y = 0$.

**19.** Points $(x, y)$ on the line satisfy

$\langle x - 7, y\rangle\cdot\langle 0, -1\rangle = 0$

$0 - y = 0$, or $y = 0$.

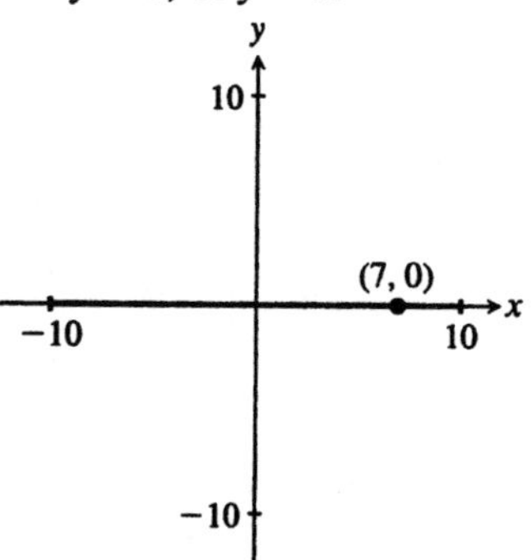


**21.** Since the line is perpendicular to $-x + 4y + 3 = 0$, which has normal vector $\langle -1, 4\rangle$, the line is in the direction of $\langle -1, 4\rangle$ and contains $(-2, 3)$. Such a line can be parametrized by

$$\mathbf{r}(t) = \langle -2, 3\rangle + \langle -1, 4\rangle t = \langle -2 - t, 3 + 4t\rangle.$$

Therefore, $x = -2 - t$ and $y = 3 + 4t$, so

$t = -x - 2 = \dfrac{y - 3}{4}$ and $\quad 4x + 8 = -y + 3$

$4x + y + 5 = 0.$

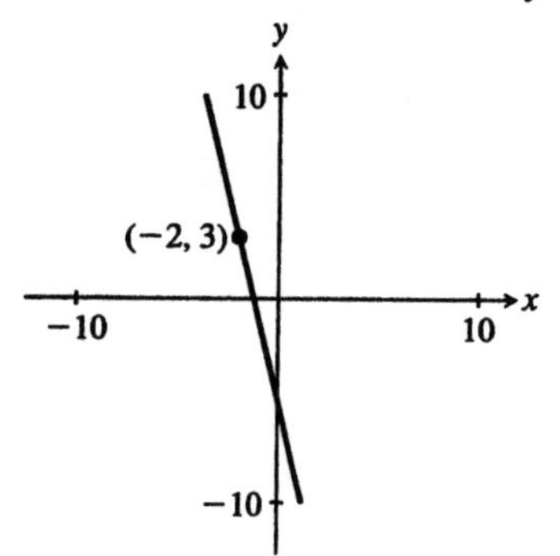

**23. a.** A vector parallel to $\ell$ is $\mathbf{v} = \langle 2, 0, -2\rangle$ and a vector normal to $\Pi$ is $\mathbf{N} = \langle 1, 4, -4\rangle$. If the line does not intersect the plane at exactly one point, it is parallel to the plane or perpendicular to the normal vector. Thus, if the dot product between **v** and **N** is not zero, $\ell$ and $\Pi$ intersect at exactly one point. Since $\mathbf{v}\cdot\mathbf{N} = 10$, $\ell$ and $\Pi$ intersect at exactly one point.

**b.** To find the point, substitute $(1 + 2t, 3, 2 - 2t)$ into the equation for $\Pi$ and solve for $t$:

$$(1+2t)+4(3)-4(2-2t)=12$$
$$10t+5=12$$
$$t=\frac{7}{10}$$

$\mathbf{r}\left\langle\frac{7}{10}\right\rangle = \left\langle\frac{12}{5}, 3, \frac{3}{5}\right\rangle$, so the point of intersection is $\left(\frac{12}{5}, 3, \frac{3}{5}\right)$.

**25. a.** $\mathbf{r}_1(t) = \langle -2, 0, -1\rangle + t\langle 3, 4, -1\rangle$ and $\mathbf{r}_2(t) = \langle 0, 7, 0\rangle + t\langle -9, -12, 3\rangle$. Hence $\ell_1$ is parallel to the vector $\mathbf{v}_1 = \langle 3, 4, -1\rangle$ and $\ell_2$ is parallel to the vector $\mathbf{v}_2 = \langle -9, -12, 3\rangle$. Since $\mathbf{v}_2 = -3\mathbf{v}_1$, $\ell_1$ and $\ell_2$ are parallel. $\mathbf{r}_1(0) = \langle -2, 0, -1\rangle$ so a point on $\ell_1$ is $(-2, 0, -1)$. If a point on $\ell_2$ has the first coordinate of $-2$, then $-9t = -2$ or $t = \frac{2}{9}$. However, since $\mathbf{r}_2\left(\frac{2}{9}\right) = \left\langle -2, \frac{13}{3}, \frac{2}{3}\right\rangle$, the only point on $\ell_2$ with the first coordinate of $-2$ is $\left(-2, \frac{13}{3}, \frac{2}{3}\right)$, so $\ell_1$ and $\ell_2$ are distinct.

**b.** $P = (-2, 0, -1)$ is a point on $\ell_1$ and $Q = (0, 7, 0)$ is a point on $\ell_2$. Thus they are also on the plane, so $\overrightarrow{PQ} = \langle 2, 7, 1\rangle$ is a vector parallel to the plane. $\mathbf{v}_1$ is also parallel to the plane because it is parallel to both $\ell_1$ and $\ell_2$. Therefore a vector perpendicular to the plane is

$$\mathbf{N} = \overrightarrow{PQ}\times\mathbf{v}_1 = \langle 2, 7, 1\rangle\times\langle 3, 4, -1\rangle = \langle -11, 5, -13\rangle.$$

An equation for a plane containing $\ell_1$ and $\ell_2$ is

$$\langle -11, 5, -13\rangle\cdot\langle x+2, y, z+1\rangle = 0$$
$$-11(x+2)+5y-13(z+1)=0,$$

and of course another is $11(x + 2) - 5y + 13(z + 1) = 0$.

**c.** To show that this plane does contain $\ell_1$ and $\ell_2$, substitute the points $(-2 + 3t, 4t, -1 - t)$ and $(-9t, 7 - 12t, 3t)$ in for $(x, y, z)$ in the equation. The plane defined by the equation contains $\ell_1$ and $\ell_2$ because $11(3t) - 5(4t) + 13(-t) = 0$ and $11(-9t + 2) - 5(7 - 12t) + 13(3t + 1) = 0$.

**27. a.** Observe that every point on $\ell_1$ has the first coordinate of 3. If a point on $\ell_2$ has the first coordinate of 3, then $-5 + 4t = 3$ or $t = 2$. Since $\mathbf{r}_2(2) = \langle 3, 0, 0\rangle$, the point $(3, 0, 0)$ is a point on $\ell_2$. Next we want to show that $(3, 0, 0)$ is a point on $\ell_1$. If a point on $\ell_1$ has the second coordinate of 0, then $3 - t = 0$ or $t = 3$. Since $\mathbf{r}_2(3) = \langle 3, 0, 0\rangle$, the point $(3, 0, 0)$ is a point on $\ell_1$. Thus $\ell_1$ and $\ell_2$ intersect.

**b.** $\mathbf{v}_1 = \langle 0, -1, 3\rangle$ is parallel to $\ell_1$ and $\mathbf{v}_2 = \langle 4, 8, -7\rangle$ is parallel to $\ell_2$. A vector parallel to the plane containing $\ell_1$ and $\ell_2$ is

$$\mathbf{N} = \mathbf{v}_1\times\mathbf{v}_2 = \langle 0, -1, 3\rangle\times\langle 4, 8, -7\rangle = \langle -17, 12, 4\rangle.$$

An equation for the plane containing $\ell_1$ and $\ell_2$ is

$$\langle -17, 12, 4\rangle\cdot\langle x-3, y, z\rangle = 0$$
$$-17(x-3)+12(y)+4(z)=0$$
$$17x-12y-4z=51.$$

**c.** The plane defined by $17x - 12y - 4z = 51$ contains $\ell_1$ and $\ell_2$ because $17(3) - 12(3 - t) - 4(-9 + 3t) = 51$ and $17(-5 + 4t) - 12(-16 + 8t) - 4(14 - 7t) = 51$.

**29.** $\mathbf{r}_1(t) = \langle 0, 4, 5\rangle + t\langle -3, 4, -1\rangle$ and $\mathbf{r}_2(t) = \langle -2, 3, 7\rangle + t\langle -1, 4, 0\rangle$. Hence $\ell_1$ is parallel to the vector $\mathbf{v}_1 = \langle -3, 4, -1\rangle$ and $\ell_2$ is parallel to the vector $\mathbf{v}_2 = \langle -1, 4, 0\rangle$. Since $\mathbf{v}_1$ and $\mathbf{v}_2$ are not scalar multiples of each other, $\ell_1$ and $\ell_2$ are not parallel. Observe that every point on $\ell_2$ has the third coordinate of 7. If a point on $\ell_1$ has the third coordinate of 7, then $5 - t = 7$ or $t = -2$. Since $\mathbf{r}_1(-2) = \langle 6, -4, 7\rangle$, the point (6, –4, 7) is the only point on $\ell_1$ with the third coordinate of 7. If a point on $\ell_2$ has the first coordinate of 6, then $-2 - t = 6$ or $t = -8$. Since $\mathbf{r}_2(-8) = \langle 6, -29, 7\rangle$, the point (6, –29, 7) is the only point on $\ell_1$ with the first coordinate of 6 and third coordinate of 7. Thus $\ell_1$ and $\ell_2$ do not intersect.

**31.** If $\ell$ does not intersect $\Pi$ at exactly one point, it must be parallel to $\Pi$, so it is also perpendicular to a vector normal to the plane. $\mathbf{v} = \langle a_1, b_1, c_1\rangle$ is a vector parallel to $\ell$. $\mathbf{N} = \langle a_2, b_2, c_2\rangle$ is a vector normal to the plane. Hence, if

$$\begin{aligned}\mathbf{v}\cdot\mathbf{N} &= \langle a_1, b_1, c_1\rangle\cdot\langle a_2, b_2, c_2\rangle \\ &= a_1a_2 + b_1b_2 + c_1c_2 \\ &= 0,\end{aligned}$$

$\ell$ is parallel to $\Pi$, so $\ell$ doesn't intersect $\Pi$ at exactly one point. Thus $\ell$ and $\Pi$ intersect at exactly one point if $a_1a_2 + b_1b_2 + c_1c_2 \neq 0$.

**33.** *Method 1*: A vector perpendicular to $\Pi$ is $\mathbf{N} = \langle 1, -1, 1\rangle$. Then an equation for the line $\ell$ through $P$ and perpendicular to $\Pi$ is

$$\begin{aligned}\mathbf{r}(t) &= \langle 0, 1, 4\rangle + t\langle 1, -1, 1\rangle \\ &= \langle t, 1-t, 4+t\rangle.\end{aligned}$$

Substitute $(t, 1 - t, 4 + t)$ into the equation for $\Pi$ and solve for $t$:

$$\begin{aligned}(t) - (1-t) + (4+t) &= 0 \\ 3t + 3 &= 0 \\ t &= -1\end{aligned}$$

$\mathbf{r}(-1) = \langle -1, 2, 3\rangle$, so $Q$ = (–1, 2, 3) is the point of intersection. Thus the perpendicular distance is $PQ = \sqrt{(-1-0)^2 + (2-1)^2 + (3-4)^2} = \sqrt{3}$.

*Method 2*: $PF = \left\|\left\|\overrightarrow{PR}\right\|\cos\theta\right\| = \left|\dfrac{1}{\|\mathbf{N}\|}\overrightarrow{PR}\cdot\mathbf{N}\right|$

Let $R$ = (1, 1, 0). Since $\|\mathbf{N}\| = \sqrt{3}$ and $\overrightarrow{PR} = \langle 1, 0, 4\rangle$, the perpendicular distance is

$$PF = \left|\frac{1}{\sqrt{3}}\langle 1, 0, -4\rangle\cdot\langle 1, -1, 1\rangle\right| = \frac{3}{\sqrt{3}} = \sqrt{3}.$$

**35.** A vector perpendicular to $\Pi$ is $\mathbf{N} = \langle a, b, c\rangle$. Then an equation for the line $\ell$ through $P$ and perpendicular to $\Pi$ is $\mathbf{r}(t) = \langle 0, 0, 0\rangle + t\langle a, b, c\rangle = \langle at, bt, ct\rangle$.

Substitute $(at, bt, ct)$ into the equation for $\Pi$ and solve for $t$:

$$a(at) + b(bt) + c(ct) = d$$
$$(a^2 + b^2 + c^2)t = d$$
$$t = \frac{d}{a^2 + b^2 + c^2}.$$

$$\mathbf{r}\left(\frac{d}{a^2+b^2+c^2}\right) = \left\langle \frac{ad}{a^2+b^2+c^2}, \frac{bd}{a^2+b^2+c^2}, \frac{cd}{a^2+b^2+c^2}\right\rangle,$$ so

$Q = \left(\frac{ad}{a^2+b^2+c^2}, \frac{bd}{a^2+b^2+c^2}, \frac{cd}{a^2+b^2+c^2}\right)$ is the point of intersection. Thus the perpendicular distance is $PQ = \sqrt{\left(\frac{ad}{a^2+b^2+c^2}\right)^2 + \left(\frac{bd}{a^2+b^2+c^2}\right)^2 + \left(\frac{cd}{a^2+b^2+c^2}\right)^2} = \frac{|d|}{\sqrt{a^2+b^2+c^2}}.$

**37.** Let $P$ be a point in $S$. $\langle 0, 3, 4\rangle$ and $\langle 1, 3, 4\rangle$ are two vectors that are not parallel to each other but are parallel to the given plane. Then $\mathbf{P} = a\langle 0, 3, 4\rangle + b\langle 1, 3, 4\rangle$ for some real numbers $a$ and $b$.

$$L(\mathbf{P}) = L\left(a\begin{pmatrix}0\\3\\4\end{pmatrix} + b\begin{pmatrix}1\\3\\4\end{pmatrix}\right)$$
$$= aL\begin{pmatrix}0\\3\\4\end{pmatrix} + bL\begin{pmatrix}1\\3\\4\end{pmatrix}$$
$$= a\begin{pmatrix}4\\-4\\-7\end{pmatrix} + b\begin{pmatrix}4\\-3\\-4\end{pmatrix}$$

Thus any point in $L(S)$ can be written in the form $a\langle 4, -4, 7\rangle + b\langle 4, -3, -4\rangle$. Since $\langle 4, -4, 7\rangle$ and $\langle 4, -3, -4\rangle$ are not parallel to each other, $L(S)$ is the plane determined by these two vectors.

**39.** Points satisfy the equation $L\begin{pmatrix}x\\y\\z\end{pmatrix} = (0)$ if and only if points satisfy the equation $4x + 7z = 0$. Thus the points in $S$ lie on the plane described by the equation $4x + 7z = 0$. Sample answer (other answers are possible): A vector normal to the plane is $\langle 4, 0, 7\rangle$ and a point on this plane is $(0, 0, 0)$.

**41.** Points satisfy the equation $T\begin{pmatrix}x\\y\\z\end{pmatrix} = (\sqrt{3})$ if and only if points satisfy the equation $4y - 3 = \sqrt{3}$. Thus the points in $S$ lie on the plane described by the equation $4y - 3 = \sqrt{3}$. Sample answer (other answers are possible): A vector normal to the plane is $\langle 0, 4, 0\rangle$ and a point on this plane is $\left(0, \frac{3+\sqrt{3}}{4}, 0\right)$.

## Section 8.7 Motion in Three Dimensions

**1.** The path is the line segment with endpoints $\mathbf{r}(0) = \langle 0, 4, -4 \rangle$ and $\mathbf{r}(3) = \langle -6, 13, 2 \rangle$.

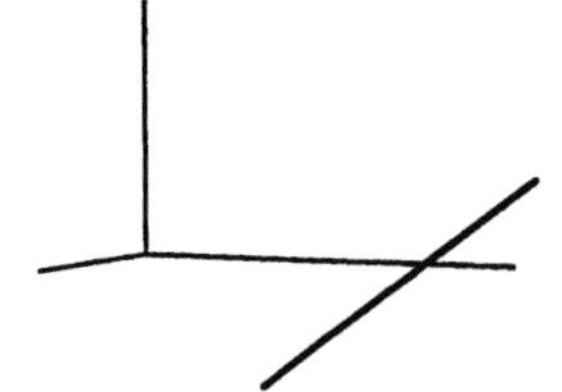

**3.** The path is a helix that when looking down the $z$-axis will look like a circle of radius 1.

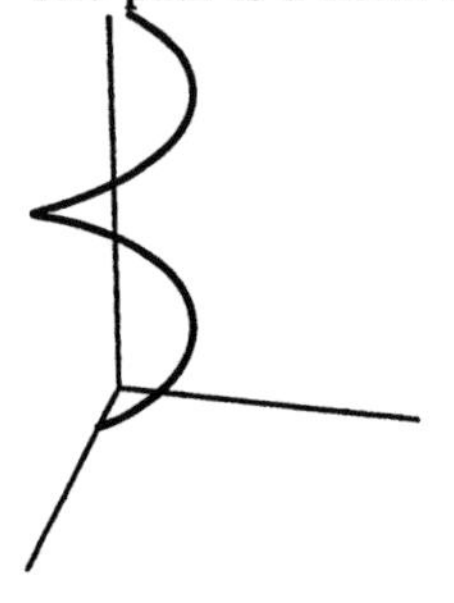

**5.** The path is a helix that when looking down the $z$-axis will look like a circle of radius 2.

**7.** The path is an elliptical helix that when looking down the $y$-axis will look like an ellipse.

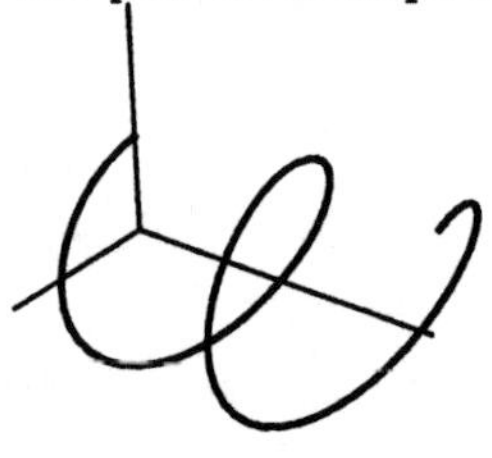

**9.** The path has a spring shape that spirals outward from the $z$-axis.

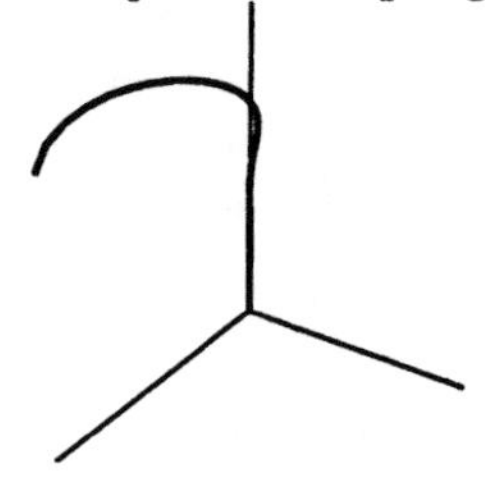

**11.** $\mathbf{r}_{xy}(t) = \langle \sin t, \cos t, 0 \rangle, \ 0 \le t \le 2\pi$

$\mathbf{r}_{xy}(t)$ describes a circle of radius 1 centered at the origin in the $(x, y)$-plane.

$\mathbf{r}_{xz}(t) = \langle \sin t, 0, 4 \rangle, \ 0 \le t \le 2\pi$

$\mathbf{r}_{xz}(t)$ describes a line segment with endpoints $(-1, 4)$ and $(1, 4)$ in the $(x, z)$-plane.

$\mathbf{r}_{yz}(t) = \langle 0, \cos t, 4 \rangle, \ 0 \le t \le 2\pi$

$\mathbf{r}_{yz}(t)$ describes a line segment with endpoints $(-1, 4)$ and $(1, 4)$ in the $(y, z)$-plane.

Thus the curve appears as in the following figure.

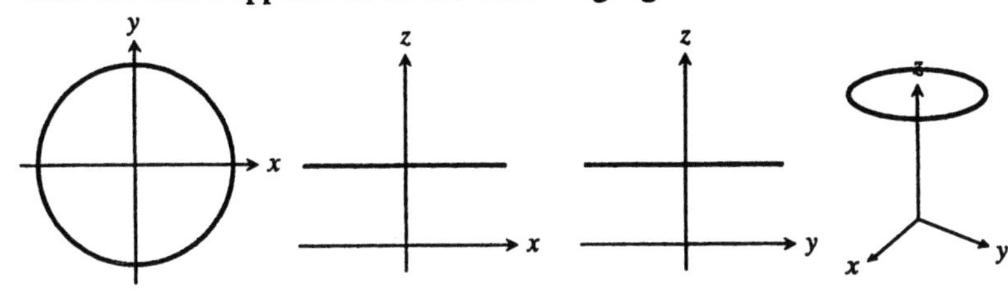


**13.** $\mathbf{r}_{xy}(t) = \left\langle t, e^t, 0 \right\rangle, \ 0 \le t \le 2\pi$

$\mathbf{r}_{xy}(t)$ describes part of the equation $y = e^x$ for $0 \le x \le 2\pi$ in the $(x, y)$-plane.

$\mathbf{r}_{xz}(t) = \left\langle t, 0, t^2 \right\rangle, \ 0 \le t \le 2\pi$

$\mathbf{r}_{xz}(t)$ describes part of the parabola $z = x^2$ for $0 \le x \le 2\pi$ in the $(x, z)$-plane.

$\mathbf{r}_{yz}(t) = \left\langle 0, e^t, t^2 \right\rangle, \ 0 \le t \le 2\pi$

$\mathbf{r}_{yz}(t)$ describes part of the equation $y = e^{\sqrt{z}}$ for $0 \le z \le 4\pi^2$ in the $(y, z)$-plane.

Thus, the curve appears as in the following figure.

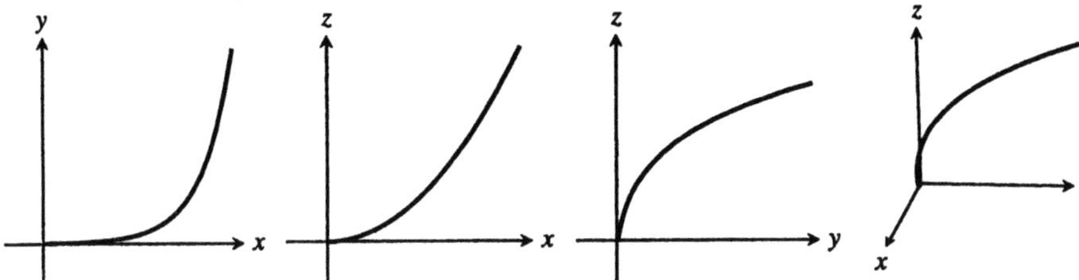


**15.** $\mathbf{v}(t) = \dfrac{d}{dt}\mathbf{r}(t) = \langle -2, 3, 2 \rangle$

$\mathbf{a}(t) = \dfrac{d}{dt}\mathbf{v}(t) = \langle 0, 0, 0 \rangle$

$\mathbf{r}(1) = \langle -2, 7, -2 \rangle$; $\mathbf{v}(1) = \langle -2, 3, 2 \rangle$;

$\mathbf{a}(1) = \langle 0, 0, 0 \rangle$

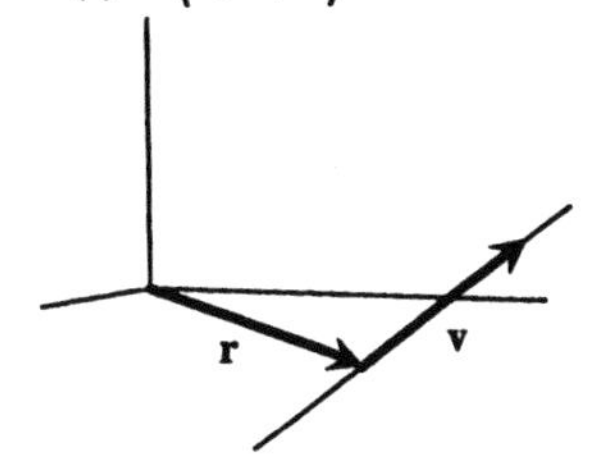

**17.** $\mathbf{v}(t) = \frac{d}{dt}\mathbf{r}(t) = \langle 1, -3\sin t, \cos t\rangle$

$\mathbf{a}(t) = \frac{d}{dt}\mathbf{v}(t) = \langle 0, -3\cos t, -\sin t\rangle$

$\mathbf{r}(1) = \langle 1, 3\cos 1, \sin 1\rangle$

$\approx \langle 1, 1.6209, 0.8145\rangle;$

$\mathbf{v}(1) = \langle 1, -3\sin 1, \cos 1\rangle$

$\approx \langle 1, -2.5244, 0.5403\rangle;$

$\mathbf{a}(1) = \langle 0, -3\cos 1, -\sin 1\rangle$

$\approx \langle 0, -1.6209, -0.8145\rangle$

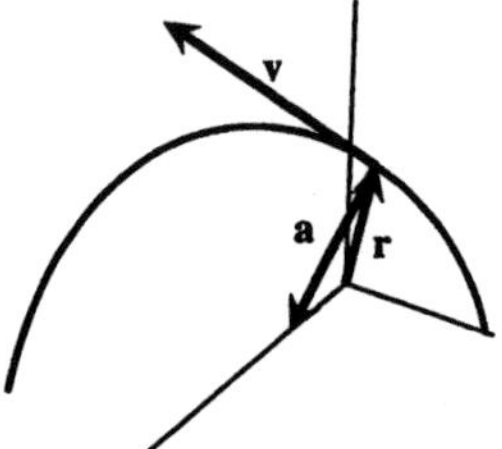


**19.** $\mathbf{v}(t) = \mathbf{r}'(t) = \langle 4t, 2t, 0\rangle$

$\mathbf{a}(t) = \mathbf{v}'(t) = \langle 4, 2, 0\rangle$

$\frac{ds}{dt} = \|\mathbf{v}(t)\| = 2\sqrt{5}t$

$\mathbf{T}(t) = \frac{1}{\|\mathbf{v}(t)\|}\mathbf{v}(t) = \left\langle \frac{2}{\sqrt{5}}, \frac{1}{\sqrt{5}}, 0\right\rangle$

$\frac{d\mathbf{T}}{ds} = \frac{d\mathbf{T}}{dt}\frac{dt}{ds} = \langle 0, 0, 0\rangle \frac{1}{2\sqrt{5}t} = \langle 0, 0, 0\rangle$

$\kappa(t) = \left\|\frac{d\mathbf{T}}{ds}\right\| = 0$

$\mathbf{N}(t) = \frac{1}{\kappa(t)}\frac{d\mathbf{T}}{ds}$ is undefined.

$\mathbf{T}(2) = \left\langle \frac{2}{\sqrt{5}}, \frac{1}{\sqrt{5}}, 0\right\rangle \approx \langle 0.894, 0.447, 0\rangle$

$\mathbf{N}(2)$ is undefined.

$\kappa(2) = 0$

$a_T(2) = \mathbf{a}(2)\cdot\mathbf{T}(2) = 2\sqrt{5} \approx 4.472$

$a_N(2)$ does not exist.

**21.** $\mathbf{v}(t) = \mathbf{r}'(t) = \langle 1, -3\sin t, \cos t\rangle$

$\mathbf{a}(t) = \mathbf{v}'(t) = \langle 0, -3\cos t, -\sin t\rangle$

$\frac{ds}{dt} = \|\mathbf{v}(t)\| = \sqrt{8\sin^2 t + 2}$

$\mathbf{T}(t) = \frac{1}{\|\mathbf{v}(t)\|}\mathbf{v}(t)$

$= \frac{1}{\sqrt{8\sin^2 t + 2}}\langle 1, -3\sin t, \cos t\rangle$

$\kappa(t) = \frac{\|\mathbf{r}'(t)\times\mathbf{r}''(t)\|}{\|\mathbf{r}'(t)\|^3} = \frac{\|\langle 3, \sin t, -3\cos t\rangle\|}{(8\sin^2 t + 2)^{3/2}}$

$= \frac{\sqrt{8\cos^2 t + 10}}{(8\sin^2 t + 2)^{3/2}}$

$\mathbf{T}(\pi) = \frac{1}{\sqrt{2}}\langle 1, 0, -1\rangle = \left\langle \frac{\sqrt{2}}{2}, 0, -\frac{\sqrt{2}}{2}\right\rangle$

$\approx \langle 0.707, 0, -0.707\rangle$

$\kappa(\pi) = \frac{\sqrt{18}}{2\sqrt{2}} = \frac{3}{2}$

$a_T(\pi) = \mathbf{a}(\pi)\cdot\mathbf{T}(\pi)$

$= \langle 0, 3, 0\rangle \cdot \left\langle \frac{\sqrt{2}}{2}, 0, -\frac{\sqrt{2}}{2}\right\rangle$

$= 0$

$a_N(\pi)\mathbf{N}(\pi) = \mathbf{a}(\pi) - a_T(\pi)\mathbf{T}(\pi)$

$= \langle 0, 3, 0\rangle - 0\mathbf{T}(\pi)$

$= \langle 0, 3, 0\rangle$

$a_N(\pi) = \|a_N(\pi)\mathbf{N}(\pi)\| = 3$

$\mathbf{N}(\pi) = \frac{1}{a_N(\pi)}a_N(\pi)\mathbf{N}(\pi) = \langle 0, 1, 0\rangle$

**23.** $\mathbf{v}(t) = \mathbf{r}'(t) = \langle 1, 1, 3t^2\rangle$

$\mathbf{a}(t) = \mathbf{v}'(t) = \langle 0, 0, 6t\rangle$

$\frac{ds}{dt} = \|\mathbf{v}(t)\| = \sqrt{9t^4 + 2}$

$T(t) = \frac{1}{\|\mathbf{v}(t)\|}\mathbf{v}(t) = \frac{1}{\sqrt{9t^4 + 2}}\langle 1, 1, 3t^2\rangle$

$\kappa(t) = \frac{\|\mathbf{r}'(t)\times\mathbf{r}''(t)\|}{\|\mathbf{r}'(t)\|^3} = \frac{\|\langle 6t, -6t, 0\rangle\|}{(9t^4 + 2)^{3/2}}$

$= \frac{6\sqrt{2}t}{(9t^4 + 2)^{3/2}}$

$\mathbf{T}(2) = \frac{1}{\sqrt{146}}\langle 1, 1, 12\rangle$

$= \left\langle \frac{1}{\sqrt{146}}, \frac{1}{\sqrt{146}}, \frac{12}{\sqrt{146}}\right\rangle$

$\approx \langle 0.083, 0.083, 0.993\rangle$

$\kappa(2) = \frac{12\sqrt{2}}{146\sqrt{146}} = \frac{6}{73\sqrt{73}} \approx 0.010$

$a_T(2) = \mathbf{a}(2)\cdot\mathbf{T}(2)$

$= \langle 0, 0, 12\rangle \cdot \left\langle \frac{1}{\sqrt{146}}, \frac{1}{\sqrt{146}}, \frac{12}{\sqrt{146}}\right\rangle$

$= \frac{144}{\sqrt{146}}$

$\approx 11.918$

$a_N(2)\mathbf{N}(2)$
$= \mathbf{a}(2) - a_T(2)\mathbf{T}(2)$
$= \langle 0, 0, 12\rangle - \frac{144}{\sqrt{146}}\left\langle \frac{1}{\sqrt{146}}, \frac{1}{\sqrt{146}}, \frac{12}{\sqrt{146}}\right\rangle$
$= \left\langle -\frac{72}{73}, -\frac{72}{73}, \frac{12}{73}\right\rangle$

$a_N(2) = \|a_N(2)\mathbf{N}(2)\| = \frac{12}{\sqrt{73}} \approx 1.404$

$\mathbf{N}(2) = \frac{1}{a_N(2)} a_N(2)\mathbf{N}(2)$
$= \left(-\frac{6}{\sqrt{73}}, -\frac{6}{\sqrt{73}}, \frac{1}{\sqrt{73}}\right)$
$= (-0.702, -0.702, 0.117)$

**25. a.** The average velocity of an object on the interval $[t_0, t_1]$ is

$$\mathbf{v}[t_0, t_1] = \frac{1}{t_1 - t_0}(\mathbf{r}(t_1) - \mathbf{r}(t_0))$$
$$= \left\langle \frac{x(t_1) - x(t_0)}{t_1 - t_0}, \frac{y(t_1) - y(t_0)}{t_1 - t_0}, \frac{z(t_1) - z(t_0)}{t_1 - t_0}\right\rangle.$$

The average velocity vector is parallel to the vector $\mathbf{r}(t_1) - \mathbf{r}(t_0)$. It gives the average velocity of each component on the interval.

**b.** The velocity, $\mathbf{v}(t_0)$, of an object at time $t_0$ is

$$\mathbf{v}(t_0) = \lim_{t_1 \to t_0} [t_0, t_1] = \lim_{t_1 \to t_0} \frac{1}{t_1 - t_0}(\mathbf{r}(t_1) - \mathbf{r}(t_0))$$
$$= \lim_{t_1 \to t_0} \left\langle \frac{x(t_1) - x(t_0)}{t_1 - t_0}, \frac{y(t_1) - y(t_0)}{t_1 - t_0}, \frac{z(t_1) - z(t_0)}{t_1 - t_0}\right\rangle$$
$$= (x'(t_0), y'(t_0), z'(t_0)).$$

We write this result as $\mathbf{v}(t_0) = \mathbf{r}'(t_0)$.

**27.** Subdivide $[a, b]$ with equally spaced points $a = t_0, t_1, \ldots, t_n = b$ and let $h = \frac{a-b}{n}$ denote the length of each subinterval. Since $x = x(t)$, $y = y(t)$, and $z = z(t)$ are differentiable, the arc length of the segment of $C$ between $\mathbf{r}(t_{i-1})$ and $\mathbf{r}(t_i)$ is approximated by

$$ds_i = \|\mathbf{r}(t_i) - \mathbf{r}(t_{i-1})\| = \|\mathbf{r}(t_{i-1} + h) - \mathbf{r}(t_{i-1})\|$$
$$= \sqrt{(x(t_{i-1} + h) - x(t_{i-1}))^2 + (y(t_{i-1} + h) - y(t_{i-1}))^2 + (z(t_{i-1} + h) - z(t_{i-1}))^2}$$
$$\approx \sqrt{(hx'(t_{i-1}))^2 + (hy'(t_{i-1}))^2 + (hz'(t_{i-1}))^2}$$
$$= \sqrt{(x'(t_{i-1}))^2 + (y'(t_{i-1}))^2 + (z'(t_{i-1}))^2}\, h.$$

We add the lengths $ds_i$ to approximate the arc length $s$ of $C$ as

$$s \approx \sum_{i=1}^{n} ds_i \approx \sum_{i=1}^{n} \sqrt{(x'(t_{i-1}))^2 + (y'(t_{i-1}))^2 + (z'(t_{i-1}))^2}\, h.$$

If we divide the interval more finely and apply the definition of the definite integral, it appears that the arc length $s$ is $s = \int_a^b \sqrt{(x'(t))^2 + (y'(t))^2 + (z'(t))^2}\, dt$.

**29.** $\mathbf{r}'(t) = \langle 2\cos t, -3\sin t, 4\rangle$

$\|\mathbf{r}'(t)\| = \sqrt{4\cos^2 t + 9\sin^2 t + 16}$

$= \sqrt{5\sin^2 t + 20}$

$s(2\pi) = \int_0^{2\pi} \sqrt{5\sin^2 t + 20}\, dt$

Using a CAS, we find that $s(2\pi) \approx 29.7807$.

**31.** Using a graphing calculator, we plot $s(t)$ and 10. After zooming on the intersection, we get $t \approx 3.019$.

**33.** $\mathbf{v}(t) = \mathbf{r}'(t) = \left(\sin t, \cos t, \sqrt{7}\right)$

$\|\mathbf{v}(t)\| = 2\sqrt{2}$

$12 = \int_0^{t_1} 2\sqrt{2}\, dt = 2\sqrt{2}t_1$

Thus $t_1 = \dfrac{12}{2\sqrt{2}} = 3\sqrt{2} \approx 4.243$.

**35.** $\mathbf{v}(t) = \mathbf{r}'(t) = \langle 2\cos t, -3\sin t, 4\rangle$

$\|\mathbf{v}(t)\| = \sqrt{20 + 5\sin^2 t}$

$s(t_1) = \int_0^{t_1} \sqrt{20 + 5\sin^2 t}\, dt$

We want to find $t_1$ such that $s(t_1) = 3\sqrt{3} \approx 5.196$. Apply Newton's method to the equation $s(t_1) - 3\sqrt{3} = 0$ to find $x_{n+1}$ where $x_{n+1} = x_n - \dfrac{s(x_n) - 3\sqrt{3}}{s'(x_n)}$. First, using a CAS, we find that $s(1) \approx 4.620$ and $s(2) \approx 9.578$, so we take our initial guess to be $x_1 = 1$. Subsequent values of $x_n$ are given in the table below.

| $n$ | $x_n$ |
|---|---|
| 1 | 1 |
| 2 | 1.1186 |
| 3 | 1.1180 |
| 4 | 1.1180 |

Thus $t_1 \approx 1.12$.

**37.** $\langle 2, 8, 5\rangle - \langle -2, 4, 3\rangle = \langle 4, 4, 2\rangle$

Let $\mathbf{r}(t) = \langle -2, 4, 3\rangle + t\langle 4, 4, 2\rangle$

$= \langle -2 + 4t, 4 + 4t, 3 + 2t\rangle$.

**39.** The position of the object can be described by $\mathbf{r}(t) = tk\langle a_1, a_2, a_3\rangle$ where $k$ is a positive constant.

$$\mathbf{v}(t) = \mathbf{r}'(t) = k\langle a_1, a_2, a_3\rangle$$

$$\|\mathbf{v}(t)\| = k\sqrt{a_1^2 + a_2^2 + a_3^2} = s$$

Hence $k = \dfrac{s}{\sqrt{a_1^2 + a_2^2 + a_3^2}}$, so the position function is

$$\mathbf{r}(t) = \left\langle \frac{a_1 st}{\sqrt{a_1^2 + a_2^2 + a_3^2}}, \frac{a_2 st}{\sqrt{a_1^2 + a_2^2 + a_3^2}}, \frac{a_3 st}{\sqrt{a_1^2 + a_2^2 + a_3^2}} \right\rangle.$$

**41.** Possible functions are of the form $\mathbf{r}(t) = \langle 5\sin(at), 5\cos(at), kt\rangle$ for nonzero constants $a$ and $k$. Observe that $\mathbf{r}(0) = \langle 0, 5, 0\rangle$.

Sample answers (other answers are possible):

If $a = 2$, then $\mathbf{v}(t) = \mathbf{r}'(t) = \langle 10\cos(2t), -10\sin(2t), k\rangle$, and so $20 = \|\mathbf{v}(t)\| = \sqrt{10^2 + k^2}$. In this case, $k^2 = 300$, so $k = \pm 10\sqrt{3}$. Choosing $k = 10\sqrt{3}$ gives the position function

$$\mathbf{r}(t) = \left\langle 5\sin(2t), 5\cos(2t), 10\sqrt{3}t\right\rangle.$$

If $a = \sqrt{12}$, then

$$\begin{aligned}\mathbf{v}(t) &= \mathbf{r}'(t) \\ &= \left\langle 5\sqrt{12}\cos\left(\sqrt{12}t\right), -5\sqrt{12}\sin\left(\sqrt{12}t\right), k\right\rangle,\end{aligned}$$

and so $20 = \|\mathbf{v}(t)\| = \sqrt{300 + k^2}$ has solutions $k = 10$, $k = -10$. Therefore, a valid position function is

$$\mathbf{r}(t) = \left\langle 5\sin\left(\sqrt{12}t\right), 5\cos\left(\sqrt{12}t\right), 10t\right\rangle.$$

If $a = 1$ or $a = -1$, then $\mathbf{r}(t) = \langle 5\sin(at), 5\cos(at), kt\rangle$ and $\mathbf{v}(t) = \mathbf{r}'(t) = \langle 5a\cos(at), -5a\sin(at), k\rangle$.

Therefore, $20 = \|\mathbf{v}(t)\| = \sqrt{25a^2 + k^2} = \sqrt{25 + k^2}$ would yield solutions $k = \pm 5\sqrt{15}$, so that

$$\mathbf{r}(t) = \left\langle 5\sin t, 5\cos t, 5\sqrt{15}t\right\rangle \text{ and}$$

$$\begin{aligned}\mathbf{r}(t) &= \left\langle 5\sin(-t), 5\cos(-t), 5\sqrt{15}t\right\rangle \\ &= \left\langle -5\sin t, 5\cos t, 5\sqrt{15}t\right\rangle\end{aligned}$$

are two more possible position functions.

**43.** By Newton's second law, $\mathbf{F} = m\mathbf{a}$.

$$\begin{aligned}\mathbf{a}(t) &= \mathbf{r}''(t) \\ &= \left(\left\langle -60\pi\sin\left(\frac{\pi t}{100}\right), 60\pi\cos\left(\frac{\pi t}{100}\right), -5\right\rangle\right)' \\ &= \left\langle -0.6\pi^2\cos\left(\frac{\pi t}{100}\right), -0.6\pi^2\sin\left(\frac{\pi t}{100}\right), 0\right\rangle,\end{aligned}$$

and so

$$\begin{aligned}\mathbf{F}(250) &= 3.6\times 10^5\left\langle -0.6\pi^2\cos(2.5\pi), -0.6\pi^2\sin(2.5\pi), 0\right\rangle \\ &= 3.6\times 10^5\left\langle 0, -0.6\pi^2, 0\right\rangle \\ &= \left\langle 0, -216{,}000\pi^2, 0\right\rangle \text{ newtons.}\end{aligned}$$

**45.** From (20), $\kappa(t)=\dfrac{\|\mathbf{r}'(t)\times\mathbf{r}''(t)\|}{\|\mathbf{r}'(t)\|^3}$. Recall from Section 8.3 that $\|\mathbf{a}\times\mathbf{b}\|^2=\|\mathbf{a}\|^2\|\mathbf{b}\|^2-(\mathbf{a}\cdot\mathbf{b})^2$, so

$\|\mathbf{r}'(t)\times\mathbf{r}''(t)\|^2=\|\mathbf{r}'(t)\|^2\|\mathbf{r}''(t)\|^2-(\mathbf{r}'(t)\cdot\mathbf{r}''(t))^2$

Hence, $\kappa(t)=\dfrac{\left(\|\mathbf{r}'(t)\|^2\|\mathbf{r}''(t)\|^2-(\mathbf{r}'(t)\cdot\mathbf{r}''(t))^2\right)^{1/2}}{\|\mathbf{r}'(t)\|^3}$.

**47.** Suppose $\mathbf{r}(t)$ is not the zero vector. Then by Exercise 46b, $\mathbf{r}(t)=e^{K(t)}\mathbf{b}$ for some constant vector $\mathbf{b}$. If there is some $t_1$ such that $\mathbf{r}(t_1)=0$, then $\mathbf{b}$ must be the zero vector since $e^{K(t)}$ is always greater than zero. Thus $\mathbf{r}(t)$ is the zero vector.

**49.** $(x, y, z)$ lies on the sphere of center $(0, 0, 0)$ with radius 1 if and only if it satisfies the equation $x^2+y^2+z^2=1$. Suppose $(x, y, z)=(\cos^2 t, \sin t\cos t, \sin t)$. Then

$$\begin{aligned}&(\cos^2 t)^2+(\sin t\cos t)^2+(\sin t)^2\\&=\cos^4 t+\sin^2 t\cos^2 t+\sin^2 t\\&=\cos^2 t(\cos^2 t+\sin^2 t)+\sin^2 t\\&=\cos^2 t+\sin^2 t\\&=1\end{aligned}$$

Thus, every point on the curve described by (19) lies on the sphere of center (0, 0, 0) and radius 1.

**51. a.** $\mathbf{v}(t)=\mathbf{r}'(t)=\langle 6\sin 3t, 6\cos 3t, 5\rangle$

$\|\mathbf{v}(t)\|=\sqrt{61}$

$s=\int_0^t \sqrt{61}\,d\tau=\sqrt{61}t$

Thus $t=\dfrac{1}{\sqrt{61}}s$, so $C$ can be described by $\mathbf{r}(s)=\left\langle -2\cos\dfrac{3}{\sqrt{61}}s, 2\sin\dfrac{3}{\sqrt{61}}s, \dfrac{5}{\sqrt{61}}s\right\rangle$.

**b.** $$\begin{aligned}\mathbf{T}(t)&=\frac{1}{\|\mathbf{r}'(t)\|}\mathbf{r}'(t)\\&=\left\langle \frac{6}{\sqrt{61}}\sin 3t, \frac{6}{\sqrt{61}}\cos 3t, \frac{5}{\sqrt{61}}\right\rangle\end{aligned}$$

$$\begin{aligned}\frac{d\mathbf{T}}{ds}&=\frac{d\mathbf{T}}{dt}\cdot\frac{dt}{ds}\\&=\left\langle \frac{18}{\sqrt{61}}\cos 3t, -\frac{18}{\sqrt{61}}\sin 3t, 0\right\rangle\frac{1}{\sqrt{61}}\\&=\left\langle \frac{18}{61}\cos 3t, -\frac{18}{61}\sin 3t, 0\right\rangle\end{aligned}$$

$$\kappa(t)=\left\|\frac{d\mathbf{T}}{ds}\right\|=\frac{18}{61}$$

**53.** If $\mathbf{r}(t) = \langle 4\cos t, -4t, 4\sin t\rangle$, then $\mathbf{r}'(t) = \langle -4\sin t, -4, 4\cos t\rangle$. Therefore, $\|\mathbf{r}'(t)\| = \sqrt{4^2 + 4^2} = 4\sqrt{2}$, and

$$\begin{aligned}\mathbf{T}(t) &= \frac{1}{\|\mathbf{r}'(t)\|}\mathbf{r}'(t)\\ &= \frac{1}{4\sqrt{2}}\langle -4\sin t, -4, 4\cos t\rangle\\ &= \left\langle -\frac{\sqrt{2}}{2}\sin t, -\frac{\sqrt{2}}{2}, \frac{\sqrt{2}}{2}\cos t\right\rangle.\end{aligned}$$

$\dfrac{ds}{dt} = \|\mathbf{r}'(t)\| = 4\sqrt{2}$ and $\dfrac{d\mathbf{T}}{dt} = \left\langle -\dfrac{\sqrt{2}}{2}\cos t, 0, -\dfrac{\sqrt{2}}{2}\sin t\right\rangle$, so

$$\begin{aligned}\frac{d\mathbf{T}}{ds} &= \frac{d\mathbf{T}}{dt}\cdot\frac{dt}{ds}\\ &= \frac{1}{4\sqrt{2}}\left\langle -\frac{\sqrt{2}}{2}\cos t, 0, -\frac{\sqrt{2}}{2}\sin t\right\rangle\\ &= \left\langle -\frac{1}{8}\cos t, 0, -\frac{1}{8}\sin t\right\rangle.\end{aligned}$$

Therefore, $\kappa(t) = \left\|\dfrac{d\mathbf{T}}{ds}\right\| = \dfrac{1}{8}$ and

$$\begin{aligned}\mathbf{N}(t) = \frac{1}{\kappa(t)}\frac{d\mathbf{T}}{ds} &= 8\left\langle -\frac{1}{8}\cos t, 0, -\frac{1}{8}\sin t\right\rangle\\ &= \langle -\cos t, 0, -\sin t\rangle.\end{aligned}$$

Therefore,

$$\begin{aligned}\mathbf{B} &= \mathbf{T}\times\mathbf{N}\\ &= \left\langle -\frac{\sqrt{2}}{2}\sin t, -\frac{\sqrt{2}}{2}, \frac{\sqrt{2}}{2}\cos t\right\rangle \times \langle -\cos t, 0, -\sin t\rangle\\ &= \left\langle \frac{\sqrt{2}}{2}\sin t, -\frac{\sqrt{2}}{2}\cos^2 t - \frac{\sqrt{2}}{2}\sin^2 t, -\frac{\sqrt{2}}{2}\cos t\right\rangle\\ &= \left\langle \frac{\sqrt{2}}{2}\sin t, -\frac{\sqrt{2}}{2}, -\frac{\sqrt{2}}{2}\cos t\right\rangle,\end{aligned}$$

so, $\mathbf{B}\left(\dfrac{\pi}{2}\right) = \left\langle \dfrac{\sqrt{2}}{2}, -\dfrac{\sqrt{2}}{2}, 0\right\rangle$.

**55.** If $\mathbf{r}(t) = \langle e^t, e^t\sin t, e^t\cos t\rangle$, then $\mathbf{r}'(t) = \langle e^t, e^t(\sin t + \cos t), e^t(\cos t - \sin t)\rangle$. Therefore,

$$\begin{aligned}\|\mathbf{r}'(t)\| &= \sqrt{e^{2t} + e^{2t}(\sin^2 t + 2\sin t\cos t + \cos^2 t) + e^{2t}(\cos^2 t - 2\sin t\cos t + \cos^2 t)}\\ &= \sqrt{e^{2t}3}\\ &= e^t\sqrt{3},\end{aligned}$$

and so

$$\begin{aligned}\mathbf{T}(t) = \frac{1}{\|\mathbf{r}'(t)\|}\mathbf{r}'(t) &= \frac{1}{\sqrt{3}e^t}\langle e^t, e^t(\sin t + \cos t), e^t(\cos t - \sin t)\rangle\\ &= \left\langle \frac{1}{\sqrt{3}}, \frac{\sin t + \cos t}{\sqrt{3}}, \frac{\cos t - \sin t}{\sqrt{3}}\right\rangle.\end{aligned}$$

$\dfrac{d\mathbf{T}}{dt} = \left\langle 0, \dfrac{\cos t - \sin t}{\sqrt{3}}, \dfrac{-\sin t - \cos t}{\sqrt{3}}\right\rangle$, so since $\dfrac{ds}{dt} = \|\mathbf{r}'(t)\| = e^t\sqrt{3}$, then since $\dfrac{d\mathbf{T}}{ds} = \dfrac{d\mathbf{T}}{dt}\dfrac{dt}{ds}$, then

$$\frac{d\mathbf{T}}{ds} = \left\langle 0, \frac{\cos t - \sin t}{\sqrt{3}}, \frac{-\sin t - \cos t}{\sqrt{3}} \right\rangle \cdot \frac{1}{e^t\sqrt{3}}$$
$$= \left\langle 0, \frac{\cos t - \sin t}{3e^t}, \frac{-\sin t - \cos t}{3e^t} \right\rangle.$$

Therefore, $\kappa(t) = \left\| \frac{d\mathbf{T}}{ds} \right\| = \sqrt{\frac{(\cos t - \sin t)^2}{9e^{2t}} + \frac{(\sin t + \cos t)^2}{9e^{2t}}} = \frac{\sqrt{2}}{3e^t}$ and

$$\mathbf{N}(t) = \frac{1}{\kappa(t)} \frac{d\mathbf{T}}{ds} = \frac{3e^t}{\sqrt{2}} \left\langle 0, \frac{\cos t - \sin t}{3e^t}, \frac{-\sin t - \cos t}{3e^t} \right\rangle$$
$$= \left\langle 0, \frac{\cos t - \sin t}{\sqrt{2}}, \frac{-\sin t - \cos t}{\sqrt{2}} \right\rangle.$$

Therefore,

$$\mathbf{B} = \mathbf{T} \times \mathbf{N} = \left\langle \frac{1}{\sqrt{3}}, \frac{\sin t + \cos t}{\sqrt{3}}, \frac{\cos t - \sin t}{\sqrt{3}} \right\rangle \times \left\langle 0, \frac{\cos t - \sin t}{\sqrt{2}}, \frac{-\sin t - \cos t}{\sqrt{2}} \right\rangle$$
$$= \left\langle \frac{-(\sin t + \cos t)^2}{\sqrt{6}} - \frac{(\cos t - \sin t)^2}{\sqrt{6}}, \frac{\sin t + \cos t}{\sqrt{6}}, \frac{\cos t - \sin t}{\sqrt{6}} \right\rangle$$
$$= \left\langle -\frac{2}{\sqrt{6}}, \frac{\sin t + \cos t}{\sqrt{6}}, \frac{\cos t - \sin t}{\sqrt{6}} \right\rangle,$$

so $\mathbf{B}(0) = \left\langle -\frac{\sqrt{6}}{3}, \frac{1}{\sqrt{6}}, \frac{1}{\sqrt{6}} \right\rangle.$

**57.** $\mathbf{r}'(t) = \langle 2t, 3, 3t^2 \rangle$

$\mathbf{r}(1) = \langle 1, 3, 0 \rangle$

$\mathbf{r}'(1) = \langle 2, 3, 3 \rangle$

An equation for the line is

$\mathbf{w}(t) = \langle 1, 3, 0 \rangle + t\langle 2, 3, 3 \rangle$

$= \langle 1 + 2t, 3 + 3t, 3t \rangle.$

**59.** $\mathbf{r}'(t) = \langle e^t + te^t, \sin t + t\cos t, \cos t - t\sin t \rangle$

$\mathbf{r}(0) = \langle 0, 0, 0 \rangle$

$\mathbf{r}'(0) = \langle 1, 0, 1 \rangle$

An equation for the line is

$\mathbf{w}(t) = \langle 0, 0, 0 \rangle + t\langle 1, 0, 1 \rangle = \langle t, 0, t \rangle.$

**61.** **a.** $\mathbf{r}'(t) = \langle e^t + te^t, 2t + 1, 4\sin t \rangle$

$\mathbf{r}(0) = \langle 0, 0, -4 \rangle$

$\mathbf{r}'(0) = \langle 1, 1, 0 \rangle$

An equation for the line is

$\mathbf{l}(t) = \langle 0, 0, -4 \rangle + t\langle 1, 1, 0 \rangle = \langle t, t, -4 \rangle.$

**b.** $\mathbf{r}(0.1) \approx \mathbf{l}(0.1) = \langle 0.1, 0.1, -4 \rangle$

**c.** The error can be described by the vector $\mathbf{l}(0.1) - \mathbf{r}(0.1)$ or perhaps its magnitude.

**63.** If we associate the two dimensional vector $\mathbf{r}(t) = \langle x(t), y(t)\rangle$ with the one in three dimensions $\mathbf{r}(t) = \langle x(t), y(t), 0\rangle$, we can use the formula for curvature

$$\kappa(t) = \frac{\|\mathbf{r}'(t)\times\mathbf{r}''(t)\|}{\|\mathbf{r}'(t)\|^3}.$$

$$\mathbf{r}'(t)\times\mathbf{r}''(t) = \langle x'(t), y'(t), 0\rangle\times\langle x''(t), y''(t), 0\rangle$$
$$= \langle 0, 0, x'(t)y''(t) - y'(t)x''(t)\rangle,$$

so

$$\|\mathbf{r}'(t)\times\mathbf{r}''(t)\|$$
$$= \sqrt{0^2 + 0^2 + (x'(t)y''(t) - y'(t)''(t))^2}$$
$$= |x'(t)y''(t) - y'(t)x''(t)|$$
$$= |x''(t)y'(t) - x'(t)y''(t)|.$$

Since

$$\|\mathbf{r}'(t)\| = \sqrt{x'(t)^2 + y'(t)^2 + 0^2}$$
$$= \sqrt{x'(t)^2 + y'(t)^2},$$

then $\|\mathbf{r}'(t)\|^3 = (x'(t)^2 + y'(t)^2)^{3/2}$. Therefore,

$$\kappa(t) = \frac{\|\mathbf{r}'(t)\times\mathbf{r}''(t)\|}{\|\mathbf{r}'(t)\|^3} = \frac{|x''(t)y'(t) - x'(t)y''(t)|}{(x'(t)^2 + y'(t)^2)^{3/2}}.$$

## Chapter 8 Review Exercises

**1.** $2\mathbf{a} - 5\mathbf{b} = 2\langle -1, 2, 3\rangle - 5\langle 3, 0, 7\rangle$
$$= \langle -2 - 15, 4 - 0, 6 - 35\rangle$$
$$= \langle -17, 4, -29\rangle$$

**3.** $\mathbf{a}\times\mathbf{c} = \det\begin{pmatrix}\mathbf{i} & \mathbf{j} & \mathbf{k}\\ -1 & 2 & 3\\ 3 & 10 & -7\end{pmatrix}$
$$= \mathbf{i}(-14 - 30) - \mathbf{j}(7 - 9) + \mathbf{k}(-10 - 6)$$
$$= -44\mathbf{i} + 2\mathbf{j} - 16\mathbf{k}$$
$$= \langle -44, 2, -16\rangle$$

**5.** $\mathbf{c}\times\mathbf{d} = \det\begin{pmatrix}\mathbf{i} & \mathbf{j} & \mathbf{k}\\ 3 & 10 & -7\\ 3 & -6 & 9\end{pmatrix}$
$$= \mathbf{i}(90 - 42) - \mathbf{j}(27 + 21) + \mathbf{k}(-18 - 30)$$
$$= 48\mathbf{i} - 48\mathbf{j} - 48\mathbf{k}$$

$$\text{area} = \|\mathbf{c}\times\mathbf{d}\| = \|\langle 48, -48, -48\rangle\|$$
$$= 48\|\langle 1, -1, -1\rangle\|$$
$$= 48\sqrt{1^2 + 1^2 + 1^2}$$
$$= 48\sqrt{3}$$

**7.** $\mathbf{c}\times\mathbf{d} = \det\begin{pmatrix}\mathbf{i} & \mathbf{j} & \mathbf{k}\\ 3 & 10 & -7\\ 3 & -6 & 9\end{pmatrix}$
$$= 48\mathbf{i} - 48\mathbf{j} - 48\mathbf{k}$$

$$\text{Volume} = |\mathbf{a}\cdot(\mathbf{c}\times\mathbf{d})|$$
$$= |\langle -1, 2, 3\rangle\cdot\langle 48, -48, -48\rangle|$$
$$= |-48 - 96 - 144|$$
$$= 288$$

**9.** Since $\|\mathbf{a}\| = \sqrt{1^2 + 2^2 + 3^2} = \sqrt{14}$, a unit vector in the direction of $\mathbf{a}$ is

$$\mathbf{u} = \frac{1}{\|\mathbf{a}\|}\mathbf{a} = \frac{1}{\sqrt{14}}\langle -1, 2, 3\rangle$$
$$= \left\langle -\frac{1}{\sqrt{14}}, \frac{2}{\sqrt{14}}, \frac{3}{\sqrt{14}}\right\rangle.$$

The projection of $\mathbf{c}$ onto $\mathbf{u}$ is given by

$$(\mathbf{c}\cdot\mathbf{u})\mathbf{u} = \left(\langle 3, 10, -7\rangle\cdot\left\langle -\frac{1}{\sqrt{14}}, \frac{2}{\sqrt{14}}, \frac{3}{\sqrt{14}}\right\rangle\right)\mathbf{u}$$
$$= \left(\frac{-3 + 20 - 21}{\sqrt{14}}\right)\mathbf{u}$$
$$= -\frac{4}{\sqrt{14}}\left\langle -\frac{1}{\sqrt{14}}, \frac{2}{\sqrt{14}}, \frac{3}{\sqrt{14}}\right\rangle$$
$$= \left\langle \frac{4}{14}, -\frac{8}{14}, -\frac{12}{14}\right\rangle$$
$$= \left\langle \frac{2}{7}, -\frac{4}{7}, -\frac{6}{7}\right\rangle.$$

**11.** $\mathbf{a}\times\mathbf{b} = \det\begin{pmatrix}\mathbf{i} & \mathbf{j} & \mathbf{k}\\ -1 & 2 & 3\\ 3 & 0 & 7\end{pmatrix} = 14\mathbf{i} + 16\mathbf{j} - 6\mathbf{k}$, and

$\mathbf{c}\times\mathbf{d} = \det\begin{pmatrix}\mathbf{i} & \mathbf{j} & \mathbf{k}\\ 3 & 10 & -7\\ 3 & -6 & 9\end{pmatrix} = 48\mathbf{i} - 48\mathbf{j} - 48\mathbf{k}$, so

$$(\mathbf{a}\times\mathbf{b})\cdot(\mathbf{c}\times\mathbf{d}) = \langle 14, 16, -6\rangle\cdot\langle 48, -48, -48\rangle$$
$$= 192.$$

**13.** $-2A = -2\begin{pmatrix}-4 & 7 & 3\\ 0 & 2 & -6\end{pmatrix} = \begin{pmatrix}8 & -14 & -6\\ 0 & -4 & 12\end{pmatrix}$

**15.** $$-B+5E=-\begin{pmatrix}7 & -3\\ 10 & -4\end{pmatrix}+5\begin{pmatrix}4 & 2\\ -3 & 7\end{pmatrix}$$
$$=\begin{pmatrix}-7 & 3\\ -10 & 4\end{pmatrix}+\begin{pmatrix}20 & 10\\ -15 & 35\end{pmatrix}$$
$$=\begin{pmatrix}-7+20 & 3+10\\ -10-15 & 4+35\end{pmatrix}$$
$$=\begin{pmatrix}13 & 13\\ -25 & 39\end{pmatrix}$$

**17.** $$\det B=\det\begin{pmatrix}7 & -3\\ 10 & -4\end{pmatrix}$$
$$=7(-4)-10(-3)$$
$$=-28+30$$
$$=2$$

**19.** Since the number of columns of $A$ does not equal the number of rows of $B$, $AB$ is undefined.

**21.** $$AC=\begin{pmatrix}-4 & 7 & 3\\ 0 & 2 & -6\end{pmatrix}\cdot\begin{pmatrix}3 & -4\\ 4 & 0\\ 5 & 11\end{pmatrix}$$
$$=\begin{pmatrix}(-4)3+7\cdot 4+3\cdot 5 & (-4)(-4)+7\cdot 0+3\cdot 11\\ 0\cdot 3+2\cdot 4+(-6)5 & 0(-4)+2\cdot 0+(-6)11\end{pmatrix}$$
$$=\begin{pmatrix}-12+28+15 & 16+0+33\\ 0+8-30 & 0+0-66\end{pmatrix}$$
$$=\begin{pmatrix}31 & 49\\ -22 & -66\end{pmatrix}$$

**23.** Since the dimension of $B$ is $2 \times 2$ and the dimension of $C$ is $3 \times 2$, the number of columns of $B$ is unequal to the number of rows of $C$, so $BC$ is undefined. Therefore, $\det(BC)$ is undefined.

**25.** **a.** $\vec{AB}=\langle 1, 5, 9\rangle-\langle 2, -3, 6\rangle=\langle -1, 8, 3\rangle$ and $\vec{AC}=\langle -2, 0, 6\rangle-\langle 2, -3, 6\rangle=\langle -4, 3, 0\rangle$ are not scalar multiples of each other, so these vectors are not parallel. Therefore, the points $A$, $B$, and $C$ are not collinear and hence form a triangle.

**b.** The area of the triangle with vertices $A$, $B$, and $C$ is one-half the area of the parallelogram formed by the vectors $\vec{AB}$ and $\vec{AC}$. That is,

$$\text{area}=\frac{1}{2}\|\langle -1, 8, 3\rangle\times\langle -4, 3, 0\rangle\|$$
$$=\frac{1}{2}\|\langle -9, -12, 29\rangle\|$$
$$=\frac{1}{2}\sqrt{1066}$$
$$\approx 16.3248.$$

**27.** $L(S)$ is the parallelogram with vertices $L\begin{pmatrix}2\\1\end{pmatrix}=\begin{pmatrix}4\\5\end{pmatrix}$, $L\begin{pmatrix}1\\4\end{pmatrix}=\begin{pmatrix}30\\13\end{pmatrix}$, $L\begin{pmatrix}4\\8\end{pmatrix}=\begin{pmatrix}56\\28\end{pmatrix}$, and $L\begin{pmatrix}3\\11\end{pmatrix}=\begin{pmatrix}82\\36\end{pmatrix}$. Since $S$ is determined by the vectors $\begin{pmatrix}4\\8\end{pmatrix}-\begin{pmatrix}2\\1\end{pmatrix}=\begin{pmatrix}2\\7\end{pmatrix}$ and $\begin{pmatrix}1\\4\end{pmatrix}-\begin{pmatrix}2\\1\end{pmatrix}=\begin{pmatrix}-1\\3\end{pmatrix}$, $L(S)$ is determined by the vectors $L\begin{pmatrix}4\\8\end{pmatrix}-L\begin{pmatrix}2\\1\end{pmatrix}=L\begin{pmatrix}2\\7\end{pmatrix}=\begin{pmatrix}52\\23\end{pmatrix}$ and $L\begin{pmatrix}-1\\3\end{pmatrix}=\begin{pmatrix}26\\8\end{pmatrix}$. Therefore, the area of parallelogram $L(S)$ is
$\left|\det\begin{pmatrix}52 & 23\\26 & 8\end{pmatrix}\right|=|-182|=182.$

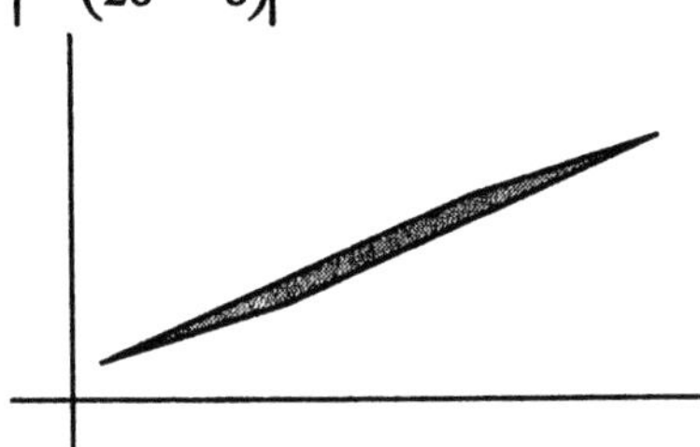

**29.** The two planes in question both have a point on the line (0, 14, 6) in common, since $2(0) - 14 + 3(6) = 4$ and $0 + 14 - 6 = 8$. Because one plane has normal vector $\langle 2, -1, 3\rangle$ and the other $\langle 1, 1, -1\rangle$, and these vectors are not parallel (because one is not a scalar multiple of the other), the two planes are distinct. The line in question is in the direction of the vector $\langle 2, -5, -3\rangle$, so since $\langle 2, -5, -3\rangle\cdot\langle 2, -1, 3\rangle = 4+5-9=0$ and $\langle 2, -5, -3\rangle\cdot\langle 1, 1, -1\rangle = 2-5+3=0$, the line is perpendicular to the two normal vectors. Therefore, the two planes each contain the line. So, since these two distinct planes both contain the line, the two planes intersect at the line.

**31.** Sample answer (other answers are possible): Find two equations of the form $a(x + 4) + by + c(z - 3) = 0$ so that $\langle a, b, c\rangle$ is normal to $\langle 1, 2, 3\rangle$ and so these two normal vectors are not parallel. Use cross products such as $\langle 6, 4, 3\rangle\times\langle 1, 2, 3\rangle=\langle 6, -15, 8\rangle$ and $\langle 6, 3, 1\rangle\times\langle 1, 2, 3\rangle=\langle 7, -17, 9\rangle$. Note that $\langle 6, -15, 8\rangle$ and $\langle 7, -17, 9\rangle$ are not scalar multiples of each other and are hence not parallel. Since they are both perpendicular to $\langle 1, 2, 3\rangle$, the equations $6(x + 4) - 15y + 8(z - 3) = 0$ and $7(x + 4) - 17y + 9(z - 3) = 0$ define planes that intersect at (–4, 0, 3), and since they are in the direction of the line, they contain the line. Since the planes are distinct, they only intersect at the line. The equations are equivalent to $6x - 15y + 8z = 0$ and $7x - 17y + 9z + 1 = 0$.

**33.** The plane is in the direction of the vectors $\mathbf{u}=\langle 1, 9, 2\rangle-\langle 0, 0, 0\rangle=\langle 1, 9, 2\rangle$ and $\mathbf{v}=\langle 8, 11, 4\rangle-\langle 0, 0, 0\rangle=\langle 8, 11, 4\rangle$, so $\mathbf{u}\times\mathbf{v}=\langle 1, 9, 2\rangle\times\langle 8, 11, 4\rangle=\langle 14, 12, -61\rangle$ is normal to the plane. Therefore, $14(x - 0) + 12(y - 0) - 61(z - 0) = 0$, or $14x + 12y - 61z = 0$ is an equation for the plane.

**35.**
$$\begin{aligned}L(S)&=\begin{pmatrix}-2 & 1 & 0\\2 & 1 & 1\\3 & 2 & -1\end{pmatrix}\begin{pmatrix}5+4t\\5t\\-2+2t\end{pmatrix}\\&=\begin{pmatrix}-10-8t+5t\\10+8t+5t-2+2t\\15+12t+10t+2-2t\end{pmatrix}\\&=\begin{pmatrix}-10-3t\\8+15t\\17+20t\end{pmatrix}\\&=\begin{pmatrix}-10\\8\\17\end{pmatrix}+t\begin{pmatrix}-3\\15\\20\end{pmatrix},\end{aligned}$$
so $L(S)$ is a line through (–10, 8, 17) in the direction of the vector $\langle -3, 15, 20\rangle$ with equation $\mathbf{r}(t)=\langle -10, 8, 17\rangle+t\langle -3, 15, 20\rangle$, or $\mathbf{r}(t)=\langle -3t-10, 15t+8, 20t+17\rangle$.

**37.** $A=\begin{pmatrix}2 & -1\\ 0 & -3\end{pmatrix}$ is the matrix associated with $L$ and $B=\begin{pmatrix}4 & 12\\ -2 & 5\end{pmatrix}$ is the matrix associated with $M$.

**a.** $(M \circ L)(S)=(BA)(S)$. Since $S$ is the set of points on and inside the square with vertices (0, 0), (1, 0), (1, 1), and (0, 1), the area of $(M \circ L)(S)$ equals $1^2|\det(BA)|$.

Since $BA=\begin{pmatrix}8 & -40\\ -4 & -13\end{pmatrix}$,

$\det(BA) = 8(-13) - (-4)(-40) = -264$.
Therefore, the area of
$(M \circ L)(S)=|-264|=264$.

**b.** $(L \circ M)(S)=(AB)(S)$. For reasons explained in part **a**, the area of
$(L \circ M)(S)=1^2|\det(AB)|$.
By properties of determinants and the commutativity of the multiplication of real numbers,

$$\begin{aligned}\det(AB) &= \det(A)\det(B)\\ &= \det(B)\det(A)\\ &= \det(BA).\end{aligned}$$

Therefore, from part **a**,
$\det(AB) = \det(BA) = -264$, so the area of
$(L \circ M)(S)=|-264|=264$.

**c.** $(M \circ M)(S)=(B^2)(S)$. For reasons cited in part **a**, the area of
$(M \circ M)(S)=1^2|\det(B^2)|$.

Since $B^2=\begin{pmatrix}-8 & 108\\ -18 & 1\end{pmatrix}$, then

$\det(B^2)=(-8)1-(-18)108=1936$.
Therefore, the area of
$(M \circ M)(S)=|1936|=1936$.

**d.** $(L \circ M \circ L)(S)=(ABA)(S)$. For reasons cited in part **a**, the area of
$(L \circ M \circ L)(S)=1^2|\det(ABA)|$. Since

$ABA=\begin{pmatrix}20 & -67\\ 12 & 39\end{pmatrix}$,

$\det(ABA) = 20 \cdot 39 - 12(-67) = 1584$.
Therefore, the area of
$(L \circ M \circ L)(S)=|1584|=1584$.

**39.** Since

$M\begin{pmatrix}x\\ y\end{pmatrix}=\det\begin{pmatrix}x & y\\ -2 & 4\end{pmatrix}=4x+2y=(4 \quad 2)\begin{pmatrix}x\\ y\end{pmatrix}$,

$M$ is a linear function with associated matrix $(4 \quad 2)$. The domain of $M$ is $R^2$ while the range of $M$ is $R$.

**41.** We know by how $L$ is defined that $L$ is a linear function with domain $R^2$ and range contained in $R^2$. To show that the range of $L$ equals $R^2$, let $\begin{pmatrix}a\\ b\end{pmatrix}$ represent any vector in $R^2$. We then need to find a vector $\begin{pmatrix}x\\ y\end{pmatrix}$ such that $L\begin{pmatrix}x\\ y\end{pmatrix}=\begin{pmatrix}a\\ b\end{pmatrix}$.

$L\begin{pmatrix}x\\ y\end{pmatrix}=\begin{pmatrix}3x-2y\\ x-4y\end{pmatrix}$, so to find $\begin{pmatrix}x\\ y\end{pmatrix}$, we need to solve the system

$$\begin{aligned}3x-2y&=a\\ x-4y&=b.\end{aligned}$$

Substitute $x = 4y + b$ in for $x$ in $3x - 2y = a$ and solve for $y$. This yields $y=\dfrac{a-3b}{10}$, so

$x=4\left(\dfrac{a-3b}{10}\right)+b=\dfrac{2a}{5}-\dfrac{b}{5}$. It can now be verified that if $x=\dfrac{2a}{5}-\dfrac{b}{5}$ and $y=\dfrac{a-3b}{10}$, then $L\begin{pmatrix}x\\ y\end{pmatrix}=\begin{pmatrix}a\\ b\end{pmatrix}$. Therefore, the range of $L$ is all of $R^2$. And since $\begin{pmatrix}x\\ y\end{pmatrix}$ is unique, given $\begin{pmatrix}a\\ b\end{pmatrix}$, the function is one-to-one.

**43.** $L\begin{pmatrix}x\\ y\end{pmatrix}=\begin{pmatrix}3x-2y\\ 3x-4y\end{pmatrix}$, so if $L(\mathbf{v}) = a\mathbf{v}$, where $a$ is a real number and $\mathbf{v}=\begin{pmatrix}x\\ y\end{pmatrix}\neq\begin{pmatrix}0\\ 0\end{pmatrix}$, then since

$a\mathbf{v}=a\begin{pmatrix}x\\ y\end{pmatrix}=\begin{pmatrix}ax\\ ay\end{pmatrix}$, this leads to the system of equations

$$\begin{aligned}3x-2y&=ax\\ 3x-4y&=ay.\end{aligned}$$

Collecting like terms leads to
$(3-a)x-2y=0$
$3x-(4+a)y=0$.
If we solve the second equation for $x$ and then substitute this for $x$ in the second

equation, we get $x = \frac{(4+a)y}{3}$ and $(3-a)\left(\frac{(4+a)y}{3}\right) - 2y = 0$. If we factor out $y$ on the left hand side of this last equation, we get $y\left(\frac{(3-a)(4+a)}{3} - 2\right) = 0$. Notice that the restriction $\begin{pmatrix} x \\ y \end{pmatrix} \neq \begin{pmatrix} 0 \\ 0 \end{pmatrix}$ prevents the possibility $y = 0$ because otherwise $x = \frac{(4+a)y}{3}$ would equal 0. Therefore, $\frac{(3-a)(4+a)}{3} - 2 = 0$, which leads to the quadratic equation $a^2 + a - 6 = 0$, which has solution $a = 2$ or $a = -3$.

**45. a.** As an aid, consider the graphs of $\mathbf{r}_{xy}(t) = \langle t\sin t,\ t\cos t,\ 0\rangle$, $\mathbf{r}_{yz}(t) = \langle 0,\ t\cos t,\ t\rangle$, and $\mathbf{r}_{xz}(t) = \langle t\sin t,\ 0,\ t\rangle$ to visualize the graph in three dimensions of a spiral of increasing radius winding around the $z$-axis.

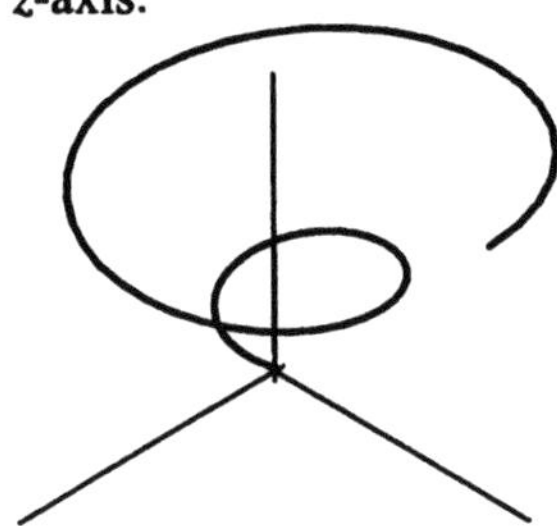

**b.** $\mathbf{r}'(t) = \langle \sin t + t\cos t,\ \cos t - t\sin t,\ 1\rangle$, so

$$\|\mathbf{r}'(t)\| = \sqrt{(\sin t + t\cos t)^2 + (\cos t - t\sin t)^2 + 1},$$

which simplifies to $\|\mathbf{r}'(t)\| = \sqrt{2+t^2}$. Therefore, the distance traveled by the object from time $t = 0$ and $t = 2$ is given by the definite integral $s = \int_0^2 \|\mathbf{r}'(t)\|\,dt = \int_0^2 \sqrt{2+t^2}\,dt$. With the help of an integral table,

$$\begin{aligned}
s &= \left(\frac{t}{2}\sqrt{t^2+2} + \ln\left|t + \sqrt{t^2+2}\right|\right)\Big|_0^2 \\
&= \sqrt{6} + \ln\left(2+\sqrt{6}\right) - \ln\sqrt{2} \\
&= \sqrt{6} + \ln\left(\frac{2+\sqrt{6}}{\sqrt{2}}\right) \\
&= \sqrt{6} + \ln\left(\sqrt{2}+\sqrt{3}\right) \\
&\approx 3.59571.
\end{aligned}$$

**c.** 
$$\begin{aligned}
\mathbf{T}(t) &= \frac{1}{\|\mathbf{r}'(t)\|}\mathbf{r}'(t) \\
&= \frac{1}{\sqrt{t^2+2}}\langle \sin t + t\cos t,\ \cos t - t\sin t,\ 1\rangle,
\end{aligned}$$

so $\mathbf{T}(0) = \frac{1}{\sqrt{2}}\langle 0, 1, 1\rangle = \left\langle 0, \frac{1}{\sqrt{2}}, \frac{1}{\sqrt{2}}\right\rangle$.

Acceleration is

$$\begin{aligned}
\mathbf{a}(t) &= \mathbf{r}(t) \\
&= \langle 2\cos t - t\sin t,\ -2\sin t - t\cos t,\ 0\rangle,
\end{aligned}$$

so the tangent component of acceleration at time $t = 0$ is

$$\begin{aligned}
a_T(0) &= \mathbf{a}(0)\cdot\mathbf{T}(0) \\
&= \langle 2, 0, 0\rangle\cdot\left\langle 0, \frac{1}{\sqrt{2}}, \frac{1}{\sqrt{2}}\right\rangle \\
&= 0.
\end{aligned}$$

Therefore, if $\mathbf{N}(t)$ is the unit normal vector at time $t$ and $a_N(t)$ is the normal component of acceleration at time $t$, then

$$\begin{aligned}
a_N(0)\mathbf{N}(0) &= \mathbf{a}(0) - a_T(0)\mathbf{T}(0) \\
&= \langle 2, 0, 0\rangle - 0\cdot\mathbf{T}(0) \\
&= \langle 2, 0, 0\rangle;
\end{aligned}$$

$a_N(0) = \|a_N(0)\mathbf{N}(0)\| = 2$; and

$$\begin{aligned}
\mathbf{N}(0) &= \frac{1}{a_N(0)}a_N(0)\mathbf{N}(0) \\
&= \frac{1}{2}\langle 2, 0, 0\rangle \\
&= \langle 1, 0, 0\rangle.
\end{aligned}$$

**d.** As discovered in part **c**, $a_T(0) = 0$ and $a_N(0) = 2$.

# Chapter 9 Functions of Several Variables

## Section 9.1 Conic Sections

**1.** This is of the form $4py = x^2$ with $p = \frac{1}{4}$ so vertex is (0, 0), focus is $\left(0, \frac{1}{4}\right)$ and directrix is $y = -\frac{1}{4}$.

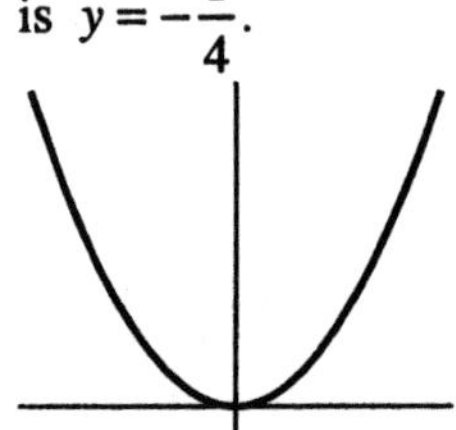

**3.** $y + 2 = 3(x+1)^2$ can be rewritten as $\frac{1}{3}(y - (-2)) = (x - (-1))^2$ with a vertex of (–1, –2), focus of $\left(-1, -\frac{23}{12}\right)$ and directrix of $y = -\frac{25}{12}$.

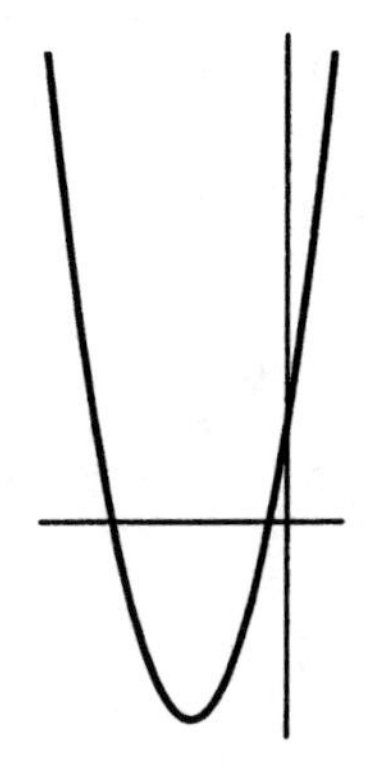

**5.** We rewrite $y = x^2 + 2x$ as $(y - (-1)) = (x - (-1))^2$ and see that the vertex is (–1, –1), focus is $\left(-1, -\frac{3}{4}\right)$ and directrix is $y = -\frac{5}{4}$.

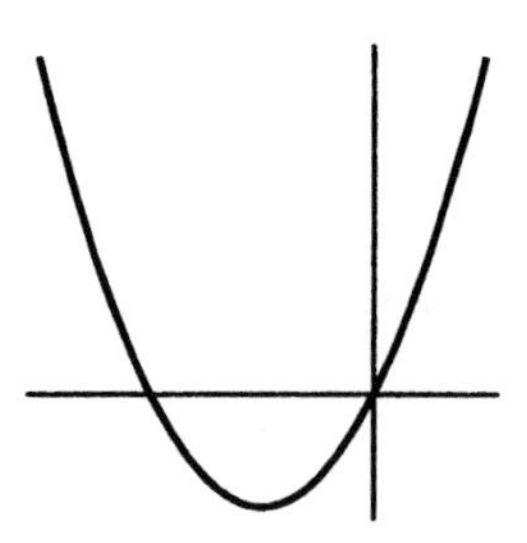

**7.** We rewrite $8x = y^2 + 2y + 25$ as $8(x-3) = (y+1)^2$ and see that the vertex is (3, –1), focus is (5, –1) and directrix is $x = 1$.

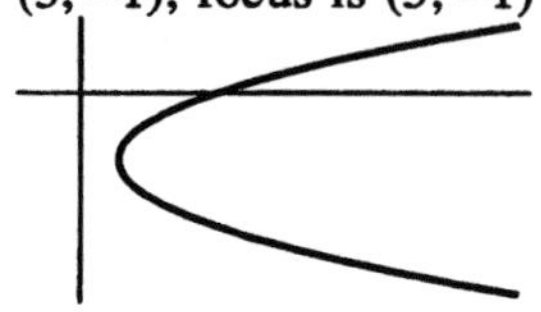

**9.** Ellipse, center: (0, 0), foci: $\left(0, \pm\sqrt{7}\right)$, vertices: (±3, 0), (0, ±4)

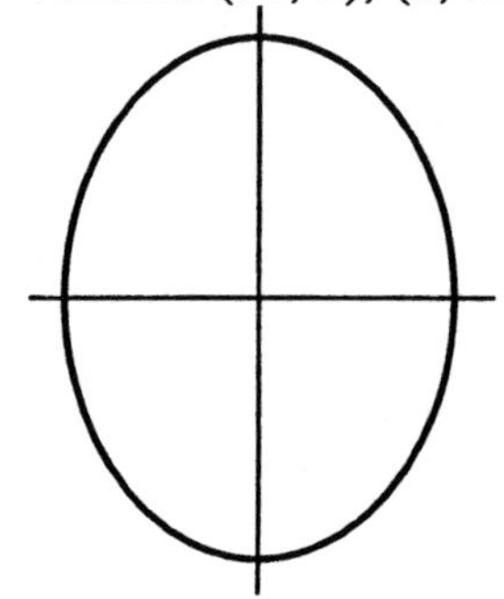

**11.** Hyperbola, center: (–1, 1), foci: $\left(-1, 1 \pm \sqrt{3}\right)$, asymptotes: $y = \pm\frac{1}{\sqrt{2}}(x+1) + 1$, vertices: (–1, 2), (–1, 0)

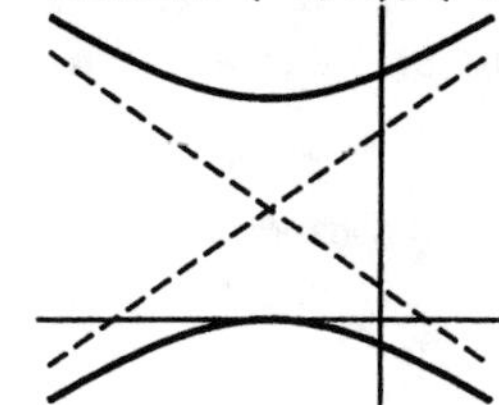

**13.** Ellipse (circle), center: $\left(\frac{1}{4}, 0\right)$, foci coincide at the center

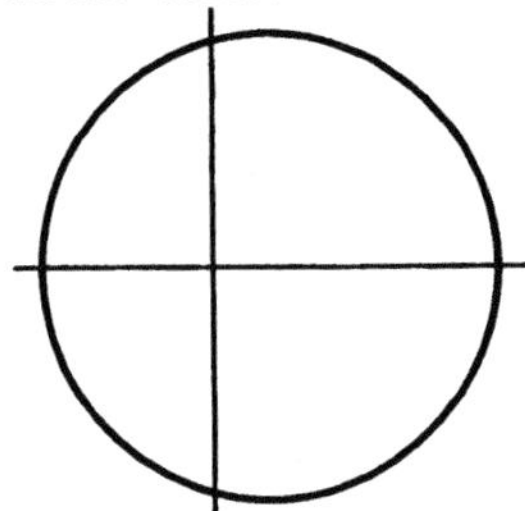

**15.** Hyperbola, center: (–3, –2), foci: $\left(-3, -2 \pm \sqrt{5}\right)$, asymptotes: $y = \pm 2(x + 3) - 2$, vertices: (–3, 0), (–3, –4)

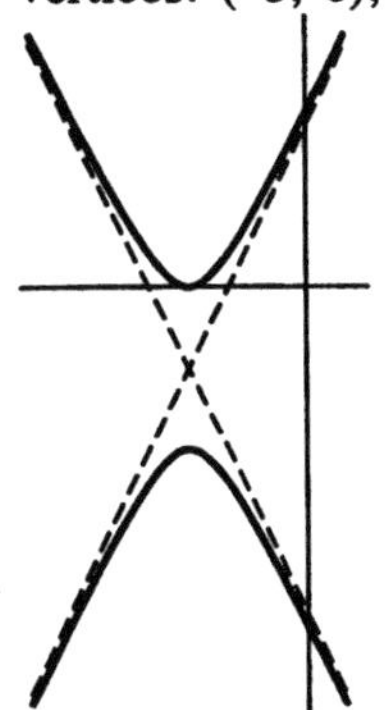

**17.** Here $c = 2$, $b = \frac{D}{2} = 4$ and $a^2 = b^2 - c^2 = 12$ so $\frac{x^2}{12} + \frac{y^2}{16} = 1.$

**19.** We want $|y - (-6)| = \sqrt{(x-2)^2 + (y-2)^2}$. Squaring and simplifying gives $16(y+2) = (x-2)^2$.

**21.** We have $a = \frac{D}{2} = 1$ and $c = \frac{(6-(-2))}{2} = 4$ so $b^2 = c^2 - a^2 = 15$. The center is at $\left(\frac{6+(-2)}{2}, \frac{1+1}{2}\right) = (2, 1)$ so $(x-2)^2 - \frac{(y-1)^2}{15} = 1.$

**23.** We begin with $\sqrt{x^2 + y^2} = 2|x+3|$. After squaring and simplifying, we obtain the hyperbola $\frac{(x+4)^2}{4} - \frac{y^2}{12} = 1.$

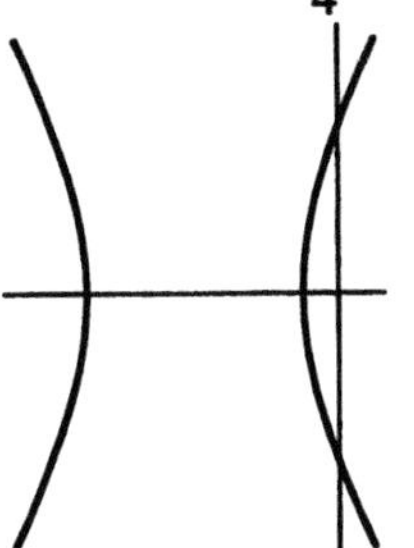

**25.** We begin with $\sqrt{x^2 + y^2} = |x+3|$. After squaring and simplifying we obtain the parabola $6\left(x + \frac{3}{2}\right) = y^2$.

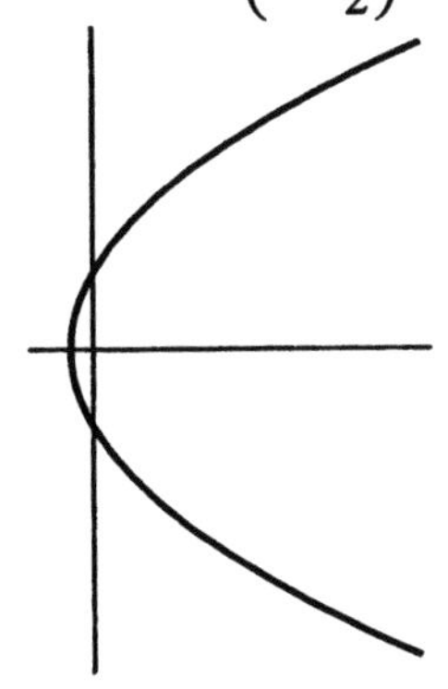

**27.** We begin with $\sqrt{(x+2)^2 + (y-3)^2} = \frac{1}{2}|y|$. After squaring and simplifying we obtain the ellipse $\frac{(x+2)^2}{3} + \frac{(y-4)^2}{4} = 1.$

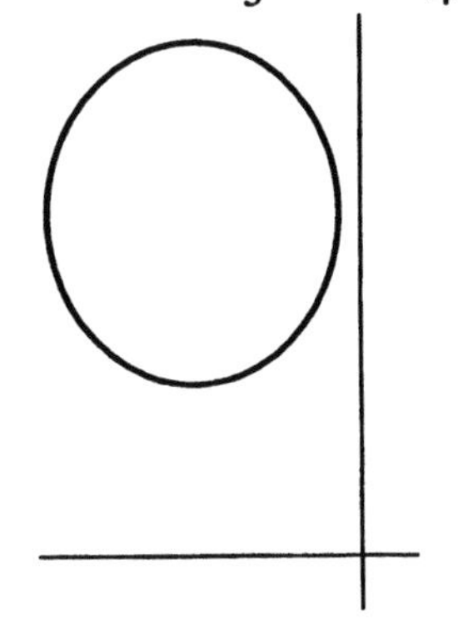

**29.** Using implicit differentiation on $\frac{x^2}{a^2}+\frac{y^2}{b^2}=1$ gives $y'=\frac{-xb^2}{ya^2}$ so a vector in the direction of $l$ may be given as $\vec{T}=\langle ya^2, -xb^2\rangle$. We assume, for now, that $a \geq b$ and therefore $\overrightarrow{F_1P}=\langle x-c, y\rangle$ and $\overrightarrow{F_2P}=\langle x+c, y\rangle$. To show these make equal angles with $l$ we will show that the cosines of the two angles made by $\vec{T}$ with $\overrightarrow{F_1P}$ and $\overrightarrow{F_2P}$ are equal with opposite signs. The cosine of the first angle is $\frac{\overrightarrow{F_1P}\cdot\vec{T}}{\left\|\overrightarrow{F_1P}\right\|\left\|\vec{T}\right\|}$ and that of the second angle is $\frac{\overrightarrow{F_2P}\cdot\vec{T}}{\left\|\overrightarrow{F_2P}\right\|\left\|\vec{T}\right\|}$. These will be equal in magnitude if and only if

$\left|\left\|\overrightarrow{F_1P}\right\|\overrightarrow{F_2P}\cdot\vec{T}\right|=\left|\left\|\overrightarrow{F_2P}\right\|\overrightarrow{F_1P}\cdot\vec{T}\right|$, which holds if and only if

$\left|\sqrt{(x-c)^2+y^2}(xya^2+cya^2-xyb^2)\right|=\left|\sqrt{(x+c)^2+y^2}(xya^2-cya^2-xyb^2)\right|$.

We replace both occurrences of $xya^2-xyb^2$ with $xyc^2$ since $c^2=a^2-b^2$ when $a>b$, and then factor out $cy$. After squaring the result and collecting terms we see that the result holds if and only if $4xc(x^2c^2+a^4)=4a^2xc(x^2+c^2+y^2)$.

Replacing $c^2$ with $a^2-b^2$ and collecting terms gives $-b^2x^2=-a^2b^2+a^2y^2$ as the resulting equation. This equality is equivalent to $\frac{x^2}{a^2}+\frac{y^2}{b^2}=1$ so our equivalent statement, that the cosines are equal in magnitude, must also be true. Thus, the angles are either equal or supplementary. If the angles are supplementary, the angle $\overrightarrow{F_1P}$ forms with $\vec{T}$ equals the angle $\overrightarrow{F_2P}$ forms with $-\vec{T}$. The case where $a<b$ is similar.

**31.** From (24) we see that

$$\begin{aligned}\hat{A}+\hat{C}&=(A\cos^2\theta+B\cos\theta\sin\theta+C\sin^2\theta)+(A\sin^2\theta-B\cos\theta\sin\theta+C\cos^2\theta)\\&=A+C\end{aligned}$$

as required.

**33.** From (24) we see that

$$\begin{aligned}\hat{B}^2-4\hat{A}\hat{C}&=((C-A)\sin2\theta+B\cos2\theta)^2-4(A\cos^2\theta+B\cos\theta\sin\theta+C\sin^2\theta)\\&\qquad\cdot(A\sin^2\theta-B\cos\theta\sin\theta+C\cos^2\theta)\\&=(C^2-2AC+A^2)\cdot4\sin^2\theta\cos^2\theta+2(BC-AB)(2\sin\theta\cos\theta)(1-2\sin^2\theta)\\&\quad+B^2(1-4\sin^2\theta+4\sin^4\theta)-4A^2\cos^2\theta\sin^2\theta+4AB\cos^3\theta\sin\theta\\&\quad-4AC\cos^4\theta-4AB\cos\theta\sin^3\theta+4B^2\sin^2\theta\cos^2\theta-4BC\cos^3\theta\sin\theta\\&\quad-4AC\sin^4\theta+4BC\cos\theta\sin^3\theta-4C^2\sin^2\theta\cos^2\theta\end{aligned}$$

Collecting our terms we have

$$\begin{aligned}&B^2(1-4\sin^2\theta+4\sin^4\theta+4\cos^2\theta\sin^2\theta)+4BC(\sin\theta\cos\theta-\sin^3\theta\cos\theta-\cos^3\theta\sin\theta)\\&\quad+4AB(-\sin\theta\cos\theta+\sin^3\theta\cos\theta+\cos^3\theta\sin\theta)+4AC(-2\sin^2\theta\cos^2\theta-\cos^4\theta-\sin^4\theta).\end{aligned}$$

Judicious use of the Pythagorean identity allows us to rewrite the above as $B^2-4AC$, as required.

**35.** Since $R = B^2 - 4AC = (-1)^2 - 4\cdot(0)\cdot(0) > 0$ we have a hyperbola.
$\theta = \frac{1}{2}\operatorname{arccot}\left(\frac{A-C}{B}\right) = \frac{1}{2}\operatorname{arccot}(0) = 45°$

**37.** Since $R = B^2 - 4AC > 0$ we have a hyperbola.
$\theta = \frac{1}{2}\operatorname{arccot}\left(\frac{3}{4}\right) \approx 26.565°$

**39.** Since $R = B^2 - 4AC = 0$ we have a parabola.
$\theta = \frac{1}{2}\operatorname{arccot}\left(-\frac{1}{\sqrt{8}}\right) \approx -35.264°$

**41.** Using $A = 1$, $B = C = D = 0$, $E = -1$, $F = 1$ and $\theta = -60°$ in (24) gives
$\frac{1}{4}x^2 + \frac{\sqrt{3}}{2}xy + \frac{3}{4}y^2 + \frac{\sqrt{3}}{2}x - \frac{1}{2}y + 1 = 0$ or
$x^2 + 2\sqrt{3}xy + 3y^2 + 2\sqrt{3}x - 2y + 4 = 0.$

**43.** Using $A = -1$, $C = 1$, $F = -3$, $B = D = E = 0$ and $\theta = \frac{\pi}{4}$ in (24) gives $2xy - 3 = 0$.

**45.** This parabola is just the parabola
$4p\left(x - \frac{\sqrt{2}}{2}\right) = y^2$ with $p = -\frac{\sqrt{2}}{2}$ rotated 45°.
Using (24) with $\theta = -45°$ gives
$\frac{1}{2}x^2 - xy + \frac{1}{2}y^2 + 2x + 2y - 2 = 0.$

**47.** If we rotated the hyperbola 45° we would have a new hyperbola, with foci on the $x$-axis, with equation $\sqrt{x^2 + y^2} = 2\left|x - \sqrt{2}\right|$ which becomes $-3x^2 + y^2 + 8\sqrt{2}x - 8 = 0$. Using (24) with $\theta = 45°$ gives $-x^2 + 4xy - y^2 + 8x - 8y - 8 = 0$.

**49.** If we rotated the hyperbola −60° we would have a new hyperbola, with foci at (±2, 0). Since $a = \frac{D}{2} = \frac{1}{2}$ and $c = 2$ it follows that $b^2 = c^2 - a^2 = \frac{15}{4}$. The equation of the new hyperbola is $4x^2 - \frac{4y^2}{15} = 1$. Using (24) with $\theta = -60°$ and then multiplying by 15, we get $12x^2 + 32\sqrt{3}xy + 44y^2 - 15 = 0$.

**51.** Equation (21):

$$\frac{\left(x - \frac{p\varepsilon^2}{1-\varepsilon^2}\right)^2}{\frac{\varepsilon^2 p^2}{(1-\varepsilon^2)^2}} + \frac{y^2}{\frac{\varepsilon^2 p^2}{(1-\varepsilon^2)}} = 1$$

For $0 < \varepsilon < 1$, $1 - \varepsilon^2 > 0$, and it follows that $\frac{\varepsilon^2 p^2}{(1-\varepsilon^2)} > 0$. Since the coefficients of the squared expressions for $x$ and $y$ are both positive, the equation represents an ellipse.

Furthermore, since $\frac{\varepsilon^2 p^2}{(1-\varepsilon^2)^2} > \frac{\varepsilon^2 p^2}{(1-\varepsilon^2)}$ when $0 < \varepsilon < 1$, the major axis of the ellipse is horizontal.

The center of the ellipse is at $\left(\frac{p\varepsilon^2}{1-\varepsilon^2}, 0\right)$ and the foci are each

$$\begin{aligned} c &= \sqrt{a^2 - b^2} \\ &= \sqrt{\frac{\varepsilon^2 p^2}{(1-\varepsilon^2)^2} - \frac{\varepsilon^2 p^2}{(1-\varepsilon^2)}} \\ &= \varepsilon p\sqrt{\frac{\varepsilon^2}{(1-\varepsilon^2)^2}} \\ &= \frac{p\varepsilon^2}{1-\varepsilon^2} \end{aligned}$$

units away, along the $x$-axis, at (0, 0) and $\left(\frac{2p\varepsilon^2}{1-\varepsilon^2}, 0\right)$.

**53.** We may rewrite the equation as $(x + 2)(y - 3) = 0$ and this describes two intersecting lines; $x = -2$ and $y = 3$.

**55.** We may rewrite the equation as $(x + 2y + 4)(x + 2y - 3) = 0$ and this describes two parallel lines; $x + 2y = -4$ and $x + 2y = 3$.

**57.** $\theta = \frac{1}{2}\operatorname{arccot}\left(\frac{A-C}{B}\right) = \frac{1}{2}\operatorname{arccot}\left(\frac{0}{-1}\right) = 45°$

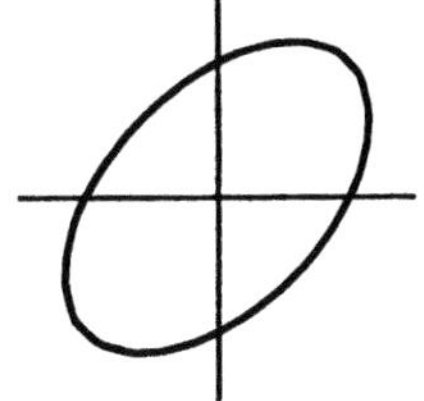

**59.** $\theta = \frac{1}{2}\operatorname{arccot}\left(\frac{A-C}{B}\right)$
$= \frac{1}{2}\operatorname{arccot}\left(\frac{\sqrt{3}-\left(-\sqrt{3}\right)}{-2}\right)$
$= -15°$

## Section 9.2 Real-World Functions

**1.** domain of $f = R^2$
$f(x, 2) = 2x(x-6) = 2x^2 - 12x$
$f(x, 2)$ is a quadratic function of $x$.

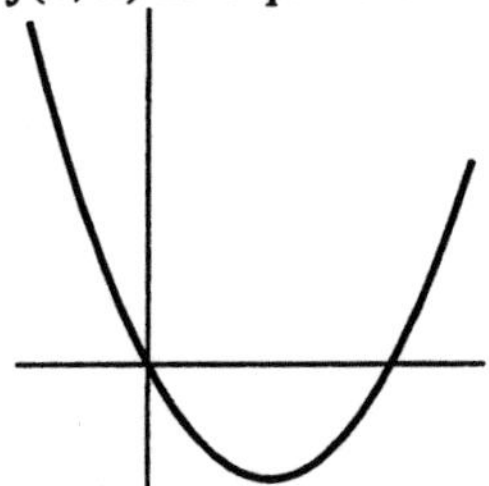

**3.** domain of $f = R^2$
$f\left(x, \frac{\pi}{2}\right) = \left(x + \frac{\pi}{2}\right)\sin\left(\frac{\pi x}{2}\right)$

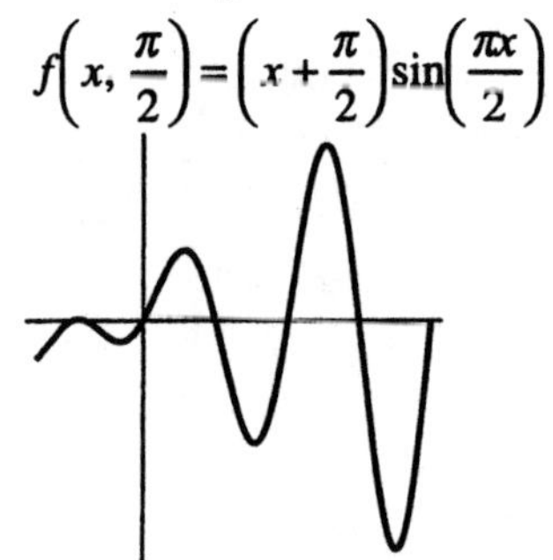

**5.** domain of $f = R^3$
$f(2, y, 1) = \sqrt{4 + 2y^2 + 4} = \sqrt{2y^2 + 8}$

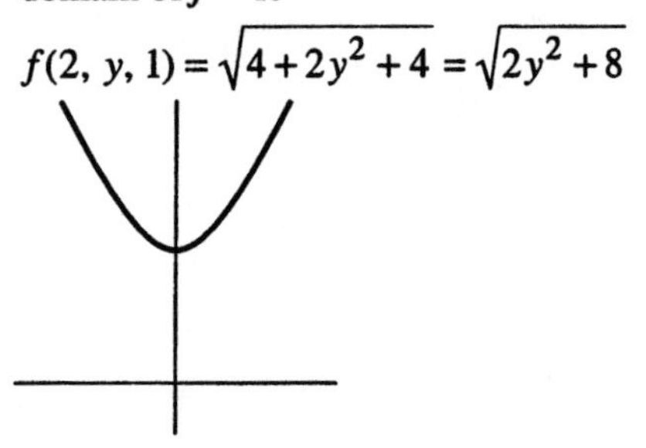

**7.** domain of $f = \{(x, y, z): x + y + z > 0\}$
$f(2, y, -4) = \ln(2 + y - 4) = \ln(y - 2)$
$f(2, y, -4)$ is a logarithmic function with domain $y > 2$.

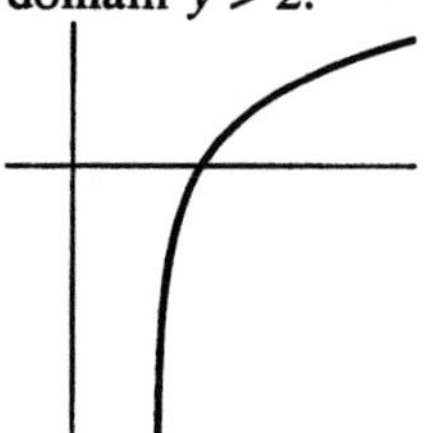

**9.** domain of $f = \{(x, y, z): x^2 + y^2 + z^2 \le 1\}$
$f\left(-\frac{1}{4}, y, \frac{1}{3}\right) = \arcsin\left(\frac{1}{16} + y^2 + \frac{1}{9}\right)$
$= \arcsin\left(y^2 + \frac{25}{144}\right)$
$f\left(-\frac{1}{4}, y, \frac{1}{3}\right)$ is a function with domain $-\frac{\sqrt{119}}{12} \le y \le \frac{\sqrt{119}}{12}$.

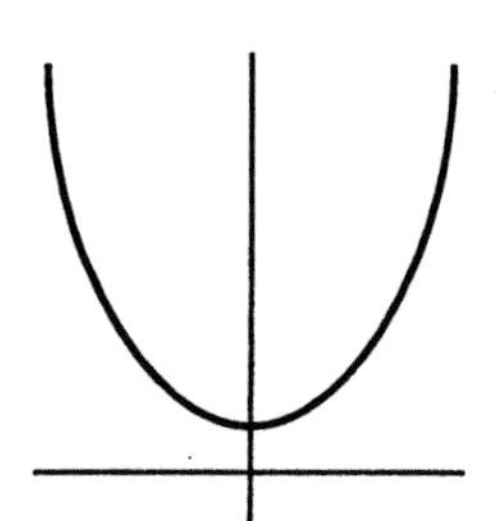

**11.** When the mass $M$ and the response height $h$ of the block are constant, the quantity $v - \sqrt{2gh}$ is inversely proportional to $m$ the mass of the bullet. When $m$ is small, $v$ is large, and as $m$ grows, $v$ decreases.

**13.** $p = p(0.9, f, 0.3)$
$= 0.9(1 - f(1 - 0.3))$
$= 0.9 - 0.63f,$
so $p$ is a linear function of $f$. The probability $p$ of being eaten decreases as the probability $f$ of the mimic fooling the predator increases, but $p$ can never be lower than 0.27 since $f$ is at most 1.

**15.** $P(0.06, t) = \frac{16t - 657}{e^{0.06t} - 1}$
$\frac{dP}{dt}(0.06, t) = \frac{e^{0.06t}(55.42 - 0.96t) - 16}{(e^{0.06t} - 1)^2}$
$\frac{dP}{dt}(0.06, t) = 0$ when

$e^{0.06t}(55.42 - 0.96t) - 16 = 0.$
From the graph in Example 5, we know that the maximum is when $t \approx 57$. Thus we use Newton's method to find the root of
$e^{0.06t}(55.42 - 0.96t) - 16 = 0$ near $t_1 = 57$.
The table shows that $t \approx 57.19$.

| $n$ | $t_n$ |
|---|---|
| 1 | 57 |
| 2 | 57.192 |
| 3 | 57.190 |
| 4 | 57.190 |

**17.** For $100 \le t \le 110$, $S$ is defined by the segment joining (100, 913) and (110, 1000). Thus, we have $S(t) = 8.7t + 43$, $100 \le t \le 110$, so
$$P(0.06, t) = \frac{8.7t + 43}{e^{0.06t} - 1}, \quad 100 \le t \le 110.$$

**19.** Using the expression in Problem 18 with $r = 0.08$, $T = 18$, and $P = 95{,}000$, the amount you want to invest is
$95{,}000e^{-(0.08)(18)} \approx 22{,}508$ dollars.

**21.** If an object starts from rest, it has velocity $v = \sqrt{2gh}$ after falling from height $h$. At this time the kinetic energy is $\text{KE} = \frac{1}{2}mv^2 = mgh.$
Hence the gravitational potential energy of the object at height $h$ is $mgh$.
The bullet-block system of Example 2 has mass $M + m$. If it rises to height $h$, then at this instant the gravitational potential energy is $(M + m)gh$. Equate this to the kinetic energy expression in (7) to get
$$(M + m)gh = \frac{1}{2}(M + m)v_1^2.$$
Solve to obtain $v_1 = \sqrt{2gh}$. Substitute this into (6) to obtain (5).

**23. a.** Since speed and mass are non-negative, $r \ge 0$, $M \ge 0$, $m > 0$, $\frac{M}{m} \ge 1$, so
domain $= \{(r, m, M): \ r \ge 0, \ M \ge m > 0\}$.

**b.** If $m = m_0$ and $M = M_0$ is fixed, then
$s = r \ln\left(\frac{M_0}{m_0}\right)$ is a linear function with respect to $r$.

**c.** $s = s(250, 1000, M) = 250 \ln\left(\frac{M}{1000}\right)$

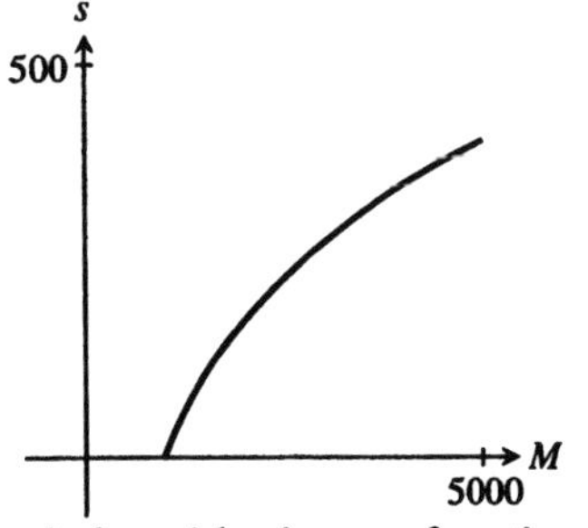


$s$ is logarithmic as a function of $M$.

**d.** $s = s(450, m, 2500) = 450 \ln\left(\frac{2500}{m}\right)$

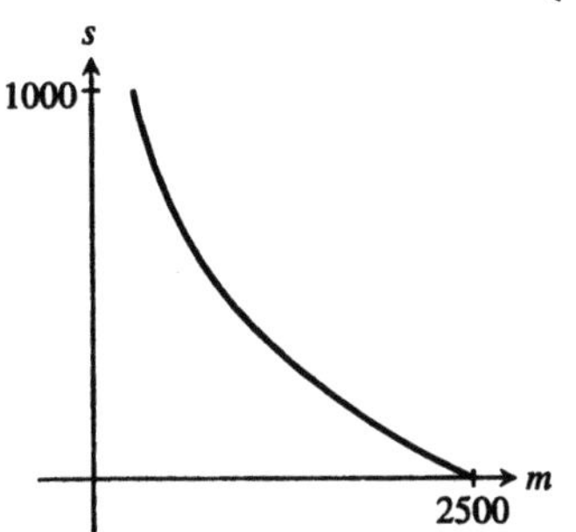


$s$ behaves like a logarithmic function of $\frac{1}{m}$.

**25. a.**

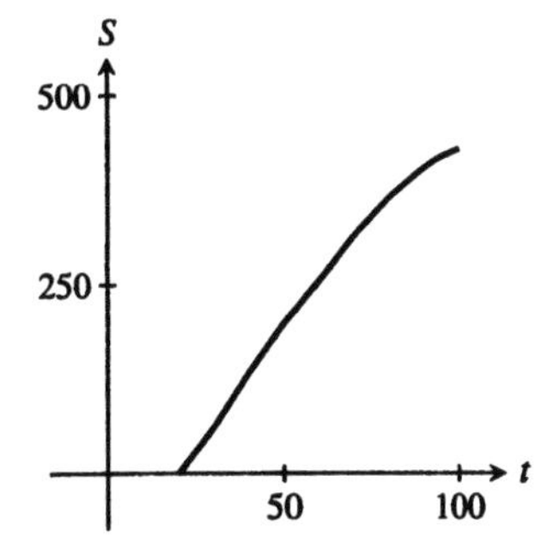


**b.**

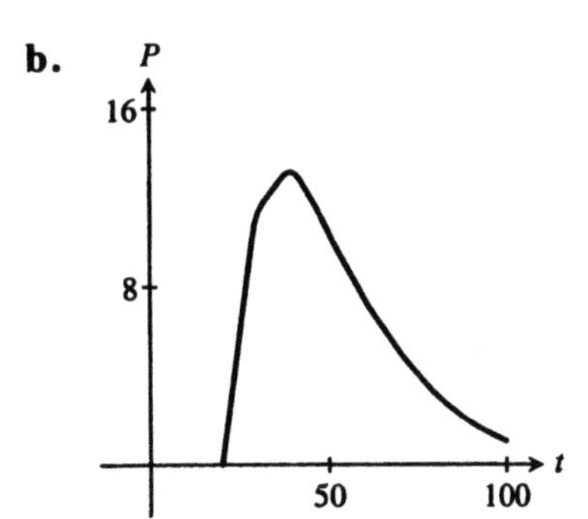

**c.** From the graph, we see that $P(0.06, t)$ has a maximum near $t = 35$. Between $t = 30$ and $t = 40$, $S(t)$ is given by $S(t) = 7.2t - 156$. Then the function $P(0.06, t) = \dfrac{S(t)}{e^{0.06t} - 1}$ has its maximum at $t \approx 36.46$.

**27. a.** The domain of $d$ is $\{(s, w): 0 \le s \le 20, 0 \le w \le 20\}$.

**b.**

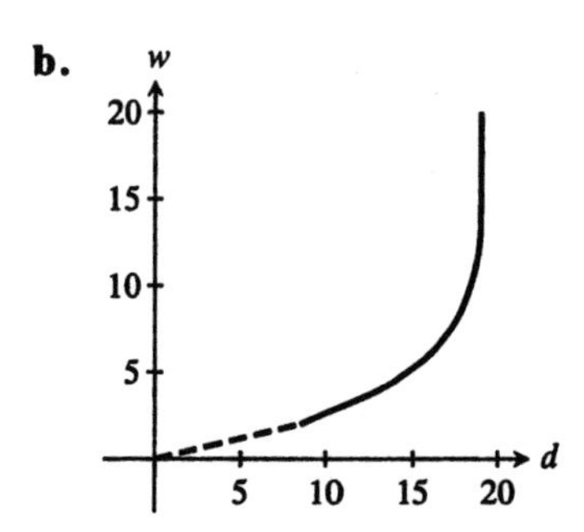


In this figure waterline $w$ is on the vertical axis and distance $d$ from center is on the horizontal axis. The graph shows the cross section of half the hull if a vertical cut is made at Station 11. For the full cross section, reflect across the $w$-axis. The dotted segment does not correspond to given data, but was filled in assuming that at Station 11 the center of the hull is at the lowest (i.e., 0) level.

**c.**

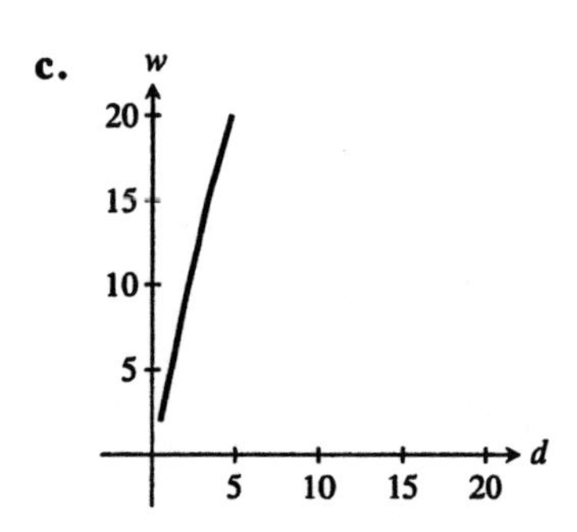


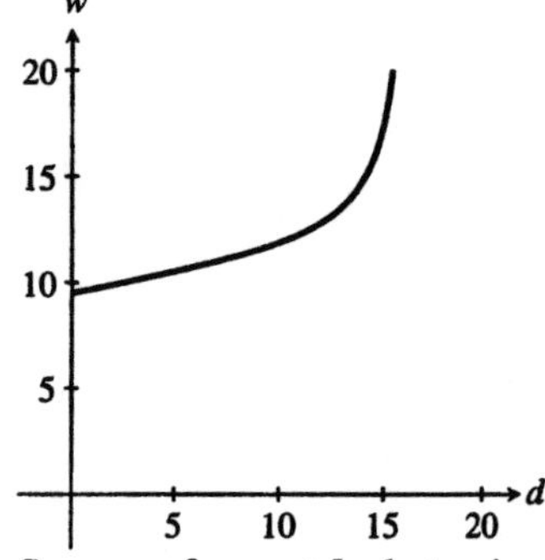


Same as for part **b**, but using Stations 1 and 19.

**d.**

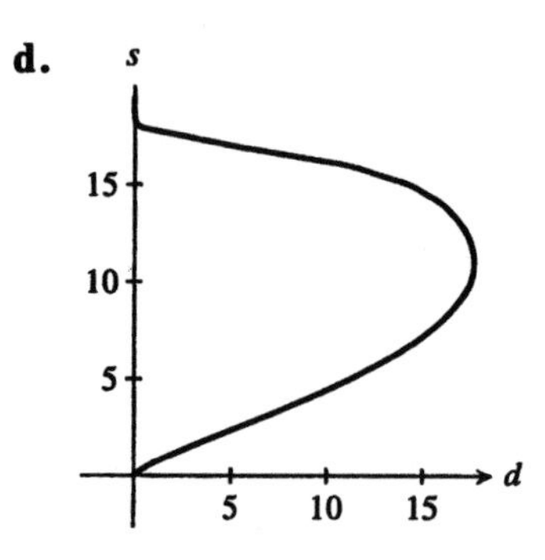


The vertical axis records stations. The horizontal axis shows the distance from center. The graph represents what a horizontal cross section of the hull at height 8 would look like. To get the full cross section, reflect about the $s$-axis. Remember that each station is 13.5 feet. Hence, for a true picture of the cross section, the illustration should be magnified by a factor of 13.5 in the vertical direction.

## Section 9.3 Graphing: Surfaces and Level Curves

**1.** Fix $z = 0$. Then $0 = 2x^2 + y^2$, so the $(x, y)$-plane cross-section is $(0, 0)$.
Fix $y = 0$. Then $z = 2x^2$, so the $(x, z)$-plane cross-section is a parabola.
Fix $x = 0$. Then $z = y^2$, so the $(y, z)$-plane cross-section is a parabola.

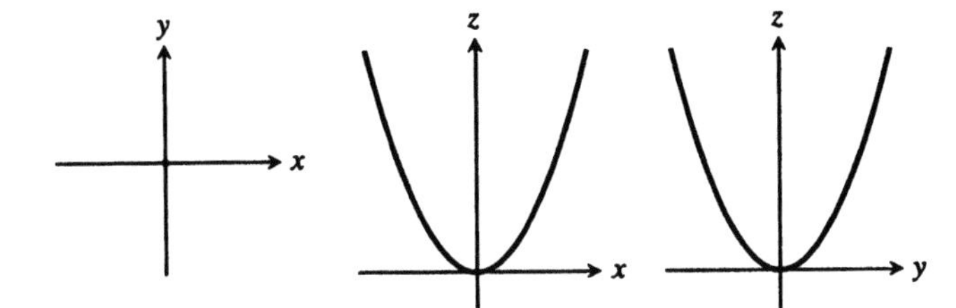


**3.** Fix $z = 0$. Then $0 = 3x - 2y$, so the $(x, y)$-plane cross-section is a line.
Fix $y = 0$. Then $z = 3x$, so the $(x, z)$-plane cross-section is a line.
Fix $x = 0$. Then $z = -2y$, so the $(y, z)$-plane cross-section is a line.

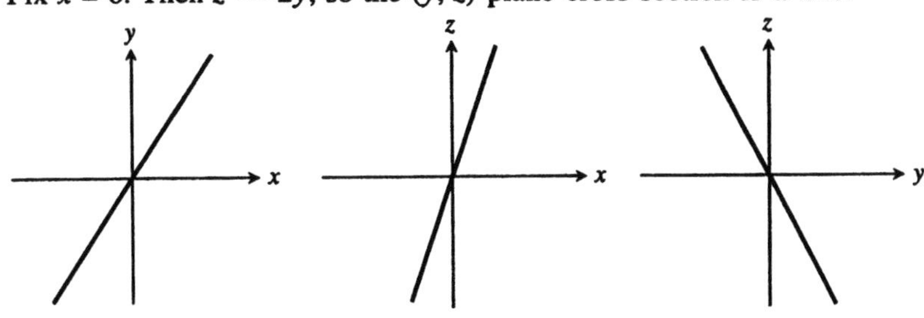


**5.** Fix $z = 0$. Then $0 = 3x^2 - y^2$ or $3x^2 = y^2$. Thus $y = \pm\sqrt{3}x$, so the $(x, y)$-plane cross-section is two lines.
Fix $y = 0$. Then $z = 3x^2$, so the $(x, z)$-plane cross-section is a parabola.
Fix $x = 0$. Then $z = -y^2$, so the $(y, z)$-plane cross-section is a parabola.

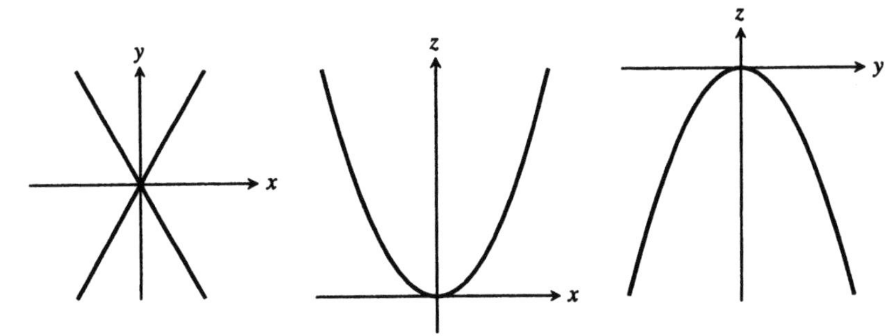


**7.** Fix $z = 0$. Then $0 = e^{xy}$ which is not possible, so there is no $(x, y)$-plane cross-section.
Fix $y = 0$. Then $z = 1$, so the $(x, z)$-plane cross-section is a line.
Fix $x = 0$. Then $z = 1$, so the $(y, z)$-plane cross-section is a line.

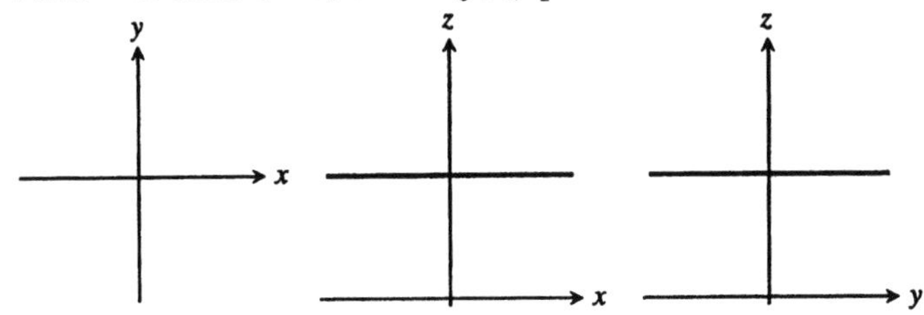

**9.** Let $c$ be a real number, and consider the set of solutions $(x, y)$ to $(3x-3y+5)^2 = c$.
If $c < 0$, there is no solution. If $c \geq 0$, $3x-3y+5 = \pm\sqrt{c}$, so the level curves are lines.

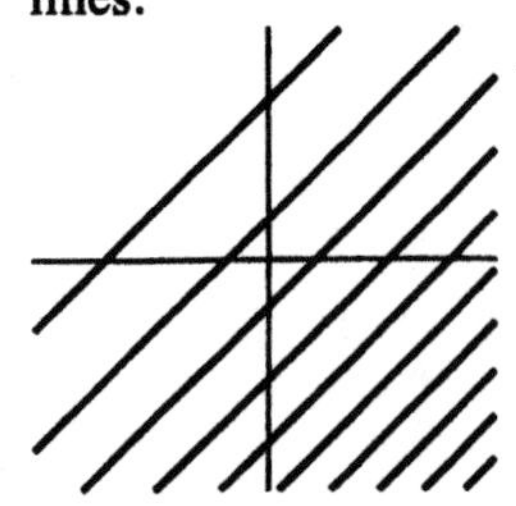

**11.** Let $c$ be a real number, and consider the set of solutions $(x, y)$ to $\frac{2}{\sqrt{x-2y^2}} = c$.
If $c \leq 0$, there is no solution. If $c > 0$, $x-2y^2 = \frac{4}{c^2}$, so the level curves are parabolas.

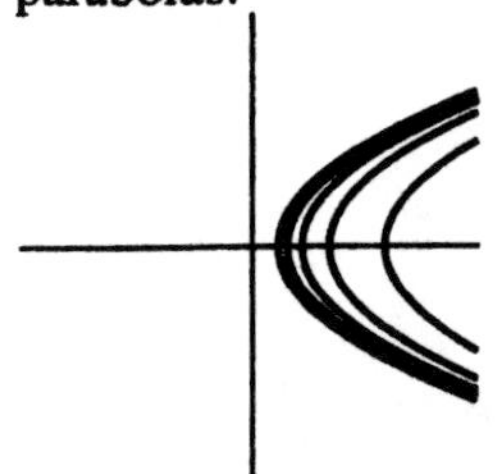

**13.** Let $c$ be a real number, and consider the set of solutions $(x, y)$ to $y^2 - x^2 = c$.
If $c \neq 0$, the level curves are hyperbolas. If $c = 0$, the level curve is two diagonal lines $y = \pm x$.

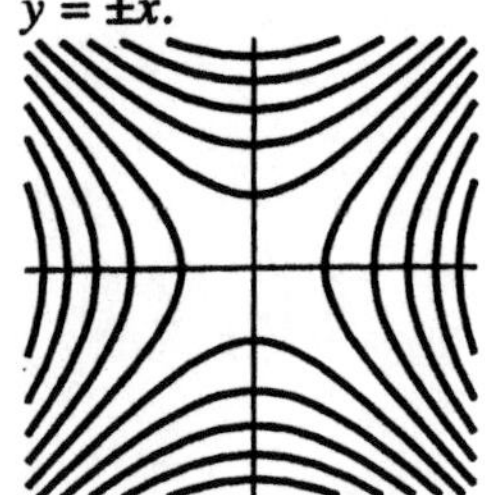

**15.** $z = 3x - y + 2$ is an equation for a plane.

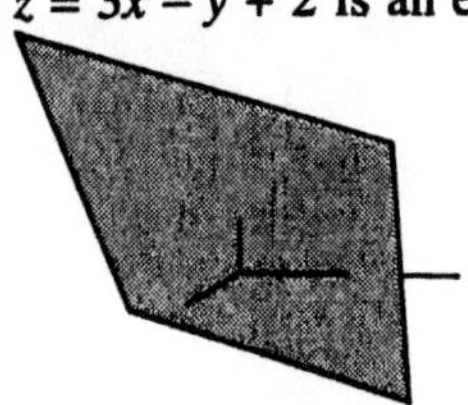

**17.** $z = \sqrt{x} + \sqrt{y}$ has a graph only above the first quadrant of the $(x, y)$-plane, rising as $x$ or $y$ grows.

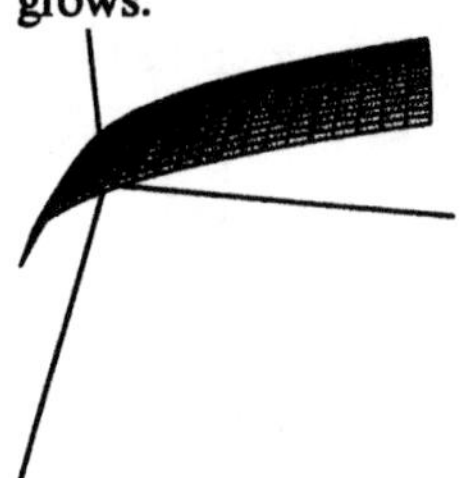

**19.** The graph of $z = \frac{1}{1+x^2+y^2}$ is symmetric about the $z$-axis and approaches the $(x, y)$-plane as $|x|$ and $|y|$ grow larger.

**21.** We want $c = \frac{1}{2}mv^2$ so $m = \frac{2c}{v^2}$ or $m = \frac{k}{v^2}$ for varying $k$. A sketch with $k$ = 1, 2, 3, is below. Since $m > 0$ and $v \geq 0$ the level curves are in quadrant I.

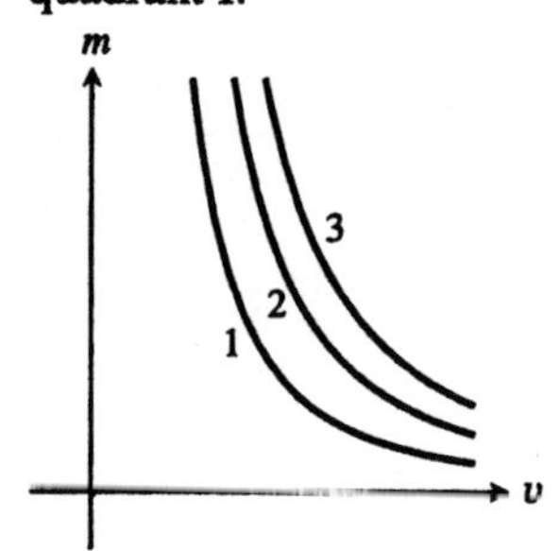


**23.** Taking $s_1$ and $s_2$ to be positive, the level curve for $f = c$ is the first-quadrant portion of the hyperbola with equation $s_2 = \frac{s_1}{cs_1 - 1}$.

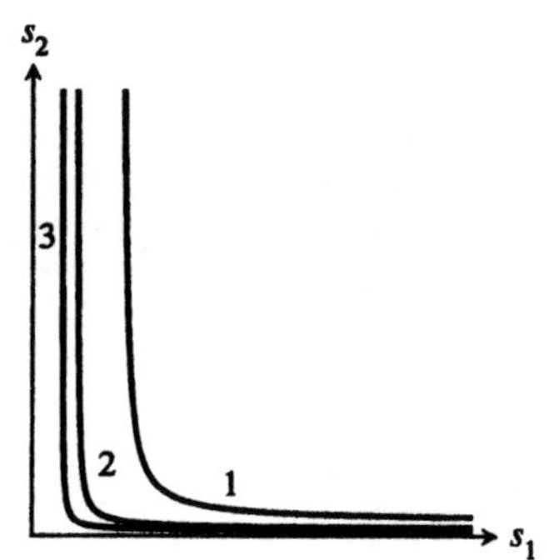

**25.** The function has as its domain locations in the United States. The range is the set of average wind powers for all locations. Given a location as input, the function returns the average wind power available at that location.

**27.** The graph of the equation $4x^2+2y^2+z^2+4=k$ is an ellipsoid for $k>4$. The figure is a level surface for $k=8$.

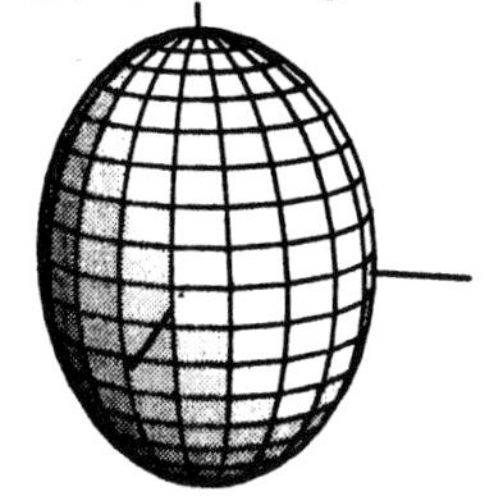

**29.** The graph of the equation $x^2-y+z^2+3=k$ is a paraboloid around the $y$-axis. The figure is a level surface for $k=0$.

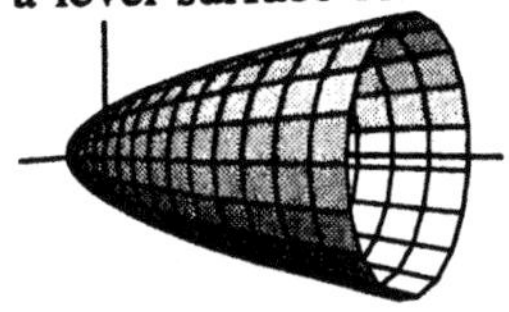

**31.** The graph of the equation $x+y^2-z^2=k$ is a hyperbolic sheet, which has both hyperbolas and parabolas as cross sections. The figure is a level surface for $k=1$.

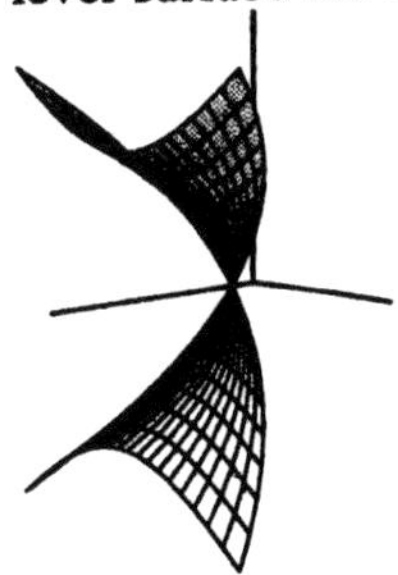

**33.** B, v

**35.** F, iii

**37.** C, iv

## Section 9.4 Graphing: Parametric Representation of Surfaces

**1.** The $(x, y, z)$-coordinates for a point on $S$ are $x=-2u+3v$, $y=u+4v$, and $z=u-v$.
Solving the first two equations for $u$ and $v$, we obtain $u=\dfrac{-4x+3y}{11}$ and $v=\dfrac{x+2y}{11}$.
Substituting these expressions into the equation for $z$, we get
$$z=\frac{-4x+3y}{11}-\frac{x+2y}{11}=\frac{-5x+y}{11}.$$
Rearranging this expression, we obtain $5x-y+11z=0$, so $5x-y+11z=0$ is an equation for $S$.
Since $5(2)-21+11(1)=0$, the point $A=(2, 21, 1)$ lies on $S$. Since $5(8)-7+11(6)=99\neq 0$, the point $B=(8, 7, 6)$ does not lie on $S$.

**3.** The $(x, y, z)$-coordinates for a point on $S$ are $x=r\cos\theta$, $y=r^2$, and $z=r\sin\theta$.
Observe that
$x^2+z^2=r^2\cos^2\theta+r^2\sin^2\theta=r^2=y$, so
$x^2+z^2=y$ is an equation for $S$.
Since $\left(-\dfrac{3}{2}\right)^2+\left(\dfrac{3\sqrt{3}}{2}\right)^2=9$, the point
$A=\left(-\dfrac{3}{2}, 9, \dfrac{3\sqrt{3}}{2}\right)$ lies on $S$.
Since $(-1)^2+0^2=1$, the point $B=(-1, 1, 0)$ lies on $S$.

**5.** The $(x, y, z)$-coordinates for a point on $S$ are $x=\cos\theta\sin\phi$, $y=\sin\theta\sin\phi$, and $z=\cos\phi$.
Observe that
$$\begin{aligned}&x^2+y^2+z^2\\&=\cos^2\theta\sin^2\phi+\sin^2\theta\sin^2\phi+\cos^2\phi\\&=\sin^2\phi+\cos^2\phi\\&=1,\end{aligned}$$
so $x^2+y^2+z^2=1$ is an equation for $S$.
Since $\left(\dfrac{\sqrt{3}}{4}\right)^2+\left(\dfrac{3}{4}\right)^2+\left(-\dfrac{1}{2}\right)^2=1$, the point
$A=\left(\dfrac{\sqrt{3}}{4}, \dfrac{3}{4}, -\dfrac{1}{2}\right)$ lies on $S$. Since
$0^2+0^2+1^2=1$, the point $B=(0, 0, 1)$ lies on $S$.

**7.** Solving the system $-3u + v = x$ and $u = y$ for $u$ and $v$, we get $u = y$ and $v = x + 3y$. Thus
$z = u + 2v = y + 2(x + 3y) = 2x + 7y.$
A scalar representation is $z = 2x + 7y$.
The first figure is plotted using a parametric plotting routine and the second figure is plotted using a standard plotting package.

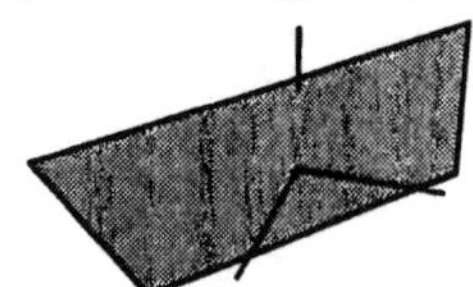
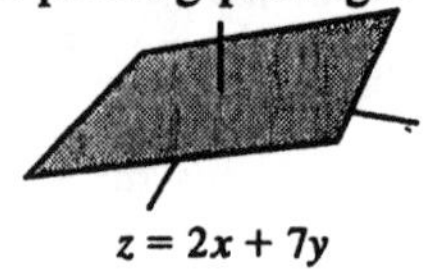

$z = 2x + 7y$

**9.** Observe that
$$x^2 + y^2 = (r\sin\theta)^2 + (r\cos\theta)^2 = r^2 = z + 2.$$
Thus a scalar representation is $z = x^2 + y^2 - 2$.
The first figure is plotted using a parametric plotting routine and the second figure is plotted using a standard plotting package.

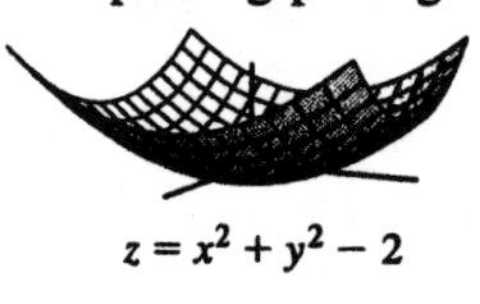

$z = x^2 + y^2 - 2$

**11.** Observe that
$$\begin{aligned} y^2 + 16z^2 &= (4r\cos\theta)^2 + 16(-r\sin\theta)^2 \\ &= 16r^2 \\ &= 16x - 32. \end{aligned}$$
Thus a scalar representation is
$$x = \frac{1}{16}y^2 + z^2 + 2.$$
The first figure is plotted using a parametric plotting routine and the second figure is plotted using a standard plotting package.

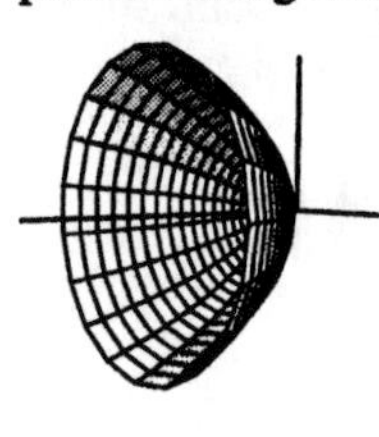
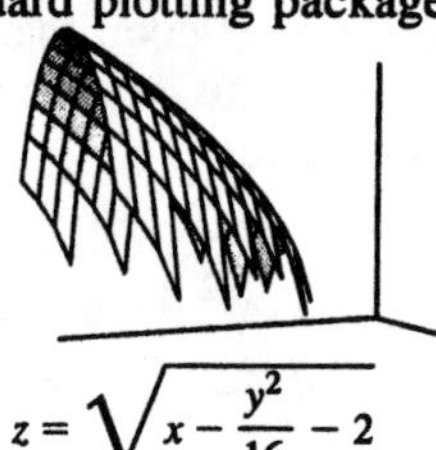

$z = \sqrt{x - \frac{y^2}{16} - 2}$

**13.** Observe that
$$\begin{aligned} 4x^2 - z^2 &= 4(2r\sec\theta)^2 - (4r\tan\theta)^2 \\ &= 16r^2 \\ &= 16y - 16. \end{aligned}$$
Thus a scalar representation is
$$y = \frac{1}{4}x^2 - \frac{1}{16}z^2 + 1.$$
The first figure is plotted using a parametric plotting routine and the second figure is plotted using a standard plotting package.

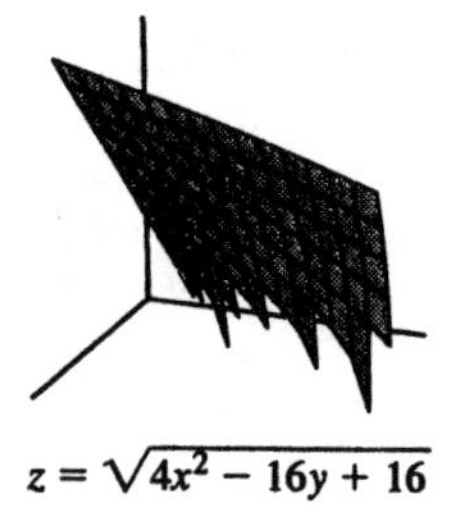

$z = \sqrt{4x^2 - 16y + 16}$

**15.** Observe that
$$e^{x+y} = e^{\ln u + \ln v} = e^{\ln uv} = uv = z.$$
Thus a scalar representation is $z = e^{x+y}$.
The first figure is plotted using a parametric plotting routine and the second figure is plotted using a standard plotting package.

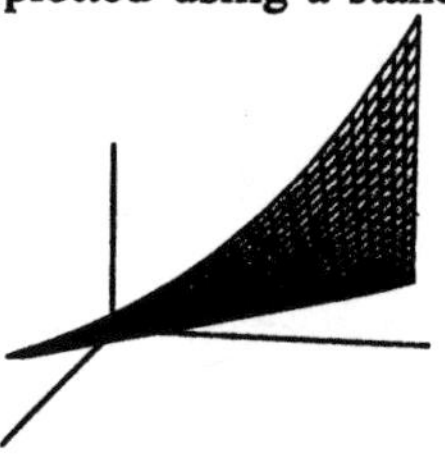
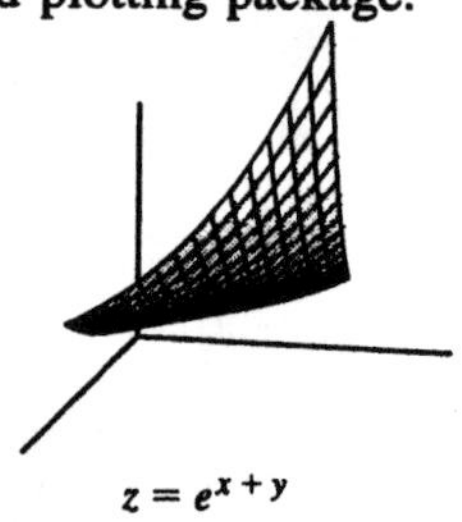

$z = e^{x+y}$

**17.** For example, let $\mathbf{f}(r, \theta) = \langle r\cos\theta, r\sin\theta, 2r^2 \rangle$, $r \ge 0$, $0 \le \theta \le 2\pi$.

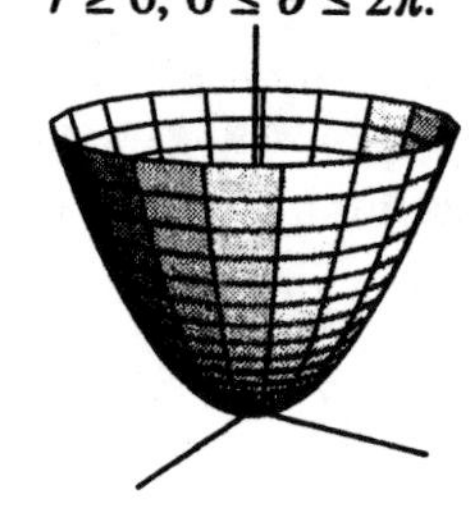

**19.** For example, using level $x$-curves, let
$$\mathbf{r}(r, \theta) = \left\langle r^2 + 1, \sqrt{2}r\cos\theta, \frac{1}{\sqrt{2}}r\sin\theta \right\rangle, \ r \ge 0,$$
$0 \le \theta \le 2\pi$.

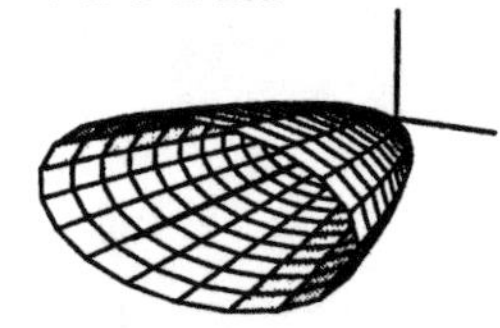

**21.** For example, using level $y$-curves, let
$\mathbf{f}(r, \theta) = \left\langle \frac{1}{\sqrt{3}} r\tan\theta, r^2, r\sec\theta \right\rangle$,
$r \geq 0, \ -\frac{\pi}{2} < \theta < \frac{\pi}{2}$.

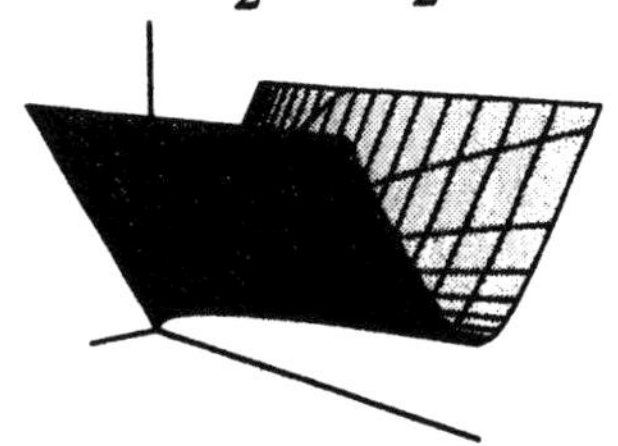

**23.** For example, let
$\mathbf{r}(u, v) = \left\langle u, -2 + \frac{2}{3}u + \frac{5}{3}v, v \right\rangle, \ -\infty < u, v < \infty.$

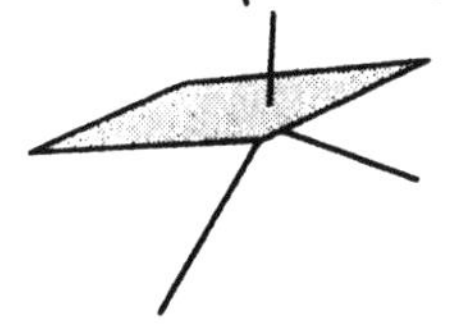

**25.** For example, using level $y$-curves, let
$\mathbf{f}(r, \theta) = \left\langle \frac{1}{\sqrt{2}} r\cos\theta, 2\ln r, \frac{1}{2} r\sin\theta \right\rangle, \ r > 0,$
$0 \leq \theta \leq 2\pi$.

**27.** Let $x = u$ and $y = v$. Then $z = \frac{1}{2}(5 - 3u + v)$, so
$\mathbf{r}(u, v) = \left\langle u, v, \frac{1}{2}(5 - 3u + v) \right\rangle$ is one representation.
Let $y = u$ and $z = v$. Then $x = \frac{1}{3}(5 + u - 2v)$, so
$\mathbf{r}(u, v) = \left\langle \frac{1}{3}(5 + u - 2v), u, v \right\rangle$ is another.
Let $x = u$ and $z = v$. Then $y = 3u + 2v - 5$, so
$\mathbf{r}(u, v) = \langle u, 3u + 2v - 5, v \rangle$ is another.

**29.** Let $x = u$ and $z = v$. Then $y = e^{u+2v}$, so
$\mathbf{r}(u, v) = \langle u, e^{u+2v}, v \rangle$ is one representation.
Let $x = u$ and $y = v$. Then $z = \frac{1}{2}(\ln v - u)$, so
$\mathbf{r}(u, v) = \left\langle u, v, \frac{1}{2}(\ln v - u) \right\rangle$ is another.
Let $y = u$ and $z = v$. Then $x = \ln u - 2v$, so
$\mathbf{r}(u, v) = (\ln u - 2v, u, v)$ is another.

**31.** Let $x = u$ and $y = v$. Then $z = f(u, v)$, so
$\mathbf{r}(u, v) = \langle u, v, f(u, v) \rangle$ is a parametric representation of the surface. This does not have a significant advantage over the Cartesian representation because they are essentially the same.

**33.** The surface $S_2$ is the reflection of the surface $S_1$ across the plane $y = z$.
The surface $S_3$ is the reflection of the surface $S_1$ across the plane $x = y$.
The surface $S_3$ is the reflection of the surface $S_2$ across the plane $x = y$ followed by a reflection across the plane $x = z$. (Note that there are other possible interpretations, such as a rotation about the vector $\langle 1, 1, 1 \rangle$ with an angle of 120° counterclockwise while looking at the origin.)

**35.** The surface $S_1$ represented by $\mathbf{r}_1$ is the reflection of the surface $S$ across the plane $y = z$.
The surface $S_2$ represented by $\mathbf{r}_2$ is the reflection of the surface $S$ across the plane $x = y$.
The surface $S_3$ represented by $\mathbf{r}_3$ is the reflection of the surface $S$ across the plane $x = y$ followed by a reflection across the plane $y = z$.
The surface $S_4$ represented by $\mathbf{r}_4$ is the reflection of the surface $S$ across the plane $x = y$ followed by a reflection across the plane $x = z$.
The surface $S_5$ represented by $\mathbf{r}_5$ is the reflection of the surface $S$ across the plane $x = z$.

**37. a.** The points on $C_1$ are the points on $S$ where $r = 3$. The points on $C_2$ are the points on $S$ where $\theta = \frac{\pi}{4}$. Observe that
$\mathbf{g}_1\left(\frac{\pi}{4}\right) = \mathbf{g}_2(3) = \mathbf{F}\left(3, \frac{\pi}{4}\right)$. Thus both $C_1$ and $C_2$ pass through the point
$\left(\frac{3\sqrt{2}}{2}, \frac{3\sqrt{2}}{2}, 9\right)$.

**b.**

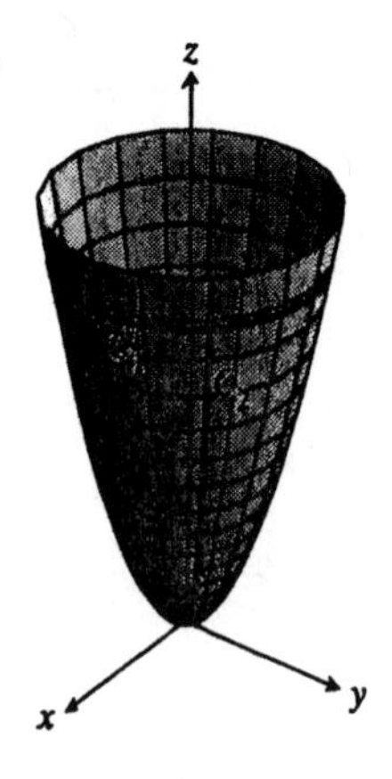


**c.** $\mathbf{g}_1'(\theta) = \langle -3\sin\theta, 3\cos\theta, 0\rangle$

The tangent vector is $g_1'\left(\frac{\pi}{4}\right) = \left\langle -\frac{3\sqrt{2}}{2}, \frac{3\sqrt{2}}{2}, 0\right\rangle.$

**d.** $\mathbf{g}_2'(r) = \left\langle \cos\frac{\pi}{4}, \sin\frac{\pi}{4}, 2r\right\rangle$

The tangent vector is $\mathbf{g}_2'(3) = \left\langle \frac{\sqrt{2}}{2}, \frac{\sqrt{2}}{2}, 6\right\rangle.$

**e.** A vector normal to the plane is $\mathbf{g}_1'\left(\frac{\pi}{4}\right)\times \mathbf{g}_2'(3) = \left\langle -\frac{3\sqrt{2}}{2}, \frac{3\sqrt{2}}{2}, 0\right\rangle \times \left\langle \frac{\sqrt{2}}{2}, \frac{\sqrt{2}}{2}, 6\right\rangle = \langle 9\sqrt{2}, 9\sqrt{2}, -3\rangle$

or the vector $\langle 3\sqrt{2}, 3\sqrt{2}, -1\rangle$. Thus an equation for the plane is

$$\langle 3\sqrt{2}, 3\sqrt{2}, -1\rangle \cdot \left\langle x - \frac{3\sqrt{2}}{2}, y - \frac{3\sqrt{2}}{2}, z - 9\right\rangle = 0$$

$$3\sqrt{2}\left(x - \frac{3\sqrt{2}}{2}\right) + 3\sqrt{2}\left(y - \frac{3\sqrt{2}}{2}\right) - 1(z-9) = 0$$

$$3\sqrt{2}x + 3\sqrt{2}y - z - 9 = 0.$$

**39. a.** The points on $C_1$ are the points on $S$ where $v = 4$. The points on $C_2$ are the points on $S$ where $u = -2$. Observe that $\mathbf{g}_1(-2) = \mathbf{g}_2(4) = \mathbf{r}(-2, 4)$. Thus both $C_1$ and $C_2$ pass through the point $(-4, 20, 5)$.

**b.**

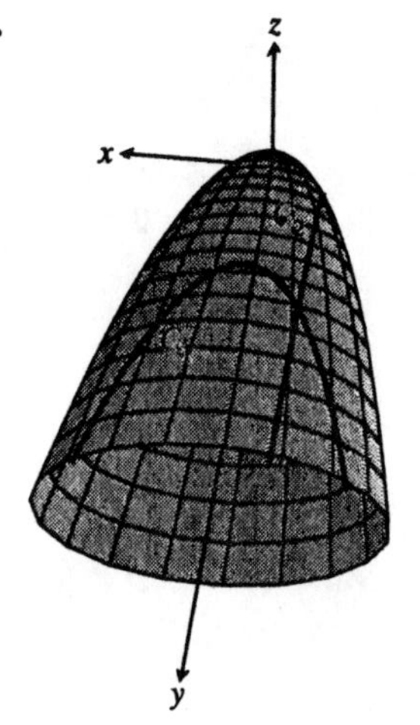

**c.** $\mathbf{g}_1'(u) = \langle 2, 2u, 0\rangle$

The tangent vector is $\mathbf{g}_1'(-2) = \langle 2, -4, 0\rangle$.

**d.** $\mathbf{g}_2'(v) = \langle 0, 2v, 1\rangle$

The tangent vector is $\mathbf{g}_2'(4) = \langle 0, 8, 1\rangle$.

**e.** A vector normal to the plane is

$$\mathbf{g}_1'(-2)\times\mathbf{g}_2'(4) = \langle 2, -4, 0\rangle \times \langle 0, 8, 1\rangle = \langle -4, -2, 16\rangle$$

or the vector $\langle -2, -1, 8\rangle$. Thus an equation for the plane is

$$\langle -2, -1, 8\rangle\cdot\langle x+4, y-20, z-5\rangle = 0$$
$$-2(x+4) - 1(y-20) + 8(z-5) = 0$$
$$-2x - y + 8z - 28 = 0.$$

**41.** Let the square be in the $(x, y)$-plane with vertices $A = (0, 1)$, $B = (-1, 0)$, $C = (0, -1)$, and $D = (1, 0)$. Note that the square is the graph of $|x| + |y| = 1$.

Let $x = u$ with $-1 \le u \le 1$. Then $y$ must range from $|x| - 1$ to $1 - |x|$, which can be accomplished by using $v$ with $-1 \le v \le 1$ but shrinking the effective range through multiplying by $1 - |x| = 1 - |u|$. So let $y = (1 - |u|)v$. Finally, for any $x$ and $y$,

$$z + |x| = 1 - |y|.$$
$$z = 1 - |x| - |y|$$
$$= 1 - |u| - |(1 - |u|)v|$$
$$= (1 - |u|) - (1 - |u|)|v|$$
$$= (1 - |u|)(1 - |v|)$$

So the top of the solid can be parametrized as $\mathbf{r}(u, v) = \langle u, (1 - |u|)v, (1 - |u|)(1 - |v|)\rangle$, for $-1 \le u, v \le 1$.

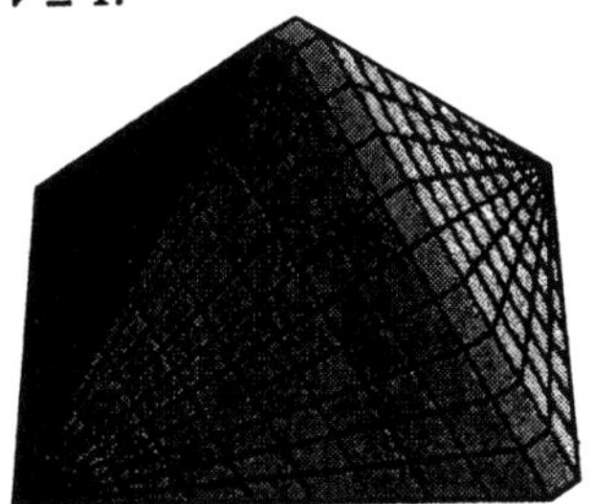

**43.** Let the equilateral triangle be in the $(x, y)$-plane with vertices $A = (0, \sqrt{3})$, $B = (-1, 0)$, and $C = (1, 0)$. Then $D = (0, 0)$. The edges of the triangle are on the lines described by $y = \sqrt{3}x + \sqrt{3}$, $y = -\sqrt{3}x + \sqrt{3}$, and $y = 0$. Consider the top of the solid. For a given $y$-coordinate, the radius of the semicircular cross-section is $1 - \frac{1}{\sqrt{3}}y$. Thus a parametric representation for the top of the solid is $\left\langle \left(1 - \frac{1}{\sqrt{3}}c\right)\cos\theta, c, \left(1 - \frac{1}{\sqrt{3}}c\right)\sin\theta\right\rangle$ for $0 \le c \le \sqrt{3}$ and $0 \le \theta \le \pi$.

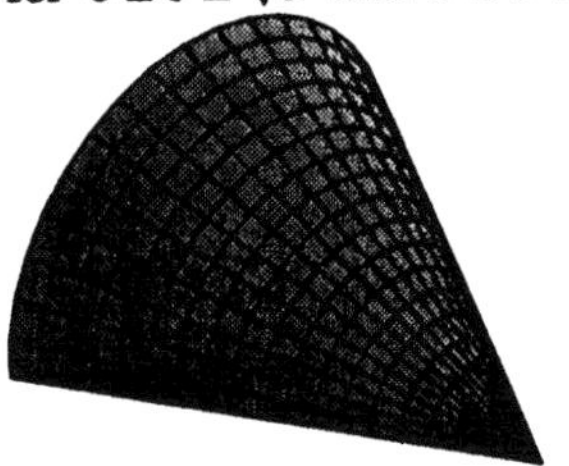

## Section 9.5 Cylindrical and Spherical Coordinates

**1.** $(x, y, z) = (1, 1, 1)$

$$r = \sqrt{1^2 + 1^2} = \sqrt{2}$$

$$\tan\theta = \frac{1}{1} = 1, \text{ so } \theta = \frac{\pi}{4}.$$

The cylindrical coordinates are $\left(\sqrt{2}, \frac{\pi}{4}, 1\right)$.

$$\rho = \sqrt{1^2 + 1^2 + 1^2} = \sqrt{3}$$

$$\phi = \arccos\left(\frac{1}{\sqrt{3}}\right)$$

$$\tan\theta = 1, \text{ so } \theta = \frac{\pi}{4}.$$

The spherical coordinates are $\left(\sqrt{3}, \frac{\pi}{4}, \arccos\left(\frac{1}{\sqrt{3}}\right)\right)$.

**3.** $(x, y, z) = \left(1, -1, -\sqrt{2}\right)$

$r = \sqrt{1^2 + (-1)^2} = \sqrt{2}$

$\tan\theta = -1$ and $x > 0$, so $\theta = -\frac{\pi}{4}$.

The cylindrical coordinates are

$\left(\sqrt{2}, -\frac{\pi}{4}, -\sqrt{2}\right)$.

$\rho = \sqrt{1^2 + (-1)^2 + \left(-\sqrt{2}\right)^2} = 2$

$\phi = \arccos\left(\frac{-\sqrt{2}}{2}\right) = \frac{3\pi}{4}$

$\tan\theta = -1$, so $\theta = -\frac{\pi}{4}$.

The spherical coordinates are $\left(2, -\frac{\pi}{4}, \frac{3\pi}{4}\right)$.

**5.** $(r, \theta, \phi) = (42, 0, 0)$
$x = 42 \cos 0 = 42$
$y = 42 \sin 0 = 0$
$z = 0$
The Cartesian coordinates are (42, 0, 0).

$\rho = \sqrt{42^2 + (0)^2 + (0)^2} = 42$

$\phi = \arccos\left(\frac{0}{42}\right) = \frac{\pi}{2}$

The spherical coordinates are $\left(42, 0, \frac{\pi}{2}\right)$.

**7.** $(r, \theta, \phi) = \left(2, \frac{\pi}{6}, -2\right)$

$x = 2\cos\frac{\pi}{6} = \sqrt{3}$

$y = 2\sin\frac{\pi}{6} = 1$

The Cartesian coordinates are $\left(\sqrt{3}, 1, -2\right)$.

$\rho = \sqrt{\left(\sqrt{3}\right)^2 + 1^2 + (-2)^2} = 2\sqrt{2}$

$\phi = \arccos\left(\frac{-2}{2\sqrt{2}}\right) = \frac{3\pi}{4}$

The spherical coordinates are $\left(2\sqrt{2}, \frac{\pi}{6}, \frac{3\pi}{4}\right)$.

**9.** $(\rho, \theta, \phi) = (42, 0, 0)$
$x = 42 \sin 0 \cos 0 = 0$
$y = 42 \sin 0 \sin 0 = 0$
$z = 42 \cos 0 = 42$
The Cartesian coordinates are (0, 0, 42).

$r = \sqrt{0^2 + 0^2} = 0$

The cylindrical coordinates are (0, 0, 42).

**11.** $(\rho, \theta, \phi) = \left(0, \frac{\pi}{4}, \frac{\pi}{4}\right)$

$x = 0\sin\frac{\pi}{4}\cos\frac{\pi}{4} = 0$; $y = 0\sin\frac{\pi}{4}\sin\frac{\pi}{4} = 0$;

$z = 0\cos\frac{\pi}{4} = 0$

The Cartesian coordinates are (0, 0, 0).
The cylindrical coordinates are (0, 0, 0).

**13.** cylindrical: $r^2 + z^2 = 16$
spherical: $\rho^2 = 16$
$\rho = 4$

**15.** cylindrical: $r\sin\theta = 5$
$r = 5\csc\theta$
spherical: $r\sin\phi\sin\theta = 5$
$r = 5\csc\phi\csc\theta$

**17.** Cartesian: $x^2 + y^2 + z^2 = 9$
spherical: $\rho^2 = 9$
$\rho = 3$

**19.** Cartesian: $r^2 = 64$
$x^2 + y^2 = 64$
spherical: $\rho^2 \sin^2\phi = 64$
$\rho = 8\csc\phi$

**21.** Cartesian: $\rho^2 = 25$
$x^2 + y^2 + z^2 = 25$
cylindrical: $r^2 + z^2 = 25$

**23.** Cartesian: $\rho\cos\phi = 3$
$z = 3$
cylindrical: $z = 3$

**25.** Let $x = r\cos\theta$ and $y = r\sin\theta$ and substitute into the equation. We get $r^2 = 5$. Then a parametric representation for $S$ is
$\mathbf{f}(\theta, t) = \left\langle\sqrt{5}\cos\theta, \sqrt{5}\sin\theta, t\right\rangle$, $0 \le \theta \le 2\pi$, $-\infty \le t \le \infty$.

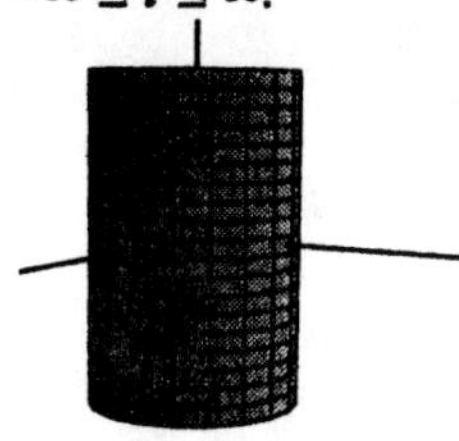

**27.** Let $x = \rho \sin \phi \cos \theta$, $y = \rho \sin \phi \sin \theta$, and $z = \rho \cos \phi$ and substitute into the equation. We get $\rho^2 - 4\rho \sin \phi \sin \theta = 0$ or $\rho = 4 \sin \phi \sin \theta$. Then a parametric representation for $S$ is $\mathbf{r}(\phi, \theta) = \left\langle 4\sin^2 \phi \sin\theta \cos\theta,\ 4\sin^2 \phi \sin^2 \theta,\ 4\cos\phi \sin\phi \sin\theta \right\rangle$, $0 \le \phi \le \pi$, $0 \le \theta \le 2\pi$.

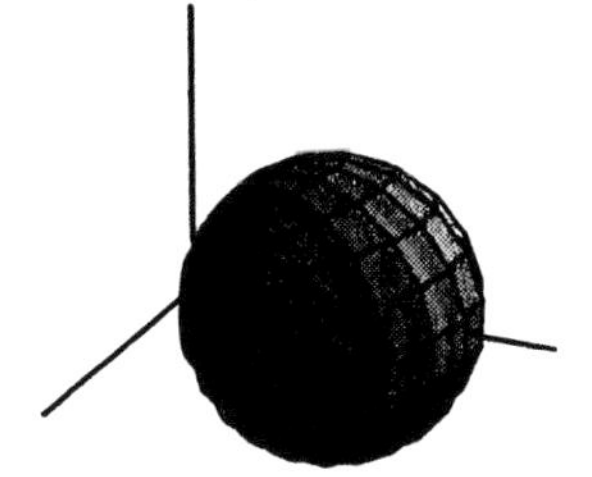

**29.** Let $x = \dfrac{1}{2} r\cos\theta$ and $y = r \sin \theta$ and substitute into the equation. We get $z = e^{-r^2}$. Then a parametric representation for $S$ is $\mathbf{r}(r, \theta) = \left\langle \dfrac{1}{2} r\cos\theta,\ r\sin\theta,\ e^{-r^2} \right\rangle$, $r \ge 0$, $0 \le \theta \le 2\pi$.

**31.** Let $x = \dfrac{1}{2}\sin\phi\cos\theta$, $y = \dfrac{1}{3}\sin\phi\sin\theta$, and $z = 2 \cos \phi$. This satisfies the given equation. Hence a parametric representation for $S$ is $\mathbf{r}(\phi, \theta) = \left\langle \dfrac{1}{2}\sin\phi\cos\theta,\ \dfrac{1}{3}\sin\phi\sin\theta,\ 2\cos\phi \right\rangle$, $0 \le \phi \le \pi$, $0 \le \theta \le 2\pi$.

**33.** $\phi = \dfrac{3\pi}{4}$, so $\cot\phi = \dfrac{z}{\sqrt{x^2+y^2}} = -1$. In cylindrical coordinates $\sqrt{x^2+y^2} = r$, so the cylindrical representation is $z = -r$. The graph of the surface is a cone below the $(x, y)$-plane. Thus the matching cylindrical representation and graph are f and III.

**35.** $\rho = 3$, so $\sqrt{x^2+y^2+z^2} = 3$ or $x^2+y^2+z^2 = 9$. In cylindrical coordinates $x^2+y^2 = r^2$, so the cylindrical representation is $r^2 + z^2 = 9$. The graph of the surface is a sphere of radius 3 centered at the origin. Thus the matching cylindrical representation and graph are b and V.

**37.** $\rho = 3 \csc \phi$ or $\rho \sin \phi = 3$, so $\sqrt{x^2+y^2} = 3$ or $x^2+y^2 = 9$. In cylindrical coordinates $x^2+y^2 = r^2$, so the cylindrical representation is $r^2 = 9$ or $r = 3$. The graph of the surface is a cylinder of radius 3 centered around the $z$-axis. Thus the matching cylindrical representation and graph are c and VI.

**39.** The graph is shaped like a double saddle.

**41.** Suppose $Q = O = (0, 0, 0)$. Then since $\rho = 4$, $x = 4\sin\phi_0\cos\theta_0 = 0$ and $y = 4\sin\phi_0\sin\theta_0 = 0$. This implies that $\sin\phi_0 = 0$ because $\cos\theta_0$ and $\sin\theta_0$ are never both 0. Thus $\phi_0 = 0$ or $\pi$. $P$ has the spherical coordinates $(4, \theta_0, 0)$ or $(4, \theta_0, \pi)$. Regardless of what $\theta_0$ is, points are included on $S_1$ as described by (11).

**43. a.**
$$\left(x+\sqrt{3}\right)^2+(y-3)^2+(z+2)^2=16$$
$$x^2+2\sqrt{3}x+3+y^2-6y+9+z^2+4z+4=16$$
$$x^2+2\sqrt{3}x+y^2-6y+z^2+4z=0$$

**b.** $r^2+2\sqrt{3}r\cos\theta-6r\sin\theta+z^2+4z=0$

**c.** $\rho^2+2\sqrt{3}\rho\sin\phi\cos\theta-6\rho\sin\phi\sin\theta+4\rho\cos\phi=0$
$$\rho=-2\sqrt{3}\sin\phi\cos\theta+6\sin\phi\sin\theta-4\cos\phi$$

**45. a.** First convert the spherical coordinates of the center to Cartesian coordinates.
$$x=10\sin\frac{\pi}{2}\cos\frac{3\pi}{4}=-5\sqrt{2},\ y=10\sin\frac{\pi}{2}\sin\frac{3\pi}{4}=5\sqrt{2},\ z=10\cos\frac{\pi}{2}=0.$$
$$\left(x+5\sqrt{2}\right)^2+\left(y-5\sqrt{2}\right)^2+z^2=100$$
$$x^2+10\sqrt{2}x+50+y^2-10\sqrt{2}y+50+z^2=100$$
$$x^2+y^2+z^2+10\sqrt{2}x-10\sqrt{2}y=0$$

**b.** $r^2+z^2+10\sqrt{2}r\cos\theta-10\sqrt{2}r\sin\theta=0$

**c.** $\rho^2+10\sqrt{2}\rho\sin\phi\cos\theta-10\sqrt{2}\rho\sin\phi\sin\theta=0$
$$\rho=10\sqrt{2}\sin\phi\sin\theta-10\sqrt{2}\sin\phi\cos\theta$$

**47.** For each point on $S$ with cylindrical coordinates $(r, \theta, z)$ such that $r \geq 1$ there is a point $(r-1, \theta, z)$ on $S^*$. Thus $S^*$ can be viewed as $S$ with a cylinder of radius 1 about the $z$-axis cut out and moved in radially by 1 unit. For example, let $g(r, \theta)=r^2$. The first graph shows the graph of $z=g(r, \theta)$ and the second graph shows the graph of $z=g(r+1, \theta)$.

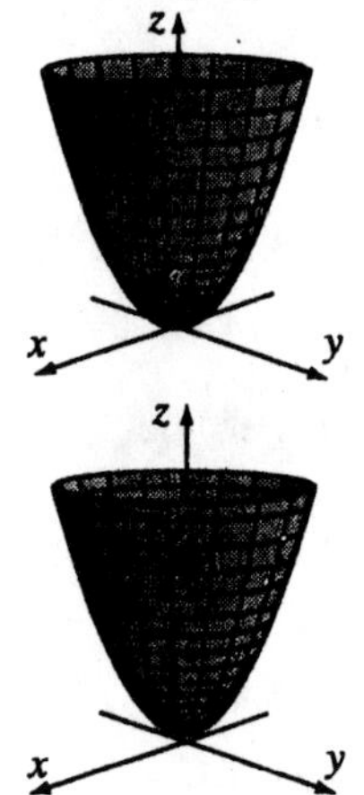

**49.** For each point on $S$ with spherical coordinate $(\rho, \theta, \phi)$ there is a point $(\rho + 1, \theta, \phi)$ on $S^*$. Thus $S^*$ can be viewed as $S$ with each point moved outward from the origin by 1 unit.
For example, let $h(\theta, \phi) = 2$. The first graph shows the graph of $\rho = h(\theta, \phi)$ and the second graph shows the graph of $\rho = h(\theta, \phi) + 1$.

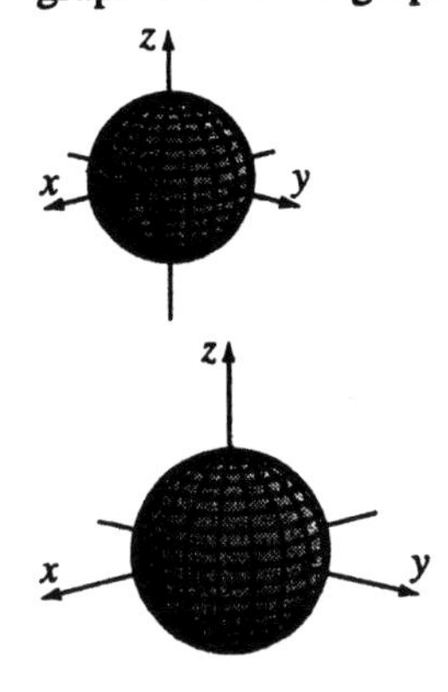

**51.** For each point on $S$ with spherical coordinate $(\rho, \theta, \phi)$ for $\frac{\pi}{6} \le \phi \le \pi$ there is a point $\left(\rho, \theta, \phi - \frac{\pi}{6}\right)$ on $S^*$. Thus $S^*$ is obtained by removing the $0 < \phi < \frac{\pi}{6}$ portion of $S$, then for the remaining points reducing $\phi$ by $\frac{\pi}{6}$.
For example, let $g(\theta, \phi) = 2 \sin \phi$. The first graph shows the graph of $r = g(\theta, \phi)$ and the second graph shows the graph of
$\rho = g\left(\theta, \phi + \frac{\pi}{6}\right)$.

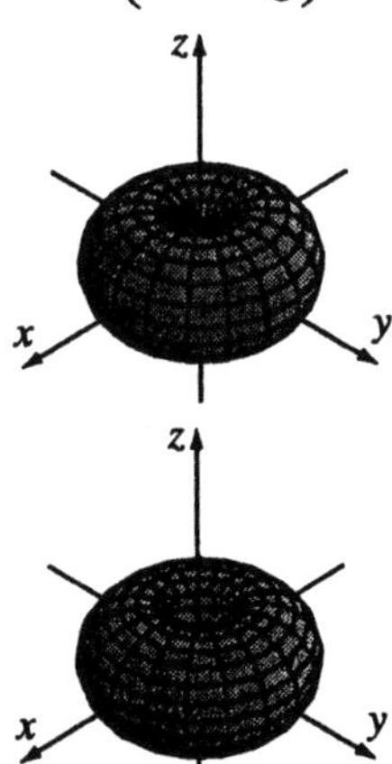

**53.** If there were no restrictions on the $\phi$-coordinate in spherical coordinates, it might be difficult to identify points on the surface.

**55. a.** If $P$ has Cartesian coordinates $(4, -4, 2)$, then
$\langle 4, -4, 2\rangle = a\langle 1, 1, 1\rangle + b\langle 4, 0, 1\rangle + c\langle 0, 0, 1\rangle$.
This equation is equivalent to the following system of three equations and three unknowns.
$a + 4b = 4$
$a = -4$
$a + b + c = 2$
Solving the system, we get $a = -4$, $b = 2$, and $c = 4$.
Thus the vector coordinates for $P$ are $\{-4, 2, 4\}$.

**b.** $Q = -2\langle 1, 1, 1\rangle + 3\langle 4, 0, 1\rangle + 4\langle 0, 0, 1\rangle$
$= \langle 10, -2, 5\rangle$
The Cartesian coordinates for $Q$ are $(10, -2, 5)$.

**c.** Let $(x, y, z)$ be the Cartesian coordinates of a point.
Thus
$\langle x, y, z\rangle = a\langle 1, 1, 1\rangle + b\langle 4, 0, 1\rangle + c\langle 0, 0, 1\rangle$,
so $x = a + 4b$, $y = a$, and $z = a + b + c$.

**d.** To express the vector coordinates $\{a, b, c\}$ in terms of the Cartesian coordinates $(x, y, z)$, we solve the system of equations
$a + 4b = x$
$a = y$
$a + b + c = z$
Thus, $a = y$, $b = \frac{1}{4}x - \frac{1}{4}y$, and
$c = -\frac{1}{4}x - \frac{3}{4}y + z$.

## Section 9.6 Limits

**1.** If $(x, y)$ is close to $(-2, 4)$, then $x$ is close to $-2$ and $y$ is close to 4. Hence
$3x^2y^2 - 5x^2y + 5x^2y + 6x - 4$ is close to
$3(-2)^2(4)^2 - 5(-2)^2(4) + 6(-2) - 4 = 96$. Thus
$$\lim_{(x, y)\to(-2, 4)} 3x^2y^2 - 5x^2y + 5x^2y + 6x - 4 = 96.$$

**3.** If $(x, y, z)$ is close to $(-2, 4, 1)$, then $x$ is close to $-2$, $y$ is close to 4, and $z$ is close to 1.
Hence $(x^2 - 2y^2 + 3x^2)^5$ is close to
$((-2)^2 - 2(4)^2 + 3(1)^2)^5 = -9{,}765{,}625$.
Thus
$$\lim_{(x, y, z)\to(-2, 4, 1)} (x^2 - 2y^2 + 3x^2)^5 = -9{,}765{,}625.$$

**5.** Let $f(x, y) = \dfrac{x^2 - y^2}{x^2 + y^2}$. We investigate the limit as $(x, y)$ approaches $(0, 0)$ along the $x$-axis and along the $y$-axis. Along the $x$-axis, points take the form $(t, 0)$ and let $t \to 0$:

$$\lim_{t\to 0} f(t, 0) = \lim_{t\to 0} \frac{t^2 - 0^2}{t^2 + 0^2} = 1$$

Along the $y$-axis, points take the form $(0, t)$ and let $t \to 0$: $\lim_{t\to 0} f(0, t) = \lim_{t\to 0} \dfrac{0^2 - t^2}{0^2 + t^2} = -1.$

Because we get different results, we conclude that $\displaystyle\lim_{(x, y)\to(0, 0)} \frac{x^2 - y^2}{x^2 + y^2}$ does not exist.

**7.** Let $f(x, y) = \dfrac{3x + 2y}{3|x| + 2|y|}$. Along the $x$-axis points take the form $(t, 0)$. Since $\displaystyle\lim_{t\to 0} f(t, 0) = \lim_{t\to 0} \frac{3x}{3|x|}$ does not exist, neither does the limit $\displaystyle\lim_{(x, y)\to(0, 0)} \frac{3x + 2y}{3|x| + 2|y|}$.

**9.** Let $f(x, y) = \dfrac{3x - 2y + 5}{|x - 1| + |y - 4|}$. Consider the behavior of $f(x, y)$ as $(x, y)$ approaches $(1, 4)$ along the line $y = 4$ for $x < 1$ and $x > 1$. Points on the line $y = 4$ have the form $(t, 4)$. First let $t \to 1^-$. This gives

$$\lim_{t\to 1^-} f(t, 4) = \lim_{t\to 1^-} \frac{3t - 3}{|t - 1|} = \lim_{t\to 1^-} \frac{3(t - 1)}{-(t - 1)} = -3.$$

Now let $t \to 1^+$. This gives

$$\lim_{t\to 1^+} f(t, 4) = \lim_{t\to 1^+} \frac{3t - 3}{|t - 1|} = \lim_{t\to 1^+} \frac{3(t - 1)}{t - 1} = 3.$$

Because we get different results as we approach $(1, 4)$ from different directions, we conclude that $\displaystyle\lim_{(x, y)\to(1, 4)} \frac{3x - 2y + 5}{|x - 1| + |y - 4|}$ does not exist.

**11.** Let $f(x, y, z) = \dfrac{\tan(x^2 + 2y^2 + 3z^2)}{x^2 + 2y^2 + 3z^2}$. Replace $x^2 + 2y^2 + 3z^2$ in $f(x, y, z)$ by $r$ and note that as $(x, y, z)$ approaches $(0, 0, 0)$, $r$ approaches 0. Hence

$$\lim_{(x, y, z)\to(0, 0, 0)} \frac{\tan(x^2 + 2y^2 + 3z^2)}{x^2 + 2y^2 + 3z^2}$$
$$= \lim_{r\to 0} \frac{\tan r}{r}$$
$$= \lim_{r\to 0} \frac{\sin r}{r} \frac{1}{\cos r}$$
$$= 1.$$

**13.** Let $f(x, y, z) = \ln(2x - 3y + z)$. Consider the behavior of $f(x, y, z)$ as $(x, y, z)$ approaches $(0, 0, 0)$ along the positive side of the $x$-axis. Points on the $x$-axis have the form $(t, 0, 0)$. As $t$ approaches 0 from the positive side, this gives $\displaystyle\lim_{t\to 0^+} f(t, 0, 0) = \lim_{t\to 0^+} \ln 2t = -\infty$. Thus we conclude that $\displaystyle\lim_{(x, y, z)\to(0, 0, 0)} \ln(2x - 3y + z)$ does not exist.

**15.** Let

$$f(x, y, z) = (2x - y + z)\sin\left(\frac{1}{4x^2 + y^2 + 8z^2}\right).$$

Let $g(x, y, z) = |2x - y + z|$.

Then $f$ and $g$ are defined for all $(x, y, z)$ in a punctured ball around $(0, 0, 0)$. Observe that $\displaystyle\lim_{(x, y, z)\to(0, 0, 0)} g(x, y, z) = 0$ and since

$$\left|\sin\left(\frac{1}{4x^2 + y^2 + 8z^2}\right)\right| \le 1,$$

$|f(x, y, z)| \le g(x, y, z)$. This gives

$$\lim_{(x, y, z)\to(0, 0, 0)} |f(x, y, z)|$$
$$\le \lim_{(x, y, z)\to(0, 0, 0)} g(x, y, z) = 0.$$

Thus we conclude that

$$\lim_{(x, y, z)\to(0, 0, 0)} (2x - y + z)\sin\left(\frac{1}{4x^2 + y^2 + 8z^2}\right) = 0.$$

**17.** The boundary is the triangular edge. The interior is the region inside of this triangle.

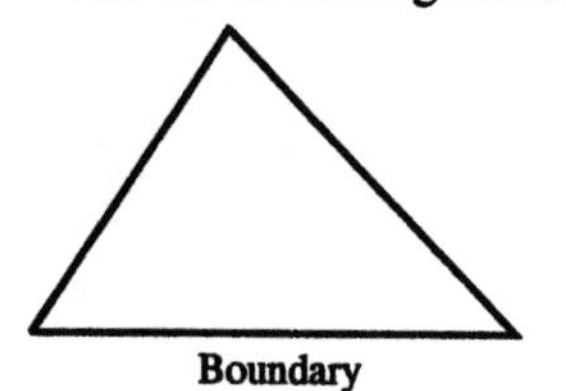
Boundary

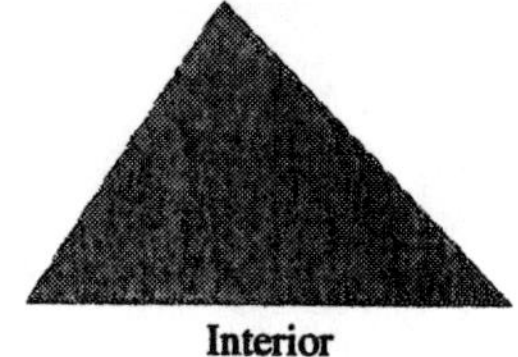
Interior

**19.** The boundary is the edge around the flower. The interior is the shaded region, and does not include any boundary points.

**21.** The boundary is the circle of center (0, 0) and radius 1. The interior is the set of all points inside of the boundary circle. $D$ is a closed set.

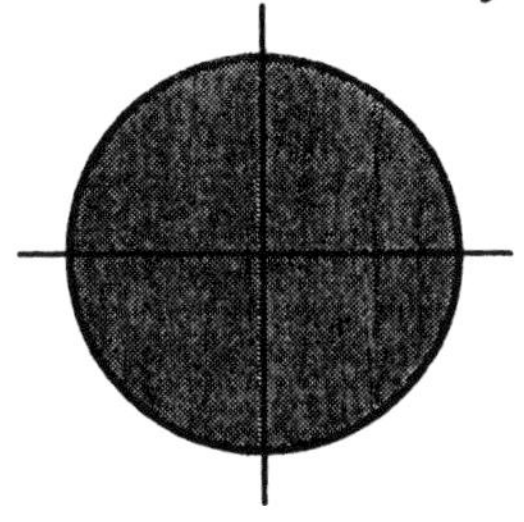

**23.** The boundary is the triangle with vertices (0, 0), (2, 0), and (0, 2). The interior is the set of all points inside of the boundary triangle. $D$ is a closed set.

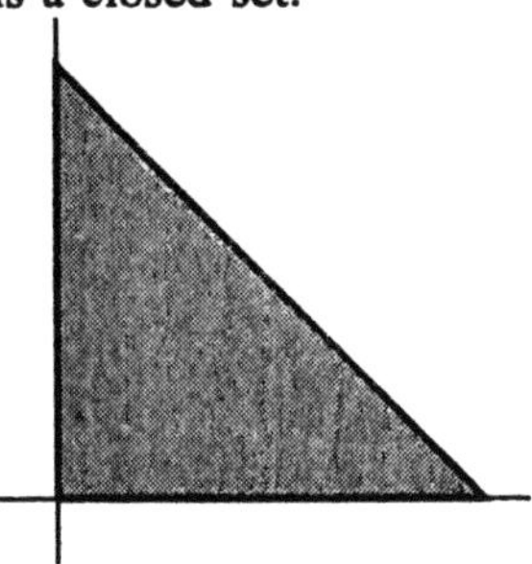

**25.** $\delta = 0.002$. Any smaller positive value is also correct.

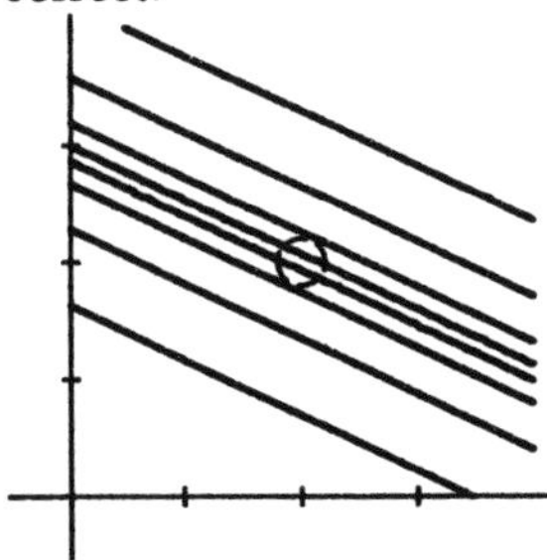

**27.** $D$ is the set of points with positive $x$-coordinates. $f$ is not continuous at points with positive $x$-coordinates on the line with equation $3x - 2y + 1 = 0$.

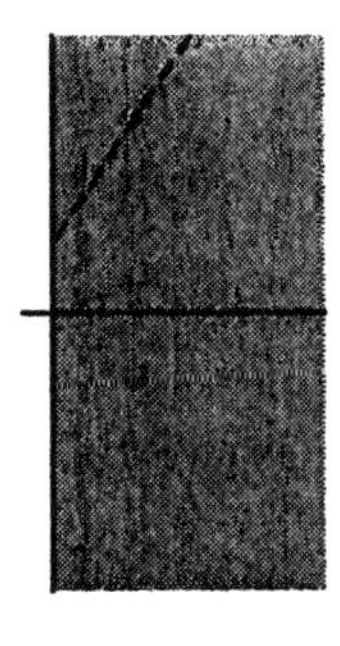

**29.** $D$ is the set of points with $x$- and $y$-coordinates less than 1 and greater than $-1$. $f$ is not continuous at points of $D$ on or to the left of the line with equation $4x - 2y + 1 = 0$.

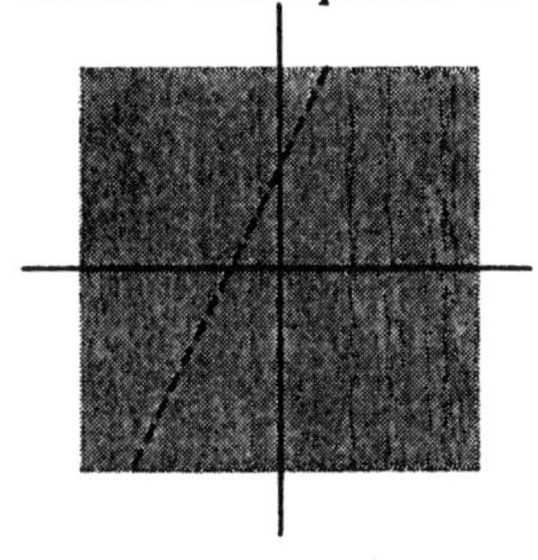

**31.** If $(x, y)$ is close to $(-2, 2)$, then $x$ is close to $-2$ and $y$ is close to 2. Hence

$3x^2y - 2xy^2 + 3x$ is close to

$3(-2)^2(2) - 2(-2)(2)^2 + 3(-2) = 34$. Thus

$$\lim_{(x, y)\to(-2, 2)} 3x^2y - 2xy^2 + 3x = 34.$$

Let $E(x, y) = |f(x, y) - 34|$

$= |3x^2y - 2xy^2 + 3x - 34|$.

Using a CAS, we plot the level curve $E(x, y) = 0.005$ for $(x, y)$ close to $(-2, 2)$. We also plot a circle of radius 0.0001 centered at $(-2, 2)$. Thus if $\delta = 0.0001$,

$|f(x, y) - 34| < 0.005$ when

$0 < \|\langle x, y\rangle - \langle -2, 2\rangle\| < \delta$.

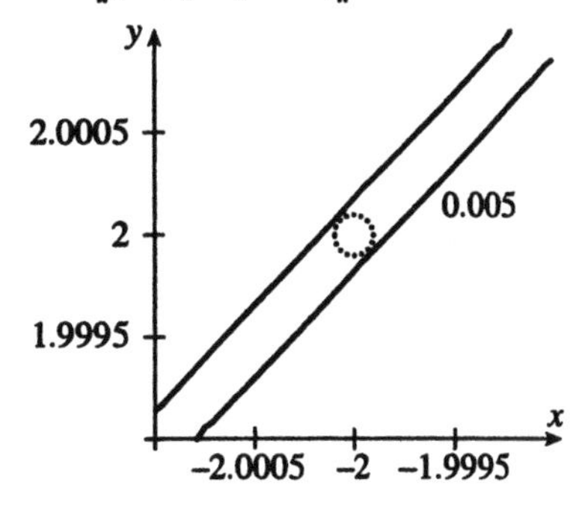

**33.** If $(x, y)$ is close to $(1, 0)$, then $x$ is close to 1 and $y$ is close to 0. Hence $\ln(2x^2 + y^3)$ is close to $\ln(2(1)^2 + 0^3) = \ln 2$. Thus, $\lim_{(x, y)\to(1, 0)} \ln(2x^2 + y^3) = \ln 2$.

Let $E(x, y) = |f(x, y) - \ln 2| = |\ln(2x^2 + y^3) - \ln 2|$. Using a CAS, we plot the level curve $E(x, y) = 0.005$ for $(x, y)$ close to $(1, 0)$. We also plot a circle of radius 0.0025 centered at $(1, 0)$. Thus if $\delta = 0.0025$, $|f(x, y) - \ln 2| < 0.005$ when $0 < \|\langle x, y\rangle - \langle 1, 0\rangle\| < \delta$.

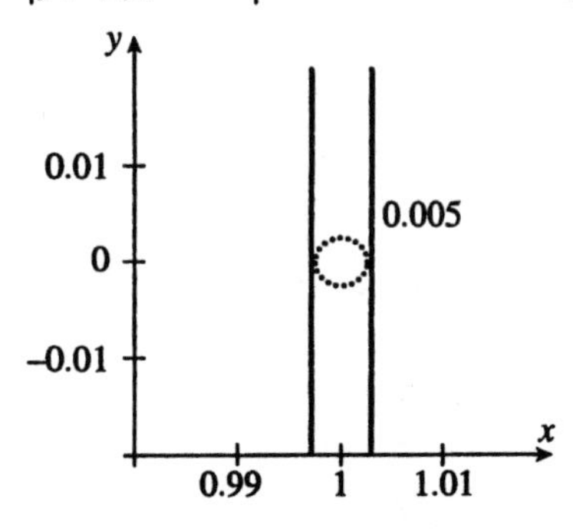


**35.** **a.** If $(x, y)$ is in $\Delta_\delta(1, 1)$ then $(x-1)^2 + (y-1)^2 < \delta^2$ which means that $|x-1| < \delta$ and $|y-1| < \delta$. Since $\delta < 1$ it also follows that $(x-1)^2 < \delta$, $(y-1)^2 < \delta$ and $|(x-1)(y-1)| < \delta$. We rewrite $x^2 = (x-1)^2 + 2(x-1) + 1$, $y^2 = (y-1)^2 + 2(y-1) + 1$ and $xy = (x-1)(y-1) + (x-1) + (y-1) + 1$ and obtain

$$\begin{aligned}|f(x, y) - 5| &= |x^2 + 2xy + 2y^2 - 5| \\ &= |(x-1)^2 + 2(x-1) + 1 + 2(x-1)(y-1) + 2(x-1) + 2(y-1) + 2 + 2(y-1)^2 + 4(y-1) + 2 - 5| \\ &= |(x-1)^2 + 4(x-1) + 2(x-1)(y-1) + 6(y-1) + 2(y-1)^2| \\ &< \delta + 4\delta + 2\delta + 6\delta + 2\delta \\ &= 15\delta, \text{ as required.}\end{aligned}$$

**b.** Let $15\delta = \varepsilon = 0.0003$ to obtain $\delta = 0.00002$.

**37.** **a.** If $(s, t)$ is in $\Delta_\delta(3, -1)$ then $(s-3)^2 + (t+1)^2 < \delta^2$ which means that $|s-3| < \delta$ and $|t+1| < \delta$. Since $\delta < 1$ it also follows that $(s-3)^2 < \delta$, $(t+1)^2 < \delta$ and $(s-3)(t+1) < \delta$.

We rewrite $s^2 = (s-3)^2 + 6(s-3) + 9$ and $t = (t+1) - 1$ and obtain

$$\begin{aligned}|h(s, t) + 9| &= |s^2 t + 9| \\ &= |((s-3)^2 + 6(s-3) + 9)((t+1) - 1) + 9| \\ &= |(s-3)^2(t+1) + 6(s-3)(t+1) + 9(t+1) - (s-3)^2 - 6(s-3) - 9 + 9| \\ &= |(s-3)^2(t+1) + 6(s-3)(t+1) + 9(t+1) - (s-3)^2 - 6(s-3)| \\ &< \delta + 6\delta + 9\delta + \delta + 6\delta \\ &= 23\delta \text{ as required.}\end{aligned}$$

**b.** Let $23\delta < \varepsilon = 0.0001$ to obtain $\delta = \dfrac{0.0001}{23} \approx 0.000004$.

**39.** We know that $\lim_{(x, y)\to(a, b)} p(x, y) = p(a, b) = L$ and $\lim_{(x, y)\to(a, b)} q(x, y) = q(a, b) = M$. Observe that

$$\begin{aligned}
\left|\frac{p(x, y)}{q(x, y)} - \frac{L}{M}\right| &= \left|\frac{p(x, y)}{q(x, y)} - \frac{p(a, b)}{q(a, b)}\right| \\
&= \left|\frac{p(x, y)q(a, b) - q(x, y)p(a, b)}{q(x, y)q(a, b)}\right| \\
&= \left|\frac{p(x, y)q(a, b) - p(a, b)q(a, b) + p(a, b)q(a, b) - q(x, y)p(a, b)}{q(x, y)q(a, b)}\right| \\
&= \left|\frac{q(a, b)(p(x, y) - p(a, b)) - p(a, b)(q(x, y) - q(a, b))}{q(x, y)q(a, b)}\right| \\
&= \left|\frac{M(p(x, y) - p(a, b)) - L(q(x, y) - q(a, b))}{q(x, y)M}\right| \\
&\le \left|\frac{M(p(x, y) - p(a, b))}{q(x, y)M}\right| + \left|\frac{L(q(x, y) - q(a, b))}{q(x, y)M}\right| \\
&= \left|\frac{p(x, y) - p(a, b)}{q(x, y)}\right| + \left|\frac{L(q(x, y) - q(a, b))}{Mq(x, y)}\right|
\end{aligned}$$

We want to show that for $(x, y)$ close to $(a, b)$, $\frac{p(x, y)}{q(x, y)}$ approaches $\frac{L}{M}$. There is a disk of some radius $r$ centered at $(a, b)$ such that $q(x, y) \neq 0$ for all $(x, y)$ in the disk. Since $q$ is polynomial function, there exists some $m > 0$ such that $|q(x, y)| \geq m$ for all $(x, y)$ in the disk. Thus for all points in that disk we have

$$\left|\frac{p(x, y)}{q(x, y)} - \frac{p(a, b)}{q(a, b)}\right| \le \left|\frac{p(x, y) - p(a, b)}{m}\right| + \left|\frac{L(q(x, y) - q(a, b))}{mM}\right|.$$

By the definition of a limit, if given $\frac{\varepsilon m}{2}$, we can find $\delta_1$ such that when $0 < \|\langle x, y\rangle - \langle a, b\rangle\| < \delta_1$ $|p(x, y) - L| < \frac{\varepsilon m}{2}$, and if given $\frac{\varepsilon mM}{2L}$, we can find $\delta_2$ such that when $0 < \|\langle x, y\rangle - \langle a, b\rangle\| < \delta_2$ $|q(x, y) - M| < \frac{\varepsilon mM}{2L}$. Let $\delta$ be the minimum among $r$, $\delta_1$, and $\delta_2$. So if given $\varepsilon$ and $0 < \|\langle x, y\rangle - \langle a, b\rangle\| < \delta$, then we know that

$$\left|\frac{p(x, y)}{q(x, y)} - \frac{L}{M}\right| \le \left|\frac{p(x, y) - p(a, b)}{m}\right| + \left|\frac{L(q(x, y) - q(a, b)}{mM}\right| < \frac{\varepsilon}{2} + \frac{\varepsilon}{2} = \varepsilon.$$

Hence $\lim_{(x, y)\to(a, b)} \frac{p(x, y)}{q(x, y)} = \frac{p(a, b)}{q(a, b)}$.

**41.** Let $D$ be a subset of $R^3$. A point $Q$ in $R^3$ is called an *interior point* of $D$ if some ball of positive radius centered at $Q$ is contained in $D$. A point $P$ in $R^3$ is called a *boundary point* of $D$ if every ball centered at $P$ contains a point in $D$ and points that are not in $D$.

**43. a.** Let $E(x, y) = \left|\sqrt{xy} - \left(\frac{1}{4}x - y\right)\right|$. Using a CAS, we plot the level curve $E(x, y) = 0.005$ for $(x, y)$ close to $(4, 1)$. We also plot a circle of radius 0.1 centered at $(4, 1)$. Thus if $\delta = 0.1$, $|E(x, y)| < 0.005$ when $0 < \|\langle x, y\rangle - \langle 4, 1\rangle\| < \delta$.

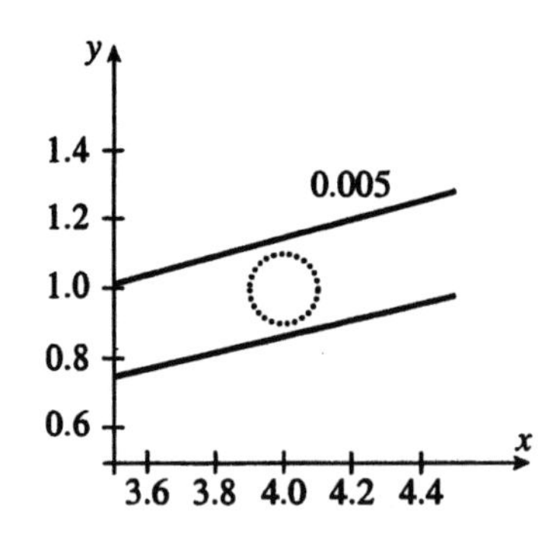


**b.** The point (3.85, 1.07) is $\sqrt{(0.15)^2+(0.07)^2} \approx 0.166$ units away from the point (4, 1). A circle of this radius is entirely contained within the level curve $E(x, y) = 0.01$. Thus the error should be less than 0.01.

$$f(3.85, 1.07) \approx \frac{1}{4}(3.85)+1.07 = 2.0325$$

$$f(3.85, 1.07) = \sqrt{(3.85)(1.07)} \approx 2.0297$$

$$\left|\sqrt{(3.85)(1.07)} - \left(\frac{1}{4}(3.85)+1.07\right)\right| \approx 0.00284$$

**45. a.** Let $\delta$ be the minimum between $1-|x_0|$ and $1-|y_0|$. Then let $r=\frac{\delta}{2}$ or any number less than $\delta$. This proves that $Q$ is an interior point of $D$.

**b.** Let $\delta$ be the minimum between $|x_1|-1$ and $|y_1|-1$. Then let $r=\frac{\delta}{2}$ or any number less than $\delta$. This proves that $R$ is neither an interior point nor a boundary point of $D$.

**47.** Suppose $(x, y)$ approaches (0, 0) along the line $y = kx$. Points on the line have the form $(t, kt)$. This gives

$$\lim_{t\to 0} F(t, kt) = \lim_{t\to 0}\frac{t^m(kt)^n}{t^{2m}+(kt)^{2n}} = \lim_{t\to 0}\frac{k^n t^{m+n}}{t^{2m}+(kt)^{2n}}.$$

If $n > m$ then we have

$$\lim_{t\to 0} F(t, kt) = \lim_{t\to 0}\frac{k^n t^{n-m}}{1+(kt)^{2n-2m}} = \frac{0}{1} = 0.$$

Similarly, if $m > n$, then we have

$$\lim_{t\to 0} F(t, kt) = \lim_{t\to 0}\frac{k^n t^{m-n}}{t^{2m-2n}+k^{2n}} = \frac{0}{k^{2n}} = 0.$$

Now suppose $(x, y)$ approaches (0, 0) along the path described by $(x, y) = (t^n, t^m)$. This gives

$$\lim_{t\to 0} F(t^n, t^m) = \lim_{t\to 0}\frac{(t^n)^m(t^m)^n}{(t^n)^{2m}+(t^m)^{2n}} = \lim_{t\to 0}\frac{t^{mn}}{2t^{2mn}} = \frac{1}{2}.$$

Hence $\lim_{(x, y)\to(0, 0)} F(x, y)$ does not exist when $m \neq n$.

If $n = m$, then $F(x, y) = \frac{x^m y^m}{x^{2m}+y^{2m}}$.

Suppose $(x, y)$ approaches (0, 0) along the path described by $(x, y) = (t, t)$. This gives

$$\lim_{t\to 0} F(t, t) = \lim_{t\to 0}\frac{t^m t^m}{t^{2m}+t^{2m}} = \lim_{t\to 0}\frac{t^{2m}}{2t^{2m}} = \frac{1}{2}.$$

Suppose $(x, y)$ approaches (0, 0) along the path described by $(x, y) = (t, 0)$.

This gives $\lim_{t\to 0} F(t, 0) = \lim_{t\to 0}\frac{0}{t^{2m}+0} = 0.$

Hence $\lim_{(x, y)\to(0, 0)} F(x, y)$ does not exist when $m = n$.

**49.** Suppose that $\lim_{t\to t_0} f(\mathbf{r}(t)) = M \neq L$. Let $\varepsilon$ be some number such that $\varepsilon = \frac{|L-M|}{2}$. By definition of limit, there is some number $\delta_0$ such that when $0 < \|\langle x, y\rangle - \langle a, b\rangle\| < \delta_0$, we have $|f(x, y) - L| < \varepsilon$. Also, since $C$ is a curve (and thus continuous), there is some number $\delta_1$ such that when $0 < |t-t_0| < \delta_1$, we have $\|\mathbf{r}(t) - \langle a, b\rangle\| < \delta_0$. By the definition of limit, there is some number $\delta_2$ such that when $0 < |t-t_0| < \delta_2$, we have $|f(\mathbf{r}(t)) - M| < \varepsilon$. Let $\delta_3$ be the minimum between $\delta_1$ and $\delta_2$. Then for $t_1$ such that $|t_1 - t_0| < \delta_3$, we have $|f(\mathbf{r}(t_1)) - M| < \varepsilon$. We also have $|f(\mathbf{r}(t_1)) - L| < \varepsilon$ since $\|\mathbf{r}(t_1) - \langle a, b\rangle\| < \delta_0$. This gives us a contradiction, since $M$ and $L$ are $2\varepsilon$ units apart, so $\lim_{t\to t_0} f(\mathbf{r}(t)) = L$.

**Section 9.7 Derivatives**

**1.** $f(x, y) = -3x + \sqrt{2}y + 4$

$$f_x(x, y) = \frac{\partial}{\partial x}\left(-3x + \sqrt{2}y + 4\right) = -3$$

$$f_y(x, y) = \frac{\partial}{\partial y}\left(-3x + \sqrt{2}y + 4\right) = \sqrt{2}$$

**3.** $f(x, y) = \dfrac{x + 2y}{(y - 2x)(y^2 + 1)}$

$$\begin{aligned}
f_x(x, y) &= \frac{\partial}{\partial x} \frac{x + 2y}{(y - 2x)(y^2 + 1)} \\
&= \frac{(y - 2x)(y^2 + 1)\frac{\partial}{\partial x}(x + 2y) - (x + 2y)\frac{\partial}{\partial x}((y - 2x)(y^2 + 1))}{((y - 2x)(y^2 + 1))^2} \\
&= \frac{(y - 2x)(y^2 + 1)(1) - (x + 2y)(-2)(y^2 + 1)}{(y - 2x)^2(y^2 + 1)^2} \\
&= \frac{(y - 2x) + 2(x + 2y)}{(y - 2x)^2(y^2 + 1)} \\
&= \frac{5y}{(y - 2x)^2(y^2 + 1)}
\end{aligned}$$

$$\begin{aligned}
f_y(x, y) &= \frac{\partial}{\partial y} \frac{x + 2y}{(y - 2x)(y^2 + 1)} \\
&= \frac{(y - 2x)(y^2 + 1)\frac{\partial}{\partial y}(x + 2y) - (x + 2y)\frac{\partial}{\partial y}((y - 2x)(y^2 + 1))}{((y - 2x)(y^2 + 1))^2} \\
&= \frac{(y - 2x)(y^2 + 1)(2) - (x + 2y)((y^2 + 1) + (y - 2x)2y)}{(y - 2x)^2(y^2 + 1)^2} \\
&= \frac{2(y - 2x)(y^2 + 1) - (x + 2y)(3y^2 - 4xy + 1)}{(y - 2x)^2(y^2 + 1)^2} \\
&= \frac{4x^2y + xy^2 - 4y^3 - 5x}{(y - 2x)^2(y^2 + 1)^2}
\end{aligned}$$

**5.** $f(x, y) = \dfrac{2e^x - xe^y}{y^2e^x - e^y}$

$$f_x(x, y) = \frac{\partial}{\partial x}\frac{2e^x - xe^y}{y^2e^x - e^y}$$

$$= \frac{(y^2e^x - e^y)\frac{\partial}{\partial x}(2e^x - xe^y) - (2e^x - xe^y)\frac{\partial}{\partial x}(y^2e^x - e^y)}{(y^2e^x - e^y)^2}$$

$$= \frac{(y^2e^x - e^y)(2e^x - e^y) - (2e^x - xe^y)(y^2e^x)}{(y^2e^x - e^y)^2}$$

$$= \frac{2y^2e^{2x} - y^2e^{x+y} - 2e^{x+y} + e^{2y} - 2y^2e^{2x} + xy^2e^{x+y}}{(y^2e^x - e^y)^2}$$

$$= \frac{e^y(xy^2e^x - y^2e^x - 2e^x + e^y)}{(y^2e^x - e^y)^2}$$

$$f_y(x, y) = \frac{\partial}{\partial y}\frac{2e^x - xe^y}{y^2e^x - e^y}$$

$$= \frac{(y^2e^x - e^y)\frac{\partial}{\partial y}(2e^x - xe^y) - (2e^x - xe^y)\frac{\partial}{\partial y}(y^2e^x - e^y)}{(y^2e^x - e^y)^2}$$

$$= \frac{(y^2e^x - e^y)(-xe^y) - (2e^x - xe^y)(2ye^x - e^y)}{(y^2e^x - e^y)^2}$$

$$= \frac{-xy^2e^{x+y} + xe^{2y} - 4ye^{2x} + 2e^{x+y} + 2xye^{x+y} - xe^{2y}}{(y^2e^x - e^y)^2}$$

$$= \frac{e^x(-xy^2e^y - 4ye^x + 2e^y + 2xye^y)}{(y^2e^x - e^y)^2}$$

**7.** $g(x, y, z) = xy^2z^3 + 2$

$$g_x(x, y, z) = \frac{\partial}{\partial x}(xy^2z^3 + 2) = y^2z^3$$

$$g_y(x, y, z) = \frac{\partial}{\partial y}(xy^2z^3 + 2) = 2xyz^3$$

$$g_z(x, y, z) = \frac{\partial}{\partial z}(xy^2z^3 + 2) = 3xy^2z^2$$

**9.** $g(x, y, z) = \dfrac{x^2 + y^2 - z^2}{x^2 - y^2 + z^2}$

$$g_x(x, y, z) = \frac{\partial}{\partial x}\left(\frac{x^2 + y^2 - z^2}{x^2 - y^2 + z^2}\right)$$

$$= \frac{(x^2 - y^2 + z^2)\frac{\partial}{\partial x}(x^2 + y^2 - z^2) - (x^2 + y^2 - z^2)\frac{\partial}{\partial x}(x^2 - y^2 + z^2)}{(x^2 - y^2 + z^2)^2}$$

$$= \frac{(x^2 - y^2 + z^2)2x - (x^2 + y^2 - z^2)2x}{(x^2 - y^2 + z^2)^2}$$

$$= \frac{4x(-y^2 + z^2)}{(x^2 - y^2 + z^2)^2}$$

$$g_y(x, y, z) = \frac{\partial}{\partial y}\left(\frac{x^2+y^2-z^2}{x^2-y^2+z^2}\right)$$
$$= \frac{(x^2-y^2+z^2)\frac{\partial}{\partial y}(x^2+y^2-z^2)-(x^2+y^2-z^2)\frac{\partial}{\partial y}(x^2-y^2+z^2)}{(x^2-y^2+z^2)^2}$$
$$= \frac{(x^2-y^2+z^2)2y-(x^2+y^2-z^2)(-2y)}{(x^2-y^2+z^2)^2}$$
$$= \frac{4x^2y}{(x^2-y^2+z^2)^2}$$
$$g_z(x, y, z) = \frac{\partial}{\partial z}\left(\frac{x^2+y^2-z^2}{x^2-y^2+z^2}\right)$$
$$= \frac{(x^2-y^2+z^2)\frac{\partial}{\partial z}(x^2+y^2-z^2)-(x^2+y^2-z^2)\frac{\partial}{\partial z}(x^2-y^2+z^2)}{(x^2-y^2+z^2)^2}$$
$$= \frac{(x^2-y^2+z^2)(-2z)-(x^2+y^2-z^2)(2z)}{(x^2-y^2+z^2)^2}$$
$$= \frac{-4x^2z}{(x^2-y^2+z^2)^2}$$

**11.** $g(x, y, z) = (x+z^2)\cos y \sin 2y$

$$g_x(x, y, z) = \frac{\partial}{\partial x}((x+z^2)\cos y \sin 2y)$$
$$= \cos y \sin 2y$$
$$g_y(x, y, z) = \frac{\partial}{\partial y}((x+z^2)\cos y \sin 2y)$$
$$= (x+z^2)(-\sin y \sin 2y + 2\cos y \cos 2y)$$
$$g_z(x, y, z) = \frac{\partial}{\partial x}((x+z^2)\cos y \sin 2y)$$
$$= 2z \cos y \sin 2y$$

**13.** $\nabla f(x, y) = \langle f_x(x, y), f_y(x, y)\rangle = \langle -3, \sqrt{2}\rangle$

$\nabla f(1, -2) = \langle -3, \sqrt{2}\rangle$

**15.** $\nabla f(x, y) = \left\langle \frac{2x+3y}{x^2+3xy+y^3}, \frac{3x+3y^2}{x^2+3xy+y^3}\right\rangle$

**17.** $\nabla g(x, y, z) = (y^2z^3, 2xyz^3, 3xy^2z^2)$

$\nabla g(1, 0, -2) = \langle 0, 0, 0\rangle$

**19.** $\nabla g(t_1, t_2, t_3) = \langle g_x(t_1, t_2, t_3), g_y(t_1, t_2, t_3), g_z(t_1, t_2, t_3)\rangle$

$$= \left\langle 2t_1\sqrt{t_2}\arctan(2t_1t_2) + \frac{2t_1^2t_2^{3/2}}{1+4t_1^2t_2^2}, \frac{t_1^2}{2\sqrt{t_2}}\arctan(2t_1t_2) + \frac{2t_1^3\sqrt{t_2}}{1+4t_1^2t_2^2}, 0\right\rangle$$

**21.** $D_{\mathbf{u}}f(\mathbf{P}) = \nabla f(1, -2)\cdot\left\langle \frac{3}{5}, -\frac{4}{5}\right\rangle = \langle -3, \sqrt{2}\rangle \cdot \left\langle \frac{3}{5}, -\frac{4}{5}\right\rangle = -\frac{9}{5} - \frac{4\sqrt{2}}{5}$

**23.** $D_{\mathbf{u}}f(\mathbf{P}) = \nabla f(a, b) \cdot \langle 1, 0\rangle = \langle f_x(a, b), f_y(a, b)\rangle \cdot \langle 1, 0\rangle = f_x(a, b) = \dfrac{2a+3b}{a^2+3ab+b^3}$

**25.** $D_{\mathbf{u}}g(\mathbf{P}) = \nabla g(1, 0, -2) \cdot \left\langle \frac{1}{3}, \frac{2}{3}, -\frac{2}{3}\right\rangle = \langle 0, 0, 0\rangle \cdot \left\langle \frac{1}{3}, \frac{2}{3}, -\frac{2}{3}\right\rangle = 0$

**27.** $D_{\mathbf{u}}g(\mathbf{P}) = \nabla g(0, 1, \pi) \cdot \langle 0, 0, 1\rangle = \langle g_x(0, 1, \pi), g_y(0, 1, \pi), g_z(0, 1, \pi)\rangle \cdot \langle 0, 0, 1\rangle = g_z(0, 1, \pi) = 0$

**29.** $f(x, y) = 2x^2 - xy^3 - x + y$

$f_x(x, y) = \frac{\partial}{\partial x}(2x^2 - xy^3 - x + y) = 4x - y^3 - 1$

$f_y(x, y) = \frac{\partial}{\partial y}(2x^2 - xy^3 - x + y) = -3xy^2 + 1$

$\nabla f(x, y) = \langle f_x(x, y), f_y(x, y)\rangle = \langle 4x - y^3 - 1, -3xy^2 + 1\rangle$

$\nabla f(1, -2) = \langle 11, -11\rangle$

$D_{\mathbf{u}}f(1, -2)$ is maximum when the angle between $\nabla f(1, -2)$ and $\mathbf{u}$ is 0, so

$\mathbf{u} = \frac{1}{\|\nabla f(1, -2)\|}\nabla f(1, -2) = \left\langle \frac{1}{\sqrt{2}}, -\frac{1}{\sqrt{2}}\right\rangle$ and $D_{\mathbf{u}}f(1, -2) = \|\langle 11, -11\rangle\|\cos 0 = 11\sqrt{2}$.

$D_{\mathbf{v}}f(1, -2)$ is minimum when the angle between $\nabla f(1, -2)$ and $\mathbf{v}$ is $\pi$, so

$\mathbf{v} = -\frac{1}{\|\nabla f(1, -2)\|}\nabla f(1, -2) = \left\langle -\frac{1}{\sqrt{2}}, \frac{1}{\sqrt{2}}\right\rangle$ and $D_{\mathbf{v}}f(1, -2) = \|\langle 11, -11\rangle\|\cos \pi = -11\sqrt{2}$.

**31.** $f(x, y) = xye^{x-y}$

$\nabla f(x, y) = \langle ye^{x-y} + xye^{x-y}, xe^{x-y} - xye^{x-y}\rangle$

$\nabla f(2, 1) = \langle 3e, 0\rangle$

$D_{\mathbf{u}}f(2, 1)$ is maximum when the angle between $\nabla f(2, 1)$ and $\mathbf{u}$ is 0,

so $\mathbf{u} = \frac{1}{\|\nabla f(2, 1)\|}\nabla f(2, 1) = \langle 1, 0\rangle$ and $D_{\mathbf{u}}f(1, -2) = \|\langle 3e, 0\rangle\|\cos 0 = 3e$.

$D_{\mathbf{v}}f(2, 1)$ is minimum when the angle between $\nabla f(2, 1)$ and $\mathbf{v}$ is $\pi$,

so $\mathbf{v} = -\frac{1}{\|\nabla f(2, 1)\|}\nabla f(2, 1) = \langle -1, 0\rangle$ and $D_{\mathbf{v}}f(2, 1) = \|\langle 3e, 0\rangle\|\cos \pi = -3e$.

**33.** $D_{\mathbf{u}}f(1, -2) = 0$ when the angle between $\nabla f(1, -2) = \langle 11, -11\rangle$ and $\mathbf{u}$ is $\frac{\pi}{2}$, so $\mathbf{u} = \pm\left\langle \frac{1}{\sqrt{2}}, \frac{1}{\sqrt{2}}\right\rangle$.

**35.** $D_{\mathbf{u}}f(2, 1) = 0$ when the angle between $\nabla f(2, 1) = \langle 3e, 0\rangle$ and $\mathbf{u}$ is $\frac{\pi}{2}$, so $\mathbf{u} = \pm\langle 0, 1\rangle$.

**37.** First we simplify the expression $I(50 + 150h, 100 + 50h) - I(50,100)$:

$I(50+150h, 100+50h) - I(50, 100)$

$$= \frac{40}{(-50+150h)^2+(-50+50h)^2+400} + \frac{40}{(-100+150h)^2+(100+50h)^2+400}$$

$$+ \frac{40}{(50+150h)^2+(100+50h)^2+400} - \frac{40}{(-50)^2+(-50)^2+400}$$

$$- \frac{40}{(-100)^2+(100)^2+400} - \frac{40}{(50)^2+(100)^2+400}$$

$$= \frac{1}{625h^2-500h+135} + \frac{1}{625h^2-500h+510} + \frac{2}{1250h^2+1250h+645} - \frac{1}{135} - \frac{1}{510} - \frac{2}{645}$$

$$= \frac{-625h^2+500h}{135(625h^2-500h+135)} + \frac{-625h^2+500h}{510(625h^2-500h+510)} + \frac{-2500h^2-2500h}{645(1250h^2+1250h+645)}$$

$$= \frac{5(-5h^2+4h)}{27(125h^2-100h+27)} + \frac{5(-5h^2+4h)}{102(125h^2-100h+102)} + \frac{-100(h^2+h)}{129(250h^2+250h+129)}$$

Thus, $\lim_{h\to 0^+} \frac{I(\mathbf{P}+h\mathbf{v})-I(\mathbf{P})}{\|h\mathbf{v}\|} = \lim_{h\to 0^+} \frac{I(50+150h, 100+50h)-I(50, 100)}{50\sqrt{10}h}$

$$= \lim_{h\to 0^+} \frac{-5h+4}{270\sqrt{10}(125h^2-100h+27)} + \lim_{h\to 0^+} \frac{-5h+4}{1020\sqrt{10}(125h^2-100h+102)}$$

$$+ \lim_{h\to 0^+} \frac{-2(h+1)}{129\sqrt{10}(250h^2+250h+129)}$$

$$= \frac{2}{3645\sqrt{10}} + \frac{1}{26{,}010\sqrt{10}} - \frac{2}{16{,}641\sqrt{10}} \approx 0.0001477$$

**39.** $\mathbf{u} = \left\langle \cos\frac{3\pi}{4}, \sin\frac{3\pi}{4} \right\rangle = \left\langle -\frac{\sqrt{2}}{2}, \frac{\sqrt{2}}{2} \right\rangle$

$\nabla g(x, y) = \left\langle (x+y+1)e^{x+2y}, (2x+2y+1)e^{x+2y} \right\rangle$

$\nabla g(2, -1) = \left\langle 2e^{-1}, 3e^{-1} \right\rangle$

$D_{\mathbf{u}}g(-2, 1) = \nabla g(2, -1)\cdot \mathbf{u} = \left\langle 2e^{-1}, 3e^{-1} \right\rangle \cdot \left\langle -\frac{\sqrt{2}}{2}, \frac{\sqrt{2}}{2} \right\rangle = \frac{\sqrt{2}}{2}e^{-1}$

**41.** $\mathbf{u} = -\frac{\mathbf{P}}{\|\mathbf{P}\|} = \frac{1}{\sqrt{5}}\langle 0, 1, -2\rangle = \left\langle 0, \frac{1}{\sqrt{5}}, -\frac{2}{\sqrt{5}} \right\rangle$

$f_x(x, y, z) = \frac{yz(x^2+y^2+z^2+1)-2x(xyz+1)}{(x^2+y^2+z^2+1)^2}$

$f_y(x, y, z) = \frac{xz(x^2+y^2+z^2+1)-2y(xyz+1)}{(x^2+y^2+z^2+1)^2}$

$f_z(x, y, z) = \frac{xy(x^2+y^2+z^2+1)-2z(xyz+1)}{(x^2+y^2+z^2+1)^2}$

$\nabla f(0, -1, 2) = \left\langle f_x(0, -1, 2), f_y(0, -1, 2), f_z(0, -1, 2) \right\rangle = \left\langle -\frac{1}{3}, \frac{1}{18}, -\frac{1}{9} \right\rangle$

$D_{\mathbf{u}}f(0, -1, 2) = \nabla f(0, -1, 2)\cdot \mathbf{u} = \left\langle -\frac{1}{3}, \frac{1}{18}, -\frac{1}{9} \right\rangle \cdot \left\langle 0, \frac{1}{\sqrt{5}}, -\frac{2}{\sqrt{5}} \right\rangle = \frac{\sqrt{5}}{18}$

**43.** Let $\nabla f(3,-7)=\langle a, b\rangle$.

$$D_{\mathbf{u}}f(3,-7)=\langle a, b\rangle\cdot\left\langle\frac{\sqrt{2}}{2}, \frac{\sqrt{2}}{2}\right\rangle$$
$$=\frac{\sqrt{2}}{2}a+\frac{\sqrt{2}}{2}b$$
$$=8$$

$$D_{\mathbf{v}}f(3,-7)=\langle a, b\rangle\cdot\left\langle-\frac{\sqrt{2}}{2}, \frac{\sqrt{2}}{2}\right\rangle$$
$$=-\frac{\sqrt{2}}{2}a+\frac{\sqrt{2}}{2}b$$
$$=-1$$

Thus we have a system of two equations and two unknowns. Solving for $a$ and $b$, we get $a=\frac{9}{\sqrt{2}}$, $b=\frac{7}{\sqrt{2}}$. Thus

$$\nabla f(3,-7)=\left\langle\frac{9}{\sqrt{2}}, \frac{7}{\sqrt{2}}\right\rangle.$$

**45.** **a.** $f_x(x, y)=\frac{\partial}{\partial x}(2x^2y-xy+1)=4xy-y$

$$f_y(x, y)=\frac{\partial}{\partial y}(2x^2y-xy+1)=2x^2-x$$
$$\nabla f(1,-1)=\langle f_x(1,-1), f_y(1,-1)\rangle=\langle -3, 1\rangle$$
$$d(\theta)=D_{\mathbf{u}_\theta}f(1,-1)$$
$$=\langle -3, 1\rangle\cdot\langle\cos\theta, \sin\theta\rangle$$
$$=-3\cos\theta+\sin\theta$$

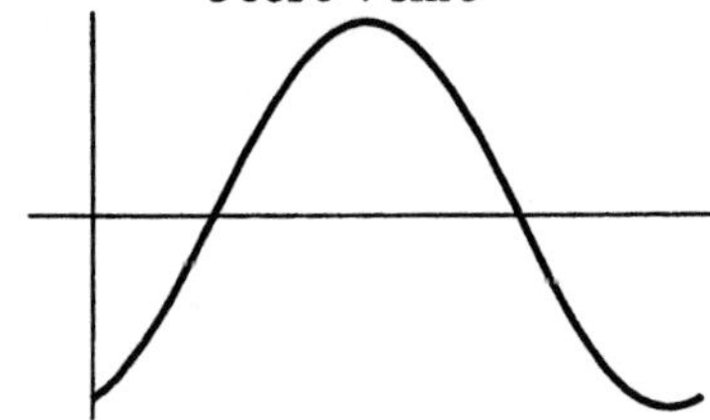

**b.** $d$ takes on a maximum value of $\sqrt{10}$ at $\theta=\pi+\arctan\left(-\frac{1}{3}\right)\approx 2.820$. $d$ takes on a minimum value of $-\sqrt{10}$ at $\theta=2\pi+\arctan\left(-\frac{1}{3}\right)\approx 5.961$. $d(\theta)=0$ when $\theta=\arctan 3\approx 1.249$ and $\theta=\pi+\arctan 3\approx 4.391$.

**c.** The direction from (1, –1) where the rate of increase is maximal is $\left\langle-\frac{3}{\sqrt{10}}, \frac{1}{\sqrt{10}}\right\rangle$ and the direction from (1, –1) where the rate of increase is minimal is $\left\langle\frac{3}{\sqrt{10}}, -\frac{1}{\sqrt{10}}\right\rangle$.

**47.** **a.** If we consider a point $(x, y, z)$ on $\beta$, we can let $x=1$ and $y=s$. Thus $z=s+1$. Then $\beta$ can be described by

$$\mathbf{r}(t)=\langle 1, s, s+1\rangle.$$
$$\mathbf{r}(2)=\langle 1, 2, 3\rangle$$

$\mathbf{r}'(t)=\langle 0, 1, 1\rangle$, so an equation for the tangent line when $s=2$ is

$$\mathbf{w}(t)=\langle 1, 2, 3\rangle+t\langle 0, 1, 1\rangle=\langle 1, 2+t, 3+t\rangle.$$

**b.** If we consider a point $(x, y, z)$ on $\alpha$, we can let $x=s$ and $y=2$. Thus $z=2s+1$. Then $\alpha$ can be described by

$$\mathbf{r}(s)=\langle s, 2, 2s+1\rangle.$$
$$\mathbf{r}(1)=\langle 1, 2, 3\rangle$$

$\mathbf{r}'(s)=\langle 1, 0, 2\rangle$, so an equation for the tangent line when $s=1$ is

$$\mathbf{w}(t)=\langle 1, 2, 3\rangle+t\langle 1, 0, 2\rangle=\langle 1+t, 2, 3+2t\rangle.$$

**c.** A vector normal to the plane is $\mathbf{N}=\langle 1, 0, 2\rangle\times\langle 0, 1, 1\rangle=\langle -2, -1, 1\rangle$. Thus an equation for the plane is

$$\langle -2, -1, 1\rangle\cdot\langle x-1, y-2, z-3\rangle=0$$
$$-2(x-1)-(y-2)+(z-3)=0$$
$$-2x-y+z+1=0.$$

**49.** **a.** If we consider a point $(x, y, z)$ on $\beta$, we can let $x=0$ and $y=s$. Thus $z=2\ln s$. Then $\beta$ can be described by

$$\mathbf{r}(s)=\langle 0, s, 2\ln s\rangle.$$
$$\mathbf{r}(1)=\langle 0, 1, 0\rangle$$

$\mathbf{r}'(s)=\left\langle 0, 1, \frac{2}{s}\right\rangle$, so $\mathbf{r}'(1)=\langle 0, 1, 2\rangle$. Thus an equation for the tangent line when $s=1$ is $\mathbf{w}(t)=\langle 0, 1, 0\rangle+t\langle 0, 1, 2\rangle=\langle 0, 1+t, 2t\rangle$.

**b.** If we consider a point $(x, y, z)$ on $\alpha$, we can let $x = s$ and $y = 1$. Thus $z = \ln(s^2 + 1)$. Then $\alpha$ can be described by $\mathbf{r}(s) = \langle s, 1, \ln(s^2 + 1)\rangle$. $\mathbf{r}(0) = \langle 0, 1, 0\rangle$ and $\mathbf{r}'(s) = \left\langle 1, 0, \frac{2s}{s^2 + 1}\right\rangle$, so $\mathbf{r}'(0) = \langle 1, 0, 0\rangle$.
Thus an equation for the tangent line when $s = 0$ is $\mathbf{w}(t) = \langle 0, 1, 0\rangle + t\langle 1, 0, 0\rangle = \langle t, 1, 0\rangle$.

**c.** A vector normal to the plane is $\mathbf{N} = \langle 0, 1, 2\rangle \times \langle 1, 0, 0\rangle = \langle 0, 2, -1\rangle$. Thus an equation for the plane is

$$\langle 0, 2, -1\rangle \cdot \langle x, y-1, z\rangle = 0$$
$$2(y-1) - (z) = 0$$
$$2y - z - 2 = 0.$$

**51. a.** $$D_{\mathbf{u}}g(0.5, 0.5) \approx \frac{g(0.6, 0.6) - g(0.5, 0.5)}{\|\langle 0.6, 0.6\rangle - \langle 0.5, 0.5\rangle\|} = \frac{0.1}{0.1\sqrt{2}} \approx 0.707$$

**b.** $$D_{\mathbf{v}}g(0.5, 0.5) \approx \frac{g(0.4, 0.6) - g(0.5, 0.5)}{\|\langle 0.4, 0.6\rangle - \langle 0.5, 0.5\rangle\|} = \frac{0.06}{0.1\sqrt{2}} \approx 0.424$$

**c.** $$g_x(0.5, 0.5) \approx \frac{g(0.6, 0.5) - g(0.5, 0.5)}{0.1} = \frac{0.01}{0.1} = 0.1$$
$$g_y(0.5, 0.5) \approx \frac{g(0.5, 0.6) - g(0.5, 0.5)}{0.1} = \frac{0.08}{0.1} = 0.8$$

**53.** Answers will vary.

**a.** Locate $(-1, -0.5)$ on the graph and draw a line segment through the point that makes an angle $\frac{\pi}{4}$ with the positive $x$-axis from the level curve of 1.29 to the level curve of 0.67. The length of this segment is approximately 0.34. Thus
$$D_{\mathbf{u}}g(-1, -0.5) \approx \frac{0.67 - 1.29}{0.34} \approx -1.82.$$

**b.** Locate $(-0.5, -0.75)$ on the graph and draw the line segment through the point that makes an angle of $\frac{2\pi}{3}$ with the positive $x$-axis from the level curve of 1.70 to the level curve of 1.29. The length of this segment is approximately 0.28. Thus
$$D_{\mathbf{v}}(-0.5, -0.75) \approx \frac{1.29 - 1.70}{0.28} \approx -1.46.$$

**c.** Locate $(-1, -1)$ on the graph. This point is on the level curve of 1.87.
$$g_x(-1, -1) \approx \frac{1.70 - 2.03}{1.5} = -0.22;$$
$$g_y(-1, -1) \approx \frac{1.70 - 2.03}{0.25} = -1.32$$
Thus $\nabla g(-1, -1) \approx \langle -0.22, -1.32\rangle$.

**d.** For the maximum, draw a vector in the same direction as $\nabla g$. For the minimum, draw a vector in the direction opposite that of $\nabla g$. Note that $\nabla g$ should be normal to the level curve of 1.87 at $(-1, -1)$.

**e.** The directions in which the directional derivative is 0 are tangent to the level curve of 1.87 at $(-1, -1)$.

**55. a.** The set of points for which $d(x, y, z) = c$ forms a sphere of radius $c$ about the point $P_0$. Thus, $\nabla d(\mathbf{Q})$ is a vector parallel to the vector $\overrightarrow{P_0Q}$. Hence $D_{\mathbf{u}}d(\mathbf{Q}) = \nabla d(\mathbf{Q}) \cdot \mathbf{u}$ is maximum when $\mathbf{u}$ is in the same direction as $\overrightarrow{P_0Q}$. $D_{\mathbf{u}}d(\mathbf{Q}) = \nabla d(\mathbf{Q}) \cdot \mathbf{u}$ is minimum when $\mathbf{u}$ is in the direction opposite $\overrightarrow{P_0Q}$.

**b.** $D_{\mathbf{v}}d(\mathbf{Q}) = 0$ when $\mathbf{v}$ is perpendicular to $\overrightarrow{P_0Q}$.

**c.** Although $\nabla d(P_0)$ is undefined, it can be shown, using the definition of the directional derivative, that the directional derivative is 1 in every direction from $P_0$.

**57. a.** The set of points for which $d(x, y, z) = c$ forms two planes parallel to $\Pi$, each $c$ units away from $\Pi$. Thus $\nabla d(\mathbf{Q})$ is a vector parallel to the vector normal to the plane pointing from $\Pi$ to Q. Hence $D_{\mathbf{u}}d(\mathbf{Q}) = \nabla d(\mathbf{Q})\cdot\mathbf{u}$ is maximum when $\mathbf{u}$ is parallel to the vector normal to the plane pointing from $\Pi$ to $Q$. $D_{\mathbf{u}}d(\mathbf{Q}) = \nabla d(\mathbf{Q})\cdot\mathbf{u}$ is minimum when $\mathbf{u}$ is parallel to the vector normal to the plane pointing from $Q$ to $\Pi$.

**b.** $D_{\mathbf{v}}d(\mathbf{Q}) = 0$ when $\mathbf{v}$ is parallel to $\Pi$.

**c.** Although $\nabla d$ is undefined on $\Pi$, the directional derivatives at a point on $\Pi$ can be calculated using the definition of the directional derivative. The directional derivative is maximal in directions perpendicular to $\Pi$ and minimal in directions parallel to $\Pi$.

**59. a.** $\dfrac{\partial z}{\partial x} = 2(x-2)(y+1)^4 + 2(x+2)(y-3)^2$

$\dfrac{\partial z}{\partial y} = 4(x-2)^2(y+1)^3 + 2(x+2)^2(y-3)$

At (1, 0), $\dfrac{\partial z}{\partial x} = 52$ and $\dfrac{\partial z}{\partial y} = -50$.

Thus at the point (1, 0, 82), the water will flow in the direction $\langle -52, 50\rangle$ as viewed from above.

**b.** The hiker should go in the direction $\langle 52, -50\rangle$ as viewed from above.

**61.** $T_p = 12.188(v - 0.00106)$; $T_v = 12.188\left(p - \dfrac{0.00007}{v^2} + \dfrac{1.484\times 10^{-7}}{v^3}\right)$

$T_p(0.6, 1.5) \approx 18.3$; $T_v(0.6, 1.5) \approx 7.31$

We should change $(p, v)$ in the "direction" $\langle -18.3, -7.31\rangle$. This means that in order to decrease the temperature as rapidly as possible, decrease the pressure by 18.3 atmospheres as the relative volume is decreased by 7.31.

**63.** $D_{\mathbf{u}}f(P_0) = \lim\limits_{t\to 0^+} \dfrac{f(P_0 + t\mathbf{u}) - f(P_0)}{t}$. Let $P_0 = (a, b)$. If we let $\mathbf{u}$ be in the direction $(h, k)$, then $\mathbf{u} = \dfrac{1}{\|(h, k)\|}(h, k)$. If we let $t$ be sufficiently small, say $\|(h, k)\|$, then

$$D_{\mathbf{u}}f(P_0) = \langle f_x(a, b), f_y(a, b)\rangle\cdot\mathbf{u} = \frac{1}{\|(h, k)\|}(f_x(a, b)h + f_y(a, b)k) \approx \frac{f(a+h, b+k) - f(a, b)}{\|(h, k)\|}$$

Thus, $f(a+h, b+k) \approx f(a, b) + f_x(a, b)h + f_y(a, b)k$. If $x = a + h$ and $y = b + k$, $f(x, y) \approx f(a, b) + f_x(a, b)(x-a) + f_y(a, b)(y-b)$.

## Chapter 9 Review Exercises

**1.** $\sqrt{(x-3)^2 + (y-1)^2} = |y-3|$

$$y - 2 = -\frac{1}{4}(x-3)^2$$

**3.** $D = 1$ so $a = \frac{1}{2}$. Center is (2, 3) so $c = 3$.

$$b^2 = c^2 - a^2 = \frac{35}{4}$$

$$\frac{-(x-2)^2}{\frac{35}{4}} + \frac{(y-3)^2}{\frac{1}{4}} = 1$$

**5.**
$$\sqrt{x^2 + y^2} = |x - 4|$$
$$y^2 = -8x + 16$$
$$x - 2 = -\frac{1}{8}y^2$$

**7.**
$$f_x(x, y) = y^3 \sec^2(xy^2)$$
$$f_y(x, y) = \tan(xy^2) + 2xy^2 \sec^2(xy^2)$$

**9.**
$$\frac{\partial \rho}{\partial \phi} = \frac{\cos\phi \sin\theta}{1 - \cos\phi\cos\theta} - \frac{\cos\theta \sin\phi(1 + \sin\phi\sin\theta)}{(1 - \cos\phi\cos\theta)^2}$$
$$\frac{\partial \rho}{\partial \theta} = \frac{\cos\theta \sin\phi}{1 - \cos\phi\cos\theta} - \frac{\cos\phi \sin\theta(1 + \sin\phi\sin\theta)}{(1 - \cos\phi\cos\theta)^2}$$

**11.**
$$h_u(u, v, w) = \frac{2\cos(u)\sin(u)}{1 + \cos(2w)^2 + \sin(u)^2 + \sin(v)^6}$$
$$h_v(u, v, w) = \frac{6\cos(v)\sin(v)^5}{1 + \cos(2w)^2 + \sin(u)^2 + \sin(v)^6}$$
$$h_w(u, v, w) = \frac{-4\cos(2w)\sin(2w)}{1 + \cos(2w)^2 + \sin(u)^2 + \sin(v)^6}$$

**13.**
$$f_x(x, y) = 6x - 2y$$
$$f_y(x, y) = -2x + 8y - 1$$
$$\nabla f(-2, 1) \cdot \mathbf{u} = \langle -14, 11 \rangle \cdot \left\langle \frac{1}{\sqrt{2}}, \frac{-1}{\sqrt{2}} \right\rangle = -\frac{25}{\sqrt{2}}$$

**15.**
$$\rho_\phi(\phi, \theta) = \frac{4\phi}{2\phi^2 + \theta^2 + 1}$$
$$\rho_\theta(\phi, \theta) = \frac{2\theta}{2\phi^2 + \theta^2 + 1}$$
$$\nabla \rho(2, -3) \cdot \mathbf{u} = \left\langle \frac{4}{9}, -\frac{1}{3} \right\rangle \cdot \langle -1, 0 \rangle = -\frac{4}{9}$$

**17.**
$$g_x(x, y, z) = -\frac{z}{(x+y)^2 + z^2}$$
$$g_y(x, y, z) = -\frac{z}{(x+y)^2 + z^2}$$
$$g_z(x, y, z) = \frac{x+y}{(x+y)^2 + z^2}$$
$$\nabla g(1, 1, 2) \cdot \mathbf{u}$$
$$= \left\langle -\frac{1}{4}, -\frac{1}{4}, \frac{1}{4} \right\rangle \cdot \left\langle \frac{1}{\sqrt{2}}, -\frac{1}{\sqrt{6}}, \frac{1}{\sqrt{3}} \right\rangle$$
$$= \frac{-3\sqrt{2} + \sqrt{6} + 2\sqrt{3}}{24}$$

**19.** Direction: $\mathbf{u} = \frac{\nabla f(-2, 1)}{\|\nabla f(-2, 1)\|} = \left\langle -\frac{14}{\sqrt{317}}, \frac{11}{\sqrt{317}} \right\rangle$

$$D_{\mathbf{u}} f(-2, 1) = \|\nabla f(-2, 1)\| = \sqrt{317}$$

**21.** Direction:

$$\mathbf{u} = \frac{\nabla f(1, 1, -1)}{\|\nabla f(1, 1, -1)\|} = \left\langle -\frac{1}{\sqrt{5}}, -\frac{2}{\sqrt{5}}, 0 \right\rangle$$
$$D_{\mathbf{u}} f(1, 1, -1) = \|\nabla f(1, 1, -1)\| = \sqrt{5}$$

**23.** Cross sections perpendicular to the $x$- and $y$-axes are parabolas. Cross sections perpendicular to the $z$-axis are circles.

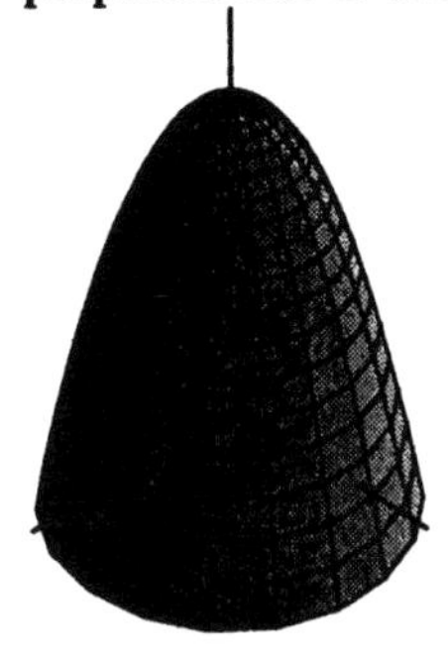

**25.** Cross sections perpendicular to the $x$- and $z$-axes are hyperbolas. Cross sections perpendicular to the $y$-axis are circles.

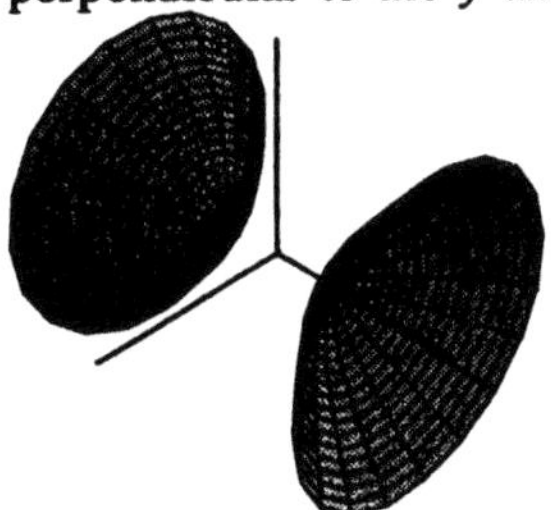

**27.** $\mathbf{r}(u, v) = \left\langle u\cos v,\ u\sin v,\ 4 - u^2\right\rangle,\ u \geq 0,$
$0 \leq v \leq 2\pi$

**29.** $\mathbf{r}(t, c) = \left\langle \frac{1}{2}t^2 + \frac{1}{2}c^2,\ c,\ t\right\rangle,\ c \geq 0, -\infty < t < \infty$

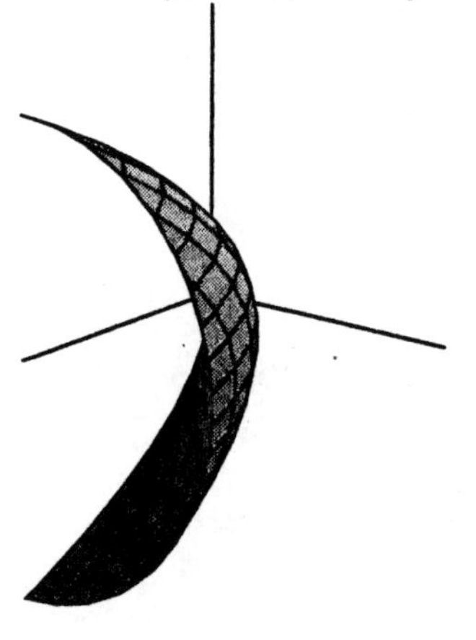

**31.** $z = 3 + r^2$
$\mathbf{w}(r, \theta) = \left\langle r\cos\theta,\ r\sin\theta,\ 3 + r^2\right\rangle$

**33.** $\rho = 2$
$\mathbf{r}(\theta, \phi) = \langle 2\cos\theta\sin\phi,\ 2\sin\theta\sin\phi,\ 2\cos\phi\rangle$

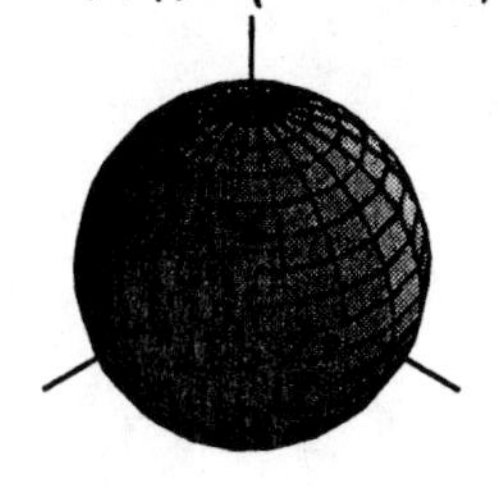

**35.** $\rho\sin\phi\cos\theta = 4$
$\rho = 4\sec\theta\csc\phi$
$\rho(\theta, \phi) = \langle 4,\ 4\tan\theta,\ 4\sec\theta\cot\phi\rangle$

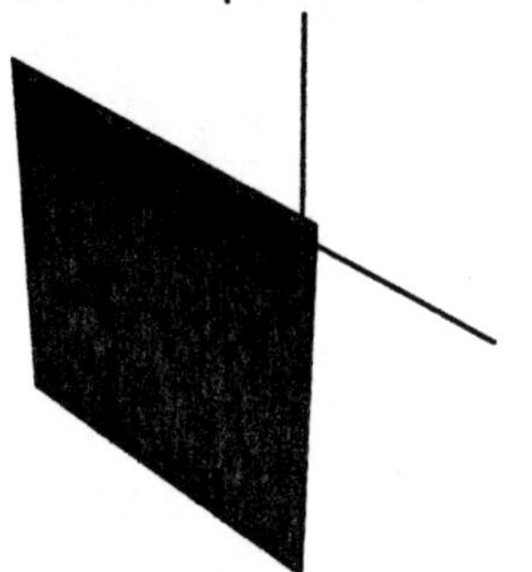

**37.** $\displaystyle\lim_{(x,\,y)\to(1,\,2)} \frac{3x^2y + 2xy^2 + 3}{4x - 3y + 4}$
$= \dfrac{3(1)^2(2) + 2(1)(2)^2 + 3}{4(1) - 3(2) + 4}$
$= \dfrac{17}{2}$

**39.** $\displaystyle\lim_{(x,\,y)\to(-1,\,2)} \frac{(x+1)(y-2)^3}{(x+1)^2 + (y-2)^2}$
$= \displaystyle\lim_{(x,\,y)\to(-1,\,2)} \frac{(x+1)(y-2)}{\left(\frac{x+1}{y-2}\right)^2 + 1}$
$= 0$

**41.** Let
$E(x, y)$
$= \left|\sqrt{1 + x + 2y} - \left[2 + \frac{1}{4}(x-1) + \frac{1}{2}(y-1)\right]\right|.$
Using a CAS, we plot the level curve $E(x, y) = 0.01$ for $(x, y)$ close to $(1, 1)$. We also plot a circle of radius 0.3 centered at $(1, 1)$. Thus, if $\delta = 0.3$, $|E(x, y)| < 0.01$ when $0 < \|\langle x, y\rangle - \langle 1, 1\rangle\| < \delta$.

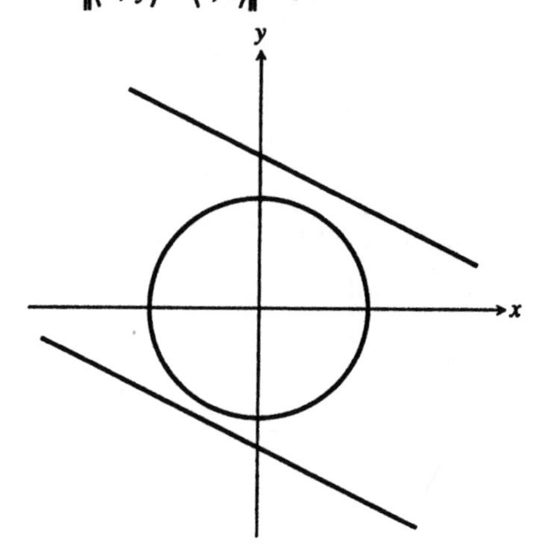

**43.** Assume distances are in centimeters, time in seconds.

$\dfrac{\partial s}{\partial x} = \dfrac{\pi}{120}\cos(20\pi t)\cos\left(\dfrac{\pi x}{30}\right) + \dfrac{\pi}{240}\cos(40\pi t)\cos\left(\dfrac{\pi x}{15}\right)$ with units of cm/cm.

$\dfrac{\partial s}{\partial t} = -5\pi\sin(20\pi t)\sin\left(\dfrac{\pi x}{30}\right) - \dfrac{5\pi}{2}\sin(40\pi t)\sin\left(\dfrac{\pi x}{15}\right)$ with units cm/s.

$\dfrac{\partial s}{\partial x}$ gives the string's slope at $(x, t)$.

$\dfrac{\partial s}{\partial t}$ gives the string's lateral speed at $(x, t)$.

**45.** Assume distances are in meters, time in seconds, and mass in kilograms.

$\dfrac{\partial F}{\partial x} = -\dfrac{2xGmM}{(x^2+y^2+z^2)^2}$ with units of newtons/m.

$\dfrac{\partial F}{\partial y} = -\dfrac{2yGmM}{(x^2+y^2+z^2)^2}$ with units of newtons/m.

$\dfrac{\partial F}{\partial z} = -\dfrac{2zGmM}{(x^2+y^2+z^2)^2}$ with units of newtons/m.

$\dfrac{\partial F}{\partial m} = \dfrac{GM}{x^2+y^2+z^2}$ with units of newtons/kg.

$\dfrac{\partial f}{\partial M} = \dfrac{Gm}{x^2+y^2+z^2}$ with units of newtons/kg.

The force decreases as $x$, $y$, or $z$ increases.
The force increases as $M$ or $m$ increases.

**47.** Assume distances are in meters, time in seconds, and mass in kilograms.

$\dfrac{\partial s}{\partial r} = \ln\left(\dfrac{M}{m}\right)$ with units of (m/s)/(m/s).

$\dfrac{\partial s}{\partial M} = \dfrac{r}{M}$ with units of (m/s)/kg.

$\dfrac{\partial s}{\partial m} = -\dfrac{r}{m}$ with units of (m/s)/kg.

The speed increases as $r$ and $M$ increase. The speed decreases as $m$ increases.

**49.** Let $P$ be in $D$.

**a.**
$$\begin{aligned}\nabla(f+g)(P) &= \left\langle (f+g)_x(P),\ (f+g)_y(P)\right\rangle \\ &= \left\langle f_x(P)+g_x(P),\ f_y(P)+g_y(P)\right\rangle \\ &= \left\langle f_x(P),\ f_y(P)\right\rangle + \left\langle g_x(P),\ g_y(P)\right\rangle \\ &= \nabla f(P) + \nabla g(P)\end{aligned}$$

**b.**
$$\begin{aligned}\nabla(cf)P &= \left\langle (cf)_x(P),\ (cf)_y(P)\right\rangle \\ &= \left\langle cf_x(P),\ cf_y(P)\right\rangle \\ &= c\nabla f(P)\end{aligned}$$

**c.** $\nabla(fg)P = \left\langle (fg)_x(P), (fg)_y(P) \right\rangle$

$= \left\langle (f_x g + f g_x)(P), (f_y g + f g_y)(P) \right\rangle$

$= \left\langle (f_x g)(P), (f_y g)(P) \right\rangle + \left\langle (fg)_x(P), (fg)_y(P) \right\rangle$

$= g\nabla f(P) + f\nabla g(P)$

**d.** $\nabla\left(\frac{f}{g}\right)P = \left\langle \left(\frac{f}{g}\right)_x P, \left(\frac{f}{g}\right)_y (P) \right\rangle$

$= \left\langle \left(\frac{f_x g - f g_x}{g^2}\right)(P), \left(\frac{f_y g - f g_y}{g^2}\right)(P) \right\rangle$

$= \left\langle \left(\frac{f_x g}{g^2}\right)(P), \left(\frac{f_y g}{g^2}\right)(P) \right\rangle - \left\langle \left(\frac{f g_x}{g^2}\right)(P), \left(\frac{f g_y}{g^2}\right)(P) \right\rangle$

$= \frac{g\nabla f(P) - f\nabla g(P)}{g^2}$

**e.** $\nabla(f^p)(P) = \left\langle (f^p)_x(P), (f^p)_y(P) \right\rangle$

$= \left\langle (pf^{p-1} f_x)(P), (pf^{p-1} f_y)(P) \right\rangle$

$= pf^{p-1}\nabla f(P)$

# Chapter 10 Differentiable Functions of Several Variables

## Section 10.1 Differentiability

**1.** $\Delta f(\mathbf{h}) = f(\mathbf{a}+\mathbf{h}) - f(\mathbf{a})$
$= 2.1(3.2)^2 - 2(3)^2$
$= 3.504$

$f_x = y^2;\ f_y = 2xy$

$df_{\mathbf{a}}(\mathbf{h}) = f_x(\mathbf{a})h_1 + f_y(\mathbf{a})h_2$
$= 3^2(0.1) + 2(2)(3)(0.2)$
$= 3.3$

$|\Delta f(\mathbf{h}) - df_{\mathbf{a}}(\mathbf{h})| = |3.504 - 3.3| = 0.204$

**3.** $\Delta f(\mathbf{h}) = f(\mathbf{a}+\mathbf{h}) - f(\mathbf{a})$
$= \ln((2.3)(1.2)) - \ln((2)(1))$
$\approx 0.322083$

$f_x = \frac{1}{x};\ f_y = \frac{1}{y}$

$df_{\mathbf{a}}(\mathbf{h}) = f_x(\mathbf{a})h_1 + f_y(\mathbf{a})h_2$
$= \frac{1}{2}(0.3) + \frac{1}{1}(0.2)$
$= 0.35$

$|\Delta f(\mathbf{h}) - df_{\mathbf{a}}(\mathbf{h})| \approx |0.322083 - 0.35|$
$= 0.0279165$

**5.** $\Delta f(\mathbf{h}) = f(\mathbf{a}+\mathbf{h}) - f(\mathbf{a})$
$= (-0.9)(0.8)(1.1)^2 - (-1)(1)(1)^2$
$= 0.1288$

$f_x = yz^2;\ f_y = xz^2;\ f_z = 2xyz$

$df_{\mathbf{a}}(\mathbf{h}) = f_x(\mathbf{a})h_1 + f_y(\mathbf{a})h_2 + f_z(\mathbf{a})h_3$
$= (1)(0.1) + (-1)(-0.2) + (-2)(0.1)$
$= 0.1$

$|\Delta f(\mathbf{h}) - df_{\mathbf{a}}(\mathbf{h})| = |0.1288 - 0.1| = 0.0288$

**7.** $f_x = 6xy + 15x^2y^2$, which is continuous near (1, 1).

$f_y = 3x^2 + 10x^3y$, which is continuous near (1, 1).

$Df(\mathbf{a}) = (f_x(\mathbf{a})\quad f_y(\mathbf{a}))$
$= (6+15\quad 3+10)$
$= (21\quad 13)$

$df(\mathbf{x}-\mathbf{a}) = (21\quad 13)\begin{pmatrix} x-1 \\ y-1 \end{pmatrix}$
$= 21(x-1) + 13(y-1)$

**9.** $f_x = \dfrac{\cos\left(\frac{x}{y}\right)}{y}$, which is continuous near $\left(\frac{\pi}{6}, 1\right)$.

$f_y = -\dfrac{x\cos\left(\frac{x}{y}\right)}{y^2}$, which is continuous near $\left(\frac{\pi}{6}, 1\right)$.

$Df(\mathbf{a}) = (f_x(\mathbf{a})\quad f_y(\mathbf{a})) = \begin{pmatrix} \frac{\sqrt{3}}{2} & -\frac{\pi\sqrt{3}}{12} \end{pmatrix}$

$df(\mathbf{x}-\mathbf{a}) = \begin{pmatrix} \frac{\sqrt{3}}{2} & \frac{\pi\sqrt{3}}{12} \end{pmatrix}\begin{pmatrix} x-\frac{\pi}{6} \\ y-1 \end{pmatrix}$
$= \frac{\sqrt{3}}{2}\left(x - \frac{\pi}{6}\right) - \frac{\pi\sqrt{3}}{12}(y-1)$

**11.** $f_x = e^{-x^2+xy}(-2x+y)$, which is continuous near (1, 0).

$f_y = e^{-x^2+xy}(x)$, which is continuous near (1, 0).

$Df(\mathbf{a}) = (f_x(\mathbf{a})\quad f_y(\mathbf{a}))$
$= (-2e^{-1}\quad e^{-1})$
$= \begin{pmatrix} -\frac{2}{e} & \frac{1}{e} \end{pmatrix}$

$df(\mathbf{x}-\mathbf{a}) = \begin{pmatrix} -\frac{2}{e} & \frac{1}{e} \end{pmatrix}\begin{pmatrix} x-1 \\ y \end{pmatrix}$
$= -\frac{2}{e}(x-1) + \frac{1}{e}y$

**13.** $f_x = \dfrac{y^2}{\sqrt{1-x^2y^4}}$, which is continuous near $\left(1, \frac{1}{\sqrt{2}}\right)$.

$f_y = \dfrac{2xy}{\sqrt{1-x^2y^4}}$, which is continuous near $\left(1, \frac{1}{\sqrt{2}}\right)$.

$$Df(\mathbf{a}) = (f_x(\mathbf{a})\ \ f_y(\mathbf{a}))$$
$$= \begin{pmatrix} \frac{\frac{1}{2}}{\sqrt{1-\frac{1}{4}}} & \frac{\sqrt{2}}{\sqrt{1-\frac{1}{4}}} \end{pmatrix}$$
$$= \begin{pmatrix} \frac{1}{\sqrt{3}} & \frac{2\sqrt{2}}{\sqrt{3}} \end{pmatrix}$$
$$df(\mathbf{x}-\mathbf{a}) = \begin{pmatrix} \frac{1}{\sqrt{3}} & \frac{2\sqrt{2}}{\sqrt{3}} \end{pmatrix}\begin{pmatrix} x-1 \\ y-\frac{1}{\sqrt{2}} \end{pmatrix}$$
$$= \frac{1}{\sqrt{3}}(x-1) + \frac{2\sqrt{2}}{\sqrt{3}}\left(y - \frac{1}{\sqrt{2}}\right)$$

**15.** $f_x = y + z$, which is continuous near $(1, -2, 1)$.
$f_y = y + z$, which is continuous near $(1, -2, 1)$.
$f_z = x + y$, which is continuous near $(1, -2, 1)$.
$$Df(\mathbf{a}) = (f_x(\mathbf{a})\ \ f_y(\mathbf{a})\ \ f_z(\mathbf{a})) = (-1\ \ 2\ \ -1)$$
$$df(\mathbf{x}-\mathbf{a}) = (-1\ \ 2\ \ -1)\begin{pmatrix} x-1 \\ y+2 \\ z-1 \end{pmatrix}$$
$$= -(x-1) + 2(y+2) - (z-1)$$

**17.** $f_x = 4x;\ f_y = -6y$
$$\mathbf{n} = \langle -f_x(\mathbf{a}),\ -f_y(\mathbf{a}),\ 1\rangle = \langle -8,\ 6,\ 1\rangle$$
The tangent plane is described by
$$z = f(\mathbf{a}) + f_x(\mathbf{a})(x-a_1) + f_y(\mathbf{a})(y-a_2)$$
$$= 5 + 8(x-2) - 6(y-1)$$
$$= 8x - 6y - 5$$

**19.** $f_x = \dfrac{e^{x/y}}{y};\ f_y = -\dfrac{xe^{x/y}}{y^2}$
$$\mathbf{n} = \langle -f_x(\mathbf{a}),\ -f_y(\mathbf{a}),\ 1\rangle = \langle -e^2,\ 2e^2,\ 1\rangle$$
The tangent plane is described by
$$z = f(\mathbf{a}) + f_x(\mathbf{a})(x-a_1) + f_y(\mathbf{a})(x-a_2)$$
$$= e^2 + e^2(x-2) - 2e^2(y-1)$$
$$= e^2x - 2e^2y + e^2$$

**21.** $f_x = 2x\cos(x^2+y^2);\ f_y = 2x\cos(x^2+y^2)$
$$\mathbf{n} = \langle -f_x(\mathbf{a}),\ -f_y(\mathbf{a}),\ 1\rangle$$
$$= \langle 6\cos 13,\ -4\cos 13,\ 1\rangle$$
The tangent plane is described by
$$z = f(\mathbf{a}) + f_x(\mathbf{a})(x-a_1) + f_y(\mathbf{a})(x-a_2)$$
$$= \sin 13 - 6\cos 13(x+3) + 4\cos 13(y-2)$$
$$= -(6\cos 13)x + (4\cos 13)y - 26\cos 13 + \sin 13$$

**23.** $\mathbf{m} = (90, 50, 30);\ \mathbf{e} = 0.02\mathbf{m} = (1.8, 1, 0.6)$
$V = lwh;\ V_l = wh;\ V_w = lh;\ V_h = lw$
$$dV_{\mathbf{m}}(\mathbf{e}) = V_l(\mathbf{m})e_l + V_w(\mathbf{m})e_2 + V_h(\mathbf{m})e_3$$
$$= (1500)(1.8) + (2700)(1) + (4500)(0.6)$$
$$= 8100$$
$$\text{approximate percentage error} = \frac{8100}{(90)(50)(30)} = 0.06 = 6\%$$
$$\Delta V(\mathbf{e}) = V(\mathbf{m}+\mathbf{e}) - V(\mathbf{m})$$
$$= (91.8)(51)(30.6) - (90)(50)(30)$$
$$= 8263.008$$
$$\text{actual percentage error} = \frac{8263.08}{(90)(50)(30)} = 0.061208 = 6.1208\%$$

**25.** Let $x$ be the perimeter of a circular cross-section perpendicular to the axis of symmetry and let $y$ be the perimeter of a cross-section lying in a plane containing the axis of symmery. Then the interior radius of the cylinder (and the hemispherical ends) in feet is $r = \dfrac{x}{2\pi}$ and the height of the cylinder is $h = \dfrac{1}{2}(y-x)$. Since the volume is $\dfrac{4}{3}\pi r^3 + \pi r^2 h$, a function for the volume of the tank is
$$f(x,y) = \frac{4}{3}\pi\left(\frac{x}{2\pi}\right)^3 + \pi\left(\frac{x}{2\pi}\right)^2\frac{1}{2}(y-x)$$
$$= \left(\frac{1}{6\pi^2} - \frac{1}{8\pi}\right)x^3 + \frac{1}{8\pi}x^2y$$
$$f_x = \left(\frac{1}{2\pi^2} - \frac{3}{8\pi}\right)x^2 + \frac{1}{4\pi}xy$$
$$f_y = \frac{1}{8\pi}x^2$$
Let $\mathbf{a} = (12.1, 37.5)$ and $\mathbf{h} = (\pm 0.05, \pm 0.1)$.
$$Df(\mathbf{a}) = (f_x(12.1, 37.5)\quad f_y(12.1, 37.5))$$
$$\approx (26.0491\quad 5.82547)$$
$$\text{error} \approx df(\mathbf{h}) = Df(\mathbf{a})\mathbf{h}$$
$$\approx (26.0491\quad 5.82547)\begin{pmatrix} \pm 0.05 \\ \pm 0.1 \end{pmatrix}$$
$$\approx \pm 1.3025 \pm 0.58255$$
Thus, error is at most 1.885 $\text{ft}^3$.

**27.** $\mathbf{r}(u, v) = \langle x, y, z\rangle = \langle u\cos v, u\sin v, u^2\rangle$

$\mathbf{r}\left(1, \frac{\pi}{4}\right) = \left\langle \frac{1}{\sqrt{2}}, \frac{1}{\sqrt{2}}, 1\right\rangle$

$x_u = \cos v, x_v = -u\sin v$

$y_u = \sin v, y_v = u\cos v$

$z_u = 2u,\ z_v = 0$

The tangent plane is described using (26):

$$x = \frac{1}{\sqrt{2}} + \frac{1}{\sqrt{2}}(u-1) - \frac{1}{\sqrt{2}}\left(v - \frac{\pi}{4}\right) = \frac{1}{\sqrt{2}}\left(1 + (u-1) - \left(v - \frac{\pi}{4}\right)\right) = \frac{1}{\sqrt{2}}\left(u - v + \frac{\pi}{4}\right)$$

$$y = \frac{1}{\sqrt{2}} + \frac{1}{\sqrt{2}}(u-1) + \frac{1}{\sqrt{2}}\left(v - \frac{\pi}{4}\right) = \frac{1}{\sqrt{2}}\left(1 + (u-1) + \left(v - \frac{\pi}{4}\right)\right) = \frac{1}{\sqrt{2}}\left(u + v - \frac{\pi}{4}\right)$$

$$z = 1 + 2(u-1) = 2u - 1$$

From these equations,
$\mathbf{r}(1.1, 0.8) \approx \langle 0.767492, 0.788143, 1.2\rangle$.
When $\mathbf{r}(1.1, 0.8)$ is calculated from the original definition, it equals approximately $\langle 0.766377, 0.789092, 1.21\rangle$.

**29.** $\mathbf{r}(u, v) = \langle x, y, z\rangle = \langle \ln u, \ln v, uv\rangle$

$\mathbf{r}(e, e) = \langle 1, 1, e^2\rangle$

$x_u = \frac{1}{u},\ x_v = 0$

$y_u = 0,\ y_v = \frac{1}{v}$

$z_u = v,\ z_v = u$

The tangent plane is described using (26):

$$x = 1 + \frac{1}{e}(u - e) = \frac{1}{e}u$$

$$y = 1 + \frac{1}{e}(v - e) = \frac{1}{e}v$$

$$z = e^2 + e(u-e) + e(v-e) = e(u + v - e)$$

From these equations,
$\mathbf{r}(2.6, 2.8) \approx \langle 0.956487, 1.03006, 7.28967\rangle$.
When $\mathbf{r}(2.6, 2.8)$ is calculated from the original definition, it equals approximately $\langle 0.955511, 1.02962, 7.28\rangle$.

**31.** $\mathbf{r}(u, v) = \langle x, y, z\rangle = \langle u\tan v, u\sec v, u^2\rangle$

$\mathbf{r}\left(2, \frac{\pi}{4}\right) = \langle 2, 2\sqrt{2}, 4\rangle$

$x_u = \tan v,\ x_v = u\sec^2 v$

$y_u = \sec v,\ y_v = u\sec v\tan v$

$z_u = 2u,\ z_v = 0$

The tangent plane is described using (26):

$$x = 2 + (u-2) + 4\left(v - \frac{\pi}{4}\right) = u + 4v - \pi$$

$$y = 2\sqrt{2} + \sqrt{2}(u-2) + 2\sqrt{2}\left(v - \frac{\pi}{4}\right) = \sqrt{2}\left(2 + (u-2) + 2\left(v - \frac{\pi}{4}\right)\right) = \sqrt{2}\left(u + 2v - \frac{\pi}{2}\right)$$

$$z = 4 + 4(u-2) = 4u - 4$$

From these equations,
$\mathbf{r}(2.2, 0.8) \approx \langle 2.25841, 3.15257, 4.8\rangle$.
When $\mathbf{r}(2.2, 0.8)$ is calculated from the original definition, it equals approximately $\langle 2.26520, 3.15771, 4.84\rangle$.

**33.** $\frac{\partial}{\partial r}(r\cos\theta) = \cos\theta;\ \frac{\partial}{\partial\theta}(r\cos\theta) = -r\sin\theta$

$\frac{\partial}{\partial r}(r\sin\theta) = \sin\theta;\ \frac{\partial}{\partial\theta}(r\sin\theta) = r\cos\theta$

All these partial derivaives are continuous over the domain, so the function is differentiable over the domain.

$$\mathbf{Df(a)} = \begin{pmatrix} \cos\frac{\pi}{6} & -(1)\sin\frac{\pi}{6} \\ \sin\frac{\pi}{6} & (1)\cos\frac{\pi}{6} \end{pmatrix} = \begin{pmatrix} \frac{\sqrt{3}}{2} & -\frac{1}{2} \\ \frac{1}{2} & \frac{\sqrt{3}}{2} \end{pmatrix}$$

$$\mathbf{df(x-a)} = \mathbf{Df(a)}\begin{pmatrix} 1.3 - 1 \\ \frac{\pi}{6} - 0.2 - \frac{\pi}{6} \end{pmatrix} = \begin{pmatrix} \frac{\sqrt{3}}{2} & -\frac{1}{2} \\ \frac{1}{2} & \frac{\sqrt{3}}{2} \end{pmatrix}\begin{pmatrix} 0.3 \\ -0.2 \end{pmatrix} \approx \begin{pmatrix} 0.359808 \\ -0.0232051 \end{pmatrix}$$

**35.** $f_x = \dfrac{e^{x/y}}{y}$; $f_y = \dfrac{xe^{x/y}}{y^2}$

$g(x, y) = f(1, 1) + f_x(1, 1)(x-1) + f_y(1, 1)(y-1) = e + e(x-1) - e(y-1)$

| point | (0.9, 0.9) | (0.9, 1.1) | (1.1, 0.9) | (1.1, 1.1) |
|---|---|---|---|---|
| $f$ | 2.7183 | 2.2663754 | 3.3947232 | 2.7183 |
| $g$ | 2.7183 | 2.1746255 | 3.2619382 | 2.7183 |
| $\lvert f-g \rvert$ | 0 | 0.0917499 | 0.132785 | 0 |

The largest error is approximately 0.132785 at (1.1, 0.9)

**37.** $\mathbf{a} = (1, 9.81)$

The worst error will occur when $L$ is too high or $g$ too low, or vice versa.

Let $\mathbf{e}_1 = (0.002, \ -0.02)$ and $\mathbf{e}_2 = (0.002, 0.02)$.

Then $\Delta T(\mathbf{e}_1) = 2\pi\sqrt{\dfrac{1.002}{9.79}} - 2\pi\sqrt{\dfrac{1}{9.81}} \approx 0.004055$ and $\Delta T(\mathbf{e}_2) = 2\pi\sqrt{\dfrac{0.998}{9.83}} - 2\pi\sqrt{\dfrac{1}{9.81}} \approx -0.004047$.

The worst error in $T$ is 0.004055 seconds.

**39.** $T_L = \pi\dfrac{1}{\sqrt{Lg}}$; $T_g = -\pi\sqrt{\dfrac{L}{g^3}}$

The function $T$ is linearized by the tangent plane, which is described by

$w(L, g) = T(1, 9.81) + T_L(1, 9.81)(L-1) + T_g(1, 9.81)(t-9.81) \approx 2.006 + 1.003(L-1) - 0.102(g-9.81)$

| point | (0.9, 9.80) | (0.9, 9.82) | (1.1, 9.80) | (1.1, 9.82) |
|---|---|---|---|---|
| $T$ | 1.90409 | 1.90215 | 2.10505 | 2.10291 |
| $w$ | 1.90672 | 1.90468 | 2.10732 | 2.10528 |
| $\lvert T-w \rvert$ | 0.00263 | 0.00253 | 0.00227 | 0.00237 |

The graph of $T$ curves more strongly along the $L$-direction than along the $g$-direction.

**41.** $f(x, y) = f(\mathbf{x}) = x^2 + 3y^3$

$\mathbf{a} = (1, 1)$

$f(\mathbf{a}) = 4$

$Df(\mathbf{a}) = (f_x(\mathbf{a}) \quad f_y(\mathbf{a})) = (2 \quad 9)$

$$df_{\mathbf{a}}(\mathbf{x}-\mathbf{a}) = (2 \quad 9)\begin{pmatrix} x-1 \\ y-1 \end{pmatrix} = 2(x-1) + 9(y-1) = 2x + 9y - 11$$

$$\begin{aligned} f(\mathbf{x}) - f(\mathbf{a}) - df_{\mathbf{a}}(\mathbf{x}-\mathbf{a}) &= x^2 + 3y^3 - 4 - 2x - 9y + 11 \\ &= x^2 - 2x + 3y^3 - 9y + 7 \\ &= x^2 - 2x + 1 + 3y^3 - 6y^2 + 3y + 6y^2 - 12y + 6 \\ &= (x^2 - 2x + 1) + 3y(y^2 - 2y + 1) + 6(y^2 - 2y + 1) \\ &= (x-1)^2 + (3y+6)(y-1)^2 \end{aligned}$$

Since $(x-1)^2 \ge 0$, $(x-1)^2 \le 12(x-1)^2$.

Near (1, 1), $0 \le y \le 2$ so $0 < 3y + 6 \le 12$.

Thus, near (1, 1), $(x-1)^2 + (3y+6)(y-1)^2 \le 12\left[(x-1)^2 + (y-1)^2\right]$.

$$\left|\frac{f(\mathbf{x})-f(\mathbf{a})-df_{\mathbf{a}}(\mathbf{x}-\mathbf{a})}{\|\mathbf{x}-\mathbf{a}\|}\right| = \left|\frac{(x-1)^2+(3y+6)(y-1)^2}{\sqrt{(x-1)^2+(y-1)^2}}\right| \le \frac{12\left[(x-1)^2+(y-1)^2\right]}{\sqrt{(x-1)^2+(y-1)^2}} = 12\sqrt{(x-1)^2+(y-1)^2}.$$

Since $\lim\limits_{\mathbf{x}\to\mathbf{a}} 12\sqrt{(x-1)^2+(y-1)^2} = 0$, it follows that $\lim\limits_{\mathbf{x}\to\mathbf{a}} \dfrac{f(\mathbf{x})-f(\mathbf{a})-df_{\mathbf{a}}(\mathbf{x}-\mathbf{a})}{\|\mathbf{x}-\mathbf{a}\|} = 0$ and $f(x, y) = x^2+3y^3$ is by definition differentiable.

**43.** Starting with (26),

$x = x_0 + x_u(u-u_0) + x_v(v-v_0)$

$y = y_0 + y_u(u-u_0) + y_v(v-v_0)$

$z = z_0 + z_u(u-u_0) + z_v(v-v_0)$,

multiply both sides by $n_1$, $n_2$ and $n_3$ respectively.

$n_1x = n_1x_0 + n_1x_u(u-u_0) + n_1x_v(v-v_0)$

$n_2y = n_2y_0 + n_2y_u(u-u_0) + n_2y_v(v-v_0)$

$n_3z = n_3z_0 + n_3z_u(u-u_0) + n_3z_v(v-v_0)$

Now add the left and right sides vertically.

$n_1x+n_2y+n_3z = n_1x_0+n_2y_0+n_3z_0+(n_1x_u+n_2y_u+n_3z_u)(u-u_0)+(n_1x_v+n_2y_v+n_3z_v)(v-v_0)$

$$\langle n_1, n_2, n_3\rangle = \begin{vmatrix} \mathbf{i} & \mathbf{j} & \mathbf{k} \\ x_u & y_u & z_u \\ x_v & y_v & z_v \end{vmatrix}$$

$$= \langle y_uz_v - y_vz_u,\ x_vz_u - x_uz_v,\ x_uy_v - x_vy_u\rangle$$

$$\begin{aligned} n_1x_u + n_2y_u + n_3z_u &= x_u(y_uz_v - y_vz_u) + y_u(x_vz_u - x_uz_v) + z_u(x_uy_v - x_vy_u) \\ &= x_uy_uz_v - x_uy_vz_u + x_vy_uz_u - x_uy_uz_v + x_uy_vz_u - x_vy_uz_u \\ &= 0 \end{aligned}$$

Similarly, $n_1x_v + n_2y_v + n_3z_v = 0$.

Then $n_1x+n_2y+n_3z = n_1x_0+n_2y_0+n_3z_0$, or equivalently, $n_1(x-x_0)+n_2(y-y_0)+n_3(z-z_0)=0$, which is (27).

**45.** $f(x, y) = -\sqrt{|xy|}$. If $xy > 0$, then $f(x, y) = \sqrt{xy}$. $f_x = -\dfrac{y}{2\sqrt{xy}}$ and $f_y = -\dfrac{x}{2\sqrt{xy}}$.

For any point in the region $\{(x, y): xy > 0\}$ the partials are continuous at and near that point.

If $xy < 0$, then $f(x, y) = -\sqrt{-xy}$. $f_x = \dfrac{y}{2\sqrt{-xy}}$ and $f_y = \dfrac{x}{2\sqrt{-xy}}$. For any point in the region $\{(x, y): xy < 0\}$ the partials are continuous at and near that point. Thus, $f$ is differentiable at points $(x, y)$ not on the coordinate axes.

**47.** In exercise 46, we showed that $g_x(0, 0) = 0$ and $g_y(0, 0) = 0$. Thus $dg_{\mathbf{a}}(\mathbf{x}-\mathbf{a}) = (0\ \ 0)\begin{pmatrix} x \\ y \end{pmatrix} = 0$.

$$\lim_{\mathbf{x}\to\mathbf{a}} \frac{g(\mathbf{x})-g(\mathbf{a})-dg_{\mathbf{a}}(\mathbf{x}-\mathbf{a})}{\|\mathbf{x}-\mathbf{a}\|} = \lim_{\mathbf{x}\to\mathbf{a}} \frac{g(x, y)}{\|\mathbf{x}-\mathbf{a}\|}$$

If $\mathbf{x}$ approaches (0, 0) along the path $(t, 0)$ for $t > 0$, then

$$\lim_{\mathbf{x}\to\mathbf{a}} \frac{g(x, y)}{\|\mathbf{x}-\mathbf{a}\|} = \lim_{t\to 0^+} \frac{g(t, 0)}{|t|} = \lim_{t\to 0^+} \frac{0}{t} = 0.$$

However, if $\mathbf{x}$ approaches (0, 0) along the path $(t, t)$ for $t > 0$, then

$\lim\limits_{\mathbf{x}\to\mathbf{a}} \dfrac{g(x, y)}{\|\mathbf{x}-\mathbf{a}\|} = \lim\limits_{t\to 0^+} \dfrac{g(t, t)}{\sqrt{2t^2}} = \lim\limits_{t\to 0^+} -\dfrac{1}{t\sqrt{2}}$ which is undefined. Thus the old-tent function is not differentiable at (0, 0).

**49.** **i.** Let $A = f_x(\mathbf{a})$ and $B = f_y(\mathbf{a})$.

**ii.** Let $\mathbf{x} - \mathbf{a} = \mathbf{h}$. Then $\mathbf{x} = \mathbf{a} + \mathbf{h}$.

**iii.** Then $df(\mathbf{h}) = (A \quad B)\begin{pmatrix} h_1 \\ h_2 \end{pmatrix} = Ah_1 + Bh_2$

$$\frac{f(\mathbf{x}) - f(\mathbf{a}) - df(\mathbf{h})}{\|\mathbf{h}\|} = \frac{f(a_1 + h_1, a_2 + h_2) - f(a_1, a_2) - Ah_1 - Bh_2}{\|\mathbf{h}\|}$$

$$= \frac{f(a_1 + h_1, a_2 + h_2) - f(a_1, a_2 + h_2) + f(a_1, a_2 + h_2) - f(a_1, a_2) - Ah_1 - Bh_2}{\|\mathbf{h}\|}$$

$$= \frac{f(a_1 + h_1, a_2 + h_2) - f(a_1, a_2 + h_2) - Ah_1}{\|\mathbf{h}\|} + \frac{f(a_1, a_2 + h_2) - f(a_1, a_2) - Bh_2}{\|\mathbf{h}\|}$$

$$= \frac{f(a_1 + h_1, a_2 + h_2) - q - Ah_1}{\|\mathbf{h}\|} + \frac{q - f(a_1, a_2) - Bh_2}{\|\mathbf{h}\|}$$

**iv.** Since $\lim_{h_1 \to 0} \frac{f(a_1 + h_1, a_2 + h_2) - f(a_1, a_2 + h_2)}{h_1} = f_x(a_1, a_2 + h_2)$,

we have that $\frac{f(a_1 + h_1, a_2 + h_2) - f(a_1, a_2 + h_2)}{h_1} = f_x(a_1, a_2 + h_2) + E_1(h_1)$ where $E_1(h_1) \to 0$ as $h_1 \to 0$.

Thus, $f(a_1 + h_1, a_2 + h_2) - f(a_1, a_2 + h_2) = h_1 f_x(a_1, a_2 + h_2) + h_1 E_1(h_1)$ and the first term in (29) can be written as $\frac{h_1(f_x(a_1, a_2 + h_2) + E_1(h_1) - A)}{\|\mathbf{h}\|} = \frac{h_1}{\|\mathbf{h}\|}(f_x(a_1, a_2 + h_2) - A + E_1(h_1))$.

Since $|h_1| \le \sqrt{h_1^2 + h_2^2}$, the ratio $\frac{h_1}{\|\mathbf{h}\|}$ lies between $-1$ and 1. $(a_1, a_2 + h_2)$ is near $\mathbf{a}$, and $f_x$ is continuous at $(a_1, a_2 + h_2)$, so $f_x(a_1, a_2 + h_2) - A = f_x(a_1, a_2 + h_2) - f_x(a_1, a_2) \to 0$ as $\|\mathbf{h}\| \to 0$.
Thus, the first term in (29) goes to 0 as $\|\mathbf{h}\| \to 0$.

Similarly, since $\lim_{h_2 \to 0} \frac{f(a_1, a_2 + h_2) - f(a_1, a_2)}{h_2} = f_y(a_1, a_2)$,
$f(a_1, a_2 + h_2) - f(a_1, a_2) = h_2 f_y(a_1, a_2) - h_2 E_2(h_2) = h_2 B - h_2 E_2(h_2)$ where $E_2(h_2) \to 0$ as $h_2 \to 0$.
the second term in (29) can be written as $\frac{h_2 E_2(h_2)}{\|\mathbf{h}\|} = \frac{h_2}{\|\mathbf{h}\|} E_2(h_2)$, which goes to 0 as $\|\mathbf{h}\| \to 0$.

## Section 10.2 The Chain Rule

**1.** $\frac{dy}{dt} = (2x_1^2 - x_2)(e^t) + (-x_1 + 2x_2)(-e^{-t}) = (2e^t - e^{-t})(e^{-t}) + (-e^t + 2e^{-t})(-e^{-t}) = 2e^{2t} - 2e^{-2t}$

Check:

$y = e^{2t} - e^t \cdot e^{-t} + e^{-2t} = e^{2t} - 1 + e^{-2t}$

$\frac{dy}{dt} = 2e^{2t} - 2e^{-2t}$

**3.** $h'(t) = \frac{\cos\left(\frac{y}{x}\right)}{y} \cdot 1 + \frac{x\cos\left(\frac{y}{x}\right)}{-y^2} \cdot \frac{t^{-1/2}}{2} = \frac{\cos\sqrt{t}}{\sqrt{t}} - \frac{t\cos\sqrt{t}}{2t\sqrt{t}} = \frac{\cos\sqrt{t}}{2\sqrt{t}}$

Check:

$h(t) = \sin\left(\frac{t}{\sqrt{t}}\right) = \sin\sqrt{t}$

$h'(t) = \cos\sqrt{t} \cdot \frac{t^{-1/2}}{2} = \frac{\cos\sqrt{t}}{2\sqrt{t}}$

**5.** $\frac{dy}{dt} = \left(\frac{1}{x_2 + \frac{x_1^2}{x_2}}\right)(2t) + \left(\frac{-x_1}{x_2^2 + x_1^2}\right)(-3t^2) = \frac{-2t}{t^3 + t} + \frac{3t^4}{t^4 + t^6} = \frac{1}{1+t^2}$

Check:

$y = \tan^{-1}\left(\frac{t^2}{-t^3}\right) = \tan^{-1}\left(-\frac{1}{t}\right)$

$\frac{dy}{dt} = \frac{1}{1 + \left(\frac{-1}{t}\right)^2} \cdot \frac{1}{t^2} = \frac{1}{1+t^2}$

**7.** $h'(t) = yz^2 \cdot 2e^{2t} - xz^2 \cdot e^{-t} + 2xyz \cdot e^t = 2e^{3t} - e^{3t} + 2e^{3t} = 3e^{3t}$

Check:

$h(t) = \left(e^{2t}\right)\left(e^{-t}\right)\left(e^t\right)^2 = e^{3t}$

$h'(t) = 3e^{3t}$

**9.** $\frac{dy}{dt} = \frac{-x_1}{\sqrt{10 - x_1^2 - x_2^2 - x_3^2}}(-\sin t) + \frac{-x_2}{\sqrt{10 - x_1^2 - x_2^2 - x_3^2}}(\cos t) + \frac{-x_3}{\sqrt{10 - x_1^2 - x_2^2 - x_3^2}}(3)$

$= \frac{\sin t \cos t}{\sqrt{9 - 9t^2}} - \frac{\sin t \cos t}{\sqrt{9 - 9t^2}} - \frac{9t}{\sqrt{9 - 9t^2}}$

$= -\frac{3t}{\sqrt{1 - t^2}}$

Check:

$y = \sqrt{10 - \cos^2 t - \sin^2 t - 9t^2} = 3\sqrt{1 - t^2}$

$\frac{dy}{dt} = \frac{3}{2\sqrt{1 - t^2}}(-2t) = -\frac{3t}{\sqrt{1 - t^2}}$

**11.** $z_u = \frac{\partial z}{\partial u} = \frac{\partial z}{\partial x}\frac{\partial x}{\partial u} + \frac{\partial z}{\partial y}\frac{\partial y}{\partial u} = -\frac{y}{x^2}\cos v + \frac{1}{x}\sin v = -\frac{u \sin v}{u^2 \cos^2 v}\cos v + \frac{1}{u \cos v}\sin v = 0$

$z_v = \frac{\partial z}{\partial v} = \frac{\partial z}{\partial x}\frac{\partial x}{\partial v} + \frac{\partial z}{\partial y}\frac{\partial y}{\partial v} = -\frac{y}{x^2}(-u \sin v) + \frac{1}{x}(u \cos v) = \frac{u^2 \sin^2 v}{u^2 \cos^2 v} + \frac{u \cos v}{u \cos v} = 1 + \tan^2 v$

**13.** $z_u = \frac{\partial z}{\partial u} = \frac{\partial z}{\partial x}\frac{\partial x}{\partial u} + \frac{\partial z}{\partial y}\frac{\partial y}{\partial u} = f_x u + f_y v = uf_x + vf_y$

$z_v = \frac{\partial z}{\partial v} = \frac{\partial z}{\partial x}\frac{\partial x}{\partial v} + \frac{\partial z}{\partial y}\frac{\partial y}{\partial v} = f_x(-v) + f_y u = -vf_x + uf_y$

**15.** $$\frac{\partial z}{\partial t_1}=\frac{\partial z}{\partial x_1}\frac{\partial x_1}{\partial t_1}+\frac{\partial z}{\partial x_2}\frac{\partial x_2}{\partial t_1}$$
$$=\frac{-x_1}{\sqrt{1-x_1^2-x_2^2}}(1)+\frac{-x_2}{\sqrt{1-x_1^2-x_2^2}}(1)$$
$$=-\frac{x_1+x_2}{\sqrt{1-x_1^2-x_2^2}}$$
$$\frac{\partial z}{\partial t_2}=\frac{\partial z}{\partial x_1}\frac{\partial x_1}{\partial t_2}+\frac{\partial z}{\partial x_2}\frac{\partial x_2}{\partial t_2}$$
$$=\frac{-x_1}{\sqrt{1-x_1^2-x_2^2}}(-1)+\frac{-x_2}{\sqrt{1-x_1^2-x_2^2}}(1)$$
$$=\frac{x_1-x_2}{\sqrt{1-x_1^2-x_2^2}}$$

When $\langle t_1, t_2\rangle=\left\langle 0, \frac{1}{2}\right\rangle$, $\langle x_1, x_2\rangle=\left\langle -\frac{1}{2}, \frac{1}{2}\right\rangle$, and so

$$\nabla z\left(0, \frac{1}{2}\right)=\left\langle \frac{dz}{dt_1}, \frac{dz}{dt_2}\right\rangle\Bigg|_{x_1=\frac{1}{2},\, x_2=\frac{1}{2}}=\langle 0, -\sqrt{2}\rangle.$$

**17.** $$\frac{d}{dt}\mathbf{y}=\begin{pmatrix}\cos x_1\cos x_2\cdot 2t+\sin x_1\sin x_2\cdot 3t^2\\ \cos x_1\sin x_2\cdot 2t-\sin x_1\cos x_2\cdot 3t^2\end{pmatrix}$$
$$=\begin{pmatrix}2t\cos t^2\cos t^3-3t^2\sin t^2\sin t^3\\ -2t\cos t^2\cos t^3-3t^2\sin t^2\sin t^3\end{pmatrix}$$

Check:
$$\mathbf{y}(t)=\left\langle \sin\left(t^2\right)\cos\left(-t^3\right), \sin\left(t^2\right)\sin\left(-t^3\right)\right\rangle$$
$$=\left\langle \sin t^2\cos t^3, -\sin t^2\sin t^3\right\rangle$$
$$\frac{d}{dt}\mathbf{y}=\begin{pmatrix}2t\cos t^2\cos t^3-3t^2\sin t^2\sin t^3\\ -2t\cos t^2\sin t^3-3t^2\sin t^2\sin t^3\end{pmatrix}$$

**19.** $$\frac{d}{dt}\mathbf{f}=\begin{pmatrix}e^{x+y}\cdot\frac{1}{t+1}+e^{x+y}\cdot\frac{1}{t}\\ e^{x-y}\cdot\frac{1}{t+1}-e^{x-y}\cdot\frac{1}{t}\end{pmatrix}$$
$$=\begin{pmatrix}\frac{e^{\ln(t^2+t)}}{t+1}+\frac{e^{\ln(t^2+t)}}{t}\\ \frac{e^{\ln\left(\frac{t+1}{t}\right)}}{t+1}-\frac{e^{\ln\left(\frac{t+1}{t}\right)}}{t}\end{pmatrix}$$
$$=\begin{pmatrix}2t+1\\ -\frac{1}{t^2}\end{pmatrix}$$

Check:
$$\mathbf{f}(t)=\left\langle e^{\ln\left(t^2+t\right)}, e^{\ln\left(\frac{t+1}{t}\right)}\right\rangle=\left\langle t^2+t, \frac{t+1}{t}\right\rangle$$
$$\frac{d}{dt}\mathbf{f}=\begin{pmatrix}2t+1\\ -\frac{1}{t^2}\end{pmatrix}$$

**21.** $$\frac{d}{dt}\mathbf{y}=\begin{pmatrix}2uw(1)+0(1)+u^2\left(\frac{1}{2\sqrt{t}}\right)\\ 0(1)+w(1)+v\left(\frac{1}{2\sqrt{t}}\right)\\ v(1)+u(1)+0\left(\frac{1}{2\sqrt{t}}\right)\end{pmatrix}$$
$$=\begin{pmatrix}2uw+\frac{u^2}{2\sqrt{t}}\\ w+\frac{v}{2\sqrt{t}}\\ u+v\end{pmatrix}$$
$$=\begin{pmatrix}\frac{5}{2}t^{3/2}\\ \frac{3}{2}t^{1/2}\\ 2t\end{pmatrix}$$

Check:
$$\mathbf{y}(t)=\left\langle t^2\sqrt{t}, t\sqrt{t}, t^2\right\rangle=\left\langle t^{5/2}, t^{3/2}, t^2\right\rangle$$
$$\frac{d}{dt}\mathbf{y}=\begin{pmatrix}\frac{5}{2}t^{3/2}\\ \frac{3}{2}t^{1/2}\\ 2t\end{pmatrix}$$

**23.** $$\frac{d}{dt}\mathbf{f}=\begin{pmatrix}-e^z\sin x\cdot 4t-0\cdot 4t+e^z\cos x\cdot 1\\ e^z\cos x\cdot 4t-0\cdot 4t+e^z\sin x\cdot 1\\ 0\cdot 4t-e^z\cdot 4t+e^z y\cdot 1\end{pmatrix}$$
$$=\begin{pmatrix}e^t\cos(2t^2)-4te^t\sin(2t^2)\\ e^t\sin(2t^2)+4te^t\cos(2t^2)\\ -2t^2e^t-4te^t\end{pmatrix}$$

Check:

$$\mathbf{f}(t)=\left(e^t\cos\left(2t^2\right),\ e^t\sin\left(2t^2\right),\ -2t^2e^t\right)$$

$$\frac{d}{dt}\mathbf{f}=\begin{pmatrix} e^t\cos(2t^2)-4te^t\sin(2t^2)\\ e^t\sin(2t^2)+4te^t\cos(2t^2)\\ -2t^2e^t-4te^t\end{pmatrix}$$

**25.** Since $u(\mathbf{r}(t))=6$ for all $t$, $C$ is a level curve of $u$.

$$\mathbf{r}'(t)=\left\langle-\sqrt{2}\sin t,\ \sqrt{3}\cos t\right\rangle$$

$$\nabla u(\mathbf{r}(t))=\langle 6x,\ 4y\rangle=\left\langle 6\sqrt{2}\cos t,\ 4\sqrt{3}\sin t\right\rangle$$

$$\nabla u(\mathbf{r}(t))\cdot\mathbf{r}'(t)=-12\sin t\cos t+12\sin t\cos t=0$$

$$\mathbf{r}\left(\frac{\pi}{4}\right)=\left\langle\frac{\sqrt{2}}{\sqrt{2}},\ \frac{\sqrt{3}}{\sqrt{2}}\right\rangle=\left\langle 1,\ \sqrt{\frac{3}{2}}\right\rangle=\mathbf{a}$$

$$\nabla u(\mathbf{a})=\left\langle 6,\ 4\sqrt{\frac{3}{2}}\right\rangle$$

$$\nabla u(\mathbf{a})\cdot\mathbf{r}'\left(\frac{\pi}{4}\right)=\left\langle 6,\ 4\sqrt{\frac{3}{2}}\right\rangle\cdot\left\langle -1,\ \sqrt{\frac{3}{2}}\right\rangle=0$$

**27.** Since $u(\mathbf{r}(t))=1$ for all $t$, $C$ is a level curve of $u$.

$$\mathbf{r}'(t)=(\sinh t,\ \cosh t)$$

$$\nabla u(\mathbf{r}(t))=(2x,\ -2y)=\langle 2\cosh t,\ -2\sinh t\rangle$$

$$\nabla u(\mathbf{r}(t))\cdot\mathbf{r}'(t)=2\cosh t\sinh t-2\cosh t\sinh t=0$$

$$\mathbf{r}\left(\ln\left(\sqrt{3}+2\right)\right)=\left\langle 2,\ \sqrt{3}\right\rangle=\mathbf{a}$$

$$\nabla u(\mathbf{a})=\left\langle 4,\ -2\sqrt{3}\right\rangle$$

$$\nabla u(\mathbf{a})\cdot\mathbf{r}'\left(\ln\left(\sqrt{3}+2\right)\right)=\left\langle 4,\ -2\sqrt{3}\right\rangle\cdot\left\langle\sqrt{3},\ 2\right\rangle=0$$

**29.**

$$T'(t)=F_x\frac{dx}{dt}+F_y\frac{dy}{dt}=e^t(\cos t-\sin t)F_x+e^t(\cos t+\sin t)F_y$$

$$T'\left(\frac{\pi}{2}\right)=-e^{\pi/2}F_x(\mathbf{p})+e^{\pi/2}F_y(\mathbf{p})\approx 18.28$$

$$T(1.5)\approx T\left(\frac{\pi}{2}\right)+T'\left(\frac{\pi}{2}\right)\left(1.5-\frac{\pi}{2}\right)=F(\mathbf{p})+T'\left(\frac{\pi}{2}\right)\left(1.5-\frac{\pi}{2}\right)\approx 148.7$$

**31.** $x+\frac{y}{2}+\frac{z}{3}=1$ implies $z=f(x,\ y)=3-3x-\frac{3y}{2}$.

$$\mathbf{r}(x,\ y)=\langle r_1,\ r_2,\ r_3\rangle=\langle x,\ y,\ f(x,\ y)\rangle$$

$$\mathbf{g}(\theta)=\langle x,\ y\rangle=\left\langle\frac{1}{3}+\frac{1}{4}\cos\theta,\ \frac{5}{8}+\frac{1}{4}\sin\theta\right\rangle$$

$$\mathbf{h}(\theta)=\langle h_1,\ h_2,\ h_3\rangle=\mathbf{r}(\mathbf{g}(\theta))$$

$$\mathbf{h}'(\theta)=\langle h_1'(\theta),\ h_2'(\theta),\ h_3'(\theta)\rangle$$

$$h_1'(\theta)=\frac{dr_1}{d\theta}=\frac{\partial h_1}{\partial x}\frac{dx}{d\theta}+\frac{\partial h_1}{\partial y}\frac{dy}{d\theta}=(1)\left(-\frac{1}{4}\sin\theta\right)+(0)\left(\frac{1}{4}\cos\theta\right)=-\frac{1}{4}\sin\theta$$

$$h_2'(\theta)=\frac{dr_2}{d\theta}=\frac{\partial h_2}{\partial x}\frac{dx}{d\theta}+\frac{\partial h_2}{\partial y}\frac{dy}{d\theta}=(0)\left(-\frac{1}{4}\sin\theta\right)+(1)\left(\frac{1}{4}\cos\theta\right)=\frac{1}{4}\cos\theta$$

$$h_3'(\theta)=\frac{dr_3}{d\theta}=\frac{\partial h_3}{\partial x}\frac{dx}{d\theta}+\frac{\partial h_3}{\partial y}\frac{dy}{d\theta}=(-3)\left(-\frac{1}{4}\sin\theta\right)+\left(-\frac{3}{2}\right)\left(\frac{1}{4}\cos\theta\right)=\frac{3}{4}\sin\theta-\frac{3}{8}\cos\theta$$

$$\mathbf{g}\left(\frac{\pi}{2}\right)=\left\langle\frac{1}{3}+\frac{1}{4}\cos\frac{\pi}{2},\ \frac{5}{8}+\frac{1}{4}\sin\frac{\pi}{2}\right\rangle=\left\langle\frac{1}{3},\ \frac{7}{8}\right\rangle$$

$$\mathbf{h}'\left(\frac{\pi}{2}\right)=\left\langle-\frac{1}{4}\sin\frac{\pi}{2},\ \frac{1}{4}\cos\frac{\pi}{2},\ \frac{3}{4}\sin\frac{\pi}{2}-\frac{3}{8}\cos\frac{\pi}{2}\right\rangle=\left\langle-\frac{1}{4},\ 0,\ \frac{3}{4}\right\rangle$$

**33.** Solving the equations relating $x$ and $y$ to $X$ and $Y$ for $X$ and $Y$, we get

$$X=(\cos\psi)x-(\sin\psi)y,$$

$$Y=(\sin\psi)x+(\cos\psi)y.$$

$$\frac{\partial X}{\partial x}=\cos\psi,\ \frac{\partial X}{\partial y}=-\sin\psi,\ \frac{\partial X}{\partial z}=0$$

$$\frac{\partial Y}{\partial x}=\sin\psi,\ \frac{\partial Y}{\partial y}=\cos\psi,\ \frac{\partial Y}{\partial z}=0$$

$$\frac{\partial Z}{\partial x}=0,\ \frac{\partial Z}{\partial y}=0,\ \frac{\partial Z}{\partial z}=1$$

$$\frac{\partial f_1}{\partial x} = \frac{\partial f_1}{\partial X}\frac{\partial X}{\partial x} + \frac{\partial f_1}{\partial Y}\frac{\partial Y}{\partial x} + \frac{\partial f_1}{\partial Z}\frac{\partial Z}{\partial x} = \frac{\partial f_1}{\partial X}\cos\psi + \frac{\partial f_1}{\partial Y}\sin\psi$$

$$\frac{\partial f_1}{\partial y} = \frac{\partial f_1}{\partial X}\frac{\partial X}{\partial y} + \frac{\partial f_1}{\partial Y}\frac{\partial Y}{\partial y} + \frac{\partial f_1}{\partial Z}\frac{\partial Z}{\partial y} = -\frac{\partial f_1}{\partial X}\sin\psi + \frac{\partial f_1}{\partial Y}\cos\psi$$

$$\frac{\partial f_1}{\partial z} = \frac{\partial f_1}{\partial X}\frac{\partial X}{\partial z} + \frac{\partial f_1}{\partial Y}\frac{\partial Y}{\partial z} + \frac{\partial f_1}{\partial Z}\frac{\partial Z}{\partial z} = \frac{\partial f_1}{\partial Z}$$

**35.** $\frac{\partial y_1}{\partial x_1} = e^{x_2};\ \frac{\partial y_1}{\partial x_2} = x_1 e^{x_2};\ \frac{\partial y_2}{\partial x_1} = x_2 e^{x_1};\ \frac{\partial y_2}{\partial x_2} = e^{x_1}$

$$\mathbf{Df}(\mathbf{g}(1,\ 0)) = \mathbf{Df}(1,\ 1) = \begin{pmatrix} \frac{\partial y_1}{\partial x_1}(1,\ 1) & \frac{\partial y_1}{\partial x_2}(1,\ 1) \\ \frac{\partial y_2}{\partial x_1}(1,\ 1) & \frac{\partial y_2}{\partial x_2}(1,\ 1) \end{pmatrix} = \begin{pmatrix} e & e \\ e & e \end{pmatrix}$$

This is the matrix associated with $\mathbf{df}_{(1,\ 1)}$.

$\frac{\partial x_1}{\partial t_1} = 1;\ \frac{\partial x_1}{\partial t_2} = 2t_2;\ \frac{\partial x_2}{\partial t_1} = t_2;\ \frac{\partial x_2}{\partial t_2} = t_1$

$$\mathbf{Dg}(1,\ 0) = \begin{pmatrix} \frac{\partial x_1}{\partial t_1}(1,\ 0) & \frac{\partial x_1}{\partial t_2}(1,\ 0) \\ \frac{\partial x_2}{\partial t_1}(1,\ 0) & \frac{\partial x_2}{\partial t_2}(1,\ 0) \end{pmatrix} = \begin{pmatrix} 1 & 0 \\ 0 & 1 \end{pmatrix}$$

This is the matrix associated with $\mathbf{dg}_{(1,\ 0)}$.

$$\mathbf{Dz}(1,\ 0) = \mathbf{Df}(\mathbf{g}(1,\ 0))\mathbf{Dg}(1,\ 0) = \begin{pmatrix} e & e \\ e & e \end{pmatrix}\begin{pmatrix} 1 & 0 \\ 0 & 1 \end{pmatrix} = \begin{pmatrix} e & e \\ e & e \end{pmatrix}$$

This is the matrix associated with $\mathbf{dz}_{(1,\ 0)}$.

$\mathbf{z}(1,\ 0) = \mathbf{f}(\mathbf{g}(1,\ 0)) = \mathbf{f}(1,\ 1) = \langle e,\ e\rangle$

$\mathbf{h} = \langle 1.05,\ 0.03\rangle - \langle 1,\ 0\rangle = \langle 0.05,\ 0.03\rangle$

$\mathbf{z}(1.05,\ 0.03) \approx \mathbf{z}(1,\ 0) + \mathbf{dz}_{(1,\ 0)}\mathbf{h} = \langle 1.08e,\ 1.08e\rangle \approx \langle 2.93574,\ 2.93574\rangle$

**37.** Half of a meridian has equation $\mathbf{m}(u,\ v_0) = \langle \sin u\cos v_0,\ \sin u\sin v_0,\ \cos u\rangle$ where $v_0$ is fixed, $0 \le v_0 < \pi$. (The other half is $\mathbf{m}(u,\ v_0 + \pi)$.)

$\mathbf{m}'(u) = \langle \cos u\cos v_0,\ \cos u\sin v_0,\ -\sin u\rangle$

The rhumb line crosses the meridian when $v = v_0$. From Example 3, when $v = v_0$,

$$\mathbf{h}'(t) = \begin{pmatrix} \cos u\cos v_0 + \dfrac{\tan(1)\sin u\sin v_0}{\sin t} \\ \cos u\sin v_0 - \dfrac{\tan(1)\sin u\cos v_0}{\sin t} \\ -\sin u \end{pmatrix}$$

where $u = t$ and $v_0 = -\tan(1)\ln\left(\tan\frac{t}{2}\right)$. Since $u = t$ and $v_0$ is fixed, we can write $\mathbf{h}'(t)$ as a function of $u$.

$$\mathbf{h}'(u) = \begin{pmatrix} \cos u\cos v_0 + \tan(1)\sin v_0 \\ \cos u\sin v_0 - \tan(1)\cos v_0 \\ -\sin u \end{pmatrix}$$

If $\theta$ is the angle between $\mathbf{h}'(u)$ and $\mathbf{m}'(u)$, $\cos\theta = \dfrac{\mathbf{m}'(u)\cdot\mathbf{h}'(u)}{\|\mathbf{m}'(u)\|\|\mathbf{h}'(u)\|}$.

$\mathbf{m}'(u)\cdot\mathbf{h}'(u) = \cos^2 u\cos^2 v_0 + \tan(1)\cos u\cos v_0\sin v_0 + \cos^2 u\sin^2 v_0 - \tan(1)\cos u\sin v_0\cos v_0 + \sin^2 u = 1$

$\|\mathbf{m}'(u)\| = 1$

$\|\mathbf{h}'(u)\| = \sqrt{1+\tan^2(1)} = \sqrt{\sec^2(1)} = \sec(1)$ since $\sec(1) > 0$.

Thus, $\cos\theta = \dfrac{1}{\sec(1)} = \cos(1)$ and $\theta = 1$.

**39.** **i.** Let $\mathbf{E_g}(t) = \dfrac{\mathbf{g}(t)-\mathbf{g}(b)-\mathbf{Dg}(b)(t-b)}{|t-b|}$. Since $\mathbf{g}$ is differentiable at $b$, $\lim\limits_{t\to b}\mathbf{E_g}(t) = \mathbf{0}$.

**ii.** Rearranging the definition of $\mathbf{E_g}(t)$ above, we get that when $t$ is near $b$,

$\mathbf{g}(t) = \mathbf{g}(b) + \mathbf{Dg}(b)(t-b) + \mathbf{E_g}(t)|t-b|$.

**iii.** Let $E_f(\mathbf{x}) = \dfrac{f(\mathbf{x})-f(\mathbf{a})-Df(\mathbf{a})(\mathbf{x}-\mathbf{a})}{\|\mathbf{x}-\mathbf{a}\|}$. Since $f$ is differentiable at $\mathbf{a}$, $\lim\limits_{\mathbf{x}\to\mathbf{a}}E_f(\mathbf{x}) = 0$.

**iv.** Rearranging the definition of $E_f(\mathbf{x})$ above, we get that when $\mathbf{x}$ is near $\mathbf{a}$,

$f(\mathbf{x}) = f(\mathbf{a}) + Df(\mathbf{a})(\mathbf{x}-\mathbf{a}) + E_f(\mathbf{x})\|\mathbf{x}-\mathbf{a}\|$.

By definition, $f\circ\mathbf{g}$ is differentiable at $b$ if there is a function $D(f\circ\mathbf{g})$ such that

$$\lim_{t\to b}\frac{(f\circ\mathbf{g})(t)-(f\circ\mathbf{g})(b)-D(f\circ\mathbf{g})(b)|t-b|}{|t-b|} = \lim_{t\to b}\left(\frac{f(\mathbf{g}(t))-f(\mathbf{g}(b))}{|t-b|}\right) - D(f\circ\mathbf{g})(b) = 0.$$

This is true just when $\lim\limits_{t\to b}\dfrac{f(\mathbf{g}(t))-f(\mathbf{g}(b))}{|t-b|}$ exists; then it equals $D(f\circ\mathbf{g})(b)$.

$$\lim_{t\to b}\frac{f(\mathbf{g}(t))-f(\mathbf{g}(b))}{|t-b|}$$

$$= \lim_{t\to b}\frac{f\left(\mathbf{g}(b)+\mathbf{Dg}(b)(t-b)+\mathbf{E_g}(t)|t-b|\right)-f(\mathbf{g}(b))}{|t-b|}$$

$$= \lim_{\substack{t\to b\\ \mathbf{x}\to\mathbf{a}}}\frac{f(\mathbf{x})-f(\mathbf{a})}{|t-b|}$$

($\mathbf{x}\to\mathbf{a}$ as $t\to b$ is guaranteed by the continuity of $\mathbf{g}$.)

$$= \lim_{\substack{t\to b\\ \mathbf{x}\to\mathbf{a}}}\frac{Df(\mathbf{a})(\mathbf{x}-\mathbf{a})+E_f(\mathbf{x})\|\mathbf{x}-\mathbf{a}\|}{|t-b|}$$

$$= Df(\mathbf{a})\lim_{\substack{t\to b\\ \mathbf{x}\to\mathbf{a}}}\frac{\mathbf{x}-\mathbf{a}}{|t-b|} + \lim_{\substack{t\to b\\ \mathbf{x}\to\mathbf{a}}}\left(E_f(\mathbf{x})\frac{\|\mathbf{x}-\mathbf{a}\|}{|t-b|}\right)$$

$$= Df(\mathbf{a})\mathbf{Dg}(b) + 0$$

$$= Df(\mathbf{g}(b))\mathbf{Dg}(b)$$

which is the desired result. The separation of one limit into two is justified by the fact that each of the two limits exists. The right limit goes to zero because $E_f(\mathbf{x})$ goes to zero and $\dfrac{\|\mathbf{x}-\mathbf{a}\|}{|t-b|}$ is finite (because $\mathbf{g}$ is differentiable).

## Section 10.3 Applications of the Chain Rule

1. $\nabla f(x_0, y_0) = \langle 2, 3 \rangle$

   Solve $f(x, y) = k$ for $y$:

   $2x + 3y + 5 = 0$

   $y = -\frac{2}{3}x - \frac{5}{3}$

   So $y' = -\frac{2}{3}$, and a tangent vector to the level curve is $\langle 3, -2 \rangle$. Since $\langle 2, 3 \rangle \cdot \langle 3, -2 \rangle = 0$, $\nabla f(x_0, y_0)$ is perpendicular to the tangent vector and thus to the level curve.

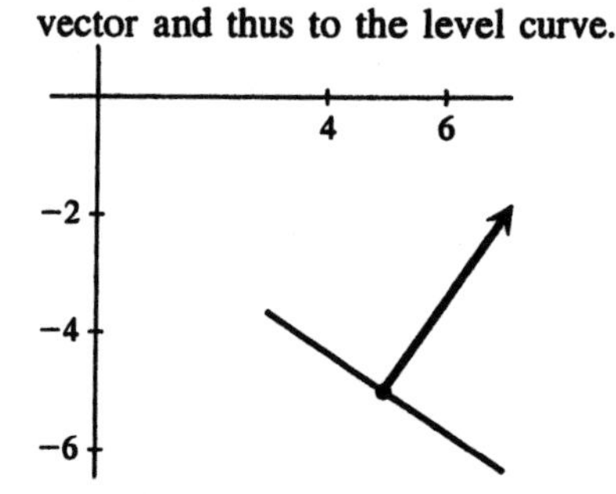


3. $\nabla f(x_0, y_0) = \langle -4x_0, 1 \rangle = \langle -4, 1 \rangle$

   Solve $f(x, y) = k$ for $y$:

   $y - 2x^2 = 1$

   $y = 2x^2 + 1$

   So $y' = 4x$, which for $x_0 = 1$ equals 4, and a tangent vector to the level curve is $\langle 1, 4 \rangle$. Since $\langle -4, 1 \rangle \cdot \langle 1, 4 \rangle = 0$, $\nabla f(x_0, y_0)$ is perpendicular to the tangent vector and thus to the level curve.

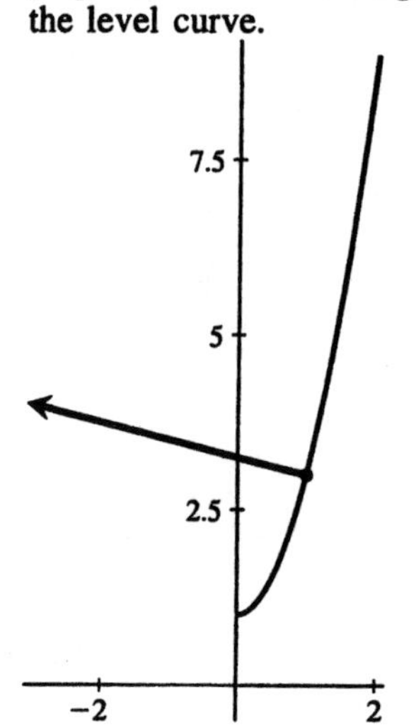


5. $\nabla f(x_0, y_0) = \langle 2x_0, 2y_0 \rangle = \langle 4, 8 \rangle$

   Solve $f(x, y) = k$ for $y$:

   $x^2 + y^2 = 20$

   $y = \sqrt{20 - x^2}$ ($y \geq 0$ portion of curve)

   So $y' = \frac{-x}{\sqrt{20 - x^2}}$, which for $x_0 = 2$ equals $-\frac{1}{2}$, and a tangent vector to the level curve is $\left\langle 1, -\frac{1}{2} \right\rangle$. Since $\langle 4, 8 \rangle \cdot \left\langle 1, -\frac{1}{2} \right\rangle = 0$, $\nabla f(x_0, y_0)$ is perpendicular to the tangent vector and thus to the level curve.

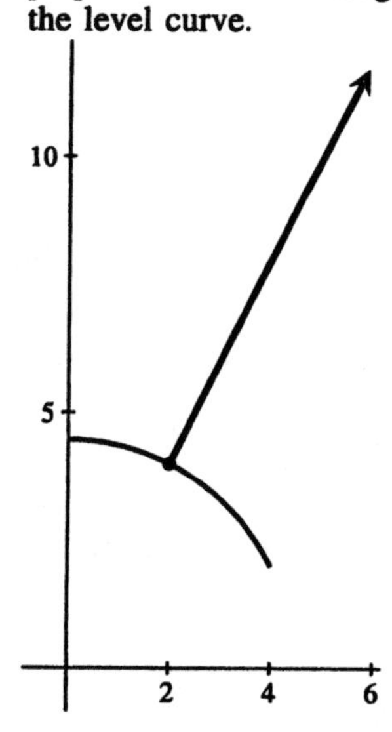


7. $F(x, y, z) = x^2 - 3xz + y^2 + 2yz - z^2$

   $F_x(x_0, y_0, z_0) = (2x - 3z)\big|_{(1, 1, 1)} = -1$

   $F_y(x_0, y_0, z_0) = (2y + 2z)\big|_{(1, 1, 1)} = 4$

   $F_z(x_0, y_0, z_0) = (-3x + 2y - 2z)\big|_{(1, 1, 1)} = -3$

   $$f_x(x_0, y_0) = -\frac{F_x(x_0, y_0, z_0)}{F_z(x_0, y_0, z_0)} = -\frac{1}{3}$$

   $$f_y(x_0, y_0) = -\frac{F_y(x_0, y_0, z_0)}{F_z(x_0, y_0, z_0)} = \frac{4}{3}$$

   So

   $$\langle -f_x(x_0, y_0), -f_y(x_0, y_0), 1 \rangle = \left\langle \frac{1}{3}, -\frac{4}{3}, 1 \right\rangle$$

This is parallel to
$\nabla F(x_0, y_0, z_0) = \langle -1, 4, -3 \rangle$ because it
equals $\left(-\frac{1}{3}\right)\nabla F(x_0, y_0, z_0)$. The tangent
plane is described by
$\nabla F(x_0, y_0, z_0) \cdot \langle x - x_0, y - y_0, z - z_0 \rangle = 0$ or
$-(x-1) + 4(y-1) - 3(z-1) = 0$.

**9.** $F(x, y, z) = e^z \cos(x+y) + z^2 - \frac{1}{2}$

$$F_x(x_0, y_0, z_0) = -e^z \sin(x+y)\Big|_{(\pi/6,\ \pi/6,\ 0)} = -\frac{\sqrt{3}}{2}$$

$$F_y(x_0, y_0, z_0) = -e^z \sin(x+y)\Big|_{(\pi/6,\ \pi/6,\ 0)} = -\frac{\sqrt{3}}{2}$$

$$F_z(x_0, y_0, z_0) = \left(e^z \cos(x+y) + 2z\right)\Big|_{(\pi/6,\ \pi/6,\ 0)} = \frac{1}{2}$$

$$f_x(x_0, y_0) = -\frac{F_x(x_0, y_0, z_0)}{F_z(x_0, y_0, z_0)} = \sqrt{3}$$

$$f_y(x_0, y_0) = -\frac{F_y(x_0, y_0, z_0)}{F_z(x_0, y_0, z_0)} = \sqrt{3}$$

So $\langle -f_x(x_0, y_0), -f_y(x_0, y_0), 1 \rangle$
$= \langle -\sqrt{3}, -\sqrt{3}, 1 \rangle$.
This is parallel to
$\nabla F(x_0, y_0, z_0) = \left\langle -\frac{\sqrt{3}}{2}, -\frac{\sqrt{3}}{2}, \frac{1}{2} \right\rangle$ because
it equals $2\nabla F(x_0, y_0, z_0)$. The tangent plane
is described by
$2\nabla F(x_0, y_0, z_0) \cdot \langle x - x_0, y - y_0, z - z_0 \rangle = 0$
or $-\sqrt{3}\left(x - \frac{\pi}{6}\right) - \sqrt{3}\left(y - \frac{\pi}{6}\right) + z = 0$.

**11.** $F(x, y, z) = x^2 + y^2 + z^2$, level surface
$F(x, y, z) = 11$. The parallel of latitude
through $P_0(3, 1, 1)$ is defined by
$C_1$: $\mathbf{r}_1(\theta) = \langle \sqrt{10}\cos\theta, \sqrt{10}\sin\theta, 1 \rangle$,
$0 \le \theta \le 2\pi$ and the meridian is defined by
$C_2$: $\mathbf{r}_2(\phi) = \left\langle 3\sqrt{\frac{11}{10}}\sin\phi, \sqrt{\frac{11}{10}}\sin\phi, \sqrt{11}\cos\phi \right\rangle$,
$0 \le \phi \le 2\pi$, because when
$\cos\phi = \frac{z_0}{r} = \frac{1}{\sqrt{11}}$, $\sin\phi = \sqrt{\frac{10}{11}}$.
$\nabla F(x_0, y_0, z_0) = \langle 2x, 2y, 2z \rangle\Big|_{(3,\ 1,\ 1)} = \langle 6, 2, 2 \rangle$
The tangent vectors to $C_1$ and $C_2$ are
$\mathbf{r}_1'(\theta) = \langle -\sqrt{10}\sin\theta, \sqrt{10}\cos\theta, 0 \rangle$
$\mathbf{r}_2'(\theta) = \left\langle 3\sqrt{\frac{11}{10}}\cos\phi, \sqrt{\frac{11}{10}}\cos\phi, -\sqrt{11}\sin\phi \right\rangle$
Since at $(3, 1, 1)$, $\theta = \arcsin\left(\frac{1}{\sqrt{10}}\right)$ and
$\phi = \arcsin\sqrt{\frac{10}{11}}$ the tangent vectors there are
$\mathbf{r}_1'\left(\arcsin\frac{1}{\sqrt{10}}\right) = \langle -1, 3, 0 \rangle$ and
$\mathbf{r}_2'\left(\arcsin\sqrt{\frac{10}{11}}\right) = \left\langle \frac{3}{\sqrt{10}}, \frac{1}{\sqrt{10}}, -\sqrt{10} \right\rangle$
or $\sqrt{10}\,\mathbf{r}_2'\left(\arcsin\sqrt{\frac{10}{11}}\right) = \langle 3, 1, -10 \rangle$.
$\nabla F(x_0, y_0, z_0)$ is perpendicular to both since
$\langle 6, 2, 2 \rangle \cdot \langle -1, 3, 0 \rangle = \langle 6, 2, 2 \rangle \cdot \langle 3, 1, -10 \rangle = 0$.

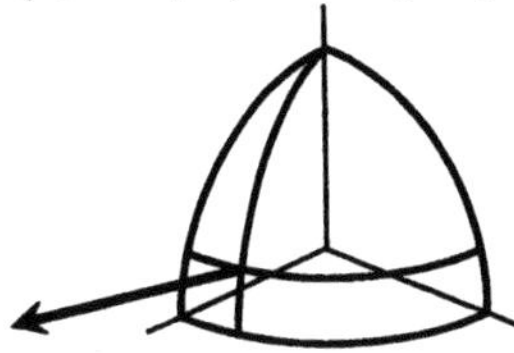

**13.** The surface $S$ is described by $x^2 - y^2 = z$ is a
level surface described by $F(x, y, z) = 0$ for
the function $F(x, y, z) = x^2 - y^2 - z$
$\nabla F(x, y, z) = \langle 2x, -2y, -1 \rangle$, so the tangent
plane at $(x_0, y_0, z_0) = (2, 1, 3)$ is described by
$\nabla F(x_0, y_0, z_0) \cdot \langle x - x_0, y - y_0, z - z_0 \rangle = 0$
$\langle 4, -2, -1 \rangle \cdot \langle x-2, y-1, z-3 \rangle = 0$ or
$4(x-2) - 2(y-1) - (z-3) = 0$

**15.** $F(x, y) = 2x^2 + xy - 3y^2 = 0$
$F_x(x_0, y_0) = (4x + y)\Big|_{(1,\ 1)} = 5$
$F_y(x_0, y_0) = (x - 6y)\Big|_{(1,\ 1)} = -5$
$$f'(x_0) = -\frac{F_x(x_0, y_0)}{F_y(x_0, y_0)} = 1$$

**17.** $F(x, y) = \arctan\left(\frac{x}{y}\right) - y + 1 = 0$

$$F_x(x_0, y_0) = \frac{y}{x^2 + y^2}\bigg|_{(0,\,1)} = 1$$

$$F_y(x_0, y_0) = \left(-\frac{x}{y^2 + x^2} - 1\right)\bigg|_{(0,\,1)} = -1$$

$$f'(x_0) = -\frac{F_x(x_0, y_0)}{F_y(x_0, y_0)} = 1$$

**19.** $F(x, y, z) = \sin(xy) + \sin(xz) - z - \sqrt{3} + 1 = 0$

$$F_x(x_0, y_0, z_0) = \left(y\cos(xy) + z\cos(xz)\right)\Big|_{(\pi/3,\,1,\,1)} = 1$$

$$F_y(x_0, y_0, z_0) = \left(x\cos(xy)\right)\Big|_{(\pi/3,\,1,\,1)} = \frac{\pi}{6}$$

$$F_z(x_0, y_0, z_0) = \left(x\cos(xz) - 1\right)\Big|_{(\pi/3,\,1,\,1)} = \frac{\pi}{6} - 1 = \frac{\pi - 6}{6}$$

$$f_x(x_0, y_0) = -\frac{F_x(x_0, y_0, z_0)}{F_z(x_0, y_0, z_0)} = -\frac{6}{\pi - 6}$$

$$f_y(x_0, y_0) = -\frac{F_y(x_0, y_0, z_0)}{F_z(x_0, y_0, z_0)} = -\frac{\pi}{\pi - 6}$$

A normal to the graph at $(x_0, y_0, z_0)$ is $\left\langle -f_x(x_0, y_0, z_0),\ -f_y(x_0, y_0, z_0),\ 1\right\rangle = \left\langle \frac{6}{\pi - 6},\ \frac{\pi}{\pi - 6},\ 1\right\rangle.$

**21.** $F(x, y, z) = \ln\left(x^2 y + z\right) + \arctan\left(\frac{z}{y}\right) - \frac{\pi}{4} = 0$

$$F_x(x_0, y_0, z_0) = \frac{2xy}{x^2 y + z}\bigg|_{(0,\,1,\,1)} = 0$$

$$F_y(x_0, y_0, z_0) = \left(\frac{x^2}{x^2 y + z} - \frac{z}{y^2 + z^2}\right)\bigg|_{(0,\,1,\,1)} = -\frac{1}{2}$$

$$F_z(x_0, y_0, z_0) = \left(\frac{1}{x^2 y + z} + \frac{y}{x^2 + y^2}\right)\bigg|_{(0,\,1,\,1)} = \frac{3}{2}$$

$$f_x(x_0, y_0) = -\frac{F_x(x_0, y_0, z_0)}{F_z(x_0, y_0, z_0)} = 0$$

$$f_y(x_0, y_0) = -\frac{F_y(x_0, y_0, z_0)}{F_z(x_0, y_0, z_0)} = \frac{1}{3}$$

A normal to the graph at $(x_0, y_0, z_0)$ is $\left\langle -f_x(x_0, y_0),\ -f_y(x_0, y_0),\ 1\right\rangle = \left\langle 0,\ -\frac{1}{3},\ 1\right\rangle.$

**23.** $f(x,y) = x^3 y^2 - xy^3$

$f_x = 3x^2 y^2 - y^3$

$f_{xx} = 6xy^2$

**25.** $f(x,y) = x^3 e^{xy}$

$f_y = x^4 e^{xy}$

$f_{yy} = x^5 e^{xy}$

**27.** $f(x,y) = x\sin(xy)$

$f_x = xy\cos(xy) + \sin(xy)$

$f_{xx} = -xy^2\sin(xy + y\cos(xy) + y\cos(xy) = -xy^2\sin(xy) + 2y\cos(xy)$

$f_{xxy} = -x^2y^2\cos(xy) - 2xy\sin(xy) - 2xy\sin(xy) + 2\cos(xy) = 2\cos(xy) - x^2y^2\cos(xy) - 4xy\sin(xy)$

**29.** $f(x,y) = x^3y^2 - xy^3$

$f_x = 3x^2y^2 - y^3$

$f_y = 2x^3y - 3xy^2$

$f_{xx} = 6xy^2$

$f_{yy} = 2x^3 - 6xy$

$f_{xy} = f_{xy} = 6x^2y - 3y^2$

$f_{xx}, f_{yy}, f_{xy},$ and $f_{yx}$ are all defined and continuous for real $x$, $y$, and $f_{xy} = f_{yx}$.

**31.** $f(x,y) = x^3e^{xy}$

$f_x = x^3ye^{xy} + 3x^2e^{xy} = x^2e^{xy}(xy+3)$

$f_y = x^4e^{xy}$

$f_{xx} = x^2ye^{xy} + \left(x^2ye^{xy} + 2xe^{xy}\right)(xy+3) = xe^{xy}\left(x^2y^2 + 6xy + 6\right)$

$f_{yy} = x^5e^{xy}$

$f_{xy} = x^3e^{xy} + x^3e^{xy}(xy+3) = x^3e^{xy}(xy+4)$

$f_{yx} = x^4ye^{xy} + 4x^3e^{xy} = x^3e^{xy}(xy+4)$

$f_{xx}, f_{yy}, f_{xy},$ and $f_{yx}$ are all defined and continuous for real $x$, $y$, and $f_{xy} = f_{yx}$.

**33.** We write $f(x,y) = x^{1/2} + y^{1/2}$.

$f_x = \frac{1}{2}x^{-1/2}y^{1/2}$; $f_{xx} = -\frac{1}{4}x^{-3/2}y^{1/2}$

$f_y = \frac{1}{2}x^{1/2}y^{-1/2}$; $f_{yy} = -\frac{1}{4}x^{1/2}y^{-3/2}$

$f_{xy} = \frac{1}{4}x^{-1/2}y^{-1/2}$

$$\begin{aligned}P(x,y) &= f(1,1) + f_x(1,1)(x-1) + f_y(1,1)(y-1) \\ &\quad + \frac{1}{2!}\left(f_{xx}(1,1)(x-1)^2 + 2f_{xy}(1,1)(x-1)(y-1) + f_{yy}(1,1)(y-1)^2\right) \\ &= 1 + \frac{1}{2}(x-1) + \frac{1}{2}(y-1) + \frac{1}{2!}\left(-\frac{1}{4}(x-1)^2 + \frac{1}{2}(x-1)(y-1) - \frac{1}{4}(y-1)^2\right) \\ &= 1 + \frac{1}{2}(x-1) + \frac{1}{2}(y-1) - \frac{1}{8}(x-1)^2 + \frac{1}{4}(x-1)(y-1) - \frac{1}{8}(y-1)^2\end{aligned}$$

**35.** $f_x = e^x \cos y;\ f_{xx} = e^x \cos y$

$f_y = -e^x \sin y;\ f_{yy} = -e^x \cos y$

$f_{xy} = -e^x \sin y$

$$P(x,y) = f(0,\ 0) + f_x(0,\ 0)(x-0) + f_y(0,\ 0)(y-0) + \frac{1}{2!}\left(f_{xx}(0,\ 0)(x-0)^2 + 2f_{xy}(0,\ 0)(x-0)(y-0) + f_{yy}(0,\ 0)(y-0)^2\right)$$

$$= 1 + x + \frac{1}{2!}\left(x^2 - y^2\right)$$

$$= 1 + x + \frac{1}{2}x^2 - \frac{1}{2}y^2$$

**37.** $f(x,\ y) = \sqrt{xy}$ (with $x,\ y > 0$)

$$f_x = \frac{\sqrt{y}}{2\sqrt{x}},\ f_y = \frac{\sqrt{x}}{2\sqrt{y}}$$

$$f_{xx} = -\frac{\sqrt{y}}{4x^{3/2}},\ f_{yy} = -\frac{\sqrt{x}}{4y^{3/2}}$$

$$f_{xxx} = \frac{3\sqrt{y}}{8x^{5/2}},\ f_{xxy} = -\frac{1}{8x^{3/2}\sqrt{y}}$$

$$f_{yyy} = \frac{3\sqrt{x}}{8y^{5/2}},\ f_{xyy} = f_{yyx} - \frac{1}{8y^{3/2}\sqrt{x}}$$

On the region where $x, y = 1 \pm 0.25$, the greatest absolute value attained by any third partial derivative is

$$\left|f_{xxx}(0.75,\ 1.25)\right| = \left|f_{yyy}(1.25,\ 0.75)\right| = \frac{3\sqrt{1.25}}{8(0.75)^{5/2}} \approx 0.86$$

This is greater than $\left|f_{xxy}(0.75,\ 0.75)\right| = \left|f_{yyx}(0.75,\ 0.75)\right| = \dfrac{1}{8(0.75)^{3/2}\sqrt{0.75}} \approx 0.22.$

So let $M = 0.9$. Then with $b = 0.5$, the error is less than $\dfrac{1}{3!}Mb^3 = \dfrac{1}{6}(0.9)(0.5)^3 = 0.01875$ for $(x, y) \in S_{0.5}$.

**39.** The first and second derivatives of $f(x,\ y) = \sqrt{\dfrac{x}{y}}$ at (1, 1) come out as stated. Then with $h_1 = x - 1$ and $h_2 = y - 1$, the Taylor polynomial inside the large parentheses in (13) becomes the polynomial given as $P(x, y)$ for (14).

**41.** Here, $u = \dfrac{x}{y}$ and $v = \dfrac{y}{x}$.

$$y' = -\frac{f_x(u,v)}{f_y(u,v)} = -\frac{f_u\dfrac{du}{dx} + f_v\dfrac{dv}{dx}}{f_u\dfrac{du}{dy} + f_v\dfrac{dv}{dy}} = -\frac{f_u\left(\dfrac{1}{y}\right) + f_v\left(-\dfrac{y}{x^2}\right)}{f_u\left(-\dfrac{x}{y^2}\right) + f_v\left(\dfrac{1}{x}\right)} = \frac{x^2yf_u - y^3f_v}{x^3f_u - xy^2f_v}$$

**43.** $F(x, y) = \sin(x-y) + x^2y = 0$

$F_y(x, y) = -\cos(x-y) + x^2 = 0$

Solve the second equation for $y$:

$\cos(x-y) = x^2$

$\sin(x-y) = \sqrt{1-x^4}$

$y = x - \arcsin\sqrt{1-x^4}$

Substitute in first equation:

$\sqrt{1-x^4} + x^2\left(x - \arcsin\sqrt{1-x^4}\right) = 0$

Solve using a CAS or Newton's method to find
$x \approx -0.703365$ and then $y \approx -1.756646$. Since the graph is symmetric about the origin, it follows that $x \approx 0.703365$, $y \approx 1.756646$ is also a solution.

This works to find the ends of the interval $I$ because $y' = -\dfrac{F_x}{F_y}$ has a vertical asymptote when $F_y = 0$.

**45.**
$$\begin{aligned} g'''(t) &= \frac{d}{dt}g''(t) \\ &= \frac{d}{dt}(f_{xx}(\mathbf{a}+t\mathbf{h})h_1^2) + \frac{d}{dt}(2f_{xy}(\mathbf{a}+t\mathbf{h})h_1h_2) + \frac{d}{dt}(f_{yy}(\mathbf{a}+t\mathbf{h})h_2^2) \\ &= (f_{xxx}h_1 + f_{xxy}h_2)h_1^2 + (2f_{xyx}h_1 + 2f_{xyy}h_2)h_1h_2 + (f_{yyx}h_1 + f_{yyy}h_2)h_2^2 \\ &= f_{xxx}h_1^3 + f_{xxy}h_1^2h_2 + 2f_{xyx}h_1^2h_2 + 2f_{xyy}h_1h_2^2 + f_{yyx}h_1h_2^2 + f_{yyy}h_2^3 \end{aligned}$$

**47.** Using the substitutions

$x = \sqrt{1+w^2}\sin\theta$, $y = \dfrac{(\sqrt{3}+1)w + (\sqrt{3}-1)\sqrt{1+w^2}\cos\theta}{2\sqrt{2}}$, and $z = \dfrac{(1-\sqrt{3})w + (1+\sqrt{3})\sqrt{1+w^2}\cos\theta}{2\sqrt{2}}$,

the equation $x^2 + yz - \frac{1}{2}\sqrt{3}(y^2 - z^2) - 1 = 0$ becomes (after some simplification)

$$(1+w^2)\sin^2\theta + \frac{2\sqrt{3}w\sqrt{1+w^2}\cos\theta + \cos^2\theta - w^2\sin^2\theta}{4}$$
$$-\frac{\sqrt{3}}{2}\left[\frac{2\sqrt{3}w^2 + 4w\sqrt{1+w^2}\cos\theta - 2\sqrt{3}(1+w^2)\cos^2\theta}{4}\right] - 1 = 0$$

and finally $0 = 0$

(This is best done on a CAS!)

**49.** Expanding the left side of $V^2\left[\left(P + \dfrac{n^2a}{V^2}\right)(V - nb) - nRT\right] = 0$ leads to

$PV^3 - n(bP + RT)V^2 + n^2aV - n^3ab = 0$ which with $P = P_0 = 5\times10^5$, $T = T_0 = 500$, $R = 8.314$, $a = 0.14$, $b = 3.91\times10^{-5}$, and $n = 100$, becomes $500{,}000V^3 - 417{,}655V^2 + 1400V - 5.474 = 0$

When displayed on a graphing calculator, the graph of the polynomial can be seen to cross the $V$-axis only once, at about 0.83196. The other two roots are therefore complex.

**51.** $f(x, y) = \begin{cases} 0 & (x, y) = 0 \\ \dfrac{xy(x^2 - y^2)}{x^2 + y^2} & (x, y) \neq 0 \end{cases}$

$f_x(0, 0)$ is found by setting $y = 0$ and allowing $x$ to approach zero. But when $y = 0, f(x, y)$ has a constant value of zero. Thus $f_x(0, 0) = 0$. By the same reasoning, $f_y(0, 0) = 0$.

For $(x, y) \neq 0$, by the quotient rule,

$$f_x(x, y) = \frac{(x^2 + y^2)(2x^2y + y(x^2 - y^2)) - 2x^2y(x^2 - y^2)}{(x^2 + y^2)^2} = \frac{y(x^4 + 4x^2y^2 - y^4)}{(x^2 + y^2)^2}$$

Also, $f_y(x, y) = \dfrac{(x^2 + y^2)(-2xy^2 + x(x^2 - y^2)) - 2xy^2(x^2 - y^2)}{(x^2 + y^2)^2} = \dfrac{x(x^4 - 4x^2y^2 - y^4)}{(x^2 + y^2)^2}$

$f_{xy}(0, 0)$ is found by setting $x = 0$ in $f_x$ and allowing $y$ to approach zero. With $x = 0$, $f_x = \dfrac{y(-y^4)}{(y^2)^2} = -y$ and the derivative of this function with respect to $y$ is a constant $-1$. To find $f_{yx}(0, 0)$, set $y = 0$ in $f_y$ and investigate the behavior with $x$ near zero. With $y = 0$, $f_y = \dfrac{x(x^4)}{(x^2)^2} = x$ and the derivative of this function with respect to $x$ is a constant 1. Thus, it is not the case that all second-order partial derivatives are continuous at (0, 0). If they were, then $f_{xy}(0, 0)$ would equal $f_{yx}(0, 0)$.

## Section 10.4 Further Applications of the Chain Rule

**1.** $u_x = 2x$; $u_{xx} = 2$
$u_y = -2y$; $u_{yy} = -2$
Thus, $u_{xx} + u_{yy} = 0$.

**3.** $u_x = 3x^2 - 3y^2$; $u_{xx} = 6x$
$u_y = -6xy$; $u_{yy} = -6x$
Thus, $u_{xx} + u_{yy} = 0$.

**5.** $u_x = \cos x \cosh y$; $u_{xx} = -\sin x \cosh y$
$u_y = \sin x \sinh y$; $u_{yy} = \sin x \cosh y$
Thus, $u_{xx} + u_{yy} = 0$.

**7.** $u_x = -\sin x \cosh y$; $u_{xx} = -\cos x \cosh y$
$u_y = \cos x \sinh y$; $u_{yy} = \cos x \cosh y$
Thus, $u_{xx} + u_{yy} = 0$.

**9.** $U_r = 2r \cos 2\theta$; $U_{rr} = 2 \cos 2\theta$
$U_\theta = -2r^2 \sin 2\theta$; $U_{\theta\theta} = -4r^2 \cos 2\theta$
Thus,

$$U_{rr} + \frac{1}{r^2}U_{\theta\theta} + \frac{1}{r}U_r = 2\cos 2\theta + \frac{1}{r^2}(-4r^2 \cos 2\theta) + \frac{1}{r}(2r \cos 2\theta) = 0.$$

**11.** $U(r, \theta) = 4r^3 \cos^3\theta - 3r^3 \cos\theta$

$U_r = 3r^2\left(4\cos^3\theta - 3\cos\theta\right) = r^2\left(12\cos^3\theta - 9\cos\theta\right)$

$U_{rr} = 2r\left(12\cos^3\theta - 9\cos\theta\right) = r\left(24\cos^3\theta - 18\cos\theta\right)$

$U_\theta = r^3\left(-12\cos^2\theta\sin\theta + 3\sin\theta\right)$

$U_{\theta\theta} = r^3\left(24\cos\theta\sin^2\theta - 12\cos^3\theta + 3\cos\theta\right)$

$= r^3\left(24\cos\theta\left(1-\cos^2\theta\right) - 12\cos^3\theta + 3\cos\theta\right)$

$= r^3\left(-36\cos^3\theta + 27\cos\theta\right)$

Thus, $U_{rr} + \frac{1}{r^2}U_{\theta\theta} + \frac{1}{r}U_r = r\left(24\cos^3\theta - 18\cos\theta\right) + r\left(-36\cos^3\theta + 27\cos\theta\right) + r\left(12\cos^3\theta - 18\cos\theta\right) = 0.$

**13.** Let $U(r, s) = u(x, y)$. $r_x = 1$; $r_y = -1$; $s_x = 1$; $s_y = 1$

Use the chain rule to obtain the following.

$u_x = U_r r_x + U_s s_x = U_r + U_s$

$u_{xx} = \frac{\partial}{\partial x}(U_r + U_s) = U_{rr}r_x + U_{rs}s_x + U_{ss}s_x + U_{sr}r_x = U_{rr} + 2U_{rs} + U_{ss}$

$u_y = U_r r_y + U_s s_y = -U_r + U_s$

$u_{yy} = \frac{\partial}{\partial y}(-U_r + U_s) = -U_{rr}r_y - U_{rs}s_y + U_{sr}r_y + U_{ss}s_y = U_{rr} - 2U_{rs} + U_{ss}$

$u_{xx} + u_{yy} = 2U_{rr} + 2U_{ss}$

Thus, the heat equation in the $(r, s)$-coordinate system is $U_{rr} + U_{ss} = 0$.

**15.** Let $U(r, s) = u(x, y)$. Differentiate the equations for $x$ and $y$, thinking of $r$ and $s$ as functions of $x$ and $y$.

$1 = r_x\cos\theta - s_x\sin\theta$; $0 = r_y\cos\theta - s_y\sin\theta$

$0 = r_x\sin\theta + s_x\cos\theta$; $1 = r_y\sin\theta + s_y\cos\theta$

Solving for $r_x$, $r_y$, $s_x$, and $s_y$, we get the following partials.

$r_x = \cos\theta$, $r_y = \sin\theta$, $s_x = -\sin\theta$, $s_y = \cos\theta$

Use the chain rule to obtain the following.

$u_x = U_r r_x + U_s s_x = (\cos\theta)U_r - (\sin\theta)U_s$

$u_{xx} = (\cos\theta)\frac{\partial}{\partial x}U_r - (\sin\theta)\frac{\partial}{\partial x}U_s$

$= (\cos\theta)(U_{rr}r_x + U_{rs}s_x) - (\sin\theta)(U_{sr}r_x + U_{ss}s_x)$

$= \left(\cos^2\theta\right)U_{rr} - (\cos\theta\sin\theta)U_{rs} - (\cos\theta\sin\theta)U_{rs} + \left(\sin^2\theta\right)U_{ss}$

$= \left(\cos^2\theta\right)U_{rr} - 2(\cos\theta\sin\theta)U_{rs} + \left(\sin^2\theta\right)U_{ss}$

$u_y = U_r r_y + U_s s_y = (\sin\theta)U_r + (\cos\theta)U_s$

$u_{yy} = (\sin\theta)\frac{\partial}{\partial y}U_r + (\cos\theta)\frac{\partial}{\partial y}U_s$

$= (\sin\theta)(U_{rr}r_y + U_{rs}s_y) + (\cos\theta)(U_{sr}r_y + U_{ss}s_y)$

$= \left(\sin^2\theta\right)U_{rr} + (\cos\theta\sin\theta)U_{rs} + (\cos\theta\sin\theta)U_{rs} + \left(\cos^2\theta\right)U_{ss}$

$= \left(\sin^2\theta\right)U_{rr} + 2(\cos\theta\sin\theta)U_{rs} + \left(\cos^2\theta\right)U_{ss}$

$u_{xx} + u_{yy} = \left(\cos^2\theta + \sin^2\theta\right)U_{rr} + \left(\sin^2\theta + \cos^2\theta\right)U_{ss} = U_{rr} + U_{ss}$

Thus, the heat equation in the $(r, s)$-coordinate system is $U_{rr} + U_{ss} = 0$.

**17.** Let $U(r, s) = u(x, y)$. From Exercise 13, we have $r_x = 1$; $r_y = -1$; $s_x = 1$; $s_y = 1$ and $u_x = U_r + U_s$. Using the chain rule,

$$u_{xy} = \frac{\partial}{\partial y}U_r + \frac{\partial}{\partial y}U_s = U_{rr}r_y + U_{rs}s_y + U_{sr}r_y + U_{ss}s_y = -U_{rr} + U_{ss}$$

To find $xy$, solve $r = x - y$, $s = x + y$ for $x$ and $y$ to obtain $x = \frac{1}{2}(s + r)$, $y = \frac{1}{2}(s - r)$. Then $xy = \frac{1}{4}\left(s^2 - r^2\right)$. So the equation $u_{xy} = xy + u_x$ becomes $-U_{rr} + U_{ss} = \frac{1}{4}\left(s^2 - r^2\right) + U_r + U_s$ or

$$-\frac{1}{4}\left(s^2 - r^2\right) - U_s + U_{ss} - U_r - U_{rr} = 0.$$

**19.** Let $U(r, s) = u(x, y)$. From Exercise 15, we have $r_x = \cos\theta$, $r_y = \sin\theta$, $s_x = -\sin\theta$, $s_y = \cos\theta$, and $u_x = (\cos\theta)U_r - (\sin\theta)U_s$.
Using the chain rule,

$$\begin{aligned} u_{xy} &= (\cos\theta)\frac{\partial}{\partial y}U_r - (\sin\theta)\frac{\partial}{\partial y}U_s \\ &= (\cos\theta)\left(U_{rr}r_y + U_{rs}s_y\right) - (\sin\theta)\left(U_{sr}r_y + U_{ss}s_y\right) \\ &= (\cos\theta\sin\theta)U_{rr} + \left(\cos^2\theta - \sin^2\theta\right)U_{rs} - (\cos\theta\sin\theta)U_{ss} \\ &= \frac{1}{2}(\sin 2\theta)(U_{rr} - U_{ss}) + (\cos 2\theta)U_{rs} \end{aligned}$$

So the equation $u_{xy} = xy + u_x$ becomes

$$\frac{1}{2}(\sin 2\theta)(U_{rr} - U_{ss}) + (\cos 2\theta)U_{rs} = (r\cos\theta - s\sin\theta)(r\sin\theta + s\cos\theta) + (\cos\theta)U_r - (\sin\theta)U_s$$

or

$$-(r\cos\theta - s\sin\theta)(r\sin\theta + s\cos\theta) + (\sin\theta)U_s + (\cos 2\theta)U_{rs} - (\cos\theta)U_r + \frac{1}{2}(\sin 2\theta)(U_{rr} - U_{ss}) = 0$$

**21.** From (5),

$$u_x^2 + u_y^2 = (U_r r_x + U_\theta\theta_x)^2 + \left(U_r r_y + U_\theta\theta_y\right)^2 = U_r^2\left(r_x^2 + r_y^2\right) + U_\theta^2\left(\theta_x^2 + \theta_y^2\right) + 2U_rU_\theta\left(r_x\theta_x + r_y\theta_y\right)$$

But because $r_x = \cos\theta$, $r_y = \sin\theta$, $\theta_x = -\frac{\sin\theta}{r}$, and $\theta_y = \frac{\cos\theta}{r}$, as stated in (8), we can write

$$\begin{aligned} u_x^2 + u_y^2 &= U_r^2 + \frac{1}{r^2}U_\theta^2 + 2U_rU_\theta\left(\frac{-\sin\theta\cos\theta}{r} + \frac{\sin\theta\cos\theta}{r}\right) \\ &= U_r^2 + \frac{1}{r^2}U_\theta^2 \end{aligned}$$

**23.** $$f'(x) = \begin{cases} 0, & -\infty < x < -1 \\ \dfrac{-\pi\sin(\pi x)}{16}, & -1 \le x \le 1 \\ 0, & 1 < x < \infty \end{cases}$$

$$f''(x) = \begin{cases} 0, & -\infty < x < -1 \\ \dfrac{-\pi^2\cos(\pi x)}{16}, & -1 \le x \le 1 \\ 0, & 1 < x < \infty \end{cases}$$

Observe that $f''(x)$ is not continuous at $-1$ and 1. In the derivation leading to (21), it was assumed that $U(w, z)$ has continuous second partial derivatives. However, since $U(w, z)$ was found to be $\frac{1}{2}f(w) + \frac{1}{2}f(z)$, this implies that $f(w)$ and $f(z)$ each have continuous second derivatives, i.e., $u(x, 0) = f(x)$ has a continuous second derivative. Thus D'Alembert's solution as presented in the text does not strictly apply to the initial shape shown in Figure 10.33.

**25.** System (6) can be written as

$$\begin{pmatrix} \cos\theta & -r\sin\theta \\ \sin\theta & r\cos\theta \end{pmatrix}\begin{pmatrix} r_x \\ \theta_x \end{pmatrix} = \begin{pmatrix} 1 \\ 0 \end{pmatrix}$$

and then by Cramer's rule,

$$r_x = \frac{\begin{vmatrix} 1 & -r\sin\theta \\ 0 & r\cos\theta \end{vmatrix}}{\begin{vmatrix} \cos\theta & -r\sin\theta \\ \sin\theta & r\cos\theta \end{vmatrix}} = \frac{r\cos\theta}{r\cos^2\theta + r\sin^2\theta} = \cos\theta$$

$$\theta_x = \frac{\begin{vmatrix} \cos\theta & 1 \\ \sin\theta & 0 \end{vmatrix}}{\begin{vmatrix} \cos\theta & -r\sin\theta \\ \sin\theta & r\cos\theta \end{vmatrix}} = \frac{-\sin\theta}{r\cos^2\theta + r\sin^2\theta} = -\frac{\sin\theta}{r}$$

Similarly, system (7) can be written as

$$\begin{pmatrix} \cos\theta & -r\sin\theta \\ \sin\theta & r\cos\theta \end{pmatrix}\begin{pmatrix} r_y \\ \theta_y \end{pmatrix} = \begin{pmatrix} 0 \\ 1 \end{pmatrix}$$

and then by Cramer's rule,

$$r_y = \frac{\begin{vmatrix} 0 & -r\sin\theta \\ 1 & r\cos\theta \end{vmatrix}}{\begin{vmatrix} \cos\theta & -r\sin\theta \\ \sin\theta & r\cos\theta \end{vmatrix}} = \frac{r\sin\theta}{r\cos^2\theta + r\sin^2\theta} = \sin\theta$$

$$\theta_y = \frac{\begin{vmatrix} \cos\theta & 0 \\ \sin\theta & 1 \end{vmatrix}}{\begin{vmatrix} \cos\theta & -r\sin\theta \\ \sin\theta & r\cos\theta \end{vmatrix}} = \frac{\cos\theta}{r\cos^2\theta + r\sin^2\theta} = \frac{\cos\theta}{r}.$$

**27.** With the calculator in radian mode, the result checks.

**29.** For $x = 0.3$,

$$g(t) = f(0.3, t) - 0.2 = e^{-t}\sin 0.3 - \frac{1}{9}e^{-9t}\sin 0.9 - 0.2.$$

then $g'(t) = -e^{-t}\sin 0.3 + e^{-9t}\sin 0.9$. Now use Newton's method, $t_{n+1} = t_n - \frac{g(t_n)}{g'(t_n)}$ starting with $t_1 = 0.3$, to find $t \approx 0.38$.

Repeating this process, using the value of $x$ as $t_1$ in each case of Newton's method, we find that some points on the 0.2 contour are (0.3, 0.38), (0.6, 1.04), (0.9, 1.37), (1.2, 1.54), (1.5, 1.61), and (1.8, 1.58).

**31.** As stated on page 894, the peak separates into two congruent pulses, one traveling to the right and the other to the left.

## Section 10.5 Optimization

**1.** $\nabla f(x, y) = \langle -(x-1)^2 - 2x(x-1),\ 4y\rangle$
$= \langle -(x-1)(3x-1),\ 4y\rangle$

$\nabla f(x, y) = \mathbf{0}$ at $(1, 0), \left(\frac{1}{3}, 0\right)$.

$f$ may have extrema at (1, 0) and $\left(\frac{1}{3}, 0\right)$

**3.** $\nabla f(x, y) = \langle 2x - y + 3,\ -x + 2y\rangle$

$\nabla f(x, y) = \mathbf{0}$ when $2x - y + 3 = 0$ and $-x + 2y = 0$.

Solving this system, we get $x = -2, y = -1$.

$f$ may have an extremum at (–2, –1).

**5.** $\nabla f(x, y) = \langle \cos y,\ -x\sin y\rangle$

$\nabla f(x, y) = \mathbf{0}$ when $\cos y = 0$ and $-x \sin y = 0$.

If $\cos y = 0$, then $y = \frac{\pi}{2}, \frac{3\pi}{2}$. If $y = \frac{\pi}{2}$ or $y = \frac{3\pi}{2}$, $\sin y \neq 0$ so $-x \sin y$ is never zero for $0 < x, y < 2\pi$.

There are no points at which $f$ may have extrema.

**7.** $\nabla f(x, y) = \langle 3x^2 - 3y,\ -3x + 3y^2\rangle$

$\nabla f(x, y) = \mathbf{0}$ when $x^2 = y$ and $y^2 = x$.

The solutions to this system are $x = 0, y = 0$ and $x = 1, y = 1$.

$f$ may have extrema at (0, 0) and (1, 1).

**9.** $\nabla f(x, y) = \langle 4x^3 - 4y,\ 4y^3 - 4x\rangle$

$\nabla f(x, y) = \mathbf{0}$ when $x^3 = y$ and $y^3 = x$.

The solutions to this system are $x = 0, y = 0$ and $x = -1, y = -1$ and $x = 1, y = 1$.

$f$ may have extrema at (0, 0), (1, 1), and (–1, –1).

**11.** $\nabla f(x, y, z) = \langle yz - 2x,\ xz - 2y,\ xy - 2z\rangle$
$\nabla f(x, y, z) = \mathbf{0}$ when $yz - 2x = 0,\ xz - 2y = 0,$
$xy - 2z = 0.$

From the first equation, $x = \frac{1}{2}yz$. Substitute into the second equation to get $\frac{1}{2}yz^2 - 2y = 0$ so $y\left(\frac{1}{2}z^2 - 2\right) = 0$. Substitute into the third equation to get $\frac{1}{2}y^2z - 2z = 0$ so $z\left(\frac{1}{2}y^2 - 2\right) = 0$. If $y = 0$, then $z = 0$ and $x = 0$.
If $y = -2$, then $z = -2$ or $z = 2$. If $y = 2$, then $z = -2$ or $z = 2$. Hence the solutions for the system are $(x, y, z) = (0, 0, 0), (2, -2, -2),$
$(-2, 2, -2), (2, 2, 2), (-2, -2, 2).$
$f$ may have extrema at $(0, 0, 0), (2, -2, -2), (-2, 2 -2), (2, 2, 2), (-2, -2, 2).$

**13.** $f(x, y) = x^3 - y^2 + 6xy$
$\nabla f(x, y) = \langle 3x^2 + 6y,\ -2y + 6x\rangle$
$\nabla f(x, y) = \mathbf{0}$ when $3x^2 + 6y = 0,\ -2y + 6x = 0.$
Solving this system, we get $(0, 0)$ and $(-6, -18)$. Only $(0, 0)$ is in the region.
$f(0, 0) = 0$. The boundary points are $(t, 0)$, $(1, t)$, and $(t, t)$ for $0 \le t \le 1$. $f(t, 0) = t^3$ is increasing on $0 \le t \le 1$ so its minimum value is $f(0, 0) = 0$ and its maximum value is $f(1, 0) = 1$.
$f(1, t) = -t^2 + 6t + 1$ is increasing on $0 \le t \le 1$ so its minimum value is $f(1, 0) = 1$ and its maximum value is $f(1, 1) = 6$.
$f(t, t) = t^3 + 5t^2$ is increasing on $0 \le t \le 1$ so its minimum value is $f(0, 0) = 0$ and its maximum value is $f(1, 1) = 6$. thus $f$ has a global minimum of 0 at $(0, 0)$ and a global maximum of 6 at $(1, 1)$.

**15.** $f(x, y) = x^2ye^{-x-y}$
$\nabla f(x, y) = \langle 2xye^{-x-y} - x^2ye^{-x-y},\ x^2e^{-x-y} - x^2ye^{-x-y}\rangle = \langle xy(2 - x)e^{-x-y},\ x^2(1 - y)e^{-x-y}\rangle$
$\nabla f(x, y) = \mathbf{0}$ at $(2, 1)$ and any point $(0, t)$.
$f(2, 1) = 4e^{-3} \approx 0.199$ and $f(0, t) = 0$.
The boundary points are $(t, 0)$, $(0, t)$, $(3, t)$, and $(t, 3)$ for $0 \le t \le 3$.
$f(t, 0) = 0$ and $f(0, t) = 0$. $f(3, t) = 9te^{-3-t}$ and $f'(3, t) = 9(1 - t)e^{-3-t}$, $f(3, 0) = 0$ and $f(3, 3) = 18e^{-6} \approx 0.045$. So $t = 1$ is a critical point.
$f(3, 1) = 9e^{-4} \approx 0.165$ and checking the endpoints, $f(t, 3) = 3t^2e^{-t-3}$ and $f'(t, 3) = (6t - 9t^2)e^{-t-3}$, so $t = 0$ and $t = \frac{3}{2}$ are critical points. $f(0, 3) = 0$, $f\left(\frac{3}{2}, 3\right) = \frac{27}{4}e^{-9/2} \approx 0.0004$, and, checking the other endpoint $f(3, 3) = 18e^{-6} \approx 0.045$.
Thus $f$ has a global minimum of 0 on the lines $x = 0$ and $y = 0$ and a global maximum of $4e^{-3}$ at $(2, 1)$.

**17.** $f(x, y) = (x - y + 1)^2$
$\nabla f(x, y) = \langle 2(x - y + 1),\ -2(x - y + 1)\rangle$
$\nabla f(x, y) = \mathbf{0}$ when $x - y + 1 = 0$. Thus any point of the form $(t, t + 1)$, $-1 \le t \le 1$ is a candidate. Note that $f(t, t + 1) = 0$ for all such $t$. Thus, points on the line $y = x + 1$ are candidates.
The boundary points are $(t - 1, 0)$, $(t - 1, 2)$, $(-1, t)$, and $(1, t)$ for $0 \le t \le 2$.
$f(t - 1, 0) = t^2$ is increasing on $0 \le t \le 2$, so its minimum value is $f(-1, 0) = 0$ and its maximum value is $f(1, 0) = 4$.
$f(t - 1, 2) = (t - 2)^2$ is decreasing on $0 \le t \le 2$, so it sminimum value is $f(1, 2) = 0$ and its maximum value is $f(-1, 2) = 4$.

$f(-1, t) = (-t)^2$ is increasing on $0 \le t \le 2$, so its minimum value is $f(-1, 0) = 0$ and its maximum value is $f(-1, 2) = 4$.

$f(1, t) = (2-t)^2$ is decreasing on $0 \le t \le 2$, so its minimum value is $f(1, 2) = 0$ and its maximum value is $f(1, 0) = 4$.

Thus $f$ has a global minimum of 0 on the line $y = x + 1$ and a global maximum of 4 at (1, 0) and (–1, 2).

**19.** $f(x, y) = x\sqrt{y} - x^2 + 9x - y$

$\nabla f(x, y) = \left\langle \sqrt{y} - 2x + 9, \ \frac{x}{2\sqrt{y}} - 1 \right\rangle$

$\nabla f(x, y) = \mathbf{0}$ when $\sqrt{y} = \frac{x}{2}$. Therefore

$\frac{x}{2} - 2x + 9 = 0$, so $x = 6$. Thus the only candidate is (6, 9).

$f(6, 9) = 27$

The boundary points are $(t, 0)$, $(0, t)$, $(t, 10)$, and $(10, t)$ for $0 \le t \le 10$. $f(t, 10) = -t^2 + 9t$ and $f'(t, 0) = -2t + 9$, so $t = \frac{9}{2}$ is a critical point.

$f\left(\frac{9}{2}, 0\right) = \frac{81}{4}$ and, checking the endpoints, $f(0, 0) = 0$ and $f(10, 0) = -10$. $f(0, t) = -t$ is decreasing on $0 \le t \le 10$, so its minimum value is $f(0, 10) = -10$ and its maximum value is $f(0, 0) = 0$.

$f(t, 10) = -t^2 + \left(9 + \sqrt{10}\right)t - 10$ and

$f'(t, 10) = -2t + 9 + \sqrt{10}$, so $t = \frac{9 + \sqrt{10}}{2}$ is a critical point.

$f\left(\frac{9+\sqrt{10}}{2}, 10\right) = \frac{51 + 18\sqrt{10}}{4} \approx 26.98$ and,

checking the endpoints, $f(0, 10) = -10$ and $f(10, 10) = 10\sqrt{10} - 20 \approx 11.62$.

$f(10, t) = -t + 10\sqrt{t} - 10$ is increasing on $0 \le t \le 10$, so its minimum value is $f(10, 0) = -10$ and its maximum value is $f(10, 10) = 10\sqrt{10} - 20 \approx 11.62$.

Thus $f$ has a global munimum of –10 at (10, 0) and (0, 10) and a global maximum of 27 at (6, 9).

**21.** $f(x, y) = 2x^2 + x + y^2 - 2$

$\nabla f(x, y) = \langle 4x + 1, \ 2y \rangle$

$\nabla f(x, y) = \mathbf{0}$ at $\left(-\frac{1}{4}, 0\right)$.

$f\left(-\frac{1}{4}, 0\right) = -\frac{17}{8}$

The boundary poins are $(2 \cos t, 2 \sin t)$ for $0 \le t \le 2\pi$.

$f(2\cos t, \ 2\sin t) = 4\cos^2 t + 2\cos t + 2$ and

$f'(2\cos t, \ 2\sin t) = -8\cos t \sin t - 2\sin t$
$= -2\sin t(4\cos t + 1)$.

On $0 \le t \le 2\pi$, the critical points are

$t = 0, \ \pi, \ \arccos\left(-\frac{1}{4}\right)$, and $2\pi \arccos\left(-\frac{1}{4}\right)$,

which are the points (2, 0), (–2, 0), $\left(-\frac{1}{2}, \frac{\sqrt{15}}{2}\right)$ and $\left(-\frac{1}{2}, -\frac{\sqrt{15}}{2}\right)$.

$f(2, 0) = 8$, $f(-2, 0) = 4$,

$f\left(-\frac{1}{2}, \frac{\sqrt{15}}{2}\right) = f\left(-\frac{1}{2}, -\frac{\sqrt{15}}{2}\right) = \frac{7}{4}$.

Thus $f$ has a global minimum of $-\frac{17}{8}$ at $\left(-\frac{1}{4}, 0\right)$ and a global maximum of 8 at (2, 0).

**23.** Let $x$, $y$, and $z$ be the dimensions of the tank where $z$ is the height, so $V = xyz$ and the restrictions are $xy + 2xz + 2yz = 12$ and $x, y, z > 0$. Hence $z = \frac{12 - xy}{2x + 2y}$ and

$V(x, y) = \frac{12xy - x^2y^2}{2x + 2y}$ for $x, y > 0$ and $xy < 12$.

$\nabla V(x, y)$

$= \left\langle \frac{-y^2\left(x^2 + 2xy - 12\right)}{2(x+y)^2}, \ \frac{-x^2\left(y^2 + 2xy - 12\right)}{2(x+y)^2} \right\rangle$

$\nabla V(x, y) = \mathbf{0}$ at (2, 2).

$V(2, 2) = 4$. As $(x, y)$ approaches the boundaries, the volume approaches zero. Thus, the absolute maximum volume is 4 $\text{ft}^3$.

**25.** The square of the distance is given by the function

$$D(x, y) = x^2 + y^2 + \left(x^2 + y^2 + \frac{1}{x^4y^2}\right) = 2x^2 + 2y^2 + \frac{1}{x^4y^2}.$$

$$\nabla D(x, y) = \left\langle 4x - \frac{4}{x^5y^2}, 4y - \frac{2}{x^4y^3}\right\rangle$$

$\nabla D(x, y) = \mathbf{0}$ at $\left(\pm 2^{1/8}, \pm 2^{-3/8}\right)$ and

$z = \pm\left(2^{5/4} - 2^{-3/4}\right)$. Since $D$ increases as $|x|$ and $|y|$ increase, the points $\left(\pm 2^{1/8}, \pm 2^{-3/8}, \pm\left(2^{5/4} + 2^{-3/4}\right)\right)$ are closest to the origin.

**27.** $V = xyz$ and the restrictions are $\frac{x}{a} + \frac{y}{b} + \frac{z}{c} = 1$ and $0 < x < a, 0 < y < b, 0 < z < c$. Hence $z = c\left(1 - \frac{x}{a} - \frac{y}{b}\right)$ and $V(x, y) = c\left(xy - \frac{x^2y}{a} - \frac{xy^2}{b}\right)$ for $0 < x < a$, $0 < y < b$.

$$\nabla V(x, y) = \left\langle cy\left(1 - \frac{2x}{a} - \frac{y}{b}\right), c\left(1 - \frac{x}{a} - \frac{2y}{b}\right)\right\rangle$$

$\nabla V(x, y) = \mathbf{0}$ at $\left(\frac{a}{3}, \frac{b}{3}\right)$, so $z = \frac{c}{3}$.

As $(x, y)$ approaches the boundaries, the volume approaches zero. Thus, the largest volume is at $\left(\frac{a}{3}, \frac{b}{3}, \frac{c}{3}\right)$.

**29.** $\nabla f(x, y) = \langle \cos x + \cos(x + y), \cos y + \cos(x + y)\rangle$

On the interval $0 \le x, y \le \pi$, $\nabla f(x, y) = \mathbf{0}$ when $\cos x = \cos y$ or $x = y$. Then $\cos x + \cos 2x = 0$ or $2\cos^2 x + \cos x - 2 = 0$, so $(2\cos x - 1)(\cos x + 1) = 0$. Therefore $x = \frac{\pi}{3}$ or $x = \pi$. The candidates are $\left(\frac{\pi}{3}, \frac{\pi}{3}\right)$ and $(\pi, \pi)$. $f\left(\frac{\pi}{3}, \frac{\pi}{3}\right) = \frac{3\sqrt{3}}{2}$ and $f(\pi, \pi) = 0$.

The boundary points are $(t, 0)$, $(0, t)$, $(t, \pi)$, and $(\pi, t)$ for $0 \le t \le \pi$.

$f(t, 0) = 2 \sin t$ has a maximum value of 2 when $t = \frac{\pi}{2}$ and a minimum value of 0 when $t = 0, \pi$.

$f(0, t) = 2 \sin t$ has a maximum value of 2 when $t = \frac{\pi}{2}$ and a minimum value of 0 when $t = 0, \pi$.

$f(t, \pi) = 0$ and $f(\pi, t) = 0$.

Thus $f$ has an absolue maximum of 2 and an absolute minimum of 0.

**31.** Let $x$, $y$, and $z$ be the three sides of the triangle, so $z = p - x - y$. From Heron's (sometimes known as Hero's) formula, the area is $\frac{1}{4}\sqrt{p(p - 2x)(p - 2y)(2x + 2y - p)}$. Then we examine the square of the area function $f(x, y) = \frac{1}{16}p(p - 2x)(p - 2y)(2x + 2y - p)$, $0 < x, y < p$.

$$\nabla f(x, y) = \frac{1}{4}\langle p(p - 2y)(p - 2x - y), p(p - 2x)(p - x - 2y)\rangle$$

$f_x = 0$ when $p = 2y$ or $p = 2x + y$ and $f_y = 0$ when $p = 2x$ or $p = 2y + x$. Note that $x, y, z > 0$.

If $p = 2y = 2x$, then $z = p - \frac{1}{2}p - \frac{1}{2}p = 0$.

If $p = 2y$ and $p = 2y + x$, then $x = 0$.

If $p = 2x + y$ and $p = 2x$, then $y = 0$.

Thus, in the region $0 < x, y < p$, $\nabla f(x, y) = \mathbf{0}$ only when $p - 2x - y = 0$ and $p - x - 2y = 0$. Solving this system, this is when $x=\frac{p}{3}, y=\frac{p}{3}$. Thus $z=\frac{p}{3}$ and $f\left(\frac{p}{3}, \frac{p}{3}\right)=\frac{1}{432}p^4$. The maximum area is $\frac{1}{12\sqrt{3}}p^2$ for an equilateral triangle. There is no minimum area since we can let one side approach 0, so the area approaches but never reaches 0.

**33.** By the regulations, $\ell+2(w+h)\le 108$. But if $\ell+2(w+h)<108$, then then the volume of the box can be increased by holding $w$ and $h$ constant and increasing $\ell$ until $\ell+2(w+h)=108$. Thus there is no way that for a box with maximum volume, $\ell+2(w+h)<108$.

**35.** $\nabla f(x, y)=\langle 6x-y-2, 4y-x\rangle$

$\mathbf{b}_1=-\nabla f(\mathbf{a}_1)=-\nabla f(0, 0)=\langle 2, 0\rangle$

$\mathbf{r}_1(t)=\mathbf{a}_1+t\mathbf{b}_1=\langle 0, 0\rangle+t\langle 2, 0\rangle=\langle 2t, 0\rangle$

$w=g_1(t)=f(2t, 0)=12t^2-4t$

The graph of $g_1$ is shown.

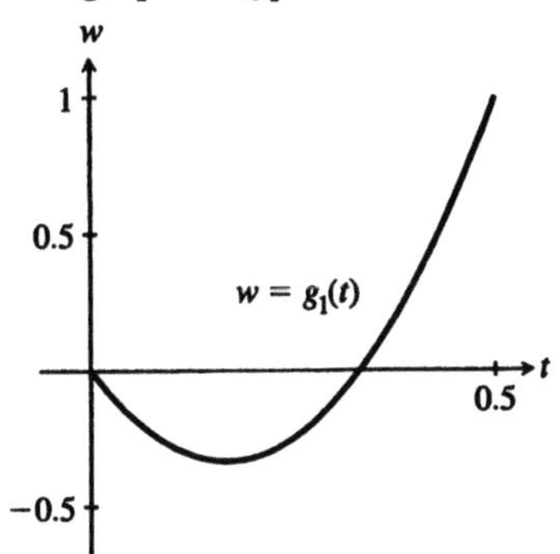


From the figure (or direct calculation), the minimum point is at $t=\frac{1}{6}\approx 0.1667$.

$\mathbf{a}_2=\mathbf{r}_1\left(\frac{1}{6}\right)=\left\langle\frac{1}{3}, 0\right\rangle$

$f(\mathbf{a}_2)=-\frac{1}{3}\approx -0.3333$

We repeat the steepest descent algorithm.

$\mathbf{b}_2=-\nabla f(\mathbf{a}_2)=-\nabla f\left(\frac{1}{3}, 0\right)=\left\langle 0, \frac{1}{3}\right\rangle$

$\mathbf{r}_2(t)=\mathbf{a}_2+t\mathbf{b}_2=\left\langle\frac{1}{3}, 0\right\rangle+t\left\langle 0, \frac{1}{3}\right\rangle=\left\langle\frac{1}{3}, \frac{1}{3}t\right\rangle$

$g_2(t)=\frac{2}{9}t^2-\frac{1}{9}t-\frac{1}{3}$

The minimum point is at $t=\frac{1}{4}$.

$\mathbf{a}_3=\mathbf{r}_2\left(\frac{1}{4}\right)=\left\langle\frac{1}{3}, \frac{1}{12}\right\rangle$

$f(\mathbf{a}_3)=-\frac{25}{72}\approx -0.3472$

Thus, $\mathbf{a}_2$ is not a minimum point for $f$.

**37.** $\nabla f(x, y, z)=\langle 4x-z, 4y+z, 2z-x+y+1\rangle$

$\mathbf{b}_1=-\nabla f(\mathbf{a}_1)=-\nabla f(0, 0, -1)=\langle -1, 1, 1\rangle$

$\mathbf{r}_1(t)=\mathbf{a}_1+t\mathbf{b}_1=\langle 0, 0, -1\rangle+t\langle -1, 1, 1\rangle=\langle -t, t, -1+t\rangle$

$w=g_1(t)=f(-t, t, -1+t)=7t^2-3t$

The graph of $g_1$ is shown.

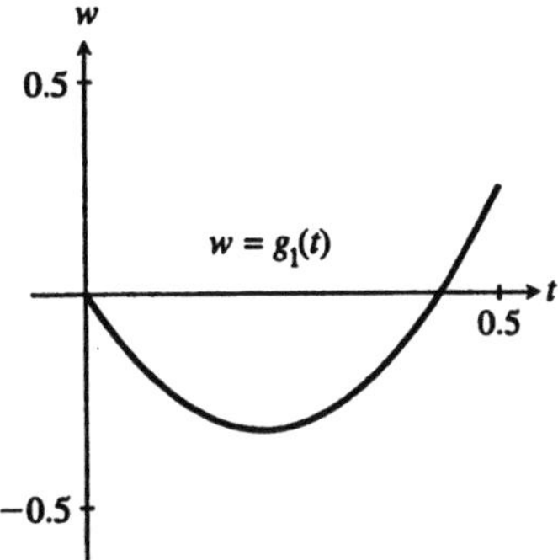


From the figure (or direct calculation), the minimum point is at $t=\frac{3}{14}\approx 0.2143$.

$\mathbf{a}_2=\mathbf{r}_1\left(\frac{3}{14}\right)=\left\langle -\frac{3}{14}, \frac{3}{14}, -\frac{11}{14}\right\rangle$

$f(\mathbf{a}_2)=-\frac{9}{28}\approx -0.3214$

We repeat the steepest descent algorithm.

$\mathbf{b}_2=-\nabla f(\mathbf{a}_2)=-\nabla f\left(-\frac{3}{14}, \frac{3}{14}, -\frac{11}{14}\right)=\left\langle\frac{1}{14}, -\frac{1}{14}, \frac{1}{7}\right\rangle$

$\mathbf{r}_2(t)=\mathbf{a}_2+t\mathbf{b}_2=\left\langle -\frac{3}{14}, \frac{3}{14}, -\frac{11}{14}\right\rangle+t\left\langle\frac{1}{14}, -\frac{1}{14}, \frac{1}{7}\right\rangle=\left\langle -\frac{3}{14}+\frac{1}{14}t, \frac{3}{14}-\frac{1}{14}t, -\frac{11}{14}+\frac{1}{7}t\right\rangle$

$g_2(t)=f\left(-\frac{3}{14}+\frac{1}{14}t, \frac{3}{14}-\frac{1}{14}t, -\frac{11}{14}+\frac{1}{7}t\right)=\frac{1}{49}t^2-\frac{3}{98}t-\frac{9}{28}$

The minimum point is at $t = \frac{3}{4}$.

$\mathbf{a}_3 = \mathbf{r}_2\left(\frac{3}{4}\right) = \left\langle -\frac{9}{56}, \frac{9}{56}, -\frac{19}{28} \right\rangle$

$f(\mathbf{a}_3) = -\frac{261}{784} \approx -0.3329$

Thus $\mathbf{a}_2$ is not a minimum point for $f$

**39.** If $a = 0$, then since $0 \le y \le \frac{a}{2}$, $y = 0$, and $f(t, y) = 0$.

If $a \ne 0$, then

$$\nabla f(t, y) = \left\langle \frac{1}{(t^2+4)^3}\left[-y^2(t^4-16) + 4a(t^2-2)y\sqrt{t^2+4} - 4a^2(3t^2-4)\right], \frac{2t}{(t^2+4)^{3/2}}\left[y\sqrt{t^2+4} - a\right] \right\rangle.$$

Since $0 < t < \infty$, $f_y = 0$ only when $y = \frac{a}{\sqrt{t^2+4}}$. Putting this into $f_t$ and solving for when $f_t = 0$, we get that $t = \frac{2}{\sqrt{3}}$. Thus $\nabla f(t, y) = \mathbf{0}$ at $\left(\frac{2}{\sqrt{3}}, \frac{a\sqrt{3}}{4}\right)$.

$f\left(\frac{2}{\sqrt{3}}, \frac{a\sqrt{3}}{4}\right) = \frac{9a^2\sqrt{3}}{128} \approx 0.1218a^2$

The boundary points are $(t, 0)$ and $\left(t, \frac{a}{2}\right)$.

$f(t, 0) = \frac{4a^2t}{(t^2+4)^2}$ and $f'(t, 0) = \frac{4a^2(4-3t^2)}{(t^2+4)^3}$

On $0 < t < \infty$, the only critical point is when $t = \frac{2}{\sqrt{3}}$.

$f\left(\frac{2}{\sqrt{3}}, 0\right) = \frac{3a^2\sqrt{3}}{32} \approx 0.1624a^2$

$f\left(t, \frac{a}{2}\right) = \frac{t}{t^2+4}\left(\frac{4a^2}{t^2+4} - \frac{a^2}{\sqrt{t^2+4}} + \frac{a^2}{4}\right)$ and $f'\left(t, \frac{a}{2}\right) = -\frac{a^2}{4(t^2+4)^3}\left(t^4 + 48t^2 - 8(t^2-2)\sqrt{t^2+4} - 80\right)$.

Using numerical methods, the only critical point on $0 < t < \infty$ is $t \approx 1.1792$.

$f\left(1.1792, \frac{a}{2}\right) \approx 0.1228a^2$

As $t \to 0$, $f(t, y) \to 0$ for all $0 \le y \le \frac{a}{2}$; as $t \to \infty$, $f(t, y) \to 0$ for all $0 \le y \le \frac{a}{2}$.

Thus, $f$ has a global maximum of $0.1624a^2$ at $\left(\frac{2}{\sqrt{3}}, 0\right)$.

## Section 10.6 Second Derivatives Test

**1.** $\nabla f(x, y) = \langle y, x \rangle$; $\nabla f(0, 0) = (0, 0)$
Thus (0, 0) is a stationary point. The point is not a local maximum since along the line $y = x$ the surface rises from 0 at (0, 0). That is, for any $t \neq 0$, $f(t, t) = t^2 > 0 = f(0, 0)$. The point is not a local minimum since along the line $y = -x$ the surface falls from 0 at (0, 0). That is, for any $t \neq 0$,
$f(t, -t) = -t^2 < 0 = f(0, 0)$.

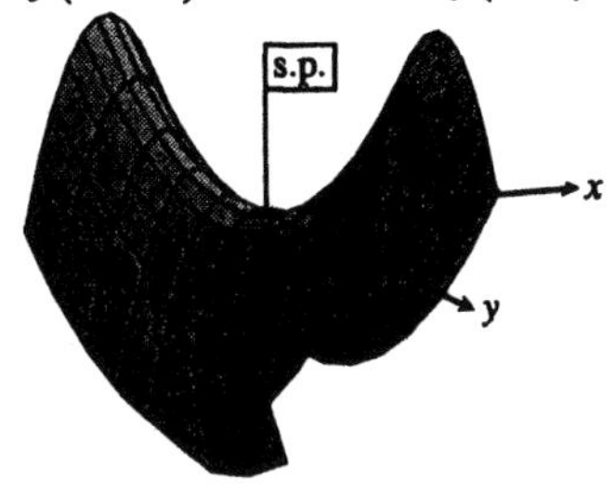


**3.** $\nabla f(x, y) = \langle 2x + 3y, 3x - 2y \rangle$
$\nabla f(x, y) = \mathbf{0}$ at (0, 0).
The only stationary point is (0, 0).
$f_{xx} = 2$; $f_{yy} = -2$; $f_{xy} = 3$
$D(0, 0) = f_{xx}(0, 0) f_{yy}(0, 0) - (f_{xy}(0, 0))^2 = -13 < 0$
Thus (0, 0) is a saddle point.

**5.** $\nabla f(x, y) = \langle 2x, -4(2y + 3) \rangle$
$\nabla f(x, y) = \mathbf{0}$ at $\left(0, -\frac{3}{2}\right)$.

The only stationary point is $\left(0, -\frac{3}{2}\right)$.

$f_{xx} = 2$; $f_{yy} = -8$; $f_{xy} = 0$

$D\left(0, -\frac{3}{2}\right) = f_{xx}\left(0, -\frac{3}{2}\right) f_{yy}\left(0, -\frac{3}{2}\right) - \left(f_{xy}\left(0, -\frac{3}{2}\right)\right)^2 = -16 < 0$

Thus $\left(0, -\frac{3}{2}\right)$ is a saddle point.

**7.** $\nabla f(x, y) = \langle 3x^2 - 3y, 3y^2 - 3x \rangle$
$\nabla f(x, y) = \mathbf{0}$ at (0, 0) and (1, 1).
The stationary points are (0, 0) and (1, 1).
$f_{xx} = 6x$; $f_{yy} = 6y$; $f_{xy} = -3$
$D(0, 0) = f_{xx}(0, 0) f_{yy}(0, 0) - (f_{xy}(0, 0))^2 = -9 < 0$
$D(1, 1) = f_{xx}(1, 1) f_{yy}(1, 1) - (f_{xy}(1, 1))^2 = 27 > 0$ and $f_{xx} = (1, 1) = 6 > 0$.
Thus (0, 0) is saddle point and (1, 1) is a local minimum.

**9.** $\nabla f(x, y) = \langle 4x^3 - 4y,\ 4y^3 - 4x \rangle$

$\nabla f(x, y) = \mathbf{0}$ at (0, 0), (1, 1) and (–1, –1). The stationary points are (0, 0), (1, 1) and (–1, –1).

$f_{xx} = 12x^2;\ f_{yy} = 12y^2;\ f_{xy} = -4$

$D(0, 0) = f_{xx}(0, 0)f_{yy}(0, 0) - (f_{xy}(0, 0))^2 = -16 < 0$

$D(1, 1) = f_{xx}(1, 1)f_{yy}(1, 1) - (f_{xy}(1, 1))^2 = 126 > 0$ and $f_{xx} = (1, 1) = 12 > 0.$

$D(-1, -1) = f_{xx}(-1, -1)f_{yy}(-1, -1) - (f_{xy}(-1, -1))^2 = 126 > 0$ and $f_{xx} = (-1, -1) = 12 > 0.$

Thus (0, 0) is saddle point and (1, 1) and (–1, –1) are local minimum.

**11.** $\nabla f(x, y) = \left\langle \frac{y-1}{x},\ \ln(xy) + 1 - \frac{1}{y} \right\rangle$

$\nabla f(x, y) = \mathbf{0}$ at (1, 1). The only stationary point is (1, 1).

$f_{xx} = \frac{-y+1}{x^2};\ f_{yy} = \frac{1}{y} + \frac{1}{y^2};\ f_{xy} = \frac{1}{x}$

$D(1, 1) = f_{xx}(1, 1)f_{yy}(1, 1) - (f_{xy}(1, 1))^2 = -1 < 0$

Thus (1, 1) is a saddle point.

**13.** It was established that $f(\mathbf{a} + \mathbf{h}) \approx f(\mathbf{a}) + \frac{1}{2}\left(Ah_1^2 + 2Bh_1h_2 + Ch_2^2\right)$, where $A = f_{xx}(\mathbf{a})$, $B = f_{xy}(\mathbf{a})$, and $C = f_{yy}(\mathbf{a})$.

If $A = 0$ it follows that $f(\mathbf{a} + \mathbf{h}) \approx f(\mathbf{a}) + Bh_1h_2 + \frac{C}{2}h_2^2$.

Now $A = 0$ implies $D(\mathbf{a}) = AC - B^2 = -B^2 \leq 0$.

This falls under case 3, if $B \neq 0$, or case 4, if $B = 0$.

Case 3: If $D(\mathbf{a}) < 0$, i.e., $B \neq 0$, then either $B > 0$ or $B < 0$. If $B > 0$, then $f(\mathbf{a} + \mathbf{h})$ can be made either greater or smaller than $f(\mathbf{a})$ by selecting $h_1$ and $h_2$ appropriately. For instance, to make $f(\mathbf{a} + \mathbf{h})$ greater, choose $h_2 > 0$ and choose $h_1$ so that

$$Bh_1h_2 + \frac{C}{2}h_2^2 > 0$$

$$h_1 > -\frac{C}{2B}h_2.$$

To make $f(\mathbf{a} + \mathbf{h}) < f(\mathbf{a})$ choose $h_1 < \frac{Ch_2}{2B}$. Similarly, if $B < 0$, we can make $f(\mathbf{a} + \mathbf{h})$ either greater or less than $f(\mathbf{a})$ by choosing $h_2 > 0$ and $h_1$ either less than or greater than $-\frac{Ch_2}{2B}$.

So whether $B > 0$ or $B < 0$, there are $(h_1, h_2)$ such that $f(\mathbf{a} + \mathbf{h}) > f(\mathbf{a})$ and $(h_1, h_2)$ such that $f(\mathbf{a} + \mathbf{h}) < f(\mathbf{a})$. Thus $\mathbf{a}$ is a saddle point, in accordance with th Second Derivatives Test Case 3.

Case 4: If $D(\mathbf{a}) = 0$, i.e., $B = 0$, then $f(\mathbf{a} + \mathbf{h}) \approx f(\mathbf{a}) + \frac{C}{2}h_2^2$

To a first approximation, $f(\mathbf{a} + \mathbf{h})$ no longer depends on $h_1$. Strictly speaking, however, it still depends on $h_1$; it is just that the analysis does not reveal how. If $C > 0$, $\mathbf{a}$ is either a local minimum or a saddle point, and if $C < 0$, $\mathbf{a}$ is either a local maximum or a saddle point. Thus the Second Derivatives Test fails.

**15.** Assume that $D(\mathbf{a}) < 0$ and $A < 0$. We first note that if we take $\mathbf{h} = \langle h_1, 0\rangle \neq \mathbf{0}$ in (3), then $f(\mathbf{a}+\mathbf{h}) < f(\mathbf{a})$. Hence, $\mathbf{a}$ is not a local minimum. Next we show that $f(\mathbf{a})$ is not a local maximum. There are two parts to this. First, if $B = 0$ and we take $\mathbf{h} = \langle 0, h_2\rangle \neq \mathbf{0}$ in (3), then (3) becomes $f(\mathbf{a}+\mathbf{h}) \approx f(\mathbf{a}) + \frac{A}{2}\cdot\frac{D(\mathbf{a})}{A^2}h_2^2 > f(\mathbf{a})$. Hence, $f(\mathbf{a})$ is not a local maximum. Secondly, if $B \neq 0$ and we take $\mathbf{h} = \left\langle h_1, -\frac{A}{B}h_1\right\rangle \neq \mathbf{0}$ in (3), then $f(\mathbf{a}+\mathbf{h}) \approx f(\mathbf{a}) + \frac{A}{2}\cdot\frac{D(\mathbf{a})}{A^2}\left(-\frac{A}{B}h_1\right)^2 > f(\mathbf{a})$, again showing that $f(\mathbf{a})$ is not a local maximum. It now follows that $\mathbf{a}$ is a saddle point.

**17.** $\nabla f(x, y) = \left\langle 3x^2\cos y + y^2, -x^3\sin y + 2xy\right\rangle$

$\nabla f(x, y) = \mathbf{0}$ implies

$3x^2\cos y + y^2 = 0$ (1)

$-x^3\sin y + 2xy = 0$ (2)

From (2), $x^2 = \dfrac{2y}{\sin y}$ since $x \neq 0$.

Substituting in (1) yields $6y\cot y + y^2 = 0$.

This can be solved numerically to yield $y \approx 1.87345$ (given $0 < y < 2$), and then $x \approx 1.98124$ (given $x > 0$).

Thus the stationary point is (1.98124, 1.87345). $f_{xx} = 6x\cos y$; $f_{yy} = -x^3\cos y + 2x$; $f_{xy} = -3x^2\sin y + 2y$

$D(1.98124, 1.87345) = f_{xx}(1.98124, 1.87345)f_{yy}(1.98124, 1.87345) - \left(f_{xy}(1.98124, 1.87345)\right)^2 \approx -78.41 < 0$

Thus (1.98124, 1.87345) is a saddle point.

**19.** The square of the distance from a point on the plane $x + \frac{1}{2}y + \frac{1}{3}z = 1$, or equivalently, $z = 3 - 3x - \frac{3}{2}y$, to the point (5, 4, 3) is $f(x, y) = (x-5)^2 + (y-4)^2 + \left(-3x - \frac{3}{2}y\right)^2$.

$\nabla f(x, y) = \left\langle 20x + 9y - 10, 9x + \frac{13}{2}y - 8\right\rangle$

$\nabla f(x, y) = \mathbf{0}$ at $\left(-\frac{1}{7}, \frac{10}{7}\right)$. The only stationary point is $\left(-\frac{1}{7}, \frac{10}{7}\right)$

$f_{xx} = 20$; $f_{yy} = \frac{13}{2}$; $f_{xy} = 9$

$D\left(-\frac{1}{7}, \frac{10}{7}\right) = f_{xx}\left(-\frac{1}{7}, \frac{10}{7}\right)f_{yy}\left(-\frac{1}{7}, \frac{10}{7}\right) - \left(f_{xy}\left(-\frac{1}{7}, \frac{10}{7}\right)\right)^2 = 20\cdot\frac{13}{2} - 9^2 = 49 > 0$ and

$f_{xx}\left(-\frac{1}{7}, \frac{10}{7}\right) = 20 > 0$.

$\left(-\frac{1}{7}, \frac{10}{7}\right)$ corresponds to a local minimum.

Thus the point on the plane closest to the point (5, 4, 3) corresponds to a local minimum of the square of a distance function.

**21.** The square of the distance from a point on the surface $z^2 = xy + 1$, or equivalently, $z = \sqrt{xy+1}$ (taking the upper portion) to the origin is

$f(x, y) = x^2 + y^2 + (xy + 1)$

$\nabla f(x, y) = \langle 2x + y, 2y + x\rangle$

$\nabla f(x, y) = \mathbf{0}$ at (0, 0). The only stationary point is (0, 0).

$f_{xx} = 2;\ f_{yy} = 2; f_{xy} = 1$

$$D(0, 0) = f_{xx}(0, 0)f_{yy}(0, 0) - \left(f_{xy}(0, 0)\right)^2$$
$$= 2 \cdot 2 - 1$$
$$= 3 > 0$$

and $f_{xx}(0, 0) = 2 > 0$.
$(x, y) = (0, 0)$ corresponds to a local minimum. (There are actually two points $(x, y, z)$ that are closest: $(0, 0, 1)$ and $(0, 0, -1)$.)
Thus, the point on the surface closest to the origin corresponds to a local minimum of the square of a distance function.

**23.** The point near (0, 0) is a local minimum and the points near (0.4, 0.8) and (–0.4, –0.8) are saddle points.

**25.** $\nabla f(x, y) = \langle 2x + y, x - 2ky + 2\rangle$

$\nabla f(x, y) = \mathbf{0}$ when $x = -\dfrac{2}{4k+1}$ and $y = \dfrac{4}{4k+1}$.

For $k \neq -\dfrac{1}{4}$, the stationary point is

$\mathbf{a} = \left\langle -\dfrac{2}{4k+1}, \dfrac{4}{4k+1}\right\rangle$. (Note that when $k = \dfrac{1}{4}$, there are no stationary points.)

$f_{xx} = 2;\ f_{xy} = 1; f_{yy} = -2k$
$D(\mathbf{a}) = -4k - 1$

$D(\mathbf{a}) < 0$ when $k > -\dfrac{1}{4}$, and $\mathbf{a}$ is a saddle point.

$D(\mathbf{a}) > 0$ when $k < -\dfrac{1}{4}$, and since $f_{xx} = 2 > 0$, $\mathbf{a}$ is a local minimum point. When $k = \dfrac{1}{4}$, there is no stationary point.

**27.** The equation of the least square lines is $y = mx + \text{b}$ where $m\sum_{i=1}^{n} x_i + bn = \sum_{i=1}^{n} y_i$.
Divide this condition by $n$ to get
$m\dfrac{1}{n}\sum_{i=1}^{n} x_i + b = \dfrac{1}{n}\sum_{i=1}^{n} y_i$. Substitute
$\bar{x} = \dfrac{1}{n}\sum_{i=1}^{n} x_i$ and $\bar{y} = \dfrac{1}{n}\sum_{i=1}^{n} y_i$ into this equation to get $\bar{y} = m\bar{x} + b$. Thus the centroid lies on the least squares line.

**29.** There are 10 data points, so $n = 10$.
$\sum_{i=1}^{10} x_i = 320; \sum_{i=1}^{10} y_i = 635; \sum_{i=1}^{10} x_i^2 = 11{,}490;$
$\sum_{i=1}^{10} x_i y_i = 21{,}275$
Solving the system of equations
$$320m + 10b = 635$$
$$11{,}490m + 320b = 21{,}275$$
we get $m = \dfrac{191}{250}$ and $b = \dfrac{9763}{250}$.
The equation of the least squares line is
$y = \dfrac{191}{250}x + \dfrac{9763}{250}$ or $y = 0.764x + 39.052$. The line and the data are shown.

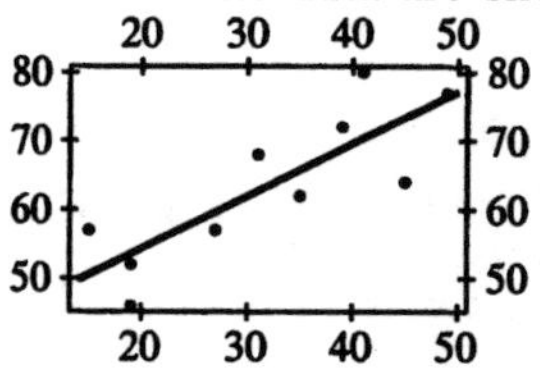


**31.** $f(x, y) = Ax^2 + 2Bxy + Cy^2$
$\nabla f(x, y) = \langle 2Ax + 2By, 2Bx + 2Cy\rangle$
$\nabla f(x, y) = \mathbf{0}$ when $2Ax + 2By = 0$ and $2Bx + 2Cy = 0$. One solution to this system is clearly (0, 0). The coefficient matrix for this system is $\begin{pmatrix} 2A & 2B \\ 2B & 2C \end{pmatrix}$.

$$\det\begin{pmatrix} 2A & 2B \\ 2B & 2C \end{pmatrix} = 4AC - 4B^2$$

If $4AC - 4B^2 \neq 0$, there is a single solution to the system, which is (0, 0). Hence there is only one stationary point, (0, 0).

**33.** In this case, $A = 1$, $B = \dfrac{1}{2}$, and $C = 1$, and $AC - B^2 = \dfrac{3}{4} > 0$. By Exercise 32, $S$ has an absolute minimum or maximum at (0, 0, 0). Since $A = 1 > 0$, (0, 0, 0) is an absolute minimum.

## Section 10.7 Optimization with Constraints

**1.** Let $f(x, y) = x^2 + y^2$ be the function for the square of the distance of the point $(x, y)$ from the origin. Let $g(x, y) = x^2y - 16$.

$\nabla f(x, y) = \langle 2x, 2y \rangle$; $\nabla g(x, y) = \langle 2xy, x^2 \rangle$

When $\nabla f = \lambda \nabla g$, we have the two equations

$2x = \lambda 2xy$

$2y = \lambda x^2$

Multiply the first equation by $x$ and the second equation by $y$.

$2x^2 = \lambda 2x^2 y$

$2y^2 = \lambda x^2 y$

Replace $x^2y$ by 16 to get

$2x^2 = 32\lambda$

$2y^2 = 16\lambda$

Solve each for $\lambda$ and set them equal to each other, so $\frac{x^2}{16} = \frac{y^2}{8}$ or $x^2 = 2y^2$. Since $x^2y = 16$, $x^2 = \frac{16}{y}$. Substitute this into the previous equation so $\frac{16}{y} = 2y^2$. Solving this we get $y = 2$. Thus $x = \pm 2\sqrt{2}$.

the points that lie closest to the origin are $\left(-2\sqrt{2}, 2\right)$ and $\left(2\sqrt{2}, 2\right)$.

**3.** Let $f(x, y) = (x-1)^2 + y^2$ be the function for the square of the distance of the point $(x, y)$ from $(1, 0)$. Let $g(x, y) = xy - 2\sqrt{2}$.

$\nabla f(x, y) = \langle 2x - 2, 2y \rangle$; $\nabla g(x, y) = \langle y, x \rangle$

When $\nabla f = \lambda \nabla g$, we have the two equations

$2x - 2 = \lambda y$

$2y = \lambda x$

Multiply the first equation by $x$ and the second by $y$, then replace every occurrence of $xy$ by $2\sqrt{2}$.

$2x^2 - 2x = 2\sqrt{2}\lambda$

$2y^2 = 2\sqrt{2}\lambda$

Solve each for $\lambda$ and set them equal.

$\frac{x^2}{\sqrt{2}} - \frac{x}{\sqrt{2}} = \frac{y^2}{\sqrt{2}}$ or $y^2 = x^2 - x$.

Since $xy = 2\sqrt{2}$, $y = \frac{2\sqrt{2}}{x}$ and $y^2 = \frac{8}{x^2}$.

Thus we have the equation $\frac{8}{x^2} = x^2 - x$.

The solutions are $x = 2$ and $x \approx -1.47797$.

When $x = 2$, $y = \sqrt{2}$ and when $x \approx -1.47797$, $y \approx -1.91373$.

$f\left(2, \sqrt{2}\right) = 3$; $f(-1.47797, -1.91373) \approx 9.80268$

The point that lies closest to $(1, 0)$ is $\left(2, \sqrt{2}\right)$.

**5.** Let $f(x, y, z) = x^2 + y^2 + z^2$ be the function for the square of the distance of the point $(x, y, z)$ from the origin. Let $g(x, y, z) = x^2y^2z - 1$.

$\nabla f(x, y, z) = \langle 2x, 2y, 2z \rangle$;

$\nabla g(x, y, z) = \langle 2xy^2z, 2x^2yz, x^2y^2 \rangle$

When $\nabla f = \lambda \nabla g$, we have the three equations

$2x = \lambda 2xy^2z$

$2y = \lambda 2x^2yz$

$2z = \lambda x^2y^2$

Multiply the first equation by $x$, the second equation by $y$, and the third equation by $z$.

$2x^2 = \lambda 2x^2y^2z$

$2y^2 = \lambda 2x^2y^2z$

$2z^2 = \lambda x^2y^2z$

Replace $x^2y^2z$ by 1 to get

$2x^2 = 2\lambda \qquad x^2 = \lambda$

$2y^2 = 2\lambda \quad \text{or} \quad y^2 = \lambda$

$2z^2 = \lambda \qquad z^2 = \frac{1}{2}\lambda$

Since $x^2y^2z = 1$, $x^4y^4z^2 = 1$. Substitute $x^4 = \lambda^2$, $y^4 = \lambda^2$, and $z^2 = \frac{1}{2}\lambda$ into the previous equation so $\frac{1}{2}\lambda^5 = 1$. Solving this, we get $\lambda = 2^{1/5}$. Thus $x = \pm 2^{1/10}$, $y = \pm 2^{1/10}$, and $z = 2^{-2/5}$ ($z$ is positive).

The points that lie closest to the origin are $\left(-2^{1/10}, -2^{1/10}, 2^{-2/5}\right)$, $\left(-2^{1/10}, 2^{1/10}, 2^{-2/5}\right)$, $\left(2^{1/10}, -2^{1/10}, 2^{-2/5}\right)$, and $\left(2^{1/10}, 2^{1/10}, 2^{-2/5}\right)$

7. Let $f(x, y, z) = (x-3)^2 + (y-1)^2 + (z+1)^2$ be the function for the square of the distance of the point $(x, y, z)$ from $(3, 1, -1)$. Let $g(x, y, z) = x^2 + y^2 + z^2 - 4$.
$\nabla f(x, y, z) = \langle 2x-6, 2y-2, 2z+2\rangle$;
$\nabla g(x, y, z) = \langle 2x, 2y, 2z\rangle$
When $\nabla f = \lambda \nabla g$, we have the three equations
$2x - 6 = \lambda 2x$
$2y - 2 = \lambda 2y$
$2z + 2 = \lambda 2z$
Solve these equations for $\lambda$ and set them equal to get $1 - \frac{3}{x} = 1 - \frac{1}{y} = 1 + \frac{1}{z}$.
Thus, $x = 3y$ and $z = -y$ and we have $4 = (3y)^2 + y^2 + (-y)^2 = 11y^2$, so $y = \pm\frac{2}{\sqrt{11}}$.
The possible points are $\left(\frac{6}{\sqrt{11}}, \frac{2}{\sqrt{11}}, -\frac{2}{\sqrt{11}}\right)$ and $\left(-\frac{6}{\sqrt{11}}, -\frac{2}{\sqrt{11}}, \frac{2}{\sqrt{11}}\right)$.
$f\left(\frac{6}{\sqrt{11}}, \frac{2}{\sqrt{11}}, -\frac{2}{\sqrt{11}}\right) = 15 - 4\sqrt{11}$ and
$f\left(-\frac{6}{\sqrt{11}}, -\frac{2}{\sqrt{11}}, \frac{2}{\sqrt{11}}\right) = 15 + 4\sqrt{11}$.
The point closest to $(3, 1, -1)$ is $\left(\frac{6}{\sqrt{11}}, \frac{2}{\sqrt{11}}, -\frac{2}{\sqrt{11}}\right)$.

9. Let $f(x, y, z) = x^4 + y^4 + z^4$ be the function to be minimized. Let $g(x, y, z) = x + y + z - 1$.
$\nabla f(x, y, z) = \langle 4x^3, 4y^3, 4z^3\rangle$;
$\nabla g(x, y, z) = \langle 1, 1, 1\rangle$
When $\nabla f = \lambda \nabla g$, we have the three equations
$4x^3 = \lambda$
$4y^3 = \lambda$
$4z^3 = \lambda$
This implies that $x = y = z$, so $3x = 1$. Thus $x = y = z = \frac{1}{3}$. Because $f$ is increasing as points on the plane move away from $\left(\frac{1}{3}, \frac{1}{3}, \frac{1}{3}\right)$, $\left(\frac{1}{3}, \frac{1}{3}, \frac{1}{3}\right)$ minimizes the function.

11. Let $V(r, h) = \pi r^2 h$ be the function for the volume of the cylinder in terms of the radius $r$ and height $h$.
Let $g(r, h) = h + 2\pi r - 108$.
$\nabla V(r, h) = \langle 2\pi rh, \pi r^2\rangle$; $\nabla g(r, h) = \langle 2\pi, 1\rangle$
When $\nabla V = \lambda \nabla g$, we have the two equations
$2\pi rh = 2\pi\lambda$ or $rh = \lambda$
$\pi r^2 = \lambda$ $\qquad \pi r^2 = \lambda$
Equating the equations for $\lambda$, we get $rh = \pi r^2$ or $h = \pi r$. Substitute this into $h + 2\pi r = 108$, so $3\pi r = 108$ or $r = \frac{36}{\pi}$. Hence $h = 36$. The cylinder is to have radius $\frac{36}{\pi}$ inches and height 36 inches.

13. Let $C(x, y) = 200x + 300y$ be the function to be minimized. The constraint is $2000K = Kx^{1/3}y^{2/3}$, so let $g(x, y) = x^{1/3}y^{2/3} - 2000$.
$\nabla C(x, y) = \langle 200, 300\rangle$;
$\nabla g(x, y) = \left\langle \frac{1}{3}x^{-2/3}y^{2/3}, \frac{2}{3}x^{1/3}y^{-1/3}\right\rangle$
When $\nabla C = \lambda \nabla g$, we have the two equations
$200 = \frac{\lambda}{3}x^{-2/3}y^{2/3}$
$300 = \frac{2\lambda}{3}x^{1/3}y^{-1/3}$
Multiply the first equation by $x$ and the second equation by $y$.
$200x = \frac{\lambda}{3}x^{1/3}y^{2/3}$
$300y = \frac{2\lambda}{3}x^{1/3}y^{2/3}$
Replace $x^{1/3}y^{2/3}$ by 2000 to get
$200x = \frac{\lambda}{3}2000$ or $\lambda = \frac{3x}{10}$
$300y = \frac{2\lambda}{3}2000$ $\qquad \lambda = \frac{9y}{40}$
So $\frac{3x}{10} = \frac{9y}{40}$ or $x = \frac{3}{4}y$. Since $x^{1/3}y^{2/3} = 2000$, $\left(\frac{3}{4}y\right)^{1/3}y^{2/3} = 2000$.
Solving this, we get $y = 2000\left(\frac{4}{3}\right)^{1/3}$, so
$x = \frac{3}{4}\cdot 2000\left(\frac{4}{3}\right)^{1/3} = 1500\left(\frac{4}{3}\right)^{1/3} = 500(6)^{2/3}$.
$500(6)^{2/3}$ units of labor and $2000\left(\frac{4}{3}\right)^{1/3}$ units of capital are needed to minimize the cost.

**15.** Let $f(x, y, z) = \frac{x}{a} + \frac{y}{b} + \frac{z}{c}$ be the function to be maximized. Let
$g(x, y, z) = x^2 + y^2 + z^2 - 1$.
$\nabla f(x, y, z) = \left\langle \frac{1}{a}, \frac{1}{b}, \frac{1}{c} \right\rangle$;
$\nabla g(x, y, z) = \langle 2x, 2y, 2z \rangle$
When $\nabla f = \lambda \nabla g$, we have the three equations
$\frac{1}{a} = 2\lambda x \qquad \frac{1}{\lambda} = 2ax$
$\frac{1}{b} = 2\lambda y \quad$ or $\quad \frac{1}{\lambda} = 2by$
$\frac{1}{c} = 2\lambda z \qquad \frac{1}{\lambda} = 2bz$
This implies that $ax = by = cz$, so $y = \frac{a}{b}x$
$z = \frac{a}{c}x$. Since $x^2 + y^2 + z^2 = 1$, we get
$x^2 + \left(\frac{a}{b}x\right)^2 + \left(\frac{a}{c}x\right)^2 = 1$. Thus
$x = \pm\frac{bc}{\sqrt{a^2b^2 + a^2c^2 + b^2c^2}}$, so
$y = \pm\frac{ac}{\sqrt{a^2b^2 + a^2c^2 + b^2c^2}}$, and
$z = \pm\frac{ab}{\sqrt{a^2b^2 + a^2c^2 + b^2c^2}}$
Let $p = \sqrt{a^2b^2 + a^2c^2 + b^2c^2}$. The function is maximized when all of the coordinates are positive, so the maximum is
$\frac{bc}{ap} + \frac{ac}{bp} + \frac{ab}{cp} = \frac{p}{abc}$ at $\left(\frac{bc}{p}, \frac{ac}{p}, \frac{ab}{p}\right)$.

**17.** Let $V(x, y, z) = xyz$ for $x, y, z > 0$ and let
$g(x, y, z) = 2x + 3y + 4z - 6$.
When $\nabla V = \lambda \nabla g$, we have the equations
$yz = 2\lambda$
$xz = 3\lambda$
$xy = 4\lambda$
Solving these equations for $\lambda$ and setting them equal we get $\frac{yz}{2} = \frac{xz}{3} = \frac{xy}{4}$.
Since $\frac{yz}{2} = \frac{xz}{3}$, $y = \frac{2x}{3}$ and since
$\frac{yz}{2} = \frac{xy}{4}$, $z = \frac{x}{2}$.

Thus $6 = 2x + 3\left(\frac{2x}{3}\right) + 4\left(\frac{x}{2}\right) = 6x$ and $x = 1$,
so $y = \frac{2}{3}$ and $z = \frac{1}{2}$.
The dimensions of the largest box are 1 unit by $\frac{2}{3}$ unit by $\frac{1}{2}$ unit.

**19.** Let $V(l, w, h) = lwh$. The constraint is
$4wh + lw + 2lh = 25$, so let
$g(l, w, h) = 4wh + lw + 2lh - 25$.
When $\nabla V = \lambda \nabla g$, we have the equations
$wh = \lambda(w + 2h)$
$lh = \lambda(4h + l)$
$lw = \lambda(4w + 2l)$
Multiply the first equation by $l$, the second by $w$, the third by $h$, set the products $lwh$ equal, and divide by $\lambda$ to get
$lw + 2lh = 4wh + lw = 4wh + 2lh$.
Thus, $2lh = 4wh$ and $lw = 2lh$ so $l = 2w$ and $w = 2h$.
Thus, $l = 2(2h) = 4h$ and for maximum volume, the length must be 4 times the height and the width must be twice the height.

**21.** The other four solutions have opposite signs on $x$ and $y$. Thus, $f(x, y, z)$ will be negative at these points. $f(x, y, z)$ is minimized on the sphere $x^2 + y^2 + z^2 = 4a^2$ at these points.

**23.** Let $f(x, y, z) = x^2 + y^2 + z^2$ be the function for the square of the distance of the point $(x, y, z)$ from the origin. (The point closest to the origin will have the distance of the plane from the origin.)
Let $g(x, y, z) = ax + by + cz + d$.
$\nabla f(x, y, z) = \langle 2x, 2y, 2z \rangle$;
$\nabla g(x, y, z) = \langle a, b, c \rangle$
When $\nabla f = \lambda \nabla g$, we have the equations
$2x = \lambda a$
$2y = \lambda b$
$2z = \lambda c$
Assuming that $a \neq 0$, $\lambda = \frac{2x}{a}$ so $y = \frac{bx}{a}$ and
$z = \frac{cx}{a}$. Thus $ax + \frac{b^2x}{a} + \frac{c^2x}{a} = -d$ and
$x = \frac{-ad}{a^2 + b^2 + c^2}$, $y = \frac{-bd}{a^2 + b^2 + c^2}$,
$z = \frac{-cd}{a^2 + b^2 + c^2}$.

At this point $f(x, y, z) = \frac{a^2d^2}{(a^2+b^2+c^2)^2} + \frac{b^2d^2}{(a^2+b^2+c^2)^2} + \frac{c^2d^2}{(a^2+b^2+c^2)^2} = \frac{d^2}{a^2+b^2+c^2}$

so the distance from the origin to the plane is $\frac{|d|}{\sqrt{a^2+b^2+c^2}}$.

**25.** Let $f(x, y) = x^2 + y^2$ be the function for the square of the distance of the point $(x, y)$ from the origin.
Let $g(x, y) = x^2 - 4xy + 5y^2 - 10$.
$\nabla f(x, y) = \langle 2x, 2y \rangle$;
$\nabla g(x, y) = \langle 2x - 4y, -4x + 10y \rangle$
When $\nabla f = \lambda \nabla g$, we have the equations
$2x = \lambda(2x - 4y)$
$2y = \lambda(-4x + 10y)$
Solving both equations for $\lambda$ and equating the results produces

$$\frac{2x}{2x-4y} = \frac{2y}{-4x+10y}$$
$$-8x^2 + 20xy = 4xy - 8y^2$$
$$xy = \frac{1}{2}\left(x^2 - y^2\right)$$

Substituting in the constraint equation gives

$$x^2 - 2\left(x^2 - y^2\right) + 5y^2 = 10$$
$$-x^2 + 7y^2 = 10$$
$$x^2 = 7y^2 - 10$$
$$x = \pm\sqrt{7y^2 - 10}$$

Resubstituting into the (original) constraint equation gives $7y^2 - 10 \pm 4y\sqrt{7y^2 - 10} + 5y^2 = 10$, which rearranges to $12y^2 - 20 = \pm 4y\sqrt{7y^2 - 10}$.
Dividing both sides by 4 and squaring,

$$\left(3y^2 - 5\right)^2 = y^2\left(7y^2 - 10\right)$$
$$2\left(y^2\right)^2 - 20y^2 + 25 = 0$$
$$y^2 = 5 \pm \frac{5}{\sqrt{2}}$$
$$y = \pm\sqrt{5 + \frac{5}{\sqrt{2}}} \text{ or } \pm\sqrt{5 - \frac{5}{\sqrt{2}}}$$
$$x = \pm\sqrt{7y^2 - 10} = \pm\sqrt{25 + \frac{35}{\sqrt{2}}} \text{ or } \pm\sqrt{25 - \frac{35}{\sqrt{2}}}$$

Note that since $xy = \frac{1}{2}(x^2 - y^2)$, $x$ and $y$ must have the same sign when $x^2 > y^2$ and $x$ and $y$ must have opposite signs when $x^2 < y^2$. Thus, we have four solutions:

$\pm\left(\sqrt{25 + \frac{35}{\sqrt{2}}}, \sqrt{5 + \frac{5}{\sqrt{2}}}\right)$ and $\pm\left(\sqrt{25 - \frac{35}{\sqrt{2}}}, -\sqrt{5 - \frac{5}{\sqrt{2}}}\right)$. For the first two,

$f(x, y) = 30 + \frac{40}{\sqrt{2}}$ is the square of the distance from the origin, and for the second two,

$f(x, y) = 30 - \frac{40}{\sqrt{2}}$ is the square of the distance from the origin. The greatest and least distances, respectively, are found by taking the square roots of these values of $f(x, y)$:

$\sqrt{30+\frac{40}{\sqrt{2}}}=\sqrt{30+20\sqrt{2}}$ and $\sqrt{30-\frac{40}{\sqrt{2}}}=\sqrt{30-20\sqrt{2}}$.

**27.** $A(a, b, c)=\sqrt{s(s-a)(s-b)(s-c)}$

Let $p$ denote the fixed perimeter. Then $s=\frac{1}{2}p$ and $A(a, b, c)=\frac{1}{4}\sqrt{p(p-2a)(p-2b)(p-2c)}$ where $p$ is a constant. The constraint is that $a+b+c=p$ so let $g(a, b, c)=a+b+c-p$. Instead of maximizing $A(a, b, c)$ we maximize $f(a, b, c)=(A(a, b, c))^2=\frac{1}{16}p(p-2a)(p-2b)(p-2c)$.

$\nabla f(a, b, c)=\left\langle -\frac{1}{8}p(p-2b)(p-2c), -\frac{1}{8}p(p-2a)(p-2c), -\frac{1}{8}p(p-2a)(p-2b)\right\rangle$; $\nabla g(a, b, c)=\langle 1, 1, 1\rangle$

When $\nabla f=\lambda\nabla g$, we have the equations

$-\frac{1}{8}p(p-2b)(p-2c)=\lambda$

$-\frac{1}{8}p(p-2a)(p-2c)=\lambda$

$-\frac{1}{8}p(p-2a)(p-2b)=\lambda$

Setting the first two expressions for $\lambda$ equal and dividing by $-\frac{1}{8}p(p-2c)$ gives $p-2b=p-2a$ so $a=b$.

Setting the second and third expressions for $\lambda$ equal and dividing by $-\frac{1}{8}p(p-2a)$ gives $p-2c=p-2b$ so $c=b$.

Thus, $a=b=c$ and the triangle of perimeter $p$ with largest area is an equilateral triangle.

**29.** Let $f(x, y, z)=xy+xz$ and $g(x, y, z)=x^2+2y^2+3z^2-1$.

$\nabla f(x, y, z)=\langle y+z, x, x\rangle$; $\nabla g(x, y, z)=\langle 2x, 4y, 6z\rangle$

When $\nabla f=\lambda\nabla g$, we have the equations

$y+z=2\lambda x$

$x=4\lambda y$

$x=6\lambda z$

From the last two equations, $y=\frac{3}{2}z$. Multiply the first equation by $x$, the second by $y$, and the third by $z$ to get

$xy+xz=2\lambda x^2$

$xy=4\lambda y^2$

$xz=6\lambda z^2$

Adding the last two equations and setting the expressions $xy+xz$ equal gives $2\lambda x^2=4\lambda y^2+6\lambda z^2$ or $x^2=2y^2+3z^2$.

Thus $1=x^2+2y^2+3z^2=2x^2$, so $x=\pm\frac{1}{\sqrt{2}}$ and $\frac{1}{2}=2y^2+3z^2$. Using $y=\frac{3}{2}z$, we get $\frac{1}{2}=\frac{9}{2}z^2+3z^2$ so $z=\pm\frac{1}{\sqrt{15}}$ and $y=\pm\frac{3}{2\sqrt{15}}$.

The possible points are $\left(\frac{1}{\sqrt{2}}, \frac{3}{2\sqrt{15}}, \frac{1}{\sqrt{15}}\right)$, $\left(-\frac{1}{\sqrt{2}}, \frac{3}{2\sqrt{15}}, \frac{1}{\sqrt{15}}\right)$, $\left(\frac{1}{\sqrt{2}}, -\frac{3}{2\sqrt{15}}, -\frac{1}{\sqrt{15}}\right)$, and $\left(-\frac{1}{\sqrt{2}}, -\frac{3}{2\sqrt{15}}, -\frac{1}{\sqrt{15}}\right)$.

$$f\left(\frac{1}{\sqrt{2}}, \frac{3}{2\sqrt{15}}, \frac{1}{\sqrt{15}}\right) = f\left(-\frac{1}{\sqrt{2}}, -\frac{3}{2\sqrt{15}}, -\frac{1}{\sqrt{15}}\right) = \frac{3}{2\sqrt{30}} + \frac{1}{\sqrt{30}} = \frac{5}{2\sqrt{30}}$$

while $f\left(-\frac{1}{\sqrt{2}}, \frac{3}{2\sqrt{15}}, \frac{1}{\sqrt{15}}\right) = f\left(\frac{1}{\sqrt{2}}, -\frac{3}{2\sqrt{15}}, -\frac{1}{\sqrt{15}}\right) = -\frac{3}{2\sqrt{30}} - \frac{1}{\sqrt{30}} = -\frac{5}{2\sqrt{30}}$.

Thus, the maximum value of $f$ on the ellipsoid is $\frac{5}{2\sqrt{30}}$ and the minimum value of $f$ on the ellipsoid is $-\frac{5}{2\sqrt{30}}$.

**31.** The constraint equation is $\cos a \sin a = \cos b(\sin b - \sin a)$.

From (18), $\sin a = -\frac{\cos 2b}{\sin b}$. Substituting,

$$\cos a\left(-\frac{\cos 2b}{\sin b}\right) = \cos b\left(\sin b + \frac{\cos 2b}{\sin b}\right)$$

Multiplying by sin $b$,

$$-\cos a \cos 2b = \cos b\left(\sin^2 b + \cos 2b\right)$$

Using $\cos 2b = \cos^2 b - \sin^2 b$ and also $\cos 2b = 2\cos^2 b - 1$,

$$\cos a\left(1 - 2\cos^2 b\right) = \cos b\left(\cos^2 b\right)$$

$$\cos a = \frac{\cos^3 b}{1 - 2\cos^2 b}$$

Squaring,

$$\cos^2 a = \frac{\cos^6 b}{\left(1 - 2\cos^2 b\right)^2}$$

$$1 - \sin^2 a = \frac{\cos^6 b}{\left(1 - 2\cos^2 b\right)^2}$$

Now using (18) again,

$$1 - \frac{\cos^2 2b}{\sin^2 b} = \frac{\cos^6 b}{\left(1 - 2\cos^2 b\right)^2}$$

$$1 - \frac{\left(2\cos^2 b - 1\right)^2}{1 - \cos^2 b} = \frac{\cos^6 b}{\left(1 - 2\cos^2 b\right)^2}$$

And now let $w = \cos^2 b$. Then $1 - \frac{(2w-1)^2}{1-w} = \frac{w^3}{(1-2w)^2}$.

Which rearranges into $w\left(15w^3 - 27w^2 + 16w - 3\right) = 0$.

**33.** We would expect the value in Exercise 38, Section 10.5 to be larger, since it does not have the additional constraint that is in Example 3.

**35.** Let $f(x, y, z) = xyz$ be the function to be maximized.

Let $g(x, y, z) = \frac{1}{3}(x + y + z) - k$.

$\nabla f(x, y, z) = \langle yz, xy, xz \rangle$

$\nabla g(x, y, z) = \left\langle \frac{1}{3}, \frac{1}{3}, \frac{1}{3} \right\rangle$

When $\nabla f = \lambda \nabla g$, we get the three equations

$yz = \frac{1}{3}\lambda$

$xz = \frac{1}{3}\lambda$

$xy = \frac{1}{3}\lambda$

These lead to $yz = xz = xy$ and from there to $x = y = z$. The maximum value of $xyz$ is attained when $x = y = z = k$. But this is also when the maximum value of $\sqrt[3]{xyz}$ is attained. And even then,

$$\sqrt[3]{xyz} = \sqrt[3]{x^3} = x \le k = \frac{x+y+z}{3} = \frac{x+x+x}{3} = x$$

So it is always the case that

$$\sqrt[3]{xyz} \le \frac{x+y+z}{3} \text{ for } x, y, z > 0.$$

## Chapter 10 Review Exercises

**1.** Let $\mathbf{x} = \langle x, y \rangle$ and $\mathbf{a} = \langle 1, 1 \rangle$. Now we calculate $\lim_{\mathbf{x} \to \mathbf{a}} \frac{f(\mathbf{x}) - f(\mathbf{a}) - df_{\mathbf{a}}(\mathbf{x} - \mathbf{a})}{\|\mathbf{x} - \mathbf{a}\|}$.

Since $f(\mathbf{x}) = 5 + x^2 + xy$, $f(\mathbf{a}) = 5 + 1 + 1 = 7$,

$Df(\mathbf{a}) = \begin{pmatrix} f_x(\mathbf{a}) & f_y(\mathbf{a}) \end{pmatrix} = \begin{pmatrix} (2x+y)\big|_{(1,1)} & x\big|_{(1,1)} \end{pmatrix} = \begin{pmatrix} 3 & 1 \end{pmatrix}$,

$\mathbf{x} - \mathbf{a} = \begin{pmatrix} x-1 \\ y-1 \end{pmatrix}$, and $\|\mathbf{x} - \mathbf{a}\| = \sqrt{(x-1)^2 + (y-1)^2}$, the limit becomes

$$\begin{aligned}
\lim_{(x,y)\to(1,1)} \frac{5 + x^2 + xy - 7 - 3(x-1) - (y-1)}{\sqrt{(x-1)^2 + (y-1)^2}} &= \lim_{(x,y)\to(1,1)} \frac{x^2 + xy - y - 3x + 2}{\sqrt{(x-1)^2 + (y-1)^2}} \\
&= \lim_{(x,y)\to(1,1)} \frac{(x-1)(x+y-2)}{\sqrt{(x-1)^2 + (y-1)^2}} \\
&= \lim_{(x,y)\to(1,1)} \left( \frac{x+y-2}{\sqrt{1 + \left(\frac{y-1}{x-1}\right)^2}} \cdot \frac{x-1}{|x-1|} \right) \\
&= 0
\end{aligned}$$

**3.** For $f(x, y, z) = z^3 - xyz^2$ and $\mathbf{a} = \langle 1, 1, 2\rangle$,
$\nabla f(x, y, z) = \langle -yz^2, -xz^2, 3z^2 - 2xyz\rangle$
$\nabla f(\mathbf{a}) = \langle -4, -4, 8\rangle$
The tangent plane is given by
$\nabla f(\mathbf{a}) \cdot (\mathbf{x} - \mathbf{a}) = \mathbf{0}$, where $\mathbf{x} = \langle x, y, z\rangle$. Thus, the equation of the tangent plane is
$-4(x-1) - 4(y-1) + 8(z-2) = 0$
which, solved for $z$, produces
$z = 2 + \frac{1}{2}(x-1) + \frac{1}{2}(y-1).$

**5.** $f(x, y) = \sqrt[3]{1 - x - y^3}$, $\mathbf{a} = \left\langle \frac{3}{4}, \frac{1}{2}\right\rangle$

$f_x = -\frac{1}{3}\left(1 - x - y^3\right)^{-2/3}$;

$f_y = -y^2\left(1 - x - y^3\right)^{-2/3}$

$f(\mathbf{a}) = \frac{1}{2}$

$f_x(\mathbf{a}) = -\frac{4}{3}$; $f_y(\mathbf{a}) = -1$

Now we write
$f(\mathbf{x}) \approx f(\mathbf{a}) + f_x(\mathbf{a})h_1 + f_y(\mathbf{a})h_2$

Which with $h_1 = x - \frac{3}{4}$ and $h_2 = y - \frac{1}{2}$ becomes

$f(x, y) \approx \frac{1}{2} - \frac{4}{3}\left(x - \frac{3}{4}\right) - \left(y - \frac{1}{2}\right)$

At (0.7, 0.4) the expression on the right equals $\frac{2}{3}$, whereas the true value, $f$(0.7, 0.4), is about 0.617975. The difference is about 7.9 percent of the true value.

**7.** A normal to the surface with equation
$F(x, y, z) = x^2 + 2y^2 + 3z^2 - 6 = 0$ is
$\nabla F(x, y, z) = \langle 2x, 4y, 6z\rangle$. At (1, 1, 1), the normal is $\langle 2, 4, 6\rangle$. This vector has magnitute
$\sqrt{2^2 + 4^2 + 6^2} = 2\sqrt{14}$, so a parallel vector with magnitude 10, representing the particle's velocity, is $\frac{10}{2\sqrt{14}}\langle 2, 4, 6\rangle = \frac{10}{\sqrt{14}}\langle 1, 2, 3\rangle$.
The motion of the particle is described by

$\mathbf{r}(t) = \langle 1, 1, 1\rangle + \frac{10}{\sqrt{14}}t\langle 1, 2, 3\rangle$ for $t \ge 0$. To find out when it crosses the surface
$x^2 + y^2 + z^2 = 10$, we solve

$$\left(1 + \frac{10}{\sqrt{14}}t\right)^2 + \left(1 + \frac{20}{\sqrt{14}}t\right)^2 + \left(1 + \frac{30}{\sqrt{14}}t\right)^2 = 10$$

Which simplifies to

$100t^2 + 60t\sqrt{\frac{2}{7}} - 7 = 0$

Which using the quadratic formula produces the positive solution

$\frac{\sqrt{469} - 3\sqrt{14}}{70} \approx 0.149021$

The particle crosses the sphere in about 0.149021 seconds.

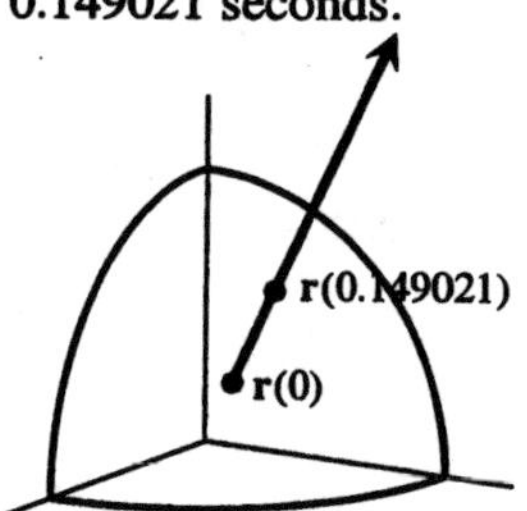


**9.** $H_s(r, s) = \frac{\partial H}{\partial s} = \frac{dH}{dx}\frac{\partial x}{\partial s}$ by the chain rule,
where $x = r^2 + s^2$. Since $\frac{dH}{dx} = f'(x)$ and $\frac{\partial x}{\partial s} = 2s$,
$H_s(r, s) = f'\left(r^2 + s^2\right)(2s)$ and
$H_s(-1, 2) = f'(5)(2 \cdot 2) = (-3)(4) = -12$

**11.** For $f(x, y) = x^2y + y^3$,
$Df(x, y) = \begin{pmatrix} f_x(x, y) & f_y(x, y)\end{pmatrix}$
$= \begin{pmatrix} 2xy & x^2 + 3y^2\end{pmatrix}$
$Df(1, 2) = \begin{pmatrix} 4 & 13\end{pmatrix}$
The linearization at (1, 2) is
$f(x, y) \approx f(1, 2) + Df(1, 2)\begin{pmatrix} x - 1 \\ y - 2\end{pmatrix}$
$= 10 + 4(x-1) + 13(y-2).$

**13.** By the chain rule, $w_t = w_x x_t + w_y y_t = f_x \frac{\partial}{\partial t}(ut) + f_y \frac{\partial}{\partial t}(t^2) = uf_x + 2tf_y$.

When $(t, u) = (2, 1)$, $(x, y) = (ut, t^2) = (2, 4)$ and $w_t = 1(5) + 2(2)(5) = 25$

Now, by the sum and product rules $w_{tu} = \frac{\partial}{\partial u} w_t = u \frac{\partial}{\partial u} f_x + f_x + 2t \frac{\partial}{\partial u} f_y$

which for $t = 2$, $u = 1$ becomes $w_{tu} = \frac{\partial}{\partial u} f_x + f_x + 4 \frac{\partial}{\partial u} f_y$.

By another application of the chain rule, $w_{tu} = f_{xx} x_u + f_{xy} y_u + f_x + 4(f_{yx} x_u + f_{yy} y_u)$

which with $x = ut$ and $y = t^2$ becomes $w_{tu} = tf_{xx} + f_x + 4(tf_{yx})$.

When $(t, u) = (2, 1)$ and $(x, y) = (2, 4)$, $w_{tu} = 2(1) + 5 + 4(2)(-2) = -9$

**15.** For $S$ given by $z = f(x, y) = 4x^2 + y^2$, the tangent plane at $(1, 1, 5)$ is given by

$z = f(1, 1) + f_x(1, 1)(x-1) + f_y(1, 1)(y-1)$.

Now we note that $f_x = 8x$, $f_y = 2y$, and so $f(1, 1) = 5$, $f_x(1, 1) = 8$, and $f_y(1, 1) = 2$. The tangent plane is given by $z = 5 + 8(x - 1) + 2(y - 1)$

A normal vector is $\langle -f_x(1, 1), -f_y(1, 1), 1 \rangle = \langle -8, -2, 1 \rangle$. A unit normal is

$$\mathbf{N} = \frac{1}{\sqrt{8^2 + 2^2 + 1^2}} \langle -8, -2, 1 \rangle = \frac{1}{\sqrt{69}} \langle -8, -2, 1 \rangle.$$

**17.** By the chain rule, with $z = f(x, y)$ and $x = \frac{1}{t}$, $y = \sqrt{t}$

$$\frac{dz}{dt} = \frac{\partial z}{\partial x}\frac{dx}{dt} + \frac{\partial z}{\partial y}\frac{dy}{dt} = f_x\left(-\frac{1}{t^2}\right) + f_y\left(\frac{1}{2\sqrt{t}}\right)$$

where $f_x$ and $f_y$ are evaluated at $(x, y) = \left(\frac{1}{t}, \sqrt{t}\right)$.

**19.** The velocity for a particle with position $\mathbf{r}(t) = \langle x(t), y(t), z(t) \rangle$ is $\mathbf{v}(t) = \mathbf{r}'(t) = \langle x'(t), y'(t), z'(t) \rangle$. Here, $\mathbf{v}(t) = \left\langle -2\sin t, 2\cos t, \frac{d}{dt} f(2\cos t, 2\sin t) \right\rangle$. We find the $z$-term using the chain rule.

$\frac{d}{dt} f(2\cos t, 2\sin t) = f_x x' + f_y y'$ where $x = 2\cos t$ and $y = 2\sin t$. We evaluate at (–2, 0 3):

$$f_x(-2, 0, 3) = -\frac{F_x(-2, 0, 3)}{F_z(-2, 0, 3)} = \frac{2.5}{-0.90} \approx 2.7778$$

$$f_y(-2, 0, 3) = -\frac{F_y(-2, 0, 3)}{F_z(-2, 0, 3)} = \frac{-1.75}{-0.90} \approx -1.9444$$

and with $t = \pi$, $x' = 0$ and $y' = -2$. So $\frac{d}{dt} f(2\cos t, 2\sin t) \approx -2(-1.9444) \approx 3.89$ and $\mathbf{v}(\pi) = \langle 0, -2, 3.89 \rangle$

**21.** For $U(r, s) = u(x, y)$ with $x = r + s$, $y = r - s$, by the chain rule

$U_r = u_x x_r + u_y y_r$

$U_s = u_x x_s + u_y y_s$

and with $x_r = 1$, $y_r = 1$, $x_s = 1$, $y_s = -1$,

$U_r = u_x + u_y$

$U_s = u_x - u_y$.

This system yields $u_x = \frac{1}{2}(U_r + U_s)$, $u_y = \frac{1}{2}(U_r - U_s)$. With these substitutions, the equation $xu_y + yu_x = x + y$ becomes $\frac{1}{2}(r+s)(U_r - U_s) + \frac{1}{2}(r-s)(U_r + U_s) = (r+s) + (r-s)$, which simplifies to $rU_r - sU_s = 2r$.

**23.** $\mathbf{y} = \mathbf{f}(\mathbf{x}) = \langle f_1(x_1, x_2), f_2(x_1, x_2) \rangle$

$\mathbf{g}(\mathbf{t}) = \langle t_1^2 + t_2^2, t_1 t_2 \rangle$

By the chain rule,

$$\mathbf{D}(\mathbf{f} \circ \mathbf{g})(\mathbf{t}) = \mathbf{Df}(\mathbf{x})\mathbf{Dg}(\mathbf{t}) = \begin{pmatrix} \frac{\partial f_1}{\partial x_1} & \frac{\partial f_1}{\partial x_2} \\ \frac{\partial f_2}{\partial x_1} & \frac{\partial f_2}{\partial x_2} \end{pmatrix} \begin{pmatrix} \frac{\partial g_1}{\partial t_1} & \frac{\partial g_1}{\partial t_2} \\ \frac{\partial g_2}{\partial t_1} & \frac{\partial g_2}{\partial t_2} \end{pmatrix} = \begin{pmatrix} \frac{\partial f_1}{\partial x_1} & \frac{\partial f_1}{\partial x_2} \\ \frac{\partial f_2}{\partial x_1} & \frac{\partial f_2}{\partial x_2} \end{pmatrix} \begin{pmatrix} 2t_1 & 2t_2 \\ t_2 & t_1 \end{pmatrix}.$$

**25.** Thinking of $H(u, v)$ as $f(x, y)$, where $x = uv$ and $y = \frac{u}{v}$, by the chain rule

$$\frac{\partial H}{\partial v} = f_x \frac{\partial x}{\partial v} + f_y \frac{\partial y}{\partial v} = f_x u - f_y \frac{u}{v^2}$$

and

$$\begin{aligned}
\frac{\partial^2 H}{\partial v^2} &= \frac{\partial}{\partial v}\left(\frac{\partial H}{\partial v}\right) \\
&= \frac{\partial}{\partial v}(f_x u) - \frac{\partial}{\partial v}\left(f_y \frac{u}{v^2}\right) \\
&= u\frac{\partial}{\partial v} f_x - f_y \frac{\partial}{\partial v}\left(\frac{u}{v^2}\right) - \frac{u}{v^2}\frac{\partial}{\partial v} f_y \\
&= u\left(f_{xx}\frac{\partial x}{\partial v} + f_{xy}\frac{\partial y}{\partial v}\right) - f_y\left(-\frac{2u}{v^3}\right) - \frac{u}{v^2}\left(f_{yx}\frac{\partial x}{\partial v} + f_{yy}\frac{\partial y}{\partial v}\right) \\
&= u\left(f_{xx}u - f_{xy}\frac{u}{v^2}\right) + f_y\frac{2u}{v^3} - \frac{u}{v^2}\left(f_{yx}u - f_{yy}\frac{u}{v^2}\right) \\
&= \frac{u^2 v^4 f_{xx} - 2u^2 v^2 f_{xy} + 2uv f_y + u^2 f_{yy}}{v^4}
\end{aligned}$$

**27.** *Method 1:*

$x^3 y^3 - 2x^2 y + 1 = 0$

Differentiate both sides with respect to $x$.

$3x^2 y^3 + 3x^3 y^2 y' - 4xy - 2x^2 y' = 0$

Solve for $y'$.

$$y' = \frac{4xy - 3x^2 y^3}{3x^3 y^2 - 2x^2}$$

Evaluate for $x = 1$, $y = 1$.

$y'(1) = 1$.

*Method 2:*

$F(x, y) = x^3y^3 - 2x^2y + 1$

$F_x(x, y) = 3x^2y^3 - 4xy$

$F_y(x, y) = 3x^3y^2 - 2x^2$

$$y' = -\frac{F_x(x, y)}{F_y(x, y)} = -\frac{3x^2y^3 - 4xy}{3x^3y^2 - 2x^2}$$

Evaluate for $x = 1$, $y = 1$.

$y'(1) = 1$.

**29.** For $g(x, y) = x^2 - 2xy + y^3 - y$,

$\nabla g(x, y) = \langle 2x - 2y, \ -2x + 3y^2 - 1 \rangle$

When $\nabla g(x, y) = \mathbf{0}$ we have the two equations

$2x - 2y = 0$

$-2x + 3y^2 - 1 = 0$

The first equation indicates that $x = y$, and the second equation then gives

$-2x + 3x^2 - 1 = (3x + 1)(x - 1) = 0$, for which the solutions are $-\frac{1}{3}$ and 1. So the two stationary points are $\left(-\frac{1}{3}, -\frac{1}{3}\right)$ and $(1, 1)$.

$g_{xx} = 2$; $g_{yy} = 6y$; $g_{xy} = -2$

$$D\left(-\frac{1}{3}, -\frac{1}{3}\right) = g_{xx}\left(-\frac{1}{3}, -\frac{1}{3}\right)g_{yy}\left(-\frac{1}{3}, -\frac{1}{3}\right) - \left(g_{xy}\left(-\frac{1}{3}, -\frac{1}{3}\right)\right)^2 = 2(-2) - (-2)^2 = -8 < 0$$

$$D(1, 1) = g_{xx}(1, 1)g_{yy}(1, 1) - \left(g_{xy}(1, 1)\right)^2 = 2(6) - (-2)^2 = 8 > 0$$

and $f_{xx} = 2 > 0$

So $\left(-\frac{1}{3}, -\frac{1}{3}\right)$ is a saddle point and $(1, 1)$ is a local minimum.

**31.** $f(x, y) = x^3 - 3xy + y^2$

$\nabla f(x, y) = \langle 3x^2 - 3y, \ -3x + 2y \rangle$

$\nabla f(x, y) = \mathbf{0}$ when

$3x^2 - 3y = 0$

$-3x + 2y = 0$

From the first equation $y = x^2$. Then the second equation yields $-3x + 2x^2 = 0$ or $x(2x - 3) = 0$, so that $x = 0$ or $x = \frac{3}{2}$. So the two candidate points are $(0, 0)$ and $\left(\frac{3}{2}, \frac{9}{4}\right)$. But only $(0, 0)$ is in the region.

$f(0, 0) = 0$

Now we examine the boundary along which $y = 0$. Let $g(x) = f(x, 0) = x^3$. $\frac{dg}{dx} = 3x^2 = 0$ only when $x = 0$, and $(0, 0)$ has already been examined. So we move to the boundary along which $x = 1$. Let $h(y) = f(1, y) = 1 - 3y + y^2$. $\frac{dh}{dy} = -3 + 2y = 0$ when $y = \frac{3}{2}$, so we have a candidate point at $\left(1, \frac{3}{2}\right)$.

$$f\left(1, \frac{3}{2}\right) = -\frac{5}{4}$$

Now we consider the boundary along which $y = 3x$. Let $p(x) = f(x, 3x) = x^3$. $\frac{dp}{dx} = 3x^2 = 0$ only when $x = 0$, and again, (0, 0) has already been considered.
Finally, we consider the corners (1, 0) and (1, 3).
$f(1, 0) = 1$
$f(1, 3) = 1$

So $f$ takes on a maximum value of 1 at (1, 0) and (1, 3) and a minimum value of $-\frac{5}{4}$ at $\left(1, \frac{3}{2}\right)$.

**33.** For $f(x, y) = xy^2 - 3x^3 + 2y$, $\nabla f(x, y) = \left\langle y^2 - 9x^2, 2xy + 2\right\rangle$.

Setting $\nabla f(x, y) = \mathbf{0}$ produces the two equations $y^2 - 9x^2 = 0$ and $2xy + 2 = 0$.
Therefore, $y = \pm 3x$ and $xy = -1$. Since $xy = -1$, $x$ and $y$ must have opposite signs and so we have $y = -3x$ and $3x^2 = 1$. The stationary points are $\left(-\frac{\sqrt{3}}{3}, \sqrt{3}\right)$ and $\left(\frac{\sqrt{3}}{3}, -\sqrt{3}\right)$.

$f_{xx} = -18x$; $f_{yy} = 2x$; $f_{xy} = 2y$

$$D\left(-\frac{\sqrt{3}}{3}, \sqrt{3}\right) = f_{xx}\left(-\frac{\sqrt{3}}{3}, \sqrt{3}\right)f_{yy}\left(-\frac{\sqrt{3}}{3}, \sqrt{3}\right) - \left(f_{xy}\left(-\frac{\sqrt{3}}{3}, \sqrt{3}\right)\right)^2 = -12 - 12 = -24 < 0$$

$$D\left(\frac{\sqrt{3}}{3}, -\sqrt{3}\right) = f_{xx}\left(\frac{\sqrt{3}}{3}, -\sqrt{3}\right)f_{yy}\left(\frac{\sqrt{3}}{3}, -\sqrt{3}\right) - \left(f_{xy}\left(\frac{\sqrt{3}}{3}, -\sqrt{3}\right)\right)^2 = -12 - 12 = -24 < 0$$

Thus both stationary points are saddle points. If $f$ has a maximum or minimum, it would appear among the stationary points; hence, $f$ has neither a maximum nor a minimum on $R^2$.

**35.** Let $f(x, y, z) = x^a y^b z^c$ be the function to be maximized. Let $g(x, y, z) = x + y + z - 100$.
$\nabla f(x, y, z) = \left\langle ax^{a-1}y^b z^c, bx^a y^{b-1} z^c, cx^a y^b z^{c-1}\right\rangle$
$\nabla g(x, y, z) = \langle 1, 1, 1\rangle$
Clearly $\nabla f = \lambda \nabla g$ when
$ax^{a-1}y^b z^c = bx^a y^{b-1} z^c$ and
$bx^a y^{b-1} z^c = cx^a y^b z^{c-1}$.
these simplify to
$ay = bx$
$bz = cy$

So $y = \frac{b}{a}x$ and $z = \frac{c}{b}y = \frac{c}{a}x$. Then we have

$x + \frac{b}{a}x + \frac{c}{a}x = 100$, which means that

$x = \frac{100a}{a+b+c}$, $y = \frac{100b}{a+b+c}$, and $z = \frac{100c}{a+b+c}$.

**37.** $T = 20(5 - x^2 + y^2) = 100 - 20(x^2 + y^2)$.

Clearly $T$ drops off in proportion to the square of the distance from the origin.

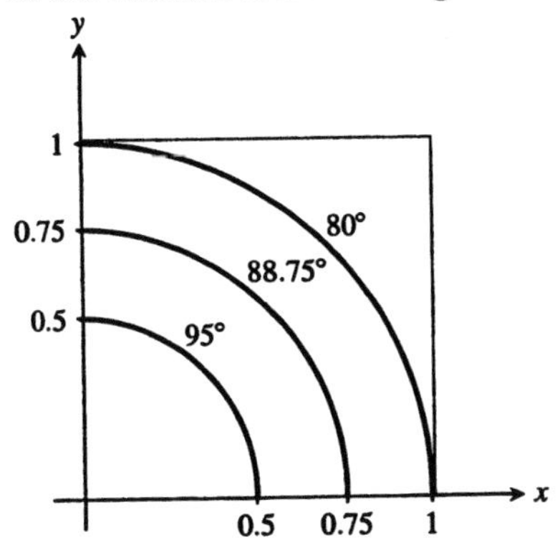


The hottest point is at (0, 0); the coldest point is at (1, 1).

**39.** The conditions are such that only two cases in the Second Derivatives Test apply. Either $D(\mathbf{r}_0) > 0$ and $f_{xx}(\mathbf{r}_0) < 0$ (Case 2, local maximum) or $D(\mathbf{r}_0) = 0$ (Case 4, test fails).

If $D(\mathbf{r}_0) = f_{xx}(\mathbf{r}_0)f_{yy}(\mathbf{r}_0) - (f_{xy}(\mathbf{r}_0))^2 > 0$ and $f_{xx}(\mathbf{r}_0) < 0$, then $f_{yy}(\mathbf{r}_0) < 0$ also—but then $f_{xx}(\mathbf{r}_0) + f_{yy}(\mathbf{r}_0) \neq 0$. So this case is ruled out.

If $D(\mathbf{r}_0) = f_{xx}(\mathbf{r}_0)f_{yy}(\mathbf{r}_0) - (f_{xy}(\mathbf{r}_0))^2 = 0$ then $f_{xx}(\mathbf{r}_0)$ and $f_{yy}(\mathbf{r}_0)$ must have the same sign unless one of them equals zero. But because $f_{xx}(\mathbf{r}_0) + f_{yy}(\mathbf{r}_0) = 0$, $f_{xx}(\mathbf{r}_0)$ and $f_{yy}(\mathbf{r}_0)$ must have opposite signs unless both of them equal zero. Thus $f_{xx}(\mathbf{r}_0) = f_{yy}(\mathbf{r}_0) = 0$, and then $D(\mathbf{r}_0) = 0$ implies $f_{xy}(\mathbf{r}_0) = 0$.

# Chapter 11 Multiple Integrals

## Section 11.1 The Double Integral on Rectangles

**1.** The area of each subregion is $1\cdot\frac{1}{2}=\frac{1}{2}$. The inscribed volume elements for each of the regions are $\frac{1}{2}f\left(\frac{1}{2},\frac{1}{4}\right)=\frac{19}{16}$,
$\frac{1}{2}f\left(\frac{1}{2},\frac{3}{4}\right)=\frac{15}{16}$, $\frac{1}{2}f\left(\frac{3}{2},\frac{1}{4}\right)=\frac{13}{16}$,
$\frac{1}{2}f\left(\frac{3}{2},\frac{3}{4}\right)=\frac{9}{16}$. Adding these numbers, the volume of the region is approximately $\frac{7}{2}$ or 3.5.

**3.** The area of each subregion is $\frac{2}{3}\cdot\frac{2}{3}=\frac{4}{9}$. The inscribed volume elements for each of the regions are $\frac{4}{9}f\left(\frac{2}{3},\frac{2}{3}\right)=\frac{176}{27}$,
$\frac{4}{9}f\left(\frac{2}{3},\frac{4}{3}\right)=\frac{16}{3}$, $\frac{4}{9}f\left(\frac{2}{3},2\right)=\frac{272}{81}$,
$\frac{4}{9}f\left(\frac{4}{3},\frac{2}{3}\right)=\frac{160}{27}$, $\frac{4}{9}f\left(\frac{4}{3},\frac{4}{3}\right)=\frac{128}{27}$,
$\frac{4}{9}f\left(\frac{4}{3},2\right)=\frac{224}{81}$, $\frac{4}{9}f\left(2,\frac{2}{3}\right)=\frac{400}{81}$,
$\frac{4}{9}f\left(2,\frac{4}{3}\right)=\frac{304}{81}$, $\frac{4}{9}f(2,2)=\frac{16}{9}$.
Adding these numbers, the volume of the region is approximately $\frac{352}{9}$ or 39.1111. Note that this is an underestimate since the evaluation is performed at points in the subregions at which $f$ takes on its minimum value.

**5.** The area of each subregion is $\frac{\pi}{8}\cdot\frac{\pi}{8}=\frac{\pi^2}{64}$. The inscribed volume elements for each of the regions are $\frac{\pi^2}{64}f\left(\frac{\pi}{16},\frac{\pi}{16}\right)\approx 0.029507$,

$\frac{\pi^2}{64}f\left(\frac{\pi}{16},\frac{3\pi}{16}\right)\approx 0.025015$,

$\frac{\pi^2}{64}f\left(\frac{\pi}{16},\frac{5\pi}{16}\right)\approx 0.016715$,

$\frac{\pi^2}{64}f\left(\frac{\pi}{16},\frac{7\pi}{16}\right)\approx 0.005869$,

$\frac{\pi^2}{64}f\left(\frac{3\pi}{16},\frac{\pi}{16}\right)\approx 0.084030$,

$\frac{\pi^2}{64}f\left(\frac{3\pi}{16},\frac{3\pi}{16}\right)\approx 0.071237$,

$\frac{\pi^2}{64}f\left(\frac{3\pi}{16},\frac{5\pi}{16}\right)\approx 0.047599$,

$\frac{\pi^2}{64}f\left(\frac{3\pi}{16},\frac{7\pi}{16}\right)\approx 0.016715$,

$\frac{\pi^2}{64}f\left(\frac{5\pi}{16},\frac{\pi}{16}\right)\approx 0.125759$,

$\frac{\pi^2}{64}f\left(\frac{5\pi}{16},\frac{3\pi}{16}\right)\approx 0.106614$,

$\frac{\pi^2}{64}f\left(\frac{5\pi}{16},\frac{5\pi}{16}\right)\approx 0.071237$,

$\frac{\pi^2}{64}f\left(\frac{5\pi}{16},\frac{7\pi}{16}\right)\approx 0.025015$,

$\frac{\pi^2}{64}f\left(\frac{7\pi}{16},\frac{\pi}{16}\right)\approx 0.148343$,

$\frac{\pi^2}{64}f\left(\frac{7\pi}{16},\frac{3\pi}{16}\right)\approx 0.125759$,

$\frac{\pi^2}{64}f\left(\frac{7\pi}{16},\frac{5\pi}{16}\right)\approx 0.084030$,

$\frac{\pi^2}{64}f\left(\frac{7\pi}{16},\frac{7\pi}{16}\right)\approx 0.029507$.

Adding these numbers, the volume over the region is approximately 1.01295.

**7.**
$$\begin{aligned}V&=\int_0^1\left(\int_0^2 3\left(1-\frac{1}{4}x-\frac{1}{3}y\right)dx\right)dy\\&=\int_0^1 3\left(x-\frac{1}{8}x^2-\frac{1}{3}xy\right)\bigg|_0^2 dy\\&=\int_0^1 3\left(\frac{3}{2}-\frac{2}{3}y\right)dy\\&=3\left(\frac{3}{2}y-\frac{1}{3}y^2\right)\bigg|_0^1\\&=\frac{7}{2}\end{aligned}$$

From Exercise 1, $R_2=\frac{7}{2}$, a percentage error of 0.

**9.** $V=\int_0^2\left(\int_0^2(16-x^2-2y^2)dx\right)dy=\int_0^2\left(16x-\frac{1}{3}x^3-2xy^2\right)\Big|_0^2 dy=\int_0^2\left(\frac{88}{3}-4y^2\right)dy=\left(\frac{88}{3}y-\frac{4}{3}y^3\right)\Big|_0^2=48$

From Exercise 3, $R_3=\frac{352}{9}$, a percentage error of about 19%.

**11.** $V=\int_0^{\pi/2}\left(\int_0^{\pi/2}(\sin x\cos y)dx\right)dy=\int_0^{\pi/2}(-\cos x\cos y)\Big|_0^{\pi/2} dy=\int_0^{\pi/2}\cos y\,dy=(\sin y)\big|_0^{\pi/2}=1$

From Exercise 5, $R_4\approx 1.01295$, a percentage error of about 1%.

**13.** 
$$\begin{aligned}\text{total rainfall}&=\int_0^{100}\left(\int_0^{100}\frac{100^2}{2x+y+200}dx\right)dy\\&=\int_0^{100}\frac{100^2}{2}\ln(2x+y+200)\Big|_0^{100} dy\\&=\frac{100^2}{2}\int_0^{100}(\ln(y+400)-\ln(y+200))dy\\&=\frac{100^2}{2}((y+400)\ln(y+400)-(y+200)\ln(y+200))\Big|_0^{100}\\&=\frac{100^2}{2}(500\ln 500-300\ln 300-400\ln 400+200\ln 200)\\&\approx 296{,}235 \text{ square mile - inches}\end{aligned}$$

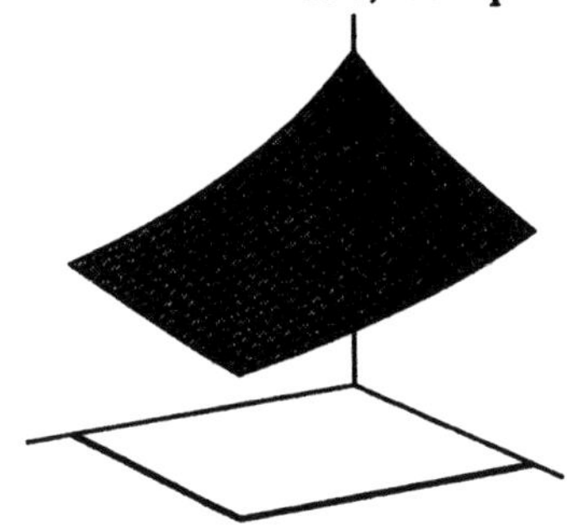

**15.** If we believe that the volume of the region beneath the graph of $r$ and above $D$ is equal to the amount of rain received by $D$, then the volume $100^2w$ of the pool is equal to $V$. The value of $w$ is equal to the average value of the rainfall function on $D$ because the product of the area of $D$ and $w$ is equal to the amount of rain received by $D$. Moreover, this fits with the definition of average value given for one-dimensional integrals, that is, $f_{\text{ave}}=\frac{1}{b-a}\int_a^b f(x)\,dx$.

$(x^*, y^*)$ is any point in $D$ such that $r(x^*, y^*)=w=\frac{V}{100^2}=100\ln 2\ln 1.5$.

For example, if $y^*=50$, then $r(x^*, 50)=\frac{100^3}{(x^*+100)(250)}=100\ln 2\ln 1.5$ and $x^*\approx 42.3250$, or $(x^*, y^*)\approx(42.3250, 50)$.

**17.** $f(2, 6) = \frac{12}{5}$; $f(0, 0) = 8$; $8 - \frac{12}{5} = \frac{28}{5}$

The volume of the base parallelepiped is $(2)(6)\left(\frac{12}{5}\right) = 28.8\text{ m}^3$. The volume of the top is $\frac{1}{2}(2)(6)\left(\frac{28}{5}\right) = 33.6\text{ m}^3$. The total volume is $62.4\text{ m}^3$.

**19.**
$$L_n = \sum_{j=1}^{n}\sum_{i=1}^{n} f\left(\frac{i-1}{n}, \frac{j-1}{n}\right)\frac{1}{n^2} = \sum_{j=0}^{n-1}\sum_{i=0}^{n-1} f\left(\frac{i}{n}, \frac{j}{n}\right)\frac{1}{n^2}$$

$$U_n = \sum_{j=1}^{n}\sum_{i=1}^{n} f\left(\frac{i}{n}, \frac{j}{n}\right)\frac{1}{n^2}$$

$$= L_n - \sum_{i=1}^{n-1} f\left(\frac{i}{n}, 0\right)\frac{1}{n^2} - \sum_{j=1}^{n-1} f\left(0, \frac{j}{n}\right)\frac{1}{n^2} - f(0, 0)\frac{1}{n^2} + \sum_{i=1}^{n-1}\left(\frac{i}{n}, 1\right)\frac{1}{n^2} + \sum_{j=1}^{n-1} f\left(1, \frac{j}{n}\right)\frac{1}{n^2} + f(1, 1)\frac{1}{n^2}$$

$$= L_n - \frac{1}{n^2}\left(f(0, 0) + \sum_{j=1}^{n-1} f\left(0, \frac{j}{n}\right) + \sum_{i=1}^{n-1} f\left(\frac{i}{n}, 0\right)\right) + \frac{1}{n^2}\left(f(1, 1) + \sum_{j=1}^{n-1} f\left(1, \frac{j}{n}\right) + \sum_{i=1}^{n-1} f\left(\frac{i}{n}, 1\right)\right)$$

Sample answer (other answers are possible): This saves redundant calculations when both $L_n$ and $U_n$ are desired and gives a maximum error, $U_n - L_n$, for the calculation of the integral.

**21.** For the cable that runs from (–2, –2, 1.5) to (2, 2, 1.5), and for all similarly oriented cables, the projection in the $(x, y)$-plane has the equation $y = x + b$ for some constant $b$. Points on the surface directly above each projection have the form

$$\left(x, x+b, \frac{1}{8}((x+b)^2 - x^2) + 1.5\right) = \left(x, x+b, \frac{1}{8}(2xb + b^2) + 1.5\right)$$

Since the $z$-component is linear, all points directly above a given projection lie in a straight line, which is the line along which the corresponding cable lies. An argument of the same form applies to cables oriented like the one that runs from (–2, 2, 1.5) to (2, –2, 1.5).

**23.**
$$\iint_I f(x, y)\,dx\,dy = \int_0^1\left(\int_0^3 f(x, y)\,dx\right)dy$$
$$= \int_0^1\left(\int_0^1 f(x, y)\,dx + \int_1^3 f(x, y)\,dx\right)dy$$
$$= \int_0^1\left(\int_0^1 f(x, y)\,dx\right)dy + \int_0^1\left(\int_1^3 f(x, y)\,dx\right)dy$$
$$= \iint_{I_1} f(x, y)\,dx\,dy + \iint_{I_2} f(x, y)\,dx\,dy$$

## Section 11.2 Extending the Double Integral and Applications

**1.**

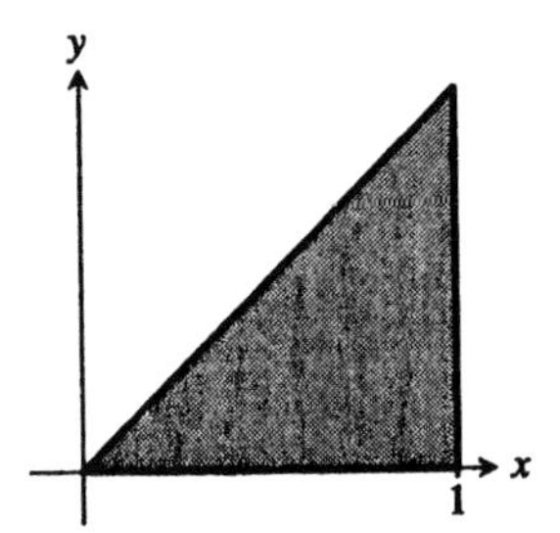


$$\int_0^1\int_0^x y\,dy\,dx = \int_0^1\left(\frac{1}{2}y^2\right)\Big|_0^x dx$$
$$= \int_0^1 \frac{1}{2}x^2 dx$$
$$= \left(\frac{1}{6}x^3\right)\Big|_0^1$$
$$= \frac{1}{6}$$

**3.**

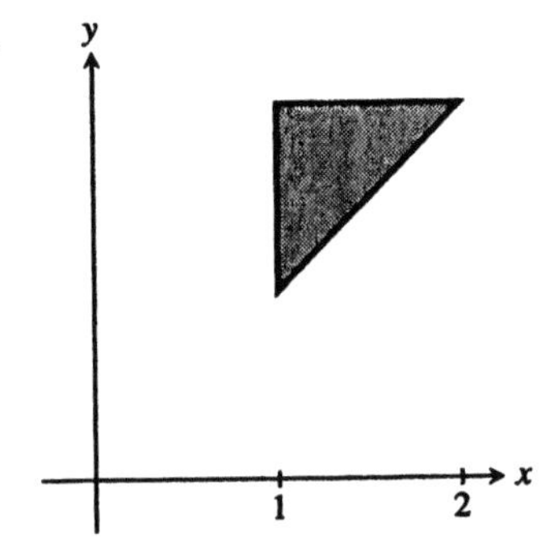


$$\int_1^2\int_1^y \frac{y}{x}dx\,dy = \int_1^2\int_x^2 \frac{y}{x}dy\,dx$$
$$= \int_1^2\left(\frac{y^2}{2x}\right)\Big|_x^2 dx$$
$$= \int_1^2\left(\frac{2}{x}-\frac{x}{2}\right)dx$$
$$= \left(2\ln x - \frac{x^2}{4}\right)\Big|_1^2$$
$$= 2\ln 2 - \frac{3}{4}$$

**5.**

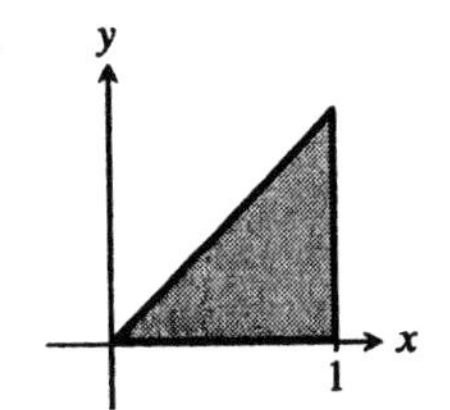


$$\int_0^1\int_0^x \sin x^2 dy\,dx = \int_0^1 (y\sin x^2)\Big|_0^x dx$$
$$= \int_0^1 x\sin x^2 dx$$
$$= \left(-\frac{1}{2}\cos x^2\right)\Big|_0^1$$
$$= \frac{1}{2} - \frac{1}{2}\cos 1$$

**7.** An equation for the line containing (1, 0) and (2, 0) is $y = 0$. An equation for the line containing (1, 0) and (1, 1) is $x = 1$. An equation for the line containing (1, 1) and (2, 0) is $y = 2 - x$. An iterated integral is
$$\int_0^1\int_0^{2-x} f(x, y)dy\,dx.$$

**9.** The region is bounded by the curves $x = y^2$ and $x = 1 + \frac{1}{\sqrt{2}}y$. When $y^2 = 1 + \frac{1}{\sqrt{2}}y$, $y = -\frac{1}{\sqrt{2}}$ or $y = \sqrt{2}$. An iterated integral is
$$\int_{-1/\sqrt{2}}^{\sqrt{2}}\int_{y^2}^{1+y/\sqrt{2}} f(x, y)dx\,dy.$$

**11.** The region is bounded by the curves $y = x^2$ and $y = 1 - x^2$. When $x^2 = 1 - x^2$, $x = -\frac{1}{\sqrt{2}}$ or $x = \frac{1}{\sqrt{2}}$. An iterated integral is
$$\int_{-1/\sqrt{2}}^{1/\sqrt{2}}\int_{x^2}^{1-x^2} f(x, y)dy\,dx.$$

**13.**

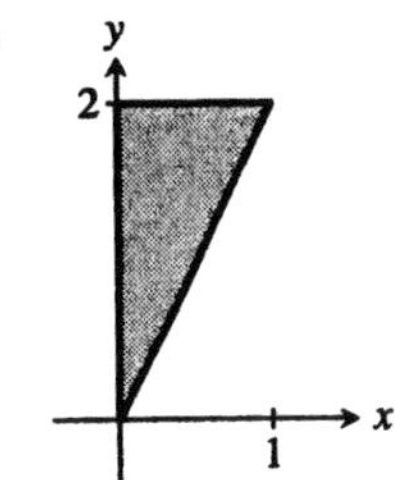

$$\int_0^1 \int_{2x}^2 xy^2\,dy\,dx = \int_0^2 \int_0^{y/2} xy^2\,dx\,dy$$
$$= \int_0^2 \left(\frac{1}{2}x^2y^2\right)\Big|_0^{y/2} dy$$
$$= \int_0^2 \left(\frac{1}{8}y^4\right)dy$$
$$= \left(\frac{1}{40}y^5\right)\Big|_0^2$$
$$= \frac{4}{5}$$

**15.**

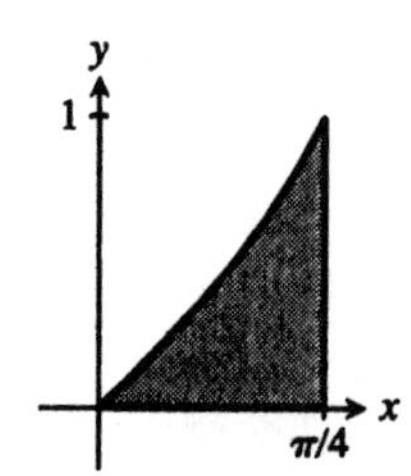


$$\int_0^1 \int_{\arctan y}^{\pi/4} y\,dx\,dy = \int_0^{\pi/4} \int_0^{\tan x} y\,dy\,dx$$
$$= \int_0^{\pi/4} \left(\frac{1}{2}y^2\right)\Big|_0^{\tan x} dy$$
$$= \int_0^{\pi/4} \left(\frac{1}{2}\tan^2 x\right)dy$$
$$= \frac{1}{2}\int_0^{\pi/4} (\sec^2 x - 1)dy$$
$$= \frac{1}{2}(\tan x - x)\Big|_0^{\pi/4}$$
$$= \frac{1}{2}\left(1 - \frac{\pi}{4}\right)$$
$$= \frac{1}{8}(4 - \pi)$$

**17.** $z = 1 - y$
An equation for the line containing (0, 0) and (1, 0) is $y = 0$. An equation for the line containing (0, 0) and (0, 1) is $x = 0$. An equation for the line containing (1, 0) and (0, 1) is $y = 1 - x$ or $x = 1 - y$.

$$V = \int_0^1 \int_0^{1-y} (1-y)dx\,dy$$
$$= \int_0^1 (1-y)x\Big|_0^{1-y} dy$$
$$= \int_0^1 (1-y)^2 dy$$
$$= -\frac{1}{3}(1-y)^3\Big|_0^1$$
$$= \frac{1}{3}$$

**19.** $z = 2 - \frac{1}{3}x - \frac{2}{3}y$
When $z = 0$, $x + 2y = 6$ or $x = 6 - 2y$. When $y = 6 - 2y$, $y = 2$.

$$V = \int_0^2 \int_y^{6-2y} \left(2 - \frac{1}{3}x - \frac{2}{3}y\right)dx\,dy$$
$$= \int_0^2 \left(2x - \frac{1}{6}x^2 - \frac{2}{3}xy\right)\Big|_y^{6-2y} dx$$
$$= \int_0^2 \left(6 - 6y + \frac{3}{2}y^2\right)dx$$
$$= \left(6y - 3y^2 + \frac{1}{2}y^3\right)\Big|_0^2$$
$$= 4$$

**21.** $z = x + y$

$$V = \int_0^1 \int_0^{x^2} (x+y)dy\,dx$$
$$= \int_0^1 \left(xy + \frac{1}{2}y^2\right)\Big|_0^{x^2} dx$$
$$= \int_0^1 \left(x^3 + \frac{1}{2}x^4\right)dx$$
$$= \left(\frac{1}{4}x^4 + \frac{1}{10}x^5\right)\Big|_0^1$$
$$= \frac{7}{20}$$

**23.** The disk can also be described by $-\sqrt{1-(x-1)^2} \le y \le \sqrt{1-(x-1)^2}$, for $0 \le x \le 2$.

$$V = \int_0^2 \int_{-\sqrt{1-(x-1)^2}}^{\sqrt{1-(x-1)^2}} (2x+1)dy\,dx$$
$$= \int_0^2 (2x+1)y\Big|_{-\sqrt{1-(x-1)^2}}^{\sqrt{1-(x-1)^2}} dx$$
$$= \int_0^2 2(2x+1)\sqrt{1-(x-1)^2}\,dx$$

To integrate this we let $x = \cos\theta + 1$, so $dx = -\sin\theta\, d\theta$. When $x = 0$, $\theta = -\pi$ and when $x = 2$, $\theta = 0$.

$$V = \int_{-\pi}^{0} 2(2\cos\theta + 3)\sin^2\theta\, d\theta$$
$$= 2\int_{-\pi}^{0} (2\cos\theta\sin^2\theta + 3\sin^2\theta)d\theta$$
$$= 2\int_{-\pi}^{0} \left(2\cos\theta\sin^2\theta + \frac{3}{2} - \frac{3}{2}\cos 2\theta\right)d\theta$$
$$= 2\left(\frac{2}{3}\sin^3\theta + \frac{3}{2}\theta - \frac{3}{4}\sin 2\theta\right)\Bigg|_{-\pi}^{0}$$
$$= 3\pi$$

**25.** By symmetry, we only need to compute the volume of the region in the first octant and then multiply the result by 8 to find the total volume. Moreover, observe that when the cylinders intersect in the first octant, $y = x$ and that the region in the first octant is symmetric with respect to the line $y = x$. Then we only need to find the volume over the region in the $(x, y)$-plane bounded by $y = x$, $y = 0$, and $x = a$ and multiply the result by 16. The height in that region is given by $z = \sqrt{a^2 - x^2}$.

$$V = 16\int_0^a \int_0^x \sqrt{a^2 - x^2}\, dy\, dx$$
$$= 16\int_0^a \sqrt{a^2 - x^2}\, y\Big|_0^x dx$$
$$= 16\int_0^a x\sqrt{a^2 - x^2}\, dx$$
$$= -\frac{16}{3}(a^2 - x^2)^{3/2}\Big|_0^a$$
$$= \frac{16}{3}a^3$$

**27.** $\delta(x, y) = k|y|$

Since $y \ge 0$ for the lamina, $\delta(x, y) = ky$. When $4 - x^2 = 0$, $x = -2$ or $x = 2$.

$$m = \int_{-2}^{2}\int_0^{4-x^2} ky\, dy\, dx$$
$$= k\int_{-2}^{2}\left(\frac{1}{2}y^2\right)\Bigg|_0^{4-x^2} dx$$
$$= \frac{k}{2}\int_{-2}^{2}(16 - 8x^2 + x^4)dx$$
$$= \frac{k}{2}\left(16x - \frac{8}{3}x^3 + \frac{1}{5}x^5\right)\Bigg|_{-2}^{2}$$
$$= \frac{256k}{15}$$

From the symmetry of the lamina and the density function, $X = 0$.

$$Y = \frac{1}{m}\int_{-2}^{2}\int_0^{4-x^2} ky^2\, dy\, dx$$
$$= \frac{15}{256}\int_{-2}^{2}\int_0^{4-x^2} y^2\, dy\, dx$$
$$= \frac{15}{256}\int_{-2}^{2}\left(\frac{1}{3}y^3\right)\Bigg|_0^{4-x^2} dx$$
$$= \frac{5}{256}\int_{-2}^{2}(64 - 48x^2 + 12x^4 - x^6)dx$$
$$= \frac{5}{256}\left(64x - 16x^3 + \frac{12}{5}x^5 - \frac{1}{7}x^7\right)\Bigg|_{-2}^{2}$$
$$= \frac{16}{7}$$

The center of mass is $\left(0, \frac{16}{7}\right)$.

**29.** $\delta(x, y) = k|x|$

When $x^2 = x + 2$, $x = -1$ or $x = 2$.

$$m = \int_{-1}^{2}\int_{x^2}^{x+2} k|x|dy\,dx$$
$$= k\int_{-1}^{2} \left(|x|y\right)\Big|_{x^2}^{x+2} dx$$
$$= k\int_{-1}^{2} \left(2|x| + x|x| - x^2|x|\right)dx$$
$$= k\int_{-1}^{0} (-2x - x^2 + x^3)dx + k\int_{0}^{2} (2x + x^2 - x^3)dx$$
$$= k\left(-x^2 - \frac{1}{3}x^3 + \frac{1}{4}x^4\right)\Big|_{-1}^{0} + k\left(x^2 + \frac{1}{3}x^3 - \frac{1}{4}x^4\right)\Big|_{0}^{2}$$
$$= \frac{5k}{12} + \frac{8k}{3}$$
$$= \frac{37k}{12}$$

$$X = \frac{1}{m}\int_{-1}^{2}\int_{x^2}^{x+2} kx|x|dy\,dx$$
$$= \frac{12}{37}\int_{-1}^{2} \left(x|x|y\right)\Big|_{x^2}^{x+2} dx$$
$$= \frac{12}{37}\int_{-1}^{2} \left(2x|x| + x^2|x| - x^3|x|\right)dx$$
$$= \frac{12}{37}\int_{-1}^{0} (-2x^2 - x^3 + x^4)dx + \frac{12}{37}\int_{0}^{2} (2x^2 + x^3 - x^4)dx$$
$$= \frac{12}{37}\left(-\frac{2}{3}x^3 - \frac{1}{4}x^4 + \frac{1}{5}x^5\right)\Big|_{-1}^{0} + \frac{12}{37}\left(\frac{2}{3}x^3 + \frac{1}{4}x^4 - \frac{1}{5}x^5\right)\Big|_{0}^{2}$$
$$= -\frac{13}{185} + \frac{176}{185}$$
$$= \frac{163}{185}$$

$$Y = \frac{1}{m}\int_{-1}^{2}\int_{x^2}^{x+2} ky|x|dy\,dx$$
$$= \frac{12}{37}\int_{-1}^{2} \left(|x|\frac{1}{2}y^2\right)\Big|_{x^2}^{x+2} dx$$
$$= \frac{6}{37}\int_{-1}^{2} \left(4|x| + 4x|x| + x^2|x| - x^4|x|\right)dx$$
$$= \frac{6}{37}\int_{-1}^{0} (-4x - 4x^2 - x^3 + x^5)dx + \frac{6}{37}\int_{0}^{2} (4x + 4x^2 + x^3 - x^5)dx$$
$$= \frac{6}{37}\left(-2x^2 - \frac{4}{3}x^3 - \frac{1}{4}x^4 + \frac{1}{6}x^6\right)\Big|_{-1}^{0} + \frac{6}{37}\left(2x^2 + \frac{4}{3}x^3 + \frac{1}{4}x^4 - \frac{1}{6}x^6\right)\Big|_{0}^{2}$$
$$= \frac{9}{74} + \frac{72}{37}$$
$$= \frac{153}{74}$$

The center of mass is $\left(\frac{163}{185}, \frac{153}{74}\right)$.

**31.** $\delta(x, y) = k\sqrt{x^2 + y^2}$

When $\sqrt{1-x^2} = 0$, $x = -1$ or $x = 1$.

$m = \frac{k\pi}{3}$

From the symmetry of the lamina and the density function, $X = 0$.

$$Y = \frac{1}{m}\int_{-1}^{1}\int_{0}^{\sqrt{1-x^2}} ky\sqrt{x^2+y^2}\,dy\,dx$$
$$= \frac{3}{\pi}\int_{-1}^{1}\int_{0}^{\sqrt{1-x^2}} y\sqrt{x^2+y^2}\,dy\,dx$$
$$= \frac{3}{\pi}\int_{-1}^{1} \frac{1}{3}(x^2+y^2)^{3/2}\Big|_{0}^{\sqrt{1-x^2}} dx$$
$$= \frac{1}{\pi}\int_{-1}^{1}\left(1-|x|^3\right)dx$$
$$= \frac{1}{\pi}\int_{-1}^{0}(1+x^3)dx + \frac{1}{\pi}\int_{0}^{1}(1-x^3)dx$$
$$= \frac{1}{\pi}\left(x+\frac{1}{4}x^4\right)\Big|_{-1}^{0} + \frac{1}{\pi}\left(x-\frac{1}{4}x^4\right)\Big|_{0}^{1}$$
$$= \frac{3}{4\pi} + \frac{3}{4\pi}$$
$$= \frac{3}{2\pi}$$

The center of mass is $\left(0, \frac{3}{2\pi}\right)$.

**33.** Let $D_1$ be the region bounded by $x = y^2$ $\left(\text{or } y = -\sqrt{x} \text{ and } y = \sqrt{x}\right)$ and $x = 1$. Let $D_2$ be the region bounded by $x = y^2$, $y = x - 2$, and $x = 1$.

$$V = \iint_D (4-x)\,dy\,dx$$
$$= \iint_{D_1}(4-x)\,dy\,dx + \iint_{D_2}(4-x)\,dy\,dx$$
$$= \int_0^1\left(\int_{-\sqrt{x}}^{\sqrt{x}}(4-x)\,dy\right)dx + \int_1^4\left(\int_{x-2}^{\sqrt{x}}(4-x)\,dy\right)dx$$

**35.** Sample answer (other answers are possible):
If the region of integration $D$, which is the quarter-circle with a square removed, is divided into three parts, $A$, $B$, and $C$ by extending the sides of the square, two of these parts resemble rounded squares and the third resembles a rounded triangle. If we denote the rounded triangle by $C$, then the other two are $A$ and $B$. Since the areas of $A$ and $B$ are equal, it would be sufficient to integrate over the regions $A$ and $C$. This would require, however, two separate integrals. Evaluating these integrals would be more complicated than those of the integral given in Example 4.

**37.** Let $D_1$ be the region bounded by $x = y$, $y = \frac{2}{5}x$, and $y = 2$. Let $D_2$ be the region bounded by $x = y$, $y = -\frac{1}{2}x + \frac{9}{2}$, and $y = 2$. $y = \frac{2}{5}x$ is equivalent to $x = \frac{5}{2}y$ and $y = -\frac{1}{2}x + \frac{9}{2}$ is equivalent to $x = -2y + 9$.

$$m = \int_0^2 \int_y^{5y/2} kx\,dm + \int_2^3 \int_y^{-2y+9} kx\,dm$$

**39.**

$$Y = \frac{k}{m}\int_{-a}^{a}\int_0^{\sqrt{a^2-x^2}} y(x^2+y^2)\,dy\,dx$$
$$= \frac{4}{\pi a^4}\int_{-a}^{a} \frac{1}{4}(x^2+y^2)^2\Big|_0^{\sqrt{a^2-x^2}}\,dx$$
$$= \frac{1}{\pi a^4}\int_{-a}^{a}(a^4 - x^4)\,dx$$
$$= \frac{1}{\pi a^4}\left(a^4x - \frac{1}{5}x^5\right)\Big|_{-a}^{a}$$
$$= \frac{8a}{5\pi}$$

**41.** Applying integration by parts to the indefinite integral several times, we obtain

$$\int\left(x^2 - \frac{2}{3}x^3\right)\sin 2x\,dx = \left(x^2 - \frac{2}{3}x^3\right)\left(-\frac{\cos 2x}{2}\right) + \int (x - x^2)\cos 2x\,dx$$
$$= -\frac{\cos 2x}{2}\left(x^2 - \frac{2}{3}x^3\right) + \frac{\sin 2x}{2}(x - x^2) - \int (1-2x)\frac{\sin 2x}{2}\,dx$$
$$= -\frac{\cos 2x}{2}\left(x^2 - \frac{2}{3}x^3\right) + \frac{\sin 2x}{2}(x - x^2) + \frac{\cos 2x}{4}(1-2x) + \int \frac{\cos 2x}{2}\,dx$$
$$= -\frac{\cos 2x}{2}\left(x^2 - \frac{2}{3}x^3 - \frac{1}{2}\right) + \frac{\sin 2x}{2}(x - x^2) - \frac{x\cos 2x}{2} + \int \frac{\cos 2x}{2}\,dx$$
$$= -\frac{\cos 2x}{2}\left(x^2 - \frac{2}{3}x^3 + x - \frac{1}{2}\right) + \frac{\sin 2x}{2}\left(x - x^2 + \frac{1}{2}\right) + \text{constant}$$

Evaluating this between the limits 0 and 1 shows that $\int_0^1\left(x^2 - \frac{2}{3}x^3\right)\sin 2x\,dx$ does in fact equal $\frac{\sin(2)}{4} - \frac{5\cos(2)}{12} - \frac{1}{4} = \frac{1}{12}(3\sin(2) - 5\cos(2) - 3) \approx 0.150719$.

**43.** The circular region is Type I, with $-1 \le x \le 1$ and $-\sqrt{1-x^2} \le y \le \sqrt{1-x^2}$. Thus,

$$V = \int_{-1}^{1}\int_{-\sqrt{1-x^2}}^{\sqrt{1-x^2}}(x^2+y^2)\,dy\,dx = \int_{-1}^{1}\left(2x^2\sqrt{1-x^2} + \frac{2}{3}(1-x^2)^{3/2}\right)dx.$$

This integral can be solved using either integral tables or trigonometric substitution and evaluates to $\frac{\pi}{2}$. It will also be easier to solve with change-of-variables techniques from Section 11.4.

**45.** The centroid of a circle is (by symmetry) at the center of the circle. Thus the centroid travels $2\pi b$. The area of the circle is $\pi r^2$. $V = (2\pi b)\pi r^2 = 2\pi^2 r^2 b$.

**47.** Substitute $x = 2$ and $y = 2$ into $\frac{x^2}{9}+\frac{y^2}{25}+\frac{z^2}{16}=1$ and solve for $z$. We get $z=\pm\frac{4\sqrt{89}}{15}$. The height of the volume element is $\frac{4\sqrt{89}}{15}$. Fix $x = 2$, and let $y=\frac{5\sqrt{5}}{3}\cos t$ and $z=\frac{4\sqrt{5}}{3}\sin t$. Then $\mathbf{r}(t)=\left(2, \frac{5\sqrt{5}}{3}\cos t, \frac{4\sqrt{5}}{3}\sin t\right)$ describes a curve that passes through the top of the volume element.

Fix $y = 2$, and let $x=\frac{3\sqrt{21}}{5}\cos t$ and $z=\frac{4\sqrt{21}}{5}\sin t$. Then $\mathbf{r}(t)=\left(\frac{3\sqrt{21}}{5}\cos t, 2, \frac{4\sqrt{21}}{5}\sin t\right)$ describes a curve that passes through the top of the volume element.

## Section 11.3 Surface Area

**1.**

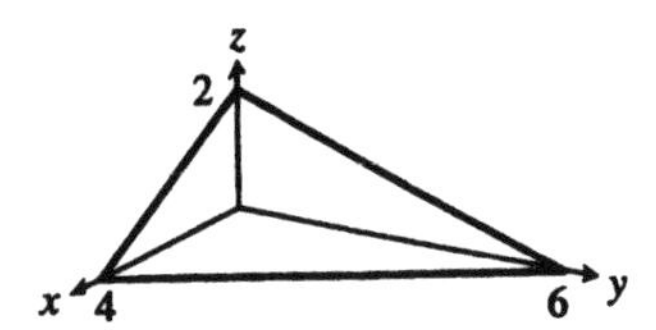


$z=2-\frac{1}{2}x-\frac{1}{3}y$

Let $\mathbf{r}(x, y)=\left(x, y, 2-\frac{1}{2}x-\frac{1}{3}y\right)$, $(x, y) \in D$ where $D$ is the region described by $0 \le x \le 4-\frac{2}{3}y$, $0 \le y \le 6$.

$$\|\mathbf{r}_x \times \mathbf{r}_y\| = \left\|\left\langle 1, 0, -\frac{1}{2}\right\rangle \times \left\langle 0, 1, -\frac{1}{3}\right\rangle\right\|$$
$$= \left\|\left\langle \frac{1}{2}, \frac{1}{3}, 1\right\rangle\right\|$$
$$= \frac{7}{6}$$

$$A=\int_0^6\int_0^{4-2y/3}\frac{7}{6}\,dx\,dy$$
$$=\frac{7}{6}\int_0^6 x\Big|_0^{4-2y/3}dy$$
$$=\frac{7}{6}\int_0^6\left(4-\frac{2}{3}y\right)dy$$
$$=\frac{7}{6}\left(4y-\frac{1}{3}y^2\right)\Big|_0^6$$
$$=14$$

*Check*: When $x = 0$ and $y = 0$, $z = 2$. When $x = 0$ and $z = 0$, $y = 6$. When $y = 0$ and $z = 0$, $x = 4$. Let $A = (4, 0, 0)$, $B = (0, 6, 0)$, and $C = (0, 0, 2)$. Hence $\overrightarrow{AB}=(-4, 6, 0)$ and $\overrightarrow{AC}=(-4, 0, 2)$.

$$\overrightarrow{AB}\times\overrightarrow{AC}=(-4, 6, 0)\times(-4, 0, 2)=(12, 8, 24)$$

$$A=\frac{1}{2}\left\|\overrightarrow{AB}\times\overrightarrow{AC}\right\|$$
$$=\frac{1}{2}\sqrt{12^2+8^2+24^2}$$
$$=14 \text{ square units}$$

**3.**

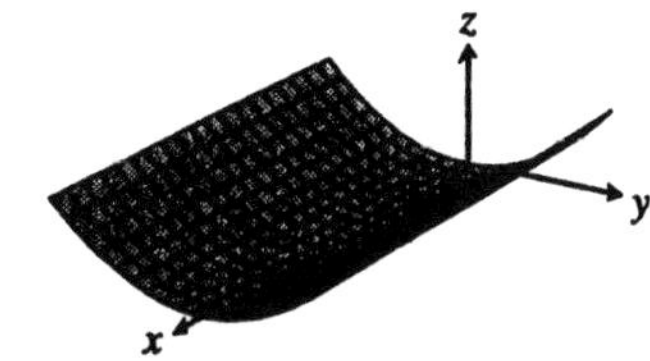


Let $\mathbf{r}(x, y)=\left(x, y, \frac{1}{2}y^2\right)$, $(x, y) \in [0, 3] \times [-1, 1]$.

$$\|\mathbf{r}_x \times \mathbf{r}_y\| = \|\langle 1, 0, 0\rangle \times \langle 0, 1, y\rangle\|$$
$$= \|\langle 0, -y, 1\rangle\|$$
$$= \sqrt{y^2+1}$$

$$A=\int_{-1}^1\int_0^3\sqrt{y^2+1}\,dx\,dy$$
$$=\int_{-1}^1\left(x\sqrt{y^2+1}\right)\Big|_0^3 dy$$
$$=3\int_{-1}^1\sqrt{y^2+1}\,dy$$

This integral is evaluated by formula (9) from

the Table of Integrals, so

$$A = 3\left(\frac{y}{2}\sqrt{y^2+1} + \frac{1}{2}\ln\left|y+\sqrt{y^2+1}\right|\right)\Bigg|_{-1}^{1}$$
$$= 3\left(\sqrt{2} + \frac{1}{2}\ln\left(1+\sqrt{2}\right) - \frac{1}{2}\ln\left(-1+\sqrt{2}\right)\right)$$
$$= 3\left(\sqrt{2} + \ln\left(1+\sqrt{2}\right)\right)$$
$$\approx 6.88676 \text{ square units}$$

**5.**

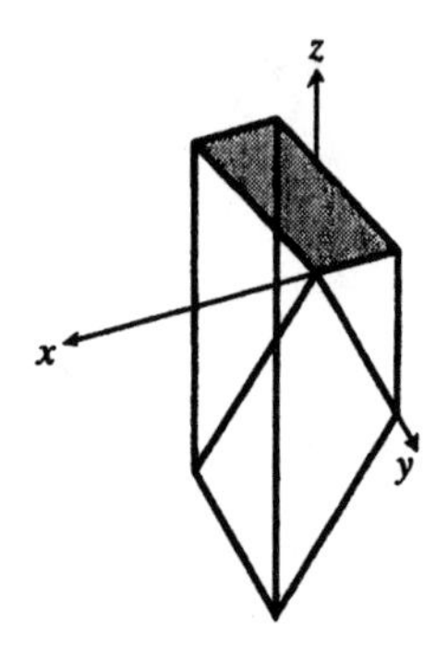


$$\|\mathbf{r}_u \times \mathbf{r}_v\| = \|\langle 1, 1, 2\rangle \times \langle 0, 1, 1\rangle\| = \|\langle -1, -1, 1\rangle\| = \sqrt{3}$$
$$A = \int_0^1\int_0^1 \sqrt{3}\,du\,dv = \sqrt{3}\int_0^1 u\Big|_0^1 dv = \sqrt{3}\int_0^1 dv = \sqrt{3}v\Big|_0^1 = \sqrt{3} \text{ square units}$$

**7.**

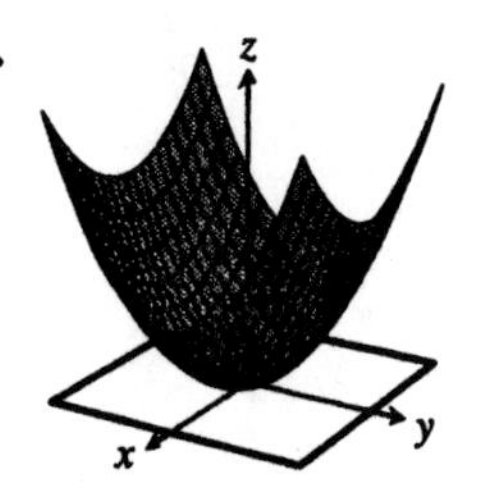


Let $\mathbf{r}(x, y) = (x, y, x^2 + 2y^2)$, $(x, y) \in [-1, 1] \times [-1, 1]$.
From symmetry, we can integrate over $[0, 1] \times [0, 1]$ and multiply by 4.

$$\|\mathbf{r}_x \times \mathbf{r}_y\| = \|\langle 1, 0, 2x\rangle \times \langle 0, 1, 4y\rangle\| = \|\langle -2x, -4y, 1\rangle\| = \sqrt{4x^2 + 16y^2 + 1}$$
$$A = 4\int_0^1\int_0^1 \sqrt{4x^2 + 16y^2 + 1}\,dx\,dy = 8\int_0^1\int_0^1 \sqrt{x^2 + \left(4y^2 + \frac{1}{4}\right)}\,dx\,dy$$

This integral is evaluated by formula (9) from the Table of Integrals, so

$$A = 8\int_0^1\left(\frac{x}{2}\sqrt{x^2 + \left(4y^2 + \frac{1}{4}\right)} + \frac{1}{2}\left(4y^2 + \frac{1}{4}\right)\ln\left|x + \sqrt{x^2 + \left(4y^2 + \frac{1}{4}\right)}\right|\right)\Bigg|_0^1 dy$$
$$= 8\int_0^1\left(\frac{1}{2}\sqrt{4y^2 + \frac{5}{4}} + \frac{1}{2}\left(4y^2 + \frac{1}{4}\right)\ln\left|1 + \sqrt{4y^2 + \frac{5}{4}}\right| + \frac{1}{2}\left(4y^2 + \frac{1}{4}\right)\ln\left|\sqrt{4y^2 + \frac{1}{4}}\right|\right)dy$$

Use numerical integration to get $A \approx 10.4670$ square units.

**9.** $\|\mathbf{r}_u \times \mathbf{r}_v\| = \left\|\langle \cos v, \sin v, 0\rangle \times \left\langle -u\sin v, u\cos v, \frac{1}{5}\right\rangle\right\| = \left\|\left\langle \frac{1}{5}\sin v, -\frac{1}{5}\cos v, u\right\rangle\right\| = \sqrt{u^2 + \frac{1}{25}}$

$$A = \int_0^{2\pi}\int_{1/2}^1 \sqrt{u^2 + \frac{1}{25}}\, du\, dv$$

**11.** Let $\mathbf{r}(x, y) = (x, y, x^2 - y^2 + 2)$.

$\|\mathbf{r}_x \times \mathbf{r}_y\| = \|\langle 1, 0, 2x\rangle \times \langle 0, 1, -2y\rangle\| = \|\langle -2x, 2y, 1\rangle\| = \sqrt{4x^2 + 4y^2 + 1}$

The circle can be described by $(x-1)^2 + y^2 = 0.1^2$. The function is symmetric with respect to the $x$-axis, so we can integrate over the region described by $0 \le y \le \sqrt{0.1^2 - (x-1)^2}$, $0.9 \le x \le 1.1$ and multiply by 2.

$$A = 2\int_{0.9}^{1.1}\int_0^{\sqrt{0.1^2-(x-1)^2}} \sqrt{4x^2 + 4y^2 + 1}\, dy\, dx$$

**13.** $x'(t) = a(1 - \cos t)$; $y'(t) = a\sin t$

$$A = 2\pi\int_0^{2\pi} a(1 - \cos t)\sqrt{a^2(1-\cos t)^2 + a^2\sin^2 t}\, dt$$
$$= 2\pi a^2\int_0^{2\pi} (1 - \cos t)\sqrt{2 - 2\cos t}\, dt$$
$$= 2\sqrt{2}\pi a^2\int_0^{2\pi} (1 - \cos t)^{3/2}\, dt$$
$$= 8\pi a^2\int_0^{2\pi} \sin^3\left(\frac{t}{2}\right) dt$$
$$= \frac{64}{3}\pi a^2 \text{ square units}$$

**15.** Let $x(t) = t$ and $y(t) = e^t$ for $0 \le t \le 1$, so $x'(t) = 1$ and $y'(t) = e^t$.

$$A = 2\pi\int_0^1 e^t\sqrt{1 + e^{2t}}\, dt$$

Let $w = e^t$, so $dw = e^t dt$. When $t = 0$, $w = 1$ and when $t = 1$, $w = e$.

$$A = 2\pi\int_1^e \sqrt{w^2 + 1}\, dw$$
$$= 2\pi\left(\frac{w}{2}\sqrt{w^2+1} + \frac{1}{2}\ln\left|w + \sqrt{w^2+1}\right|\right)\Bigg|_1^e$$
$$= 2\pi\left(\frac{e}{2}\sqrt{e^2+1} + \frac{1}{2}\ln\left(e + \sqrt{e^2+1}\right) - \frac{1}{2}\sqrt{2} - \frac{1}{2}\ln\left(1 + \sqrt{2}\right)\right)$$
$$\approx 22.9430 \text{ square units}$$

**17.** $\|\mathbf{r}_u \times \mathbf{r}_v\| = \|\langle a\cos v, a\sin v, b\rangle \times \langle -au\sin v, au\cos v, 0\rangle\|$

$$= \|\langle -abu\cos v, -abu\sin v, a^2u\rangle\|$$
$$= au\sqrt{a^2 + b^2}$$

$$m = \int_0^{2\pi}\int_0^1 a\delta u\sqrt{a^2+b^2}\, du\, dv = a\delta\sqrt{a^2+b^2}\int_0^{2\pi} \frac{1}{2}u^2\Big|_0^1 dv = \frac{1}{2}a\delta\sqrt{a^2+b^2}\int_0^{2\pi} dv = a\delta\pi\sqrt{a^2+b^2} \text{ kg}$$

An equation in Cartesian coordinates for the surface is $z = \frac{b}{a}\sqrt{x^2+y^2}$, $x^2+y^2 \le a$. By symmetry, the center of mass is on the $z$-axis. Thus if $(X, Y, Z)$ is the center of mass $X = 0$, $Y = 0$, and

$$Z = \frac{1}{m}\int_0^{2\pi}\int_0^1 \delta abu^2\sqrt{a^2+b^2}\,du\,dv = \frac{b}{\pi}\int_0^{2\pi}\int_0^1 u^2\,dv = \frac{b}{\pi}\int_0^{2\pi}\frac{1}{3}u^3\Big|_0^1 dv = \frac{b}{3\pi}\int_0^{2\pi}dv = \frac{2b}{3} \text{ meters.}$$

$$\mathbf{R} = \left\langle 0, 0, \frac{2b}{3}\right\rangle$$

**19.** $\|\mathbf{r}_u \times \mathbf{r}_v\| = \|\langle\cos v, \sin v, 2u\rangle \times \langle -u\sin v, u\cos v, 0\rangle\| = \|\langle -2u^2\cos v, -2u^2\sin v, u\rangle\| = u\sqrt{4u^2+1}$,

$$m = \int_0^{2\pi}\int_0^{3/2}\delta u\sqrt{4u^2+1}\,du\,dv = \frac{\pi}{6}\delta\left(-1+10\sqrt{10}\right) \approx 16.034\delta \text{ kg},$$

while

$$Z = \frac{1}{m}\int_0^{2\pi}\int_0^{3/2}\delta u\sqrt{4u^2+1}u^2\,du\,dv = \frac{2\pi\delta}{m}\int_0^{3/2}u^3\sqrt{4u^2+1}\,du \approx 1.29408 \text{ m.}$$

**21.** Let $\mathbf{r}(\phi_1, \theta_1) = \mathbf{r}(\phi_2, \theta_2)$ for $(\phi_1, \theta_1)$ and $(\phi_2, \theta_2)$ in $D$. Thus
$\langle a\cos\theta_1\sin\phi_1, a\sin\theta_1\sin\phi_1, a\cos\phi_1\rangle = \langle a\cos\theta_2\sin\phi_2, a\sin\theta_2\sin\phi_2, a\cos\phi_2\rangle$.

If $a\cos\phi_1 = a\cos\phi_2$, then $\phi_1 = \phi_2$ since $\phi_1, \phi_2 \in \left[\varepsilon, \frac{\pi}{2}\right]$ and $a \ne 0$. If $a\cos\theta_1\sin\phi_1 = a\cos\theta_2\sin\phi_2$ and $a\sin\theta_1\sin\phi_1 = a\sin\theta_2\sin\phi_2$, then $\cos\theta_1 = \cos\theta_2$ and $\sin\theta_1 = \sin\theta_2$ since $\sin\phi_1 = \sin\phi_2$ is never zero because $\phi_1, \phi_2 \in \left[\varepsilon, \frac{\pi}{2}\right]$, so $\theta_1 = \theta_2$ since $\theta_1, \theta_2 \in \left[0, \frac{\pi}{2}\right]$. Thus $(\phi_1, \theta_1) = (\phi_2, \theta_2)$ and $\mathbf{r}$ is one-to-one.

$$A_\varepsilon = \int_0^{\pi/2}\int_\varepsilon^{\pi/2} a^2\sin\phi\,d\phi\,d\theta = \int_0^{\pi/2} a^2(-\cos\phi)\Big|_\varepsilon^{\pi/2}d\theta = \int_0^{\pi/2} a^2\cos\varepsilon\,d\theta = \frac{1}{2}\pi a^2\cos\varepsilon$$

$$\lim_{\varepsilon\to 0} A_\varepsilon = \lim_{\varepsilon\to 0}\frac{1}{2}\pi a^2\cos\varepsilon = \frac{1}{2}\pi a^2$$

**23.** The length of the curve is $2\pi a$ and the distance traveled by its centroid is $2\pi b$, thus
$A = (2\pi a)(2\pi b) = 4\pi^2 ab$.

**25.** If we let $x = t$ and $y = f(t)$, then equation (14) becomes

$$A = 2\pi\int_a^b f(t)\sqrt{1+f'(t)}\,dt = 2\pi\int_a^b f(x)\sqrt{1+f'(x)}\,dx.$$

**27.** The visible area is described by the lamina $\mathbf{r} = 6378\langle\cos\theta, \sin\theta, \sin\psi\rangle$, $0 \le \theta \le 2\pi$, $\sin^{-1}\left(\frac{6378}{6478}\right) \le \psi \le \frac{\pi}{2}$. Since $\|\mathbf{r}_\theta \times \mathbf{r}_\psi\| = 6378^2\cos\psi$, it follows that

$$A = \int_0^{2\pi}\int_{\sin^{-1}\left(\frac{6378}{6478}\right)}^{\pi/2} 6378^2\cos\psi\,d\psi\,d\theta = 2\pi\cdot(6378)^2\cdot\frac{100}{6478},$$

so the percentage of surface area seen is $\frac{A}{4\pi(6378)^2} = \frac{50}{6478} \approx 0.77\%$.

**29.** Let the earth have polar radius $b$ and equatorial radius $a$, $b < a$. Consider the half-ellipse in the $(x, y)$-plane $y = \frac{a}{b}\sqrt{b^2 - x^2}$. If we rotate the portion between $x_0$ and $x_0 + d$ about the $x$-axis and compute the surface area, we get the area between two parallels of latitude whose planes are distance $d$ apart.
Using the formula derived in Exercise 25, the surface area is

$$A = 2\pi \int_{x_0}^{x_0+d} \frac{a}{b}\sqrt{b^2 - x^2}\sqrt{1 + \frac{a^2x^2}{b^2(b^2 - x^2)}}\, dx$$

$$= \frac{2\pi a}{b^2}\sqrt{a^2 - b^2} \int_{x_0}^{x_0+d} \sqrt{x^2 + \frac{b^4}{a^2 - b^2}}\, dx.$$

Since the integrand varies with $x$, and for small $d$, $A \approx \frac{2\pi a}{b^2}\sqrt{a^2 - b^2}\sqrt{x_0^2 + \frac{b^4}{a^2 - b^2}} \cdot d$, it follows that $A$ depends not just on $d$ but also on $x_0$, and so the property does not hold.

**31.** Let $D_1$ be the region bounded by the triangle with vertices (50, 0), (0, 30) and (–65, 0). The approximate volume $V$ above $D_1$ is $\frac{2}{3}\iint_{D_1} \sqrt{1 + f_x^2 + f_y^2}\, dx\, dy$.

Since $f(x, y) = f(x, -y)$ and $D$ is symmetric about the $x$-axis, $V = \frac{4}{3}\iint_{D_1} \sqrt{1 + f_x^2 + f_y^2}\, dx\, dy$.

$D_1$ is bounded by the lines described by $x = \frac{13}{6}y - 65$, $x = -\frac{5}{3}y + 50$, and $y = 0$.

With $f_x = \frac{8x}{1725}$, $f_y = -\frac{7y}{15{,}525}$, $V = \frac{4}{3}\int_0^{30}\int_{13y/6-65}^{-5y/3+50} \sqrt{1 + \left(\frac{8x}{1725}\right)^2 + \left(\frac{7y}{15{,}525}\right)^2}\, dx\, dy$, which evaluates numerically to about 2314 cubic feet of aggregate.

**33.** Suppose that $\mathbf{r}(x_1, y_1) = \mathbf{r}(x_2, y_2)$ for some $(x_1, y_1), (x_2, y_2) \in D$. Then $\left\langle x_1, y_1, \sqrt{a^2 - x_1^2 - y_1^2}\right\rangle = \left\langle x_2, y_2, \sqrt{a^2 - x_2^2 - y_2^2}\right\rangle$, from which it is clear that $x_1 = x_2$ and $y_1 = y_2$.

$\mathbf{r}_x$ and $\mathbf{r}_y$ are not continuous on $D$; in particular, discontinuity occurs when $x^2 + y^2 = a^2$, or at the boundary of the disk.
To complete the details leading up to equation (9), we verify the norm $\|\mathbf{r}_x \times \mathbf{r}_y\|$:

$$\|\mathbf{r}_x \times \mathbf{r}_y\| = \sqrt{\frac{x^2}{a^2 - x^2 - y^2} + \frac{y^2}{a^2 - x^2 - y^2} + 1} = \sqrt{\frac{x^2 + y^2 + (a^2 - x^2 - y^2)}{a^2 - x^2 - y^2}} = \frac{a}{\sqrt{a^2 - x^2 - y^2}}.$$

Straightforward application of (6) now leads to the result.

**35.** Here $x(\theta) = a\cos\theta$ and $y(\theta) = a\sin\theta$, so $\sqrt{x'(\theta)^2 + y'(\theta)^2} = a$. Application of (14) now gives $A = 2\pi\int_0^{\pi} a^2 \sin\theta\, d\theta = 4\pi a^2$.

**37.** If the line with equation $y = \frac{r}{h}x$, $0 \le x \le h$, is rotated about the $x$-axis to form a conical segment with height $h$ and base radius $r$, its area can be calculated via (14) as

$A = 2\pi\int_0^h \frac{r}{h}x\sqrt{1 + \left(\frac{r}{h}\right)^2}\, dx = \pi r\sqrt{r^2 + h^2} = \pi r l$, where $l = \sqrt{r^2 + h^2}$ is the slant height of the cone.
Applying this result to the case of the triangle $ABC$ being rotated about side $AB$, we see that the area of the generated surface is the sum of the areas of two cones, one with slant height $a$ and height $p$ and one with slant height $b$ and height $p$. The surface area is thus $\pi pa + \pi pb = \pi(a + b)p$.

## Section 11.4 Change-of-Variables Formula for Double Integrals

**1.** $\dfrac{\partial g_1}{\partial u} = 2u;\ \dfrac{\partial g_1}{\partial v} = -2v$

$\dfrac{\partial g_2}{\partial u} = v;\ \dfrac{\partial g_2}{\partial v} = u$

$\mathbf{Dg}(u, v) = \begin{pmatrix} 2u & -2v \\ v & u \end{pmatrix}$

$\dfrac{\partial(x, y)}{\partial(u, v)} = \begin{vmatrix} 2u & -2v \\ v & u \end{vmatrix} = 2u^2 + 2v^2$

**3.** $\dfrac{\partial g_1}{\partial u} = 2u\cos v;\ \dfrac{\partial g_1}{\partial v} = -u^2 \sin v$

$\dfrac{\partial g_2}{\partial u} = 2u\sin v;\ \dfrac{\partial g_2}{\partial v} = u^2 \cos v$

$\mathbf{Dg}(u, v) = \begin{pmatrix} 2u\cos v & -u^2 \sin v \\ 2u\sin v & u^2 \cos v \end{pmatrix}$

$\dfrac{\partial(x, y)}{\partial(u, v)} = \begin{vmatrix} 2u\cos v & -u^2 \sin v \\ 2u\sin v & u^2 \cos v \end{vmatrix} = 2u^3$

**5.** $\dfrac{\partial g_1}{\partial u} = \dfrac{2}{3}u^{-1/3}v^{-2/3};\ \dfrac{\partial g_1}{\partial v} = -\dfrac{2}{3}u^{2/3}v^{-5/3}$

$\dfrac{\partial g_2}{\partial u} = \dfrac{4}{3}u^{1/3}v^{-1/3};\ \dfrac{\partial g_2}{\partial v} = -\dfrac{1}{3}u^{4/3}v^{-4/3}$

$\mathbf{Dg}(u, v) = \begin{pmatrix} \frac{2}{3}u^{-1/3}v^{-2/3} & -\frac{2}{3}u^{2/3}v^{-5/3} \\ \frac{4}{3}u^{1/3}v^{-1/3} & -\frac{1}{3}u^{4/3}v^{-4/3} \end{pmatrix}$

$\dfrac{\partial(x, y)}{\partial(u, v)} = \begin{vmatrix} \frac{2}{3}u^{-1/3}v^{-2/3} & -\frac{2}{3}u^{2/3}v^{-5/3} \\ \frac{4}{3}u^{1/3}v^{\ 1/3} & -\frac{1}{3}u^{4/3}v^{-4/3} \end{vmatrix}$

$= \dfrac{2}{3}uv^{-2}$

**7.** $\mathbf{g}(D) = [0, 2] \times [0, 3]$

**9.** Consider $(u, v) = (t, 1)$ for $1 \le t \le 2$. Then $t = \dfrac{y}{\sqrt{x}}$ and $1 = \dfrac{y}{x^2}$. Solving for $x$ and $y$ in terms of $t$, we get $(t^{2/3}, t^{4/3})$, $1 \le t \le 2$ as one of the boundaries of $\mathbf{g}(D)$. Consider $(u, v) = (2, t)$ for $1 \le t \le 3$. Then $2 = \dfrac{y}{\sqrt{x}}$ and $t = \dfrac{y}{x^2}$. Solving for $x$ and $y$ in terms of $t$, we get $\left(\left(\dfrac{t}{2}\right)^{-2/3}, 2\left(\dfrac{t}{2}\right)^{-1/3}\right)$, $1 \le t \le 3$ as one of the boundaries of $\mathbf{g}(D)$.
Consider $(u, v) = (t, 3)$ for $1 \le t \le 2$. Then $t = \dfrac{y}{\sqrt{x}}$ and $3 = \dfrac{y}{x^2}$. Solving for $x$ and $y$ in terms of $t$, we get $\left(\left(\dfrac{t}{3}\right)^{2/3}, 3\left(\dfrac{t}{3}\right)^{4/3}\right)$, $1 \le t \le 2$ as one of the boundaries of $\mathbf{g}(D)$.
Consider $(u, v) = (1, t)$ for $1 \le t \le 3$. Then $1 = \dfrac{y}{\sqrt{x}}$ and $t = \dfrac{y}{x^2}$. Solving for $x$ and $y$ in terms of $t$, we get $(t^{-2/3}, t^{-1/3})$, $1 \le t \le 3$ as one of the boundaries of $\mathbf{g}(D)$.
Plotting these boundaries, we sketch $\mathbf{g}(D)$.

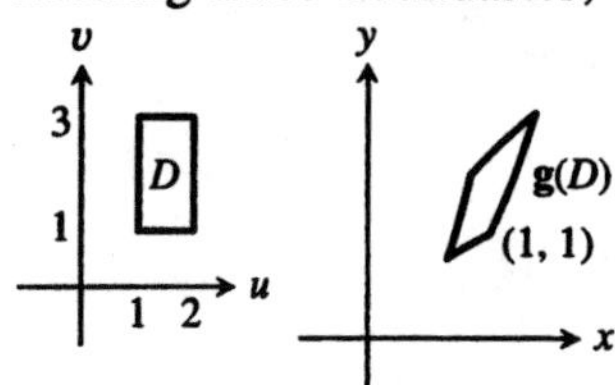


**11.** Consider $(u, v) = (t, 0)$ for $0 \le t \le 1$. Then $(x, y) = (t, t)$, $0 \le t \le 1$ is one of the boundaries of $\mathbf{g}(D)$.
Consider $(u, v) = (1, t)$ for $0 \le t \le 1$. Then $(x, y) = (1 - t, 1 + t)$, $0 \le t \le 1$ is one of the boundaries of $\mathbf{g}(D)$.
Consider $(u, v) = (t, 1)$ for $0 \le t \le 1$. Then $(x, y) = (t - 1, t + 1)$, $0 \le t \le 1$ is one of the boundaries of $\mathbf{g}(D)$.
Consider $(u, v) = (0, t)$ for $0 \le t \le 1$. Then $(x, y) = (-t, t)$, $0 \le t \le 1$ is one of the boundaries of $\mathbf{g}(D)$. Plotting these boundaries, we sketch $\mathbf{g}(D)$.

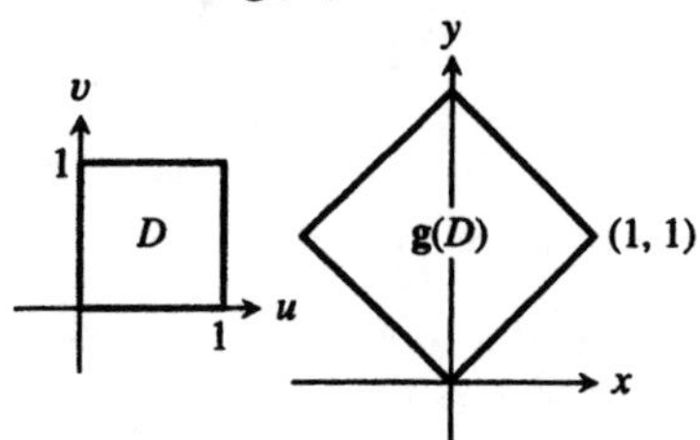

**13.** Let $u = x + y$ and $v = 2x - y$. Thus the region in $D$ is $[0, 1] \times [0, 3]$. Solving for $x$ and $y$ in terms of $u$ and $v$, we get $x = \frac{1}{3}u + \frac{1}{3}v$ and $y = \frac{2}{3}u - \frac{1}{3}v$, so $\mathbf{g}(u, v) = \left\langle \frac{1}{3}u + \frac{1}{3}v, \frac{2}{3}u - \frac{1}{3}v \right\rangle$.

$\frac{\partial g_1}{\partial u} = \frac{1}{3}$; $\frac{\partial g_1}{\partial v} = \frac{1}{3}$; $\frac{\partial g_2}{\partial u} = \frac{2}{3}$; $\frac{\partial g_2}{\partial v} = -\frac{1}{3}$

$$\left|\frac{\partial(x, y)}{\partial(u, v)}\right| = \left|\left(\frac{1}{3}\right)\left(-\frac{1}{3}\right) - \left(\frac{1}{3}\right)\left(\frac{2}{3}\right)\right| = \frac{1}{3}$$

Let $f(x, y) = \sqrt{x + y}$, so $f(\mathbf{g}(u, v)) = \sqrt{u}$.

$$\iint_T \sqrt{x+y}\, dx\, dy = \iint_D \frac{1}{3}\sqrt{u}\, du\, dv$$
$$= \int_0^3 \int_0^1 \frac{1}{3}\sqrt{u}\, du\, dv$$
$$= \int_0^3 \left(\frac{2}{9}u^{3/2}\right)\Big|_0^1 dv$$
$$= \frac{2}{9}\int_0^3 dv$$
$$= \frac{2}{9}(v)\Big|_0^3$$
$$= \frac{2}{3}$$

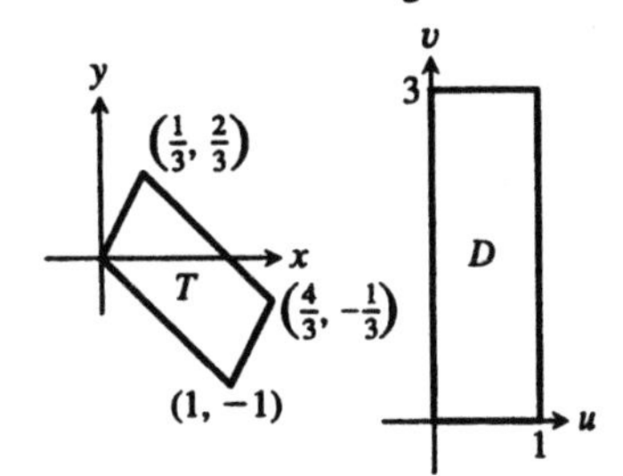


**15.** Let $u = xy$ and $v = \frac{y^2}{x}$. Thus the region in $D$ is $[2, 4] \times [1, 3]$. Solving for $x$ and $y$ in terms of $u$ and $v$, we get $x = u^{2/3}v^{-1/3}$ and $y = u^{1/3}v^{1/3}$, so $\mathbf{g}(u, v) = \left\langle u^{2/3}v^{-1/3}, u^{1/3}v^{1/3} \right\rangle$.

$\frac{\partial g_1}{\partial u} = \frac{2}{3}u^{-1/3}v^{-1/3}$; $\frac{\partial g_1}{\partial v} = -\frac{1}{3}u^{2/3}v^{-4/3}$

$\frac{\partial g_2}{\partial u} = \frac{1}{3}u^{-2/3}v^{1/3}$; $\frac{\partial g_2}{\partial v} = \frac{1}{3}u^{1/3}v^{-2/3}$

$$\left|\frac{\partial(x, y)}{\partial(u, v)}\right| = \left|\frac{2}{9}v^{-1} + \frac{1}{9}v^{-1}\right| = \frac{1}{3}v^{-1}$$

Let $f(x, y) = x^2y^2$, so $f(\mathbf{g}(u, v)) = u^2$.

$$\iint_T x^2y^2\, dx\, dy = \iint_D \frac{1}{3}u^2v^{-1}\, du\, dv$$
$$= \int_1^3 \int_2^4 \frac{1}{3}u^2v^{-1}\, du\, dv$$
$$= \int_1^3 \left(\frac{1}{9}u^3v^{-1}\right)\Big|_2^4 dv$$
$$= \frac{56}{9}\int_1^3 v^{-1}\, dv$$
$$= \frac{56}{9}(\ln v)\Big|_1^3$$
$$= \frac{56}{9}\ln 3$$

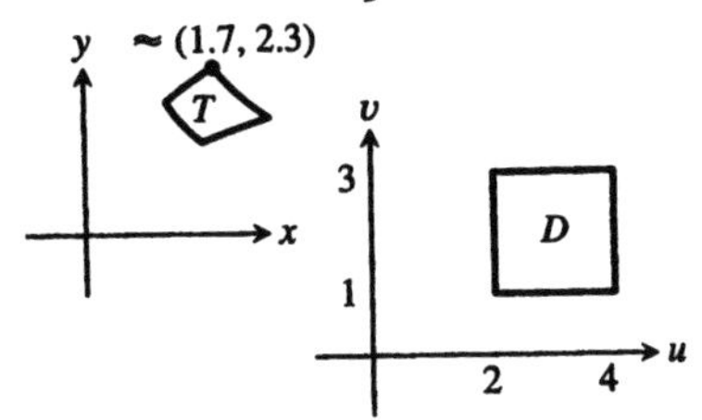


**17.** Substitute $x = \frac{1}{2}u + \frac{1}{\sqrt{6}}v$ and $y = -\frac{1}{2}u + \frac{1}{\sqrt{6}}v$ into the equation of the ellipse and simplify to get $u^2 + v^2 = 6$, so $D$ is bounded by the circle centered at the origin with radius $\sqrt{6}$.

$$\mathbf{g}(u, v) = \left\langle \frac{1}{2}u + \frac{1}{\sqrt{6}}v, -\frac{1}{2}u + \frac{1}{\sqrt{6}}v \right\rangle$$

$\frac{\partial g_1}{\partial u} = \frac{1}{2}$; $\frac{\partial g_1}{\partial v} = \frac{1}{\sqrt{6}}$; $\frac{\partial g_2}{\partial u} = -\frac{1}{2}$; $\frac{\partial g_2}{\partial v} = \frac{1}{\sqrt{6}}$

$$\left|\frac{\partial(x, y)}{\partial(u, v)}\right| = \left|\frac{1}{2\sqrt{6}} + \frac{1}{2\sqrt{6}}\right| = \frac{1}{\sqrt{6}}$$

Let $f(x, y) = xy$, so $f(\mathbf{g}(u, v)) = -\frac{1}{4}u^2 + \frac{1}{6}v^2$.

$$\iint_T xy\, dx\, dy = \frac{1}{\sqrt{6}}\iint_D \left(-\frac{1}{4}u^2 + \frac{1}{6}v^2\right) du\, dv$$

Since $D$ is a disk, we evaluate the integral by changing to polar coordinates. Let $u = r\cos\theta$, $v = r\sin\theta$, and $du\, dv = r\, dr\, d\theta$.

$$\iint_T xy\,dx\,dy = \frac{1}{\sqrt{6}}\int_0^{2\pi}\int_0^{\sqrt{6}}\left(-\frac{1}{4}r^2\cos^2\theta+\frac{1}{6}r^2\sin^2\theta\right)r\,dr\,d\theta$$
$$= \frac{1}{\sqrt{6}}\int_0^{2\pi}\int_0^{\sqrt{6}}\left(-\frac{1}{4}\cos^2\theta+\frac{1}{6}\sin^2\theta\right)r^3dr\,d\theta$$
$$= \frac{1}{\sqrt{6}}\int_0^{2\pi}\left(-\frac{1}{4}\cos^2\theta+\frac{1}{6}\sin^2\theta\right)\frac{1}{4}r^4\bigg|_0^{\sqrt{6}}d\theta$$
$$= \frac{9}{\sqrt{6}}\int_0^{2\pi}\left(-\frac{1}{4}\cos^2\theta+\frac{1}{6}\sin^2\theta\right)d\theta$$
$$= -\frac{9}{\sqrt{6}}\int_0^{2\pi}\left(\frac{1}{24}+\frac{5}{24}\cos 2\theta\right)d\theta$$
$$= -\frac{9}{\sqrt{6}}\left(\frac{1}{24}\theta+\frac{5}{48}\sin 2\theta\right)\bigg|_0^{2\pi}$$
$$= -\frac{3\pi}{4\sqrt{6}}$$
$$= -\frac{\pi}{4}\sqrt{\frac{3}{2}}$$

$5x^2 + 2xy + 5y^2 = 12$

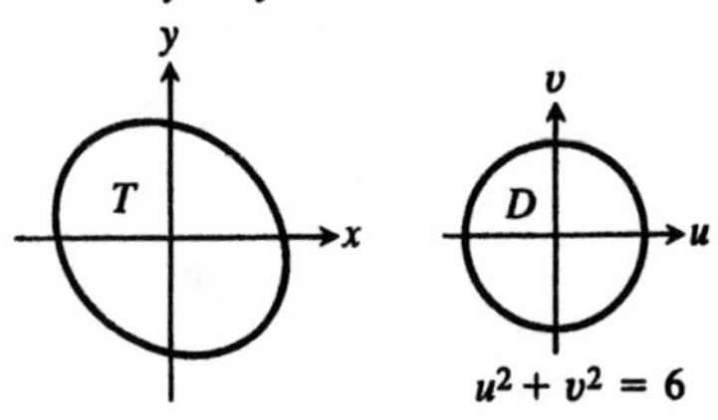


**19.** Let $x = r\cos\theta$ and $y = r\sin\theta$, so $\mathbf{g}(r, \theta) = \langle r\cos\theta, r\sin\theta\rangle$. Let $D$ be $[0, 1]\times\left[0, \frac{\pi}{4}\right]$.

$$\left|\frac{\partial(x, y)}{\partial(r, \theta)}\right| = |r|$$

Let $f(x, y) = \sqrt{x^2+y^2}$, so $f(\mathbf{g}(r, \theta)) = |r|$.

$$\iint_T \sqrt{x^2+y^2}\,dx\,dy = \iint_D r^2 dr\,d\theta = \int_0^{\pi/4}\int_0^1 r^2 dr\,d\theta = \int_0^{\pi/4}\left(\frac{1}{3}r^3\right)\bigg|_0^1 d\theta = \frac{1}{3}\int_0^{\pi/4}d\theta = \frac{1}{3}(\theta)\bigg|_0^{\pi/4} = \frac{\pi}{12}$$

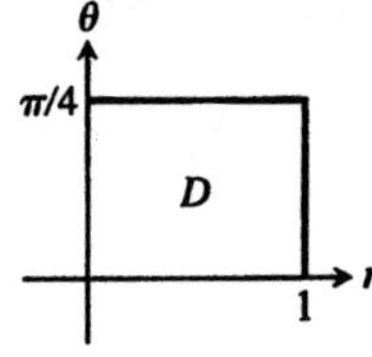


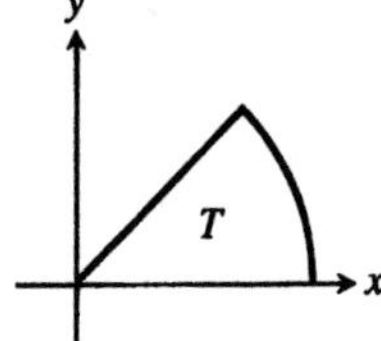

**21.** Describe the surface by
$\mathbf{r}(u, v) = \left\langle u\cos v, u\sin v, \sqrt{1-u^2}\right\rangle$, $0 \le u \le 0.1$,
$0 \le v \le 2\pi$.

$$\|\mathbf{r}_u \times \mathbf{r}_v\|$$
$$= \left\|\left\langle \cos v, \sin v, \frac{-u}{\sqrt{1-u^2}}\right\rangle \times \langle -u\sin v, u\cos v, 0\rangle\right\|$$
$$= \left\|\left\langle \frac{u^2}{\sqrt{1-u^2}}\cos v, \frac{u^2}{\sqrt{1-u^2}}\sin v, u\right\rangle\right\|$$
$$= \frac{u}{\sqrt{1-u^2}}$$
$$A = \int_0^{2\pi}\int_0^{0.1} \frac{u}{\sqrt{1-u^2}}\,du\,dv$$
$$= \int_0^{2\pi}\left(-\sqrt{1-u^2}\right)\Big|_0^{0.1} dv$$
$$= \int_0^{2\pi}\left(1-\frac{3\sqrt{11}}{10}\right)dv$$
$$= \left(1-\frac{3\sqrt{11}}{10}\right)(v)\Big|_0^{2\pi}$$
$$= 2\pi\left(1-\frac{3\sqrt{11}}{10}\right) \text{ square units}$$

Note that you can first describe the surface by
$\mathbf{r}(x, y) = \left\langle x, y, \sqrt{1-x^2-y^2}\right\rangle$, $x^2+y^2 \le 0.1^2$,
and then perform a change of variables into polar coordinates.

**23.** Let $T$ be the disk with boundary
$x^2 + y^2 - 2y = 0$.

$$V = \iint_T \left(2\sqrt{x^2+y^2}\right)dx\,dy$$

Make a change of variables into polar coordinates. The equation $x^2+y^2-2y=0$ can be written in polar coordinates as $r = 2\sin\theta$. So let $D$ be the region defined by $0 \le r \le 2\sin\theta$, $0 \le \theta \le \pi$. Thus,

$$V = \iint_D (2r)r\,dr\,d\theta$$
$$= 2\int_0^{\pi}\int_0^{2\sin\theta} r^2\,dr\,d\theta$$
$$= 2\int_0^{\pi}\left(\frac{1}{3}r^3\right)\Big|_0^{2\sin\theta} d\theta$$
$$= \frac{16}{3}\int_0^{\pi}\sin^3\theta\,d\theta$$
$$= \frac{16}{3}\int_0^{\pi}\sin\theta(1-\cos^2\theta)d\theta$$
$$= \frac{16}{3}\left(-\cos\theta + \frac{1}{3}\cos^3\theta\right)\Big|_0^{\pi}$$
$$= \frac{64}{9} \text{ cubic units}$$

**25.** $\displaystyle\int_0^{2\pi}\sin^2\theta\cos^2\theta\,d\theta$

$$= \left(-\frac{\sin\theta\cos^3\theta}{4}\right)\Big|_0^{2\pi} + \frac{1}{4}\int_0^{2\pi}\cos^2\theta\,d\theta$$
$$= \frac{1}{4}\int_0^{2\pi}\cos^2\theta\,d\theta$$
$$= \frac{1}{8}\int_0^{2\pi}(1+\cos 2\theta)d\theta$$
$$= \frac{1}{8}\left(\theta + \frac{1}{2}\sin 2\theta\right)\Big|_0^{2\pi}$$
$$= \frac{\pi}{4}$$

**27.** One change of variable is given by the equations $X = r\cos\theta$ and $Y = r\sin\theta$ and the other is given by $x = aX$ and $y = bY$. The combination of the two changes is $(x, y) = (ar\cos\theta, br\sin\theta)$.

$$\frac{\partial(x, y)}{\partial(r, \theta)} = \begin{vmatrix} a\cos\theta & b\sin\theta \\ -ar\sin\theta & br\cos\theta \end{vmatrix} = abr$$

The Jacobian of the combination is the product of the two original Jacobians.

**29.** Suppose $g$ is strictly increasing on $[a, b]$. If $S_n$ consists of the points
$x_0 = a$, $x_1 = a+\delta$, $x_2 = a+2\delta$, …, $x_n = b$,
where $\delta = \dfrac{b-a}{n}$, then $x_i < x_{i+1}$ for $0 \le i < n$, and consequently $g(x_i) < g(x_{i+1})$, $0 \le i < n$. Therefore, the set $\{g(x_i)\}$ is a subdivision of $[c, d]$, assuming $g(a) = c$ and $g(b) = d$. To see

that $G_n$ need not be regular, let $g(x) = x^2$, $[a, b] = [c, d] = [0, 1]$ and $n \geq 2$; then

$$S_n = \left\{\frac{i}{n}\right\}_{i=0}^{n} \text{ is regular, while } G_n = \left\{\left(\frac{i}{n}\right)^2\right\}_{i=0}^{n}$$

is not. The fact that the lengths of the subintervals of $G_n$ go to zero as $n \to \infty$ follows from the fact that from the Mean-value Theorem $g(x_{i+1}) - g(x_i) = g'(c_i)(x_{i+1} - x_i)$ for some $c_i \in [x_i, x_{i+1}]$, so that with $g'(c_i)$ bounded due to the continuity of $g$, $g(x_{i+1}) - g(x_i)$ goes to zero as $x_{i+1} - x_i$ goes to zero. A similar argument applies if $g$ is strictly decreasing.

**31.** First we wish to find the radius of the circle for the congruent arcs. Let $r$ be the radius of the circle. Then $\left(r - \frac{h-a}{2}\right)^2 + \left(\frac{w}{2}\right)^2 = r^2$.

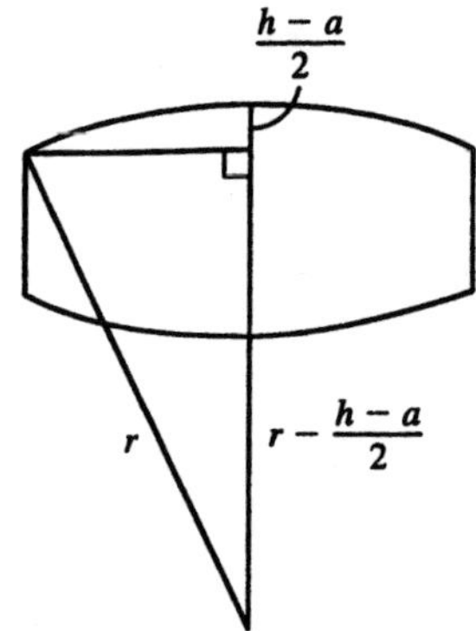


Substitute $a = 2$, $h = 4$, and $w = 10$ into this equation and solve for $r$ to get $r = 13$. Let $S$ be the upper half of the sphere $x^2 + y^2 + z^2 = 169$ inside the cylinder $x^2 + y^2 = 25$. Then the surface area of $S$ is the surface area of the top and bottom of the aspirin tablet. Let $D$ be the region $\{(x, y)\text{: } x^2 + y^2 \leq 25\}$ and $f(x, y) = \sqrt{169 - x^2 - y^2}$ describes the upper half of the sphere.

$$f_x = -\frac{x}{\sqrt{169 - x^2 - y^2}} \text{ and}$$

$$f_y = -\frac{y}{\sqrt{169 - x^2 - y^2}}.$$

$$S = \iint_D \sqrt{1 + f_x^2 + f_y^2}\, dx\, dy$$

$$= \iint_D \frac{13}{\sqrt{169 - x^2 - y^2}}\, dx\, dy$$

Use polar coordinates.

$$S = \int_0^{2\pi}\int_0^5 \frac{13}{\sqrt{169 - r^2}}\, r\, dr\, d\theta$$

$$= \int_0^{2\pi} (-13(169 - r^2)^{1/2})\Big|_0^5\, d\theta$$

$$= \int_0^{2\pi} 13\, d\theta$$

$$= 26\pi$$

The area of the cylindrical side is $\pi wa = 20\pi$. Thus the total surface area of the tablet is $2(26\pi) + 20\pi = 72\pi$ mm$^2$.

**33.** The triangular region is defined by $T = \{(x, y)\text{: } 0 \leq y \leq x, 0 \leq x \leq 1\}$. Using polar coordinates, let $x = r\cos\theta$ and $y = r\sin\theta$, so when $y = x$, $\theta = \frac{\pi}{4}$ and when $x = 1$, $r = \frac{1}{\cos\theta}$. Then

$$D = \left\{(r, \theta)\text{: } 0 \leq r \leq \frac{1}{\cos\theta}, 0 \leq \theta \leq \frac{\pi}{4}\right\}.$$

$$\iint_T (1 + x^2 + y^2)^{-2}\, dx\, dy$$

$$= \iint_D (1 + r^2)^{-2}\, r\, dr\, d\theta$$

$$= \int_0^{\pi/4}\int_0^{1/\cos\theta} r(1 + r^2)^{-2}\, dr\, d\theta$$

$$= \int_0^{\pi/2}\left(-\frac{1}{2}(1 + r^2)^{-1}\right)\Bigg|_0^{1/\cos\theta} d\theta$$

$$= -\frac{1}{2}\int_0^{\pi/4}\left(\frac{\cos^2\theta}{1 + \cos^2\theta} - 1\right)d\theta$$

$$= \frac{1}{2}\int_0^{\pi/4}\left(-\frac{\cos^2\theta}{1 + \cos^2\theta} + 1\right)d\theta$$

$$= \frac{1}{2}\int_0^{\pi/4}\frac{1}{1 + \cos^2\theta}\, d\theta$$

$$\approx 0.217605$$

**35.** Using polar coordinates, let $x = r\cos\theta$ and $y = r\sin\theta$, so when $y = x$, $\theta = \frac{\pi}{4}$ and when $y = x^2$, $r = \frac{\sin\theta}{\cos^2\theta}$.

Then $D = \left\{(r, \theta)\text{: } 0 \leq r \leq \frac{\sin\theta}{\cos^2\theta}, 0 \leq \theta \leq \frac{\pi}{4}\right\}$.

$\int_0^1 \int_{x^2}^x (x^2+y^2)^{-1/2}\,dx\,dy$

$= \int_0^{\pi/4} \int_0^{\sin\theta/\cos^2\theta} dr\,d\theta$

$= \int_0^{\pi/2} (r)\Big|_0^{\sin\theta/\cos^2\theta} d\theta$

$= \int_0^{\pi/4} \left(\frac{\sin\theta}{\cos^2\theta}\right) d\theta$

$= \left(\frac{1}{\cos\theta}\right)\Big|_0^{\pi/4}$

$= \sqrt{2}-1$

$\approx 0.4142$

**37.** Let $x = r\cos\theta$ and $y = r\sin\theta$. An equation for the cylinder in polar coordinates is $r = 2\sin\theta$. Thus, we can describe the surface by $\mathbf{r}(r,\theta) = \langle r\cos\theta, r\sin\theta, 2r\rangle$, $0 \le r \le 2\sin\theta$, $0 \le \theta \le \pi$.

$\|\mathbf{r}_r \times \mathbf{r}_\theta\| = \|\langle\cos\theta, \sin\theta, 2\rangle \times \langle -r\sin\theta, r\cos\theta, 0\rangle\| = \|\langle -2r\cos\theta, -2r\sin\theta, r\rangle\| = \sqrt{5}r$

$A = \int_0^{\pi} \int_0^{2\sin\theta} \sqrt{5}r\,dr\,d\theta$

$= \int_0^{\pi} \left(\frac{1}{2}\sqrt{5}r^2\right)\Big|_0^{2\sin\theta} d\theta$

$= 2\sqrt{5}\int_0^{\pi} \sin^2\theta\,d\theta$

$= 2\sqrt{5}\left(\frac{1}{2}\theta - \frac{1}{4}\sin 2\theta\right)\Big|_0^{\pi}$

$= \sqrt{5}\pi$ square units

## Section 11.5 Triple Integrals

**1.** Each subregion has a volume of $\frac{1}{4}$ units$^3$. Function values at the evaluation set are $\delta\left(\frac{5}{4}, \frac{3}{2}, \frac{5}{4}\right) = \frac{225}{64}$, $\delta\left(\frac{5}{4}, \frac{3}{2}, \frac{7}{4}\right) = \frac{315}{64}$, $\delta\left(\frac{5}{4}, \frac{5}{2}, \frac{5}{4}\right) = \frac{625}{64}$, $\delta\left(\frac{5}{4}, \frac{5}{2}, \frac{7}{4}\right) = \frac{875}{64}$, $\delta\left(\frac{7}{4}, \frac{3}{2}, \frac{5}{4}\right) = \frac{315}{64}$, $\delta\left(\frac{7}{4}, \frac{3}{2}, \frac{7}{4}\right) = \frac{441}{64}$, $\delta\left(\frac{7}{4}, \frac{5}{2}, \frac{5}{4}\right) = \frac{875}{64}$, $\delta\left(\frac{7}{4}, \frac{5}{2}, \frac{7}{4}\right) = \frac{1225}{64}$.

Adding these elements and multiplying by $\frac{1}{4}$ yields $m \approx 19.125$. This compares with

$$\text{true mass} = \int_1^2 \int_1^3 \int_1^2 xy^2 z\,dx\,dy\,dz$$
$$= \int_1^2 \int_1^3 \left(\frac{1}{2}x^2y^2z\right)\Big|_1^2 dy\,dz$$
$$= \int_1^2 \int_1^3 \frac{3}{2}y^2 z\,dy\,dz$$
$$= \int_1^2 \left(\frac{1}{2}y^3 z\right)\Big|_1^3 dz$$
$$= \int_1^2 13z\,dz$$
$$= \left(\frac{13}{2}z^2\right)\Big|_1^2$$
$$= \frac{39}{2}$$
$$= 19.5$$

3. An analogue of the iterated integral theorem for Type II regions, including the definition of such a region, is as follows:
Assume that $f$ is continuous on the Type II region
$B = \{(x, y, z) \in R^3: (x, z) \in D,\ p(x, z) \le y \le q(x, z)\}$, where $p$ and $q$ are bounded, continuous functions defined on the bounded region $D$ of the $(x, z)$-plane. Then $f$ is integrable on $B$ and
$$\iiint_B f(x, y, z)dV = \iint_D \left(\int_{p(x,z)}^{q(x,z)} f(x, y, z)dy\right)dx\,dz.$$

5.

$$V = \iint_D \left(\int_{\frac{1}{2}\sqrt{1-y}}^{\sqrt{1-y}} 1\,dz\right)dx\,dy$$
$$= \iint_D \frac{1}{2}\sqrt{1-y}\,dx\,dy$$
$$= \int_0^1 \int_0^{1+y^2} \frac{1}{2}\sqrt{1-y}\,dx\,dy$$
$$= \int_0^1 \left(\frac{1}{2}x\sqrt{1-y}\right)\Big|_0^{1+y^2} dy$$
$$= \frac{1}{2}\int_0^1 (1+y^2)\sqrt{1-y}\,dy$$

Note that this last integral is $\frac{1}{2}$ the integral in Example 1, so $V = \frac{43}{105}$ cubic units.

7. The region $D$ should be in the $(y, z)$-plane and limits in $x$ should be in terms of $y$ and $z$. Here, this can be accomplished by simply reordering the equations: $D$ is defined by $0 \le y \le 1$ and $\frac{1}{2}\sqrt{1-y} \le z \le \sqrt{1-y}$, while $0 \le x \le y^2 + 1$.
The corresponding integral is
$$V = \int_0^1 \int_{\frac{1}{2}\sqrt{1-y}}^{\sqrt{1-y}} \int_0^{y^2+1} dx\,dz\,dy$$
$$= \frac{43}{105} \text{ cubic units.}$$

9. $$V = \int_0^1 \int_0^{2(1-x)} \int_0^{2(1-x-y/2)} dz\,dy\,dx$$
$$= \int_0^1 \int_0^{2(1-x)} 2\left(1 - x - \frac{1}{2}y\right)dy\,dx$$
$$= 2\int_0^1 \left(y - xy - \frac{1}{4}y^2\right)\Big|_0^{2(1-x)} dx$$
$$= 2\int_0^2 (1-x)^2 dx$$
$$= -\frac{2}{3}(1-x)^3\Big|_0^1$$
$$= \frac{2}{3} \text{ cubic units}$$

**11.** $V = \iint_D \left( \int_0^{\sqrt{1-y}} 1\,dz \right) dx\,dy$

$= \iint_D \sqrt{1-y}\,dx\,dy$

$= \int_0^1 \int_0^{y+1} \sqrt{1-y}\,dx\,dy$

$= \int_0^1 \left( x\sqrt{1-y} \right)\Big|_0^{y+1} dy$

$= \int_0^1 (y+1)\sqrt{1-y}\,dy$

Make the substitution $w = 1 - y$, so $dw = -dy$.

$V = -\int_1^0 (-w+2)\sqrt{w}\,dw$

$= \int_0^1 (-w^{3/2} + 2w^{1/2})dw$

$= \left( -\frac{2}{5}w^{5/2} + \frac{4}{3}w^{3/2} \right)\Big|_0^1$

$= \frac{14}{15}$ cubic units

**13.** Equations for the boundary of $D$ are $x = 0$, $y = 0$, and $x + y = 1$

$V = \iint_D \int_0^{x^2+2y^2} 1\,dz\,dx\,dy$

$= \int_0^1 \int_0^{1-y} x^2 + 2y^2 dx\,dy$

$= \int_0^1 \left( \frac{1}{3}x^3 + 2xy^2 \right)\Big|_0^{1-y} dy$

$= \int_0^1 \left( -\frac{7}{3}y^3 + 3y^2 - y + \frac{1}{3} \right) dy$

$= \left( -\frac{7}{12}y^4 + y^3 - \frac{1}{2}y^2 + \frac{1}{3}y \right)\Big|_0^1$

$= \frac{1}{4}$ cubic units

**15.** Two vectors parallel to the upper plane are $\langle 2, 1, -3 \rangle$ and $\langle 0, 2, -3 \rangle$. Thus a vector normal to the plane is $\langle 2, 1, 3 \rangle \times \langle 0, 2, -3 \rangle = \langle 3, 6, 4 \rangle$. An equation for the plane is

$3x + 6y + 4(z - 3) = 0$ or $z = \frac{3}{4}(4 - x - 2y)$.

The region in the $(x, y)$-plane is bounded by the lines $x = 0$, $y = \frac{1}{2}x$, and $y = -\frac{1}{2}x + 2$.

$V = \int_0^2 \int_{x/2}^{-x/2+2} \int_0^{3(4-x-2y)/4} dz\,dy\,dx$

$= \int_0^2 \int_{x/2}^{-x/2+2} \frac{3}{4}(4 - x - 2y)dy\,dx$

$= \frac{3}{4}\int_0^2 (4y - xy - y^2)\Big|_{x/2}^{-x/2+2} dx$

$= \frac{3}{4}\int_0^2 (x^2 - 4x + 4)dx$

$= \frac{3}{4}\left( \frac{1}{3}x^3 - 2x^2 + 4x \right)\Big|_0^2$

$= 2$ cubic units

**17.** $V = \int_{-2}^2 \int_{y^2/2}^2 \int_0^{2-z} dx\,dz\,dy$

$= \int_{-2}^2 \int_{y^2/2}^2 (2 - z)dz\,dy$

$= \int_{-2}^2 \left( 2z - \frac{1}{2}z^2 \right)\Big|_{y^2/2}^2 dy$

$= \int_{-2}^2 \left( \frac{1}{8}y^4 - y^2 + 2 \right) dy$

$= \left( \frac{1}{40}y^5 - \frac{1}{3}y^3 + 2y \right)\Big|_{-2}^2$

$= \frac{64}{15}$ cubic units

**19.** Using a CAS, we find that, with $\delta = k(c - z)$,

$$m = \int_0^a \int_0^{b(1-x/a)} \int_0^{c(1-x/a-y/b)} k(c-z)\,dz\,dy\,dx = \frac{1}{8}kabc^2,$$

while the center of mass is

$$\mathbf{R} = \frac{1}{m}\int_0^a \int_a^{b(1-x/a)} \int_0^{c(1-x/a-y/b)} k(c-z)\langle x, y, z\rangle dz\,dy\,dx = \left\langle \frac{4a}{15}, \frac{4b}{15}, \frac{c}{5}\right\rangle = \frac{1}{15}\langle 4a, 4b, 3c\rangle.$$

**21.**
$$\begin{aligned}\int_0^1\int_0^{1-x}\int_0^{1-x-y} xy\,dz\,dy\,dx &= \int_0^1\int_0^{1-x}(xy - x^2y - xy^2)dy\,dx \\ &= \int_0^1 \left(\frac{1}{2}xy^2 - \frac{1}{2}x^2y^2 - \frac{1}{3}xy^3\right)\Bigg|_0^{1-x} dx \\ &= \int_0^1 \left(\frac{1}{6}x - \frac{1}{2}x^2 + \frac{1}{2}x^3 - \frac{1}{6}x^4\right)dx \\ &= \left(\frac{1}{12}x^2 - \frac{1}{6}x^3 + \frac{1}{8}x^4 - \frac{1}{30}x^5\right)\Bigg|_0^1 \\ &= \frac{1}{120}\end{aligned}$$

**23.**
$$\begin{aligned}\int_0^3\int_0^{1-z/3}\int_0^2 x\,dx\,dy\,dz &= \int_0^3\int_0^{1-z/3}\left(\frac{1}{2}x^2\right)\Bigg|_0^2 dy\,dz \\ &= \int_0^3\int_0^{1-z/3} 2\,dy\,dz \\ &= 2\int_0^3 (y)\big|_0^{1-z/3} dz \\ &= 2\int_0^3\left(1 - \frac{1}{3}z\right)dz \\ &= 2\left(z - \frac{1}{6}z^2\right)\Bigg|_0^3 \\ &= 3\end{aligned}$$

**25.**
$$\begin{aligned}\int_0^\pi\int_0^{\pi-x}\int_0^{xy} \sin x\,dz\,dy\,dx &= \int_0^\pi\int_0^{\pi-x}(z\sin x)\big|_0^{xy} dy\,dx \\ &= \int_0^\pi\int_0^{\pi-x} xy\sin x\,dy\,dx \\ &= \int_0^\pi\left(\frac{1}{2}xy^2\sin x\right)\Bigg|_0^{\pi-x} dx \\ &= \int_0^\pi\left(\frac{1}{2}x(\pi-x)^2\sin x\right)dx \\ &= \frac{1}{2}\int_0^\pi (x^3 - 2\pi x^2 + \pi^2 x)\sin x\,dx\end{aligned}$$

Use a CAS or formula (25) and (26) from the Table of Integrals to get $\int_0^\pi\int_0^{\pi-x}\int_0^{xy}\sin x\,dz\,dy\,dx = \pi.$

**27.** The cone $B$ can be defined by the equations $-a \le x \le a$, $-\sqrt{a^2-x^2} \le y \le \sqrt{a^2-x^2}$, and $z = h\left(1 - \dfrac{\sqrt{x^2+y^2}}{a}\right)$. The density is proportional to $z$, $\delta = kz$. Therefore, the mass of the cone is

$$m = \int_{-a}^{a}\int_{-\sqrt{a^2-x^2}}^{\sqrt{a^2-x^2}}\int_0^{h\left(1-\sqrt{x^2+y^2}/a\right)} kz\,dz\,dy\,dx$$

and its center of mass is

$$\mathbf{R} = \frac{1}{m}\int_{-a}^{a}\int_{-\sqrt{a^2-x^2}}^{\sqrt{a^2-x^2}}\int_0^{h\left(1-\sqrt{x^2+y^2}/a\right)} kz\langle x,\ y,\ z\rangle\,dz\,dy\,dx = \left\langle 0,\ 0,\ \frac{2h}{5}\right\rangle$$

This integral is more easily evaluated using cylindrical coordinates, as described in the next section.

**29.** When $4 - x^2 = 2 + x^2$, $x = -1$ or $x = 1$.

$$V = \int_{-1}^{3}\int_{-1}^{1}\int_{2+x^2}^{4-x^2} dz\,dx\,dy = \int_{-1}^{3}\int_{-1}^{1}(2-2x^2)\,dx\,dy = \int_{-1}^{3}\left(2x - \frac{2}{3}x^3\right)\Bigg|_{-1}^{1} dy = \int_{-1}^{3}\frac{8}{3}\,dy = \left(\frac{8}{3}y\right)\Bigg|_{-1}^{3} = \frac{32}{3} \text{ cubic units}$$

**31.** $$W = g\delta\int_{-a}^{a}\int_{-\sqrt{a^2-x^2}}^{\sqrt{a^2-x^2}}\int_0^{h}(h-z)\,dz\,dy\,dx = \frac{a^2h^2\pi g\delta}{2} \text{ Joules}$$

**33.** Let $P = (p_1, p_2, p_3)$. The force needed to pump a particle $dV$ at $(x, y, z)$ is $\langle 0, 0, g\rangle\delta\,dV$ where $g$ is the acceleration due to gravity. Thus work is $\langle |p_1 - x|, |p_2 - y|, |p_3 - z|\rangle \cdot \langle 0, 0, g\rangle\delta dV = (p_3 - z)g\delta dV$.

The total work over $B$ is $\iiint\limits_B (p_3 - z)g\delta dV = gp_3\iiint\limits_B \delta dV - g\iiint\limits_B z\delta dV$.

$\iiint\limits_B \delta\,dV = \delta V$ is the mass of the region and $Z = \dfrac{1}{\delta V}\iiint\limits_B z\delta\,dV$, so $Z\delta V = \iiint\limits_B z\delta\,dV$.

Thus, the total work over $B$ is $gp_3\delta V - gZ\delta V = g(p_3 - Z)\delta V$.

This is the work required to lift a particle of mass $\delta V$ from the center of mass of $B(X, Y, Z)$ to $P$.

**35.** Over the subset $D_1$ in the $(x, z)$-plane, $B$ is between the graphs of $y = 0$ and $y = 1 - z^2$. $D_1$ is bounded by the curves $x = 0$, $x = 1$, $z = 0$, and $z = 1$. Over the subregion $D_2$ in the $(x, z)$-plane, $B$ is between the graphs of $y = \sqrt{x-1}$ and $y = 1 - z^2$. When $\sqrt{x-1} = 1 - z^2$, $z = \sqrt{1-\sqrt{x-1}}$. $D_2$ is the region bounded by the curves $x = 1$, $z = \sqrt{1-\sqrt{x-1}}$, and $z = 0$.

$$V = \iint\limits_{D_1}\int_0^{1-z^2} dy\,dz\,dx + \iint\limits_{D_2}\int_{\sqrt{x-1}}^{1-z^2} dy\,dz\,dx = \int_0^1\int_0^1\int_0^{1-z^2} dy\,dz\,dx + \int_1^2\int_0^{\sqrt{1-\sqrt{x-1}}}\int_{\sqrt{x-1}}^{1-z^2} dy\,dz\,dx$$

## Section 11.6 Change-of-Variables Formula for Triple Integrals

**1.** $\mathbf{g}(1, 0, 0) = \langle 1, 1, 0\rangle$; $\mathbf{g}(0, 1, 0) = \langle 1, 1, 1\rangle$; $\mathbf{g}(0, 0, 1) = \langle 1, 0, 1\rangle$

The volume is

$$|(\mathbf{g}(1, 0, 0) \times \mathbf{g}(0, 1, 0)) \cdot \mathbf{g}(0, 0, 1)|$$
$$= |(\langle 1, 1, 0\rangle \times \langle 1, 1, 1\rangle) \cdot \langle 1, 0, 1\rangle|$$
$$= |\langle 1, -1, 0\rangle \cdot \langle 1, 0, 1\rangle|$$
$$= 1.$$

$$\frac{\partial x}{\partial u} = 1;\ \frac{\partial x}{\partial v} = 1;\ \frac{\partial x}{\partial w} = 1$$
$$\frac{\partial y}{\partial u} = 1;\ \frac{\partial y}{\partial v} = 1;\ \frac{\partial y}{\partial w} = 0$$
$$\frac{\partial z}{\partial u} = 0;\ \frac{\partial z}{\partial v} = 1;\ \frac{\partial z}{\partial w} = 1$$

$$\left|\frac{\partial(x, y, z)}{\partial(u, v, w)}\right| = \begin{vmatrix} 1 & 1 & 1 \\ 1 & 1 & 0 \\ 0 & 1 & 1 \end{vmatrix} = 1$$

Then

$$\iiint_{\mathbf{g}(D)} 1\,dx\,dy\,dz = \iiint_{D} \left|\frac{\partial(x, y, z)}{\partial(u, v, w)}\right| du\,dv\,dw = 1.$$

**3.** $$V = \int_0^1 \int_0^{\pi^2 r/9} \int_{\sqrt{z/r}}^{\pi/3} r\,d\theta\,dz\,dr$$
$$= \int_0^1 \int_0^{\pi^2 r/9} (r\theta)\Big|_{\sqrt{z/r}}^{\pi/3} dz\,dr$$
$$= \int_0^1 \int_0^{\pi^2 r/9} \left(\frac{\pi}{3} r - \sqrt{rz}\right) dz\,dr$$
$$= \int_0^1 \left(\frac{\pi}{3} rz - \frac{2}{3}\sqrt{r}(z)^{3/2}\right)\Bigg|_0^{\pi^2 r/9} dr$$
$$= \int_0^1 \left(\frac{\pi^3}{27} r^2 - \frac{2\pi^3}{81} r^2\right) dr$$
$$= \int_0^1 \frac{\pi^3}{81} r^2\,dr$$
$$= \left(\frac{\pi^3}{243} r^3\right)\Bigg|_0^1$$
$$= \frac{\pi}{243}$$

**5.** $$V = 2\int_0^{\pi/2} \int_0^{a\cos\theta} \int_{-\sqrt{a^2-r^2}}^{\sqrt{a^2-r^2}} r\,dz\,dr\,d\theta$$
$$= 4\int_0^{\pi/2} \int_0^{a\cos\theta} r\sqrt{a^2 - r^2}\,dr\,d\theta$$
$$= -\frac{4}{3}\int_0^{\pi/2} a^3((1-\cos^2\theta)\sin\theta - 1)\,d\theta$$
$$= \frac{2a^3}{9}(3\pi - 4) \text{ cubic units}$$

**7.** The surfaces $z = r^2$ and $z = 8 - r^2$ intersect when $r = 2$. Therefore, the region has volume

$$V = \int_0^{2\pi} \int_0^2 \int_{r^2}^{8-r^2} r\,dz\,dr\,d\theta$$
$$= \int_0^{2\pi} \int_0^2 8r - 2r^3\,dr\,d\theta$$
$$= 16\pi \text{ cubic units.}$$

**9.** $$m = \int_0^h \int_0^{2\pi} \int_0^a kr \cdot r\,dr\,d\theta\,dz$$
$$= k\int_0^h \int_0^{2\pi} \int_0^a r^2 dr\,d\theta\,dz$$
$$= k\int_0^h \int_0^{2\pi} \left(\frac{1}{3} r^3\right)\Bigg|_0^a d\theta\,dz$$
$$= k\int_0^h \int_0^{2\pi} \frac{1}{3} a^3 d\theta\,dz$$
$$= \frac{2}{3} ka^3 h\pi \text{ mass units}$$

**11.** From Example 5, $m = V = \frac{2}{3}\pi a^3(1-\cos b)$.

By symmetry, $X = Y = 0$.

$$Z = \frac{1}{m}\int_0^{2\pi}\int_0^b\int_0^a (\rho\cos\phi)\rho^2 \sin\phi\, d\rho\, d\phi\, d\theta$$
$$= \frac{1}{m}\int_0^{2\pi}\int_0^b\int_0^a \rho^3 \cos\phi\sin\phi\, d\rho\, d\phi\, d\theta$$
$$= \frac{1}{m}\int_0^{2\pi}\int_0^b \left(\frac{1}{4}\rho^4\cos\phi\sin\phi\right)\Big|_0^a d\phi\, d\theta$$
$$= \frac{1}{m}\int_0^{2\pi}\int_0^b \frac{1}{4}a^4\cos\phi\sin\phi\, d\phi\, d\theta$$
$$= \frac{a^4}{4m}\int_0^{2\pi}\left(\frac{1}{2}\sin^2\phi\right)\Big|_0^b d\theta$$
$$= \frac{a^4}{8m}\int_0^{2\pi}\sin^2 b\, d\theta$$
$$= \frac{\pi a^4 \sin^2 b}{4m}$$
$$= \frac{\pi a^4(1-\cos^2 b)}{4m}$$
$$= \frac{3\pi a^4(1-\cos^2 b)}{8\pi a^3(1-\cos b)}$$
$$= \frac{3}{8}a(1+\cos b) \text{ length units}$$

**13.** $$\iiint_B e^{-(x^2+y^2+z^2)^{3/2}}\,dx\,dy\,dz$$
$$= \int_0^{2\pi}\int_0^{\pi}\int_0^1 e^{-\rho^3}\rho^2\sin\phi\, d\rho\, d\phi\, d\theta$$
$$= \int_0^{2\pi}\int_0^{\pi}\left(-\frac{1}{3}\sin\phi\, e^{-\rho^3}\right)\Big|_0^1 d\phi\, d\theta$$
$$= \frac{1}{3}(1-e^{-1})\int_0^{2\pi}\int_0^{\pi}\sin\phi\, d\phi\, d\theta$$
$$= \frac{1}{3}(1-e^{-1})\int_0^{2\pi}(-\cos\phi)\Big|_0^{\pi} d\theta$$
$$= \frac{2}{3}(1-e^{-1})\int_0^{2\pi} d\theta$$
$$= \frac{4\pi}{3}(1-e^{-1})$$

**15.** $$\iiint_B x\,dx\,dy\,dz$$
$$= \int_0^{\pi/2}\int_0^{\pi/2}\int_0^1 (\rho\sin\phi\cos\theta)(\rho^2\sin\phi)d\rho\, d\phi\, d\theta$$
$$= \int_0^{\pi/2}\int_0^{\pi/2}\int_0^1 (\rho^3\sin^2\phi\cos\theta)d\rho\, d\phi\, d\theta$$
$$= \int_0^{\pi/2}\int_0^{\pi/2}\left(\frac{1}{4}\rho^4\sin^2\phi\cos\theta\right)\Big|_0^1 d\phi\, d\theta$$
$$= \frac{1}{4}\int_0^{\pi/2}\int_0^{\pi/2}\sin^2\phi\cos\theta\, d\phi\, d\theta$$
$$= \frac{1}{4}\int_0^{\pi/2}\int_0^{\pi/2}\left(\frac{1}{2}-\frac{1}{2}\cos 2\phi\right)\cos\theta\, d\phi\, d\theta$$
$$= \frac{1}{4}\int_0^{\pi/2}\left(\left(\frac{1}{2}\phi-\frac{1}{4}\sin 2\phi\right)\cos\theta\right)\Big|_0^{\pi/2} d\theta$$
$$= \frac{1}{4}\int_0^{\pi/2}\frac{\pi}{4}\cos\theta\, d\theta$$
$$= \frac{\pi}{16}(\sin\theta)\Big|_0^{\pi/2}$$
$$= \frac{\pi}{16}$$

**17.** $$\frac{\partial(x, y, z)}{\partial(r, \theta, z)} = \begin{vmatrix} a & 0 & 0 \\ 0 & b & 0 \\ 0 & 0 & c \end{vmatrix} = abc$$

$$V = \iiint_{\mathbf{g}(D)} 1\,dx\,dy\,dz = \iiint_D abc\,du\,dv\,dw$$

Observe that $u^2 + v^2 + w^2 = 1$. Change to spherical coordinates. The line $x = y$ is $au = bv$ after a change of variable. In spherical coordinates, $a\rho \sin\phi \cos\theta = b\rho \sin\phi \sin\theta$ or $\tan\theta = \frac{a}{b}$.

$$V = \int_{\arctan(a/b)}^{\pi/2}\int_0^{\pi/2}\int_0^1 abc\rho^2\sin\phi\, d\rho\, d\phi\, d\theta$$
$$= abc\int_{\arctan(a/b)}^{\pi/2}\int_0^{\pi/2}\left(\frac{1}{3}\rho^3\sin\phi\right)\Big|_0^1 d\phi\, d\theta$$
$$= \frac{abc}{3}\int_{\arctan(a/b)}^{\pi/2}\int_0^{\pi/2}\sin\phi\, d\phi\, d\theta$$
$$= \frac{abc}{3}\int_{\arctan(a/b)}^{\pi/2}(-\cos\phi)\Big|_0^{\pi/2} d\theta$$
$$= \frac{abc}{3}\int_{\arctan(a/b)}^{\pi/2} d\theta$$
$$= \frac{abc}{3}(\theta)\Big|_{\arctan(a/b)}^{\pi/2}$$
$$= \frac{abc}{6}\left(\pi - 2\arctan\left(\frac{a}{b}\right)\right) \text{ cubic units}$$

**19.** Using cylindrical coordinates, the region is bounded above by $z = r\sin\theta$ and below by $z = r^2$. When $y = x^2 + y^2$, $r = 0$, or $r = \sin\theta$. Thus $0 \le r \le \sin\theta$ and $0 \le \theta \le \pi$.

$$\iiint_B dx\,dy\,dz = \int_0^{\pi}\int_0^{\sin\theta}\int_{r^2}^{r\sin\theta} r\,dz\,dr\,d\theta$$
$$= \int_0^{\pi}\int_0^{\sin\theta}\left(rz\Big|_{r^2}^{r\sin\theta}\right)dr\,d\theta$$
$$= \int_0^{\pi}\int_0^{\sin\theta}(r^2\sin\theta - r^3)\,dr\,d\theta$$
$$= \int_0^{\pi}\left(\frac{1}{3}r^3\sin\theta - \frac{1}{4}r^4\right)\Bigg|_0^{\sin\theta} d\theta$$
$$= \int_0^{\pi}\frac{1}{12}\sin^4\theta\,d\theta$$
$$= \frac{1}{48}\int_0^{\pi}(1-\cos 2\theta)^2\,d\theta$$
$$= \frac{1}{48}\int_0^{\pi}(1 - 2\cos 2\theta + \cos^2 2\theta)\,d\theta$$
$$= \frac{1}{48}\int_0^{\pi}\left(\frac{3}{2} - 2\cos 2\theta + \frac{1}{2}\cos 4\theta\right)d\theta$$
$$= \frac{1}{48}\left(\frac{3}{2}\theta - \sin 2\theta + \frac{1}{8}\sin 4\theta\right)\Bigg|_0^{\pi}$$
$$= \frac{\pi}{32} \text{ cubic units}$$

**21.**
$$V = \int_0^{2\pi}\int_0^{\pi/3}\int_0^{4\sin\phi}\rho^2\sin\phi\,d\rho\,d\phi\,d\theta$$
$$= \int_0^{2\pi}\int_0^{\pi/3}\frac{64}{3}\sin^4\phi\,d\phi\,d\theta$$
$$= \frac{16}{3}\int_0^{2\pi}\int_0^{\pi/3}(1-\cos 2\phi)^2\,d\phi\,d\theta$$
$$= \frac{16}{3}\int_0^{2\pi}\int_0^{\pi/3}\left(\frac{3}{2} - 2\cos 2\phi + \frac{1}{2}\cos 4\phi\right)d\phi\,d\theta$$
$$= \frac{16}{3}\int_0^{2\pi}\left(\frac{\pi}{2} - \frac{9\sqrt{3}}{16}\right)d\theta$$
$$= \frac{32\pi}{3}\left(\frac{\pi}{2} - \frac{9\sqrt{3}}{16}\right)$$
$$= \frac{2\pi}{3}\left(8\pi - 9\sqrt{3}\right) \text{ cubic units}$$

**23.** By symmetry $X = Y = 0$.

From geometry $m = \frac{1}{2}\left(\frac{4}{3}\pi a^3\right) = \frac{2}{3}\pi a^3$.

$$Z = \frac{1}{m}\int_0^{2\pi}\int_0^{a}\int_0^{\sqrt{a^2-r^2}} rz\,dz\,dr\,d\theta = \frac{1}{m}\int_0^{2\pi}\int_0^{a}\frac{1}{2}(a^2r - r^3)\,dr\,d\theta = \frac{1}{m}\int_0^{2\pi}\frac{1}{8}a^4\,d\theta = \frac{3}{8}a \text{ units}$$

**25.** Let $u = \rho^3$ and $dv = \rho e^{-\rho^2}d\rho$, so $du = 3\rho^2 d\rho$ and $v = -\frac{1}{2}e^{-\rho^2}$. Then using integration by parts,

$$\int_0^1 \rho^4 e^{-\rho^2}d\rho = \left(-\frac{1}{2}\rho^3 e^{-\rho^2}\right)\Bigg|_0^1 + \int_0^1\frac{3}{2}\rho^2 e^{-\rho^2}d\rho = -\frac{1}{2}e^{-1} + \frac{3}{2}\int_0^1\rho^2 e^{-\rho^2}d\rho$$

Now let $u = \rho$ and $dv = \rho e^{-\rho^2} d\rho$ and $v = -\frac{1}{2}e^{-\rho^2}$.

$$\int_0^1 \rho^4 e^{-\rho^2} d\rho = -\frac{1}{2}e^{-1} + \frac{3}{2}\left[\left(-\frac{1}{2}\rho^2 e^{-\rho^2}\right)\Big|_0^1 + \int_0^1 \frac{1}{2}e^{-\rho^2} d\rho\right]$$
$$= -\frac{5}{4}e^{-1} + \frac{3}{4}\int_0^1 e^{-\rho^2} d\rho$$
$$= -\frac{5}{4}e^{-1} + \frac{3}{8}\sqrt{\pi}\,\mathrm{erf}(1)$$

Applying this result in Example 6, we find that $I = \frac{4\pi}{3}\left[-\frac{5}{4e} + \frac{3\sqrt{\pi}}{8}\mathrm{erf}(1)\right] = \frac{\pi}{6e}[-10 + 3e\sqrt{\pi}\,\mathrm{erf}(1)]$

Using the given value of erf(1), we find $I \approx 0.420005$, which agrees with the result from Example 6.

**27.** Since $\delta$ is a function of the distance from the center, there exists $g$ such that $\delta(x, y, z) = g(\rho)$.

$$\mathbf{F} = G\iiint_B \frac{\delta(x, y, z)}{(x^2 + y^2 + (z-q)^2)^{3/2}}\langle x, y, z - q\rangle dV$$
$$= G\int_0^a \int_0^\pi \int_0^{2\pi} \frac{g(\rho)}{(\rho^2 - 2\rho q\cos\phi + q^2)^{3/2}}\langle \rho\sin\phi\cos\theta, \rho\sin\phi\sin\theta, \rho\cos\phi - q\rangle dV$$

where $dV = \rho^2 \sin\phi\, d\theta\, d\phi\, d\rho$. Since $\int_0^{2\pi}\cos\theta\, d\theta = 0$ and $\int_0^{2\pi}\sin\theta\, d\theta = 0$, $F_1 = 0$ and $F_2 = 0$.

$$F_3 = G\int_0^a \int_0^\pi \int_0^{2\pi} \frac{g(\rho)(\rho\cos\phi - q)}{(\rho^2 - 2\rho q\cos\phi + q^2)^{3/2}}\rho^2 \sin\phi\, d\theta\, d\phi\, d\rho$$

Make the substitution $w = \rho^2 - 2\rho q\cos\phi + q^2$, $dw = 2\rho q \sin\phi\, d\phi$, and $\rho\cos\phi - q = \frac{\rho^2 - q^2 - w}{2q}$.

$\left(\text{Note that } \rho^2 \sin\phi\, d\phi = \frac{\rho}{2q}dw.\right)$

$$F_3 = G\int_0^a \int_{(\rho-q)}^{(\rho+q)^2} \int_0^{2\pi} \frac{g(\rho)}{w^{3/2}}\frac{(\rho^2 - q^2 - w)}{2q}\frac{\rho}{2q}\, d\theta\, dw\, d\rho$$
$$= \frac{G\pi}{2q^2}\int_0^a \int_{(\rho-q)^2}^{(\rho+q)^2} \rho g(\rho)((\rho^2 - q^2)w^{-3/2} - w^{-1/2})dw\, d\rho$$
$$= \frac{G\pi}{2q^2}\int_0^a (\rho g(\rho)(-2(\rho^2 - q^2)w^{-1/2} - 2w^{1/2}))\Big|_{(\rho-q)^2}^{(\rho+q)^2} d\rho$$
$$= \frac{G\pi}{2q^2}\int_0^a (-2\rho g(\rho))\left(\frac{(\rho^2 - q^2)}{|\rho + q|} + |\rho + q| - \frac{(\rho^2 - q^2)}{|\rho - q|} - |\rho - q|\right)d\rho$$

Since $q > \rho$, we get

$$F_3 = \frac{G\pi}{2q^2}\int_0^a (-8\rho g(\rho))(\rho)d\rho = \frac{-4\pi G}{q^2}\int_0^a \rho^2 g(\rho)d\rho.$$

The mass of the sphere is

$$M = \iiint_B g(\rho)dV = \int_0^a \int_0^\pi \int_0^{2\pi} g(\rho)\rho^2 \sin\phi\, d\theta\, d\phi\, d\rho = 4\pi\int_0^a \rho^2 g(\rho)\, d\rho.$$

Thus $F_3 = -\frac{GM}{q^2}$, so the main conclusion remains true.

**29.** The two balls have equations $x^2+y^2+z^2=1$ and $x^2+y^2+\left(z-\frac{7}{5}\right)^2=\frac{9}{16}$. Their surfaces intersect when $z=\frac{137}{160}$ and $\sqrt{x^2+y^2}=\frac{3\sqrt{759}}{160}$. Thus, we can separate the station in spherical coordinates into two regions, one for which $0\le\phi\le\cos^{-1}\left(\frac{137}{160}\right)$ and the other for which $\cos^{-1}\left(\frac{137}{160}\right)\le\phi\le\pi$. In the second region, $0\le\theta\le 2\pi$ and $0\le\rho\le 1$. In the first region, $\rho$ extends from 0 to the second surface $x^2+y^2+\left(z-\frac{7}{5}\right)^2=\frac{9}{16}$. Rewriting this as $x^2+y^2+\left(z-\frac{7}{5}\right)^2-\frac{9}{16}=x^2+y^2+z^2-\frac{14}{5}z+\frac{49}{25}-\frac{9}{16}=\rho^2-\frac{14}{5}\rho\cos\phi+\frac{559}{400}=0$, we see that in fact we will need $0\le\rho\le\frac{\frac{14}{5}\cos\phi-\sqrt{\left(\frac{14}{5}\cos\phi\right)^2-\frac{559}{100}}}{2}$, where we take the negative root to stay on the appropriate surface.

Therefore, with $\delta=k\rho$, we find that the mass of the station is

$$m=\int_0^{2\pi}\int_{\cos^{-1}\left(\frac{137}{160}\right)}^{\pi}\int_0^1 k\rho^3\sin\phi\,d\rho\,d\phi\,d\theta+\int_0^{2\pi}\int_0^{\cos^{-1}\left(\frac{137}{160}\right)}\int_0^{\frac{7}{5}\cos\phi-\frac{1}{2}\sqrt{\left(\frac{14}{5}\cos\phi\right)^2-\frac{559}{100}}} k\rho^3\sin\phi\,d\rho\,d\phi\,d\theta$$

$\approx 3.0062k$ mass units

**31.** $x^2+13y^2+2z^2-4xy-6yz=(x^2-4xy+4y^2)+(9y^2-6yz+z^2)+z^2=(x-2y)^2+(3y-z)^2+z^2$

Solve $u=x-2y$, $v=3y-z$, and $w=z$ for $x$, $y$, and $z$.

$x=u+\frac{2}{3}(v+w)$

$y=\frac{1}{3}(v+w)$

$z=w$

$\mathbf{g}(u,v,w)=\left\langle u+\frac{2}{3}v+\frac{2}{3}w,\ \frac{1}{3}v+\frac{1}{3}w,\ w\right\rangle$

Thus $\mathbf{g}$ is a linear transformation. $D$ is determined by $\langle 1,0,0\rangle$, $\langle 0,1,0\rangle$, and $\langle 0,0,1\rangle$.

$\mathbf{g}(1,0,0)=\langle 1,0,0\rangle$; $\mathbf{g}(0,1,0)=\left\langle \frac{2}{3},\frac{1}{3},0\right\rangle$; $\mathbf{g}(0,0,1)=\left\langle \frac{2}{3},\frac{1}{3},1\right\rangle$

Since $B$ is determined by $\langle 1,0,0\rangle$, $\left\langle \frac{2}{3},\frac{1}{3},0\right\rangle$ and $\left\langle \frac{2}{3},\frac{1}{3},1\right\rangle$ and $\mathbf{g}(0,0,0)=\langle 0,0,0\rangle$, $B=\mathbf{g}(D)$.

$$\frac{\partial(x,y,z)}{\partial(u,v,w)}=\begin{vmatrix}1 & \frac{2}{3} & \frac{2}{3}\\ 0 & \frac{1}{3} & \frac{1}{3}\\ 0 & 0 & 1\end{vmatrix}=\frac{1}{3}$$

Thus,

$$I=\iiint_B e^{-(x^2+13y^2+2z^2-4xy-6yz)}dV$$
$$=\iiint_B e^{-(x-2y)^2}e^{-(3y-z)^2}e^{-z^2}\,dV$$
$$=\iiint_D e^{-u^2}e^{-v^2}e^{-w^2}\frac{1}{3}du\,dv\,dw$$
$$=\frac{1}{3}\int_0^1\int_0^1\int_0^1 e^{-u^2}e^{-v^2}e^{-w^2}du\,dv\,dw$$
$$=\frac{1}{3}\left(\int_0^1 e^{-u^2}du\right)\left(\int_0^1 e^{-v^2}dv\right)\left(\int_0^1 e^{-w^2}dw\right)$$
$$=\frac{1}{3}\left(\int_0^1 e^{-t^2}dt\right)^3$$

## Chapter 11 Review Exercises

**1.** The area of each subregion is $\frac{1}{4}\cdot\frac{1}{4}=\frac{1}{16}$. The inscribed volume elements of $\frac{1}{16}(x+2y^2)$ for each of the regions are as follows:

| $x$ \ $y$ | $\frac{1}{8}$ | $\frac{3}{8}$ | $\frac{5}{8}$ | $\frac{7}{8}$ |
|---|---|---|---|---|
| $\frac{1}{8}$ | $\frac{5}{512}$ | $\frac{13}{512}$ | $\frac{29}{512}$ | $\frac{53}{512}$ |
| $\frac{3}{8}$ | $\frac{13}{512}$ | $\frac{21}{512}$ | $\frac{37}{512}$ | $\frac{61}{512}$ |
| $\frac{5}{8}$ | $\frac{21}{512}$ | $\frac{29}{512}$ | $\frac{45}{512}$ | $\frac{69}{512}$ |
| $\frac{7}{8}$ | $\frac{29}{512}$ | $\frac{37}{512}$ | $\frac{53}{512}$ | $\frac{77}{512}$ |

Summing these elements gives the volume estimate $V\approx\frac{37}{32}=1.15625$. The actual volume is $V=\int_0^1\int_0^1 x+2y^2\,dy\,dx=\frac{7}{6}$, so the estimate has an error of about 0.89%.

**3.** $\int_0^1\int_x^1 e^{x/y}dy\,dx=\int_0^1\int_0^y e^{x/y}dx\,dy.$
$$=\int_0^1 y(e-1)dy$$
$$=\frac{1}{2}(e-1)$$

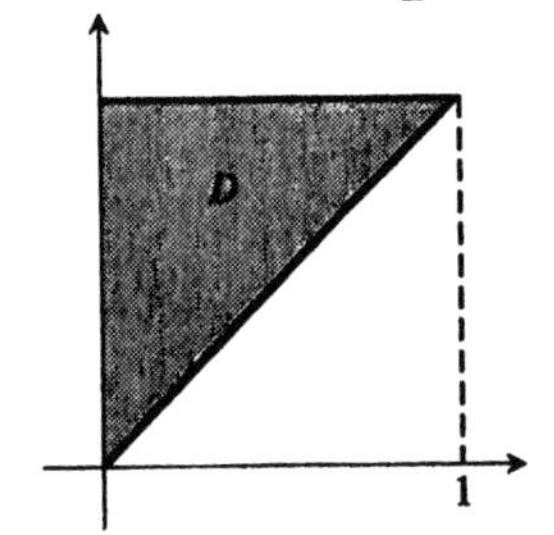


**5.** The projection of the given triangle into the $(x, y)$-plane is the triangle with vertices (0, 0), (2, 2) and (0, 2). The plane containing the upper triangle is given by the equation $x+y+z=6$, or $z=6-x-y$. Therefore, the volume of $K$ is

$$V=\int_0^2\int_0^y(6-x-y)dx\,dy=8\text{ units}^3.$$

**7.** The region of integration in the $(x, y)$-plane is $[0, 2]\times[0, 1]$, with integration in the $z$-direction between $z=0$ and $z=1+x^2+2y^2$. Thus, the mass of the solid is

$$m=\int_0^1\int_0^2\int_0^{1+x^2+2y^2}(2+z)dz\,dx\,dy\text{ grams}.$$

**9.** Since $\delta=kr$, we have $m=k\int_0^{\pi/2}\int_0^{\theta}r^2dr\,d\theta$,

$$X=\frac{k}{m}\int_0^{\pi/2}\int_0^{\theta}(r\cos\theta)r^2\,dr\,d\theta,$$
$$Y=\frac{k}{m}\int_0^{\pi/2}\int_0^{\theta}(r\sin\theta)r^2\,dr\,d\theta.$$

**11.** The curves bounding $D$ intersect at $x=\frac{5}{2}\pm\sqrt{6}$. Therefore, the mass of $K$ is

$$m=\int_{\frac{5}{2}-\sqrt{6}}^{\frac{5}{2}+\sqrt{6}}\int_{\frac{(x-\frac{1}{2})^2}{4}}^{x}\int_0^{8-x}2z\,dz=\frac{629\sqrt{6}}{10}\text{ kg}.$$

**13.** Reversing the order of integration, we find

$$\int_0^1\int_x^1 \sin(y^2)\,dy\,dx = \int_0^1\int_0^y \sin(y^2)\,dx\,dy = \int_0^1 y\sin(y^2)\,dy = \frac{1-\cos(1)}{2}.$$

We choose this method because symbolic integration with respect to $y$ is possible for $y\sin(y^2)$ but not for $\sin(y^2)$.

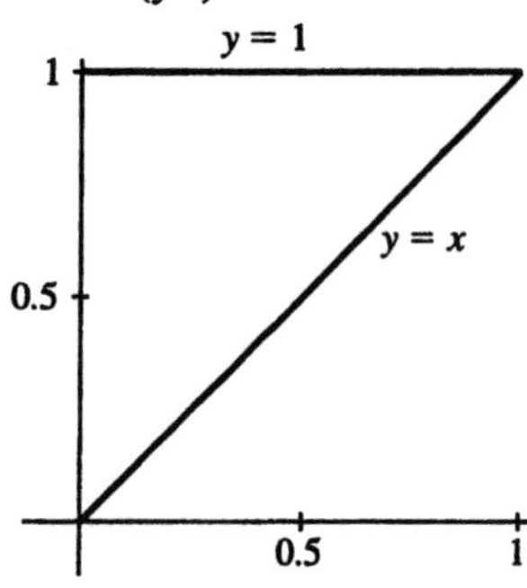


**15.** The bounding surfaces in $z$ are $z=0$ and $z=3-x-y$. The region of integration in the $(x, y)$-plane is given by the equations $1 \le y \le 2,\ 0 \le x \le \frac{y}{2}$. The volume is thus

$$V = \int_1^2\int_0^{y/2}\int_0^{3-x-y} dz\,dx\,dy = \frac{19}{24} \text{ cubic units.}$$

**17.** A normal vector to $\mathbf{r}$ at a point $(u, v)$ will be $(\mathbf{r}_u \times \mathbf{r}_v)(u, v)$. Since $\mathbf{r}_u = \langle 2u, 0, 1\rangle$ and $\mathbf{r}_v = \langle 0, 2v, 2\rangle$, $\mathbf{r}_u \times \mathbf{r}_v = \langle -2v, -4u, 4uv\rangle$. The point (4, 4, 6) corresponds to $(u, v) = (2, 2)$, so $(\mathbf{r}_u \times \mathbf{r}_v)(2, 2) = \langle -4, -8, 16\rangle$; the unit normal at (4, 4, 6) is thus $\dfrac{\langle -1, -2, 4\rangle}{\sqrt{21}}$. The area of $S$ is

$$\iint \|\mathbf{r}_u \times \mathbf{r}_v\|\,dv\,du = \int_0^4\int_0^4 2\sqrt{4u^2+v^2+4u^2v^2}\,du\,dv.$$

**19.** $A = \displaystyle\int_0^{2\sqrt{2}}\int_0^y \sqrt{1+y^2}\,dx\,dy = \frac{26}{3}$ square units.

**21.** The integral is either $\displaystyle\int_{-2}^{2}\int_{-\sqrt{4-x^2}}^{\sqrt{4-x^2}} \sqrt{1+4x^2+4y^2}\,dy\,dx$ in rectangular coordinates or $\displaystyle\int_0^{2\pi}\int_0^2 r\sqrt{1+4r^2}\,dr\,d\theta$ in polar coordinates.

**23.** Note that the surface can be rewritten as $y = \sqrt{z+x^2}$ in the given region. The surface area will then be $A = \iint \sqrt{1+{y_x}^2+{y_z}^2}\,dz\,dx$. Since $y_x = \dfrac{x}{\sqrt{z+x^2}}$ and $y_z = \dfrac{1}{2\sqrt{z+x^2}}$, this becomes

$$A = \frac{1}{2}\int_0^a\int_0^{\sqrt{a^2-x^2}} \sqrt{\frac{1+8x^2+4z}{x^2+z}}\,dz\,dx \text{ square units.}$$

**25.** Here $\mathbf{r}_u = \langle -\sin u\cos v, -\sin u \sin v, \cos u\rangle$ and $\mathbf{r}_v = \langle -\sin v(R+\cos u), \cos v(R+\cos u), 0\rangle$, so $\mathbf{r}_u \times \mathbf{r}_v = \langle -\cos u\cos v(R+\cos u), -\cos u \sin v(R+\cos u), -\sin u(R+\cos u)\rangle$. Using a CAS, we verify that $A = \displaystyle\int_0^{2\pi}\int_0^{2\pi} \|\mathbf{r}_u \times \mathbf{r}_v\|\,du\,dv = 4\pi^2 R$ square units.

**27.** The region being integrated over is the eigth-circle described by $0 \le \theta \le \frac{\pi}{4}$, $0 \le r \le 2$. Therefore, the integral is $\int_0^{\pi/4} \int_0^2 \frac{r}{1+r^2} \, dr \, d\theta = \frac{\pi \ln 5}{8}$.

**29.** The vent pipe's base can be described by the inequalities $-\frac{\pi}{2} \le \theta \le \frac{\pi}{2}$, $0 \le r \le 2\cos\theta$ in polar coordinates. Meanwhile, if the cone's base is centered at the origin, the cone can be described by the equation $\frac{x^2}{20^2} + \frac{y^2}{20^2} = \frac{(12-z)^2}{12^2}$, or $z = 12\left(1 - \frac{1}{20}r\right)$ in cylindrical coordinates. Therefore, the volume of grain in the pile is

$V(\text{cone}) - V(\text{space occupied by pipe}) = \frac{1}{3}\pi(20)^2(12) - 2\int_0^{\pi/2} \int_0^{2\cos\theta} 12\left(1 - \frac{1}{20}r\right) r \, dr \, d\theta$ cubic units.

**31.** The region $D$ is the parallelogram with vertices (0, 0), $\left(\frac{1}{3}, \frac{2}{3}\right)$, $\left(\frac{4}{3}, -\frac{1}{3}\right)$ and (1, −1), and is the image of $[0, 1] \times [0, 1]$ under the change of variables $x = u + \frac{1}{3}v$, $y = -u + \frac{2}{3}v$. The Jacobian of this change is 1, so

$$\iint_D (x+y)^2 \, dy \, dx = \int_0^1 \int_0^1 \left[\left(u + \frac{1}{3}v\right) + \left(-u + \frac{2}{3}v\right)\right]^2 dv \, du = \int_0^1 \int_0^1 v^2 \, dv \, du = \frac{1}{3}.$$

**33.** The two planes intersect when their
$z$-coordinates match, which happens along the line $x = 3 - 3y$ in the $(x, y)$-plane. The region of integration is therefore $0 \le y \le 1$, $0 \le x \le 3 - 3y$. The top plane can be written as $z = 4 - \frac{4}{3}x - 2y$ and the bottom plane as $z = 3 - x - y$, so the volume of the solid is

$V = \int_0^1 \int_0^{3-3y} \int_{3-x-y}^{4(1-x/3-y/2)} dz \, dx \, dy$ cubic units.

**35.** The region of integration in the $(x, y)$-plane is bounded by $x = 0$, $y = x^2$ and $y = 1 - x$ and can be described by the inequalities $0 \le x \le \frac{-1+\sqrt{5}}{2}$ and $x^2 \le y \le 1 - x$. Therefore, an integral for the volume is

$V = \int_0^{-1/2+\sqrt{5}/2} \int_{x^2}^{1-x} \int_0^{1-x-y} dz \, dy \, dx$ cubic units.

**37.** The cylindrical coordinate limits $0 \le \theta \le 2\pi$, $0 \le r \le R$ correspond to rectangular coordinate limits $-R \le x \le R$, $-\sqrt{R^2 - x^2} \le y \le \sqrt{R^2 - x^2}$. Recalling that $r^2 = x^2 + y^2$ and $dx \, dy = r \, dr \, d\theta$, we see that the integral is $\int_{-R}^{R} \int_{-\sqrt{R^2-x^2}}^{\sqrt{R^2-x^2}} \int_0^H \frac{\sqrt{x^2+y^2}}{1+x^2+y^2} \, dz \, dy \, dx$ in rectangular coordinates.

**39.** From the first two equations we get the limits of integration $a \le \rho \le b$. From the equation of the cone we find $0 \le \phi \le \frac{\pi}{4}$, $0 \le \theta \le 2\pi$. Therefore, the volume of the solid is

$V = \int_0^{2\pi} \int_0^{\pi/4} \int_a^b \rho^2 \sin\phi \, d\rho \, d\phi \, d\theta = \frac{\pi}{3}(b^3 - a^3)\left(2 - \sqrt{2}\right)$ cubic units.

**41.** The curves $x^2+y^2+z^2=4$ and $z^2=3(x^2+y^2)$ intersect when $x^2+y^2=1$, which gives cylindrical coordinate limits of $0\le\theta\le 2\pi$, $0\le r\le 1$, $r\sqrt{3}\le z\le\sqrt{4-r^2}$ and an integral

i) $I=\int_0^{2\pi}\int_0^1\int_{r\sqrt{3}}^{\sqrt{4-r^2}} rz\,dz\,dr\,d\theta$.

In spherical coordinates, the equation of the cone leads to the $\phi$ limits $0\le\phi\le\frac{\pi}{6}$ which along with $0\le\rho\le 2$ and $0\le\theta\le 2\pi$ leads to

ii) $I=\int_0^{2\pi}\int_0^{\pi/6}\int_0^2 \rho^3\sin\phi\,d\rho\,d\phi\,d\theta$.

**43.** The given curves in cylindrical coordinates are $r+\frac{z}{3}=1$ $(z=3(1-r))$ and $z=r$. These intersect when $z=r=\frac{3}{4}$; consequently, the solid has volume

$V=\int_0^{2\pi}\int_0^{3/4}\int_r^{3(1-r)} r\,dz\,dr\,d\theta$ cubic units.

# Chapter 12 Line and Surface Integrals

## Section 12.1 The Line Integral

**1.** $C$ is the line segment from (0, 0, 0) to (2, 4, 6).

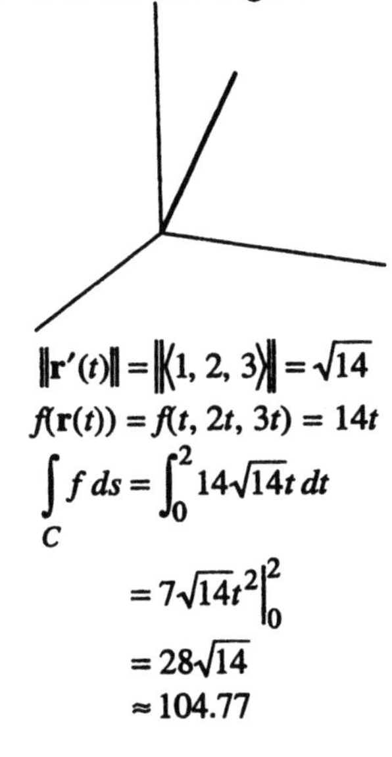

$\|\mathbf{r}'(t)\| = \|\langle 1, 2, 3\rangle\| = \sqrt{14}$

$f(\mathbf{r}(t)) = f(t, 2t, 3t) = 14t$

$$\int_C f\,ds = \int_0^2 14\sqrt{14}t\,dt$$
$$= 7\sqrt{14}t^2\Big|_0^2$$
$$= 28\sqrt{14}$$
$$\approx 104.77$$

**3.** $C$ is a circular helix.

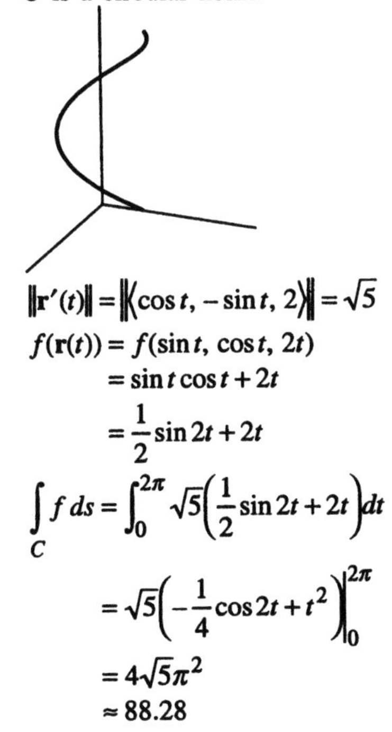

$\|\mathbf{r}'(t)\| = \|\langle \cos t, -\sin t, 2\rangle\| = \sqrt{5}$

$$f(\mathbf{r}(t)) = f(\sin t, \cos t, 2t)$$
$$= \sin t\cos t + 2t$$
$$= \frac{1}{2}\sin 2t + 2t$$

$$\int_C f\,ds = \int_0^{2\pi} \sqrt{5}\left(\frac{1}{2}\sin 2t + 2t\right)dt$$
$$= \sqrt{5}\left(-\frac{1}{4}\cos 2t + t^2\right)\Big|_0^{2\pi}$$
$$= 4\sqrt{5}\pi^2$$
$$\approx 88.28$$

**5.** $C$ can be described by $\mathbf{r}(t) = \langle 1 + 3t, -1 - t, 2 - t\rangle$, $0 \le t \le 1$.

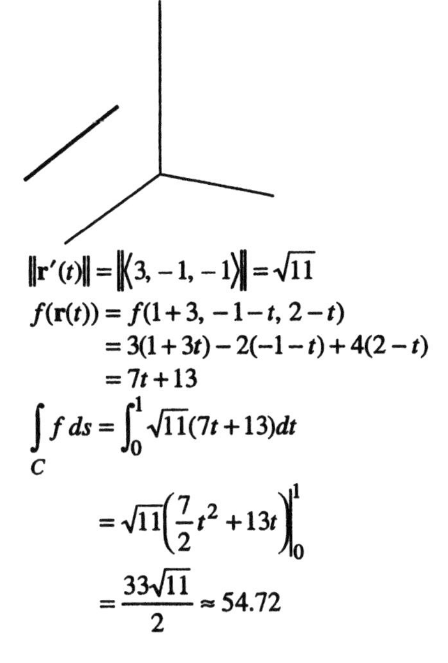

$\|\mathbf{r}'(t)\| = \|\langle 3, -1, -1\rangle\| = \sqrt{11}$

$$f(\mathbf{r}(t)) = f(1+3, -1-t, 2-t)$$
$$= 3(1+3t) - 2(-1-t) + 4(2-t)$$
$$= 7t + 13$$

$$\int_C f\,ds = \int_0^1 \sqrt{11}(7t+13)dt$$
$$= \sqrt{11}\left(\frac{7}{2}t^2 + 13t\right)\Big|_0^1$$
$$= \frac{33\sqrt{11}}{2} \approx 54.72$$

**7.** $C$ can be described by $\mathbf{r}(t) = \langle 2\cos t, 0, 2\sin t\rangle$, $0 \le t \le \pi$.

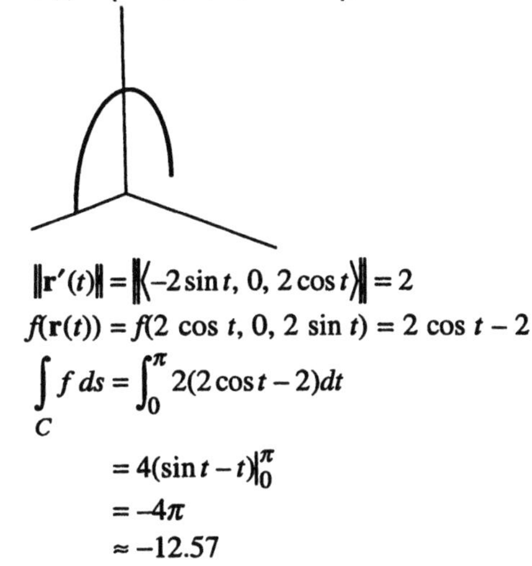

$\|\mathbf{r}'(t)\| = \|\langle -2\sin t, 0, 2\cos t\rangle\| = 2$

$f(\mathbf{r}(t)) = f(2\cos t, 0, 2\sin t) = 2\cos t - 2$

$$\int_C f\,ds = \int_0^{\pi} 2(2\cos t - 2)dt$$
$$= 4(\sin t - t)\Big|_0^{\pi}$$
$$= -4\pi$$
$$\approx -12.57$$

**9.** Let $\phi(\tau) = \arcsin \tau$, $\frac{1}{2} \le \tau \le \frac{\sqrt{3}}{2}$. $\phi$ is increasing with a continuous, nonzero derivative on the interval and $\phi\left(\frac{1}{2}\right) = \frac{\pi}{6}$, $\phi\left(\frac{\sqrt{3}}{2}\right) = \frac{\pi}{3}$. Thus

$\hat{\mathbf{r}}(\tau) = \mathbf{r}(\phi(\tau)) = \left\langle \sqrt{1-\tau^2}, \tau, \arcsin \tau \right\rangle$ is reparameterization of $C$.

$$\|\mathbf{r}'(t)\| = \|\langle -\sin t, \cos t, 1\rangle\| = \sqrt{2}$$

$$f(\mathbf{r}(t)) = f(\cos t, \sin t, t) = 1 + t^2$$

$$\int_C f\,ds = \int_{\pi/6}^{\pi/3} \sqrt{2}(1+t^2)\,dt$$

$$= \sqrt{2}\left(t + \frac{1}{3}t^3\right)\Bigg]_{\pi/6}^{\pi/3}$$

$$= \frac{\pi(108+7\pi^2)}{324\sqrt{2}} \approx 1.21$$

$$\|\hat{\mathbf{r}}'(t)\| = \left\|\left\langle -\frac{\tau}{\sqrt{1-\tau^2}}, 1, \frac{1}{\sqrt{1-\tau^2}}\right\rangle\right\| = \frac{\sqrt{2}}{\sqrt{1-\tau^2}}$$

$$f(\mathbf{r}(t)) = f\left(\sqrt{1-\tau^2}, \tau, \arcsin \tau\right)$$

$$= 1 + \arcsin^2 \tau$$

$$\int_C f\,ds = \int_{1/2}^{\sqrt{3}/2} \frac{\sqrt{2}}{\sqrt{1-\tau^2}}(1+\arcsin^2 \tau)\,dt$$

$$= \sqrt{2}\left(\arcsin \tau + \frac{1}{3}\arcsin^3 \tau\right)\Bigg]_{1/2}^{\sqrt{3}/2}$$

$$= \frac{\pi(108+7\pi^2)}{324\sqrt{2}} \approx 1.21$$

**11.** We divide the wire into the intervals $0 = t_0 < t_1 < t_2 < \cdots < t_{10} = \frac{\pi}{2}$ where $t_k = k\Delta t$, $k = 0, 1, \ldots, 10$ and $\Delta t = \frac{\pi}{20}$.

$$\|\mathbf{r}'(t)\| = \|\langle 1, 3\cos t, -3\sin t\rangle\| = \sqrt{10}$$

$$\Delta s_k = \int_{t_{k-1}}^{t_k} \|\mathbf{r}'(t)\|dt$$

$$= \int_{t_{k-1}}^{t_k} \sqrt{10}\,dt$$

$$= \sqrt{10}\Delta t$$

$$= \frac{\pi}{2\sqrt{10}}$$

Let $d_k$ be the minimum value of the density $\delta$ on the $k$th piece of the wire, and $D_k$ be the maximum value of $\delta$ on this piece.

$$\delta(\mathbf{r}(t)) = 9\cos^2 t + 1$$

$$d_k = 9\cos^2 \frac{k\pi}{20} + 1 \text{ for } k = 1, 2, \ldots, 10$$

$$D_k = 9\cos^2 \frac{(k-1)\pi}{20} + 1 \text{ for } k = 1, 2, \ldots, 10$$

$$L_{10} = \sum_{k=1}^{10} d_k \Delta s_k \approx 25.0848$$

$$U_{10} = \sum_{k=1}^{10} D_k \Delta s_k \approx 29.5554$$

$$25.0848 \le \int_0^{\pi/2} \sqrt{10}(9\cos^2 t + 1)\,dt \le 29.5554$$

**13.** $\|\mathbf{r}'(t)\| = \|\langle 1, 2t, 3t^2\rangle\|$

$$= \sqrt{1+4t^2+9t^4}$$

$$\delta(\mathbf{r}(t)) = \rho(t, t^2, t^3)$$

$$= 1 + t^2 + t^6$$

$$\text{mass} = \int_0^1 (1+t^2+t^6)\sqrt{1+4t^2+9t^4}\,dt$$

$$\approx 3.17 \text{ grams}$$

(We used numerical integration on a CAS to evaluate the integral.)

**15.** Let $C_1$ be the line segment from (0, 0, 0) to (1, 1, 0), $C_2$ the line segment from (1, 1, 0) to $\left(1, 1, \sqrt{2}\right)$, $C_3$ the line segment from $\left(1, 1, \sqrt{2}\right)$ to $\left(0, 0, \sqrt{2}\right)$, and $C_4$ the line segment from $\left(0, 0, \sqrt{2}\right)$ to (0, 0, 0). Parametrically describe the curves by $\mathbf{r}_1(t) = \langle t, t, 0\rangle$, $\mathbf{r}_2(t) = \langle 1, 1, \sqrt{2}t\rangle$, $\mathbf{r}_3(t) = \langle 1-t, 1-t, \sqrt{2}\rangle$, and $\mathbf{r}_4(t) = \langle 0, 0, \sqrt{2} - \sqrt{2}t\rangle$, respectively, for $0 \le t \le 1$.

$\|\mathbf{r}_1'(t)\| = \sqrt{2}$, $\|\mathbf{r}_2'(t)\| = \sqrt{2}$, $\|\mathbf{r}_3'(t)\| = \sqrt{2}$, $\|\mathbf{r}_4'(t)\| = \sqrt{2}$.

$\delta(\mathbf{r}_1(t)) = 1$, $\delta(\mathbf{r}_2(t)) = 1 + \sqrt{2}t$, $\delta(\mathbf{r}_3(t)) = 1 + \sqrt{2} - 2\sqrt{2}t + \sqrt{2}t^2$, $\delta(\mathbf{r}_4(t)) = 1$

$$\begin{aligned}
\text{mass} &= \int_C \delta\, ds = \int_{C_2} \delta\, ds + \int_{C_2} \delta\, ds + \int_{C_3} \delta\, ds + \int_{C_4} \delta\, ds \\
&= \int_0^1 \sqrt{2}\, dt + \int_0^1 \sqrt{2}\left(1 + \sqrt{2}t\right) dt + \int_0^1 \sqrt{2}\left(1 + \sqrt{2} - 2\sqrt{2}t + \sqrt{2}t^2\right) dt + \int_0^1 \sqrt{2}\, dt \\
&= \sqrt{2}\int_0^1 \left(4 + \sqrt{2} - \sqrt{2}t + \sqrt{2}t^2\right) dt \\
&= \sqrt{2}\left(4t + \sqrt{2}t - \frac{\sqrt{2}}{2}t^2 + \frac{\sqrt{2}}{3}t^3\right)\Bigg|_0^1 \\
&= \frac{5}{3} + 4\sqrt{2} \approx 7.32
\end{aligned}$$

**17.** Divide the path using the intervals $a = t_0 < t_1 < t_2 < \cdots < t_n = b$ where $t_k = a + k\Delta t$, $k = 0, 1, \ldots, n$ and $\Delta t = \frac{b-a}{n}$. Then the points $\mathbf{r}(t_0), \mathbf{r}(t_1), \ldots, \mathbf{r}(t_n)$ determine a subdivision of $C$. The $k$th piece of the path is the piece between $\mathbf{r}(t_{k-1})$ and $\mathbf{r}(t_k)$. The length of the $k$th piece is $\Delta s_k = \int_{t_{k-1}}^{t_k} \|\mathbf{r}'(t)\| dt$.
Let $\mathbf{c}_k = \mathbf{r}(\tilde{t}_k)$ where $t_{k-1} < \tilde{t}_k < t_k$. Then $f(\mathbf{c}_k)$ gives the height of the wall at that point in the $k$th piece. For sufficiently small $\Delta s_k$, $f(\mathbf{c}_k)\Delta s_k$ is approximately the area of the wall above the $k$th piece. Then the approximate area of the wall is $\sum_{k=1}^{n} f(\mathbf{c}_k)\Delta s_k$. Moreover, if we choose $d_k$ such that $d_k$ is the minimum height of the wall above the $k$th piece and $D_k$ such that $D_k$ is the maximum height, $L_n = \sum_{k=1}^{n} d_k \Delta s_k$, $U_n = \sum_{k=1}^{n} D_k \Delta s_k$. The area is number $I$ such that $L_n \le I \le U_n$ for every subdivision over the path $C$. Thus area $= \int_C f\, ds$.

**19.** $\|\mathbf{r}'(t)\| = \left\|\left(\frac{1}{25}t, 1\right)\right\| = \frac{1}{25}\sqrt{t^2 + 625}$

$f(\mathbf{r}(t)) = 2 + \sin\left(\frac{t^2 + 50t}{200}\right)$

$$\text{area} = \int_C f\, ds = \int_{-50}^{50} \frac{1}{25}\sqrt{t^2 + 625}\left(2 + \sin\left(\frac{t^2 + 50t}{200}\right)\right) dt$$

Using a CAS, we numerically estimate the area to be approximately 258.23.

**21. a.** Divide the wire using the intervals $a = t_0 < t_1 < t_2 < \cdots < t_n = b$ where $t_k = a + k\Delta t$, $k = 0, 1, \ldots, n$ and $\Delta t = \frac{b-a}{n}$. Then the points $\mathbf{r}(t_0), \mathbf{r}(t_1), \ldots, \mathbf{r}(t_n)$ determine a subdivision of $C$. The $k$th piece of the path is the piece between $\mathbf{r}(t_{k-1})$ and $\mathbf{r}(t_k)$. The length of the $k$th piece is

$$\Delta s_k = \int_{t_{k-1}}^{t_k} \|\mathbf{r}'(t)\| dt.$$

Let $\mathbf{r}_k = \mathbf{r}(\tilde{t}_k)$ where $t_{k-1} < \tilde{t}_k < t_k$. Then $\delta(\mathbf{r}_k)$ gives the density at that point in the $k$th piece. For sufficiently small $\Delta s_k$, $\Delta m_k \approx \delta(\mathbf{r}_k)\Delta s_k$ is approximately the mass of the $k$th piece.

**b.** If we think of the mass of the wire as a collection of points $\mathbf{r}_k$ with mass $\Delta m_k$, the center of mass of the wire is approximately the center of mass of the set of points, which is $\frac{\sum_{k=1}^{n} \Delta m_k \mathbf{r}_k}{\sum_{k=1}^{n} \Delta m_k}$ by the definition in Section 6.5. Note that $\sum_{k=1}^{n} \Delta m_k$ is the mass of the set of points. Let $\mathbf{r}_k = \langle x(t_k), y(t_k), z(t_k)\rangle$, so

$$(X_n, Y_n, Z_n) = \frac{\sum_{k=1}^{n} \langle x(\tilde{t}_k), y(\tilde{t}_k), z(\tilde{t}_k)\rangle \Delta m_k}{\sum_{k=1}^{n} \Delta m_k}.$$

Then $X_n = \frac{\sum_{k=1}^{n} x(\tilde{t}_k)\Delta m_k}{\sum_{k=1}^{n} \Delta m_k}$

$$Y_n = \frac{\sum_{k=1}^{n} y(\tilde{t}_k)\Delta m_k}{\sum_{k=1}^{n} \Delta m_k}$$

$$Z_n = \frac{\sum_{k=1}^{n} z(\tilde{t}_k)\Delta m_k}{\sum_{k=1}^{n} \Delta m_k}$$

**c.** As the wire is cut into smaller pieces by letting $n \to \infty$, $\sum_{k=1}^{n} \Delta m_k$ approaches $M$ and $\sum_{k=1}^{n} x(t_k)\Delta m_k$ approaches $\int_C x(t)\delta(\mathbf{r}(t))\, ds$.

Thus $X_n$ approaches $X = \frac{1}{M}\int_C x(t)\delta(\mathbf{r}(t))\, ds$. Similarly, $Y_n$ approaches $Y = \frac{1}{M}\int_C y(t)\delta(\mathbf{r}(t))\, ds$ and $Z_n$ approaches $Z = \frac{1}{M}\int_C z(t)\delta(\mathbf{r}(t))\, ds$.

Hence the center of mass of the wire is $(X, Y, Z)$

**23.** $\|\mathbf{r}'(t)\| = \left\|\langle 1, 2e^t, 2e^{2t}\rangle\right\| = 2e^{2t} + 1$

$\delta(\mathbf{r}(t)) = 2e^{3t}$

$$\text{mass} = \int_0^1 (4e^{5t} + 2e^{3t})dt$$

$$= \left(\frac{4}{5}e^{5t} + \frac{2}{3}e^{3t}\right)\Bigg|_0^1$$

$$= \frac{4}{5}e^5 + \frac{2}{3}e^3 - \frac{22}{15}$$

$$\approx 130.654 \text{ grams}$$

$\|\mathbf{r}'(t)\| = \left\|\langle 1, 2e^t, 2e^{2t}\rangle\right\| = 1 + 2e^{2t}$

$\delta(\mathbf{r}(t)) = 2e^{3t}$

$$\int_C x(t)\delta(\mathbf{r}(t))\, ds = \int_0^1 (2te^{3t} + 4te^{5t})\, dt$$

$$= \frac{86 + 100e^3 + 144e^5}{225}$$

$$\approx 104.294$$

$$\int_C y(t)\delta(\mathbf{r}(t))\, ds = \int_0^1 (4e^{4t} + 8e^{6t})\, dt$$

$$= \frac{-7 + 3e^4 + 4e^6}{3}$$

$$\approx 590.170$$

$$\int_C z(t)\delta(\mathbf{r}(t))\, ds = \int_0^1 (2e^{5t} + 4e^{7t})\, dt$$

$$= \frac{-34 + 14e^5 + 20e^7}{35}$$

$$\approx 685.041$$

$$X = \frac{1}{M}\int_C x(t)\delta(\mathbf{r}(t))\, dt \approx 0.798,$$

$$Y = \frac{1}{M}\int_C y(t)\delta(\mathbf{r}(t))\, dt \approx 4.517,$$

$$Z = \frac{1}{M}\int_C z(t)\delta(\mathbf{r}(t))\, dt \approx 5.243$$

**25.** $\|\mathbf{r}'(t)\| = \|\langle 1, 2\cos t, -2\sin t\rangle\| = \sqrt{5}$

$\delta(\mathbf{r}(t)) = 4 + 4\cos t \sin t$

$$M = \int_C \delta(\mathbf{r}(t))ds = \int_0^{2\pi} \sqrt{5}(4 + 4\cos t \sin t)dt = \sqrt{5}(4t + 2\sin^2 t)\Big|_0^{2\pi} = 8\sqrt{5}\pi$$

$$X = \frac{1}{M}\int_0^{2\pi} x(t)\delta(\mathbf{r}(t))ds = \frac{1}{M}\int_0^{2\pi} \sqrt{5}(4t + 4t\cos t \sin t)dt = \frac{1}{8\pi}\int_0^{2\pi} (4t + 4t\cos t \sin t)dt = \frac{1}{4}(4\pi - 1)$$

$$Y = \frac{1}{M}\int_0^{2\pi} y(t)\delta(\mathbf{r}(t))ds = \frac{1}{M}\int_0^{2\pi} \sqrt{5}(8\sin t + 8\cos t \sin^2 t)dt = 0$$

$$Z = \frac{1}{M}\int_0^{2\pi} z(t)\delta(\mathbf{r}(t))ds = \frac{1}{M}\int_0^{2\pi} \sqrt{5}(8\cos t + 8\cos^2 t \sin t)dt = 0$$

## Section 12.2 Vector Fields, Work and Flow

**1.** $\mathbf{F}(1, 0) = \left\langle \frac{1}{2}, 0 \right\rangle$; $\mathbf{F}(1, 1) = \left\langle \frac{1}{2}, -\frac{1}{2} \right\rangle$; $\mathbf{F}\left(-1, \frac{1}{2}\right) = \left\langle -\frac{1}{2}, -\frac{1}{4} \right\rangle$; $\mathbf{F}(-1, -1) = \left\langle -\frac{1}{2}, \frac{1}{2} \right\rangle$

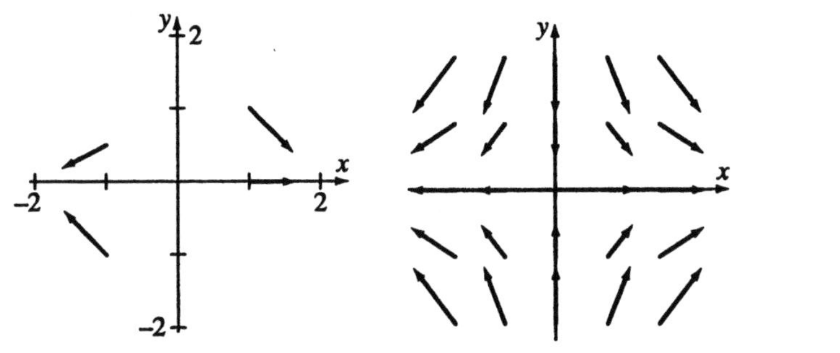


**3.** $\mathbf{F}(1, 1) = \langle \sqrt{2}, \sqrt{2} \rangle$; $\mathbf{F}(-2, 4) = \langle 2\sqrt{5}, 2\sqrt{5} \rangle$; $\mathbf{F}(0, -2) = \langle 2, 2 \rangle$; $\mathbf{F}\left(-\frac{\sqrt{2}}{2}, \frac{\sqrt{2}}{2}\right) = \langle 1, 1 \rangle$

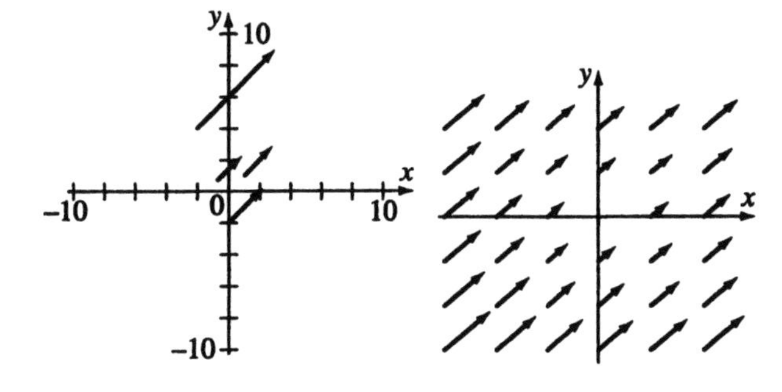

**5.** $\mathbf{F}(1, 1, 1) = \langle 1, 1, 1\rangle$; $\mathbf{F}(-2, 0, 1) = \langle 0, -2, 0\rangle$; $\mathbf{F}(0, -3, 0) = \langle 0, 0, 0\rangle$

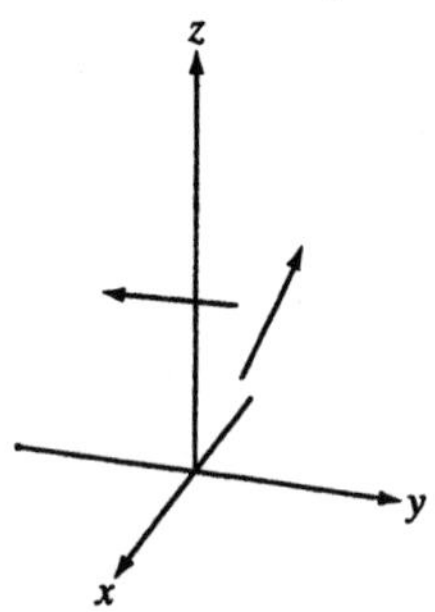


The second graph shows several vectors on the sphere of center (0, 0, 0) and radius 1. The third graph shows several vectors on the plane $z = 2$.

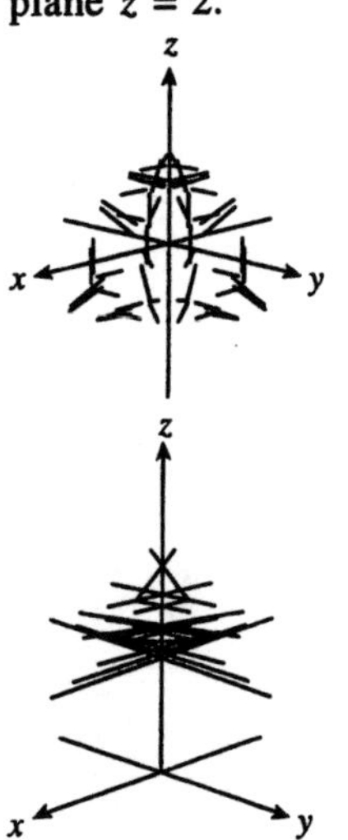


**7.** $\mathbf{F}(1, 2, 3) = \left\langle \frac{1}{\sqrt{14}}, \frac{2}{\sqrt{14}}, \frac{3}{\sqrt{14}} \right\rangle$;

$\mathbf{F}(-1, 4, 0) = \left\langle -\frac{1}{\sqrt{17}}, \frac{4}{\sqrt{17}}, 0 \right\rangle$;

$\mathbf{F}\left(\frac{1}{\sqrt{3}}, \frac{1}{\sqrt{3}}, \frac{1}{\sqrt{3}}\right) = \left\langle \frac{1}{\sqrt{3}}, \frac{1}{\sqrt{3}}, \frac{1}{\sqrt{3}} \right\rangle$

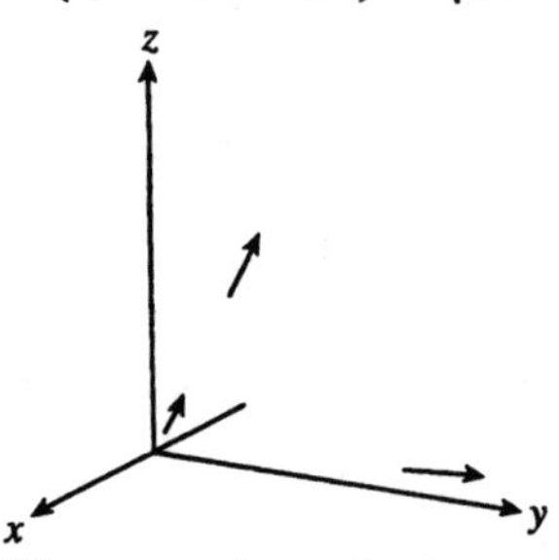


The second graph shows several vectors on the sphere of center (0, 0, 0) and radius 1. The third graph shows several vectors on the plane $z = 2$.

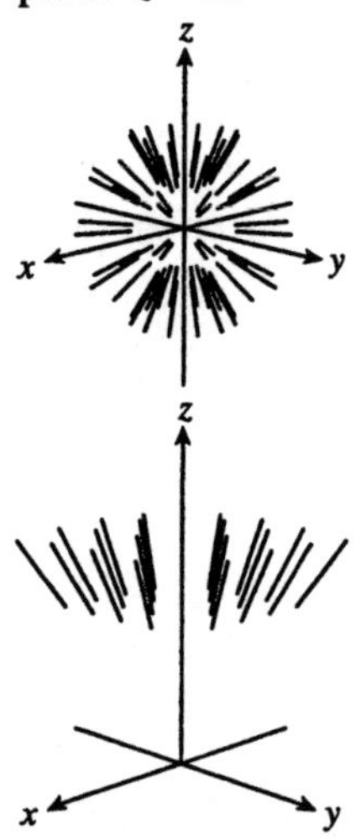


**9.** $C$ can be described by $\mathbf{r}(t) = \langle 3t, 4t\rangle$ for $0 \le t \le 1$.

$\mathbf{r}'(t) = \langle 3, 4\rangle$; $\mathbf{F}(\mathbf{r}(t)) = \langle 7t, -t\rangle$

$$\int_C (\mathbf{F}\cdot\mathbf{T})\,ds = \int_C \mathbf{F}\cdot d\mathbf{r} = \int_0^1 \langle 7t, -t\rangle\cdot\langle 3, 4\rangle\,dt$$

$$= \int_0^1 17t\,dt$$

$$= \frac{17}{2}t^2\Big|_0^1$$

$$= \frac{17}{2}$$

**11.** $C$ can be described by $\mathbf{r}(t) = \langle 1-t, 2-2t, 3-3t\rangle$ for $0 \le t \le 2$.

$\mathbf{r}'(t) = \langle -1, -2, -3\rangle$; $\mathbf{F}(\mathbf{r}(t)) = \langle e^{3-3t}, 3-6t+3t^2, 2-2t\rangle$

$$\int_C (\mathbf{F}\cdot\mathbf{T})\,ds = \int_C \mathbf{F}\cdot d\mathbf{r} = \int_0^2 \langle e^{3-3t}, 3-6t+3t^2, 2-2t\rangle\cdot\langle -1, -2, -3\rangle\,dt$$

$$= \int_0^2 (-e^{3-3t} - 6t^2 + 18t - 12)\,dt$$

$$= \left(\frac{1}{3}e^{3-3t} - 2t^3 + 9t^2 - 12t\right)\Bigg|_0^2$$

$$= -\frac{1}{3}e^3 + \frac{1}{3}e^{-3} - 4$$

$$\approx -10.68$$

**13.** $C$ can be described by $\mathbf{r}(t) = \langle -3t, 4t\rangle$ for $0 \le t \le 1$.

$\mathbf{r}'(t) = \langle -3, 4\rangle$; $\mathbf{T}(\mathbf{r}(t)) = \dfrac{1}{\|\mathbf{r}'(t)\|}\mathbf{r}'(t) = \dfrac{1}{5}\langle -3, 4\rangle$; $\mathbf{N}(\mathbf{r}(t)) = \dfrac{1}{5}\langle 4, 3\rangle$; $\mathbf{v}(\mathbf{r}(t)) = \langle 5t, -10t\rangle$

$$\int_C (\mathbf{v}\cdot\mathbf{N})\,ds = \int_0^1 \mathbf{v}(\mathbf{r}(t))\cdot\mathbf{N}(\mathbf{r}(t))\|\mathbf{r}'(t)\|dt = \int_0^1 \langle 5t, -10t\rangle\cdot\langle 4, 3\rangle\,dt = \int_0^1 -10t\,dt = -5t^2\Big|_0^1 = -5$$

(Choosing $\mathbf{N}(\mathbf{r}(t)) = \frac{1}{5}\langle -4, -3\rangle$ instead would lead to a final answer of 5.)

**15.** $\mathbf{r}'(t) = \langle 1, 2t, 3t^2\rangle$

$$\int_C 2xyz\,dx + x^2z\,dy + x^2y\,dz = \int_0^2 [2t^6 + 2t^6 + 3t^6]\,dt = \int_0^2 7t^6\,dt = t^7\Big|_0^2 = 128$$

**17.** $C$ can be described by $\mathbf{r}(t) = \langle -t, t^2\rangle$ for $0 \le t \le 2$.

$\mathbf{r}'(t) = \langle -1, 2t\rangle$; $\mathbf{F}(\mathbf{r}(t)) = \left\langle -t\sqrt{t^2+t^4}, t^2\sqrt{t^2+t^4}\right\rangle$

$$\int_C \mathbf{F}\cdot d\mathbf{r} = \int_0^2 \left\langle -t\sqrt{t^2+t^4}, t^2\sqrt{t^2+t^4}\right\rangle\cdot\langle -1, 2t\rangle\,dt$$

$$= \int_0^2 (t+2t^3)\sqrt{t^2+t^4}\,dt$$

$$= \frac{1}{3}(t^2+t^4)^{3/2}\Big|_0^2$$

$$= \frac{40\sqrt{5}}{3} \approx 29.81$$

**19.** The distance of the object from the center of the earth is $\sqrt{x^2+y^2+z^2}$. Thus $\|\mathbf{F}\| = \dfrac{GM}{x^2+y^2+z^2}$.

Since $\mathbf{F}$ points from $(x, y, z)$ towards the center of the earth, it is parallel to the vector $\langle -x, -y, -z\rangle = -\langle x, y, z\rangle$ so $\mathbf{F}$ is parallel to the unit vector $-\dfrac{1}{\sqrt{x^2+y^2+z^2}}\langle x, y, z\rangle$. Hence

$$\mathbf{F} = \|\mathbf{F}\|\left(-\frac{1}{\sqrt{x^2+y^2+z^2}}\langle x, y, z\rangle\right) = -\frac{GM}{(x^2+y^2+z^2)^{3/2}}\langle x, y, z\rangle$$

**21.** $C$ can be described by $\mathbf{r}(t) = \langle t, t^2 \rangle$ for $-2 \le t \le 1$.

$\mathbf{r}'(t) = \langle 1, 2t \rangle$; $\mathbf{T}(\mathbf{r}(t)) = \dfrac{1}{\|\mathbf{r}'(t)\|}\mathbf{r}'(t) = \dfrac{1}{\|\mathbf{r}'(t)\|}\langle 1, 2t \rangle$; $\mathbf{N}(\mathbf{r}(t)) = \dfrac{1}{\|\mathbf{r}'(t)\|}\langle 2t, -1 \rangle$; $\mathbf{v}(\mathbf{r}(t)) = \langle 2t^4 - 1, 2t - t^2 \rangle$

$$\text{flux across } C = \int_C \mathbf{v}\cdot\mathbf{N}\,ds = \int_{-2}^{1} \mathbf{v}(\mathbf{r}(t))\cdot\mathbf{N}(\mathbf{r}(t))\|\mathbf{r}'(t)\|dt = \int_{-2}^{1}(4t^5 + t^2 - 4t)dt = \left(\frac{2}{3}t^6 + \frac{1}{3}t^3 - 2t^2\right)\Big|_{-2}^{1} = -33 \text{ g/s}$$

**23.** $\mathbf{r}'(t) = \langle -\pi \sin \pi t, \pi \cos \pi t, 1 \rangle$

$$\mathbf{T}(\mathbf{r}(t)) = \frac{1}{\|\mathbf{r}'(t)\|}\mathbf{r}'(t) = \frac{1}{\|\mathbf{r}'(t)\|}\langle -\pi \sin \pi t, \pi \cos \pi t, 1 \rangle$$

$$\mathbf{F}(\mathbf{r}(t)) = \left\langle \sin \pi t \ln(t+1), \cos \pi t \ln(t+1), \frac{\cos \pi t \sin \pi t}{t+1} \right\rangle$$

$$\begin{aligned}\text{work} = \int_C \mathbf{F}\cdot\mathbf{T}\,ds = \int_0^1 \mathbf{F}(\mathbf{r}(t))\cdot\mathbf{T}(\mathbf{r}(t))\|\mathbf{r}'(t)\|dt &= \int_0^1\left(-\pi \sin^2 \pi t \ln(t+1) + \pi \cos^2 \pi t \ln(t+1) + \frac{\cos \pi t \sin \pi t}{t+1}\right)dt \\ &= \cos \pi t \sin \pi t \ln(t+1)\Big|_0^1 \\ &= 0\end{aligned}$$

**25.** **a.** There appears to be neither a sink nor a source inside the triangle.

**b.** Let $C_1$ be the segment from (0, 0) to (3, 0), $C_2$ the segment from (3, 0) to (0, 4), and $C_3$ the segment from (0, 4) to (0, 0).

We can describe $C_1$ by $\mathbf{r}_1(t) = \langle t, 0 \rangle$ $0 \le t \le 3$, $C_2$ by $\mathbf{r}_2(t) = \langle 3 - 3t, 4t \rangle$, $0 \le t \le 1$, and $C_3$ by $\mathbf{r}_3(t) = \langle 0, 4 - t \rangle$, $0 \le t \le 4$.

Thus $\mathbf{r}_1'(t) = \langle 1, 0 \rangle$, $\mathbf{r}_2'(t) = \langle -3, 4 \rangle$, and $\mathbf{r}_3'(t) = \langle 0, -1 \rangle$. $\mathbf{v}(\mathbf{r}_1(t)) = \langle t, t \rangle$, $\mathbf{v}(\mathbf{r}_2(t)) = \langle 3 + t, 3 - 7t \rangle$, and $\mathbf{v}(\mathbf{r}_3(t)) = \langle 4 - t, -4 + t \rangle$.

$$\begin{aligned}\text{flux} = \int_C \mathbf{v}\cdot\mathbf{N}\,ds &= \int_{C_1} \mathbf{v}\cdot\mathbf{N}\,ds + \int_{C_2} \mathbf{v}\cdot\mathbf{N}\,ds + \int_{C_3} \mathbf{v}\cdot\mathbf{N}\,ds \\ &= \int_0^3 \langle t, t \rangle\cdot\langle 0, -1 \rangle dt + \int_0^1 \langle 3+t, 3-7t \rangle\cdot\langle 4, 3 \rangle dt + \int_0^4 \langle 4-t, -4+t \rangle\cdot\langle -1, 0 \rangle dt \\ &= \int_0^3 (-t)\,dt + \int_0^1 (21 - 17t)\,dt + \int_0^4 (-4 + t)\,dt \\ &= -\frac{1}{2}t^2\Big|_0^3 + \left(21t - \frac{17}{2}t^2\right)\Big|_0^1 + \left(-4t + \frac{1}{2}t^2\right)\Big|_0^4 \\ &= 0\end{aligned}$$

**c.**

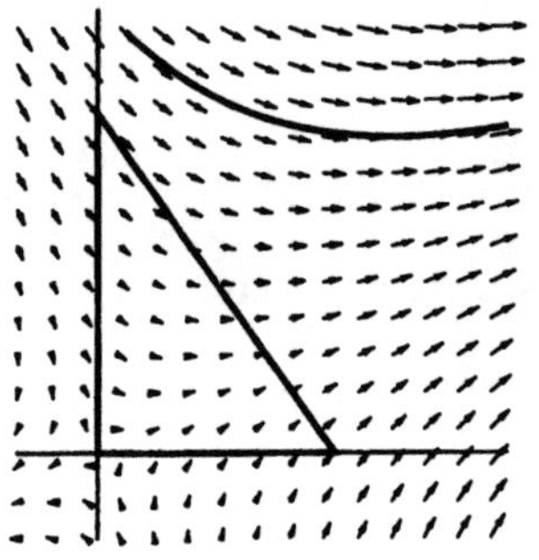

**27. a.** As $a \le t \le b$, $a \le a + b - t \le b$. Thus $\mathbf{r}_1(t)$ traces points of $C$. $\mathbf{r}_1(a) = \mathbf{r}(b) = \mathbf{B}$ and $\mathbf{r}_1(b) = \mathbf{r}(a) = \mathbf{A}$, so the path $-C$ goes from $\mathbf{B}$ to $\mathbf{A}$.

**b.** $$\int_{-C} (\mathbf{F} \cdot \mathbf{T})\, ds = \int_a^b \mathbf{F}(\mathbf{r}_1(t)) \cdot \mathbf{r}_1'(t)\, dt = \int_a^b \mathbf{F}(\mathbf{r}(a+b-t)) \cdot \mathbf{r}'(a+b-t)\, dt$$

We make a change of variables by letting $u = a + b - t$, so $du = -dt$ and when $t = a$ and $t = b$, $u = b$ and $u = a$, respectively.

Then $$\int_{-C} (\mathbf{F} \cdot \mathbf{T}) ds = \int_b^a \mathbf{F}(\mathbf{r}(u)) \cdot \mathbf{r}'(u)\, du = -\int_a^b \mathbf{F}(\mathbf{r}(u)) \cdot \mathbf{r}'(u)\, du = -\int_C \mathbf{F} \cdot \mathbf{T}\, ds$$

**29.** Let $C$ be parameterized by $\mathbf{r}(t) = \langle x(t), y(t), z(t) \rangle$, $a \le t \le b$ with $\mathbf{r}(a) = \mathbf{A}$ and $\mathbf{r}(b) = \mathbf{B}$.

$$\int_C \mathbf{F} \cdot d\mathbf{r} = \int_a^b \mathbf{F}(\mathbf{r}(t)) \cdot \mathbf{r}'(t)\, dt = \int_a^b \left( y(t) \ln z(t) x'(t) + x(t) \ln z(t) y'(t) + \frac{x(t)y(t)}{z(t)} z'(t) \right) dt$$

$$= \int_a^b \frac{d}{dt}(x(t)y(t) \ln z(t))\, dt$$

$$= x(b)y(b) \ln z(b) - x(a)y(a) \ln z(a)$$

Let $f(x, y, z) = xy \ln z$. Then $\int_C \mathbf{F} \cdot d\mathbf{r} = f(\mathbf{B}) - f(\mathbf{A})$.

**31.** Let $C$ be parameterized by $\mathbf{r}(t) = \langle x(t), y(t), z(t) \rangle$, $a \le t \le b$ with $\mathbf{r}(a) = \mathbf{A}$ and $\mathbf{r}(b) = \mathbf{B}$.

$$\nabla f(x, y, z) = \left\langle 2x \sin(xy + yz) + x^2 y \cos(xy + yz),\ x^2(x + z) \cos(xy + yz),\ x^2 y \cos(xy + yz) \right\rangle$$

$$\int_C (\mathbf{F} \cdot \mathbf{T})\, ds = \int_a^b \nabla f(\mathbf{r}(t)) \cdot \mathbf{r}'(t)\, dt$$

Before going on, observe that $\frac{d}{dt} = (f(\mathbf{r}(t))) = \nabla f(\mathbf{r}(t)) \cdot \mathbf{r}'(t)$

Thus $\int_C (\mathbf{F} \cdot \mathbf{T}) ds = \int_a^b \frac{d}{dt}(f(\mathbf{r}(t))\, dt = f(\mathbf{r}(b)) - f(\mathbf{r}(a)) = f(\mathbf{B}) - f(\mathbf{A})$.

Hence for $C$ any simple closed smooth curve, parameterized by $\mathbf{r}(t) = \langle x(t), y(t), z(t) \rangle$, $a \le t \le b$ with $\mathbf{r}(a) = \mathbf{r}(b)$, $\int_C (\mathbf{F} \cdot \mathbf{T})\, ds = 0$. Note that the precise nature of $\nabla f(x, y, z)$ is irrelevant.

## Section 12.3 The Fundamental Theorem of Line Integrals

**1.** $$\frac{\partial f}{\partial x} = \frac{1}{2}\sqrt{y}\frac{1}{\sqrt{x}} + 2x = \frac{1}{2}\sqrt{\frac{y}{x}} + 2x; \quad \frac{\partial f}{\partial y} = \frac{1}{2}\sqrt{x}\frac{1}{\sqrt{y}} = \frac{1}{2}\sqrt{\frac{x}{y}}$$

Thus $\nabla f(x, y) = \left\langle \frac{1}{2}\sqrt{\frac{y}{x}} + 2x, \frac{1}{2}\sqrt{\frac{x}{y}} \right\rangle = \mathbf{F}(x, y)$.

$\mathbf{r}(0) = \langle 1, 1 \rangle$ and $\mathbf{r}(3) = \langle 28, 2 \rangle$, so by the Fundamental Theorem of Line Integrals,

$$\int_C (\mathbf{F} \cdot \mathbf{T})\, ds = f(28, 2) - f(1, 1) = \left(\sqrt{56} + 783\right) - 1 = 782 + 2\sqrt{14} \approx 789.48$$

**3.** $$\frac{\partial f}{\partial x} = e^{yz} - yz; \quad \frac{\partial f}{\partial y} = xze^{yz} - xz; \quad \frac{\partial f}{\partial y} = xye^{yz} - xy$$

Thus $\nabla f(x, y, z) = \left\langle e^{yz} - yz, xze^{yz} - xz, xye^{yz} - xy \right\rangle = \mathbf{F}(x, y)$.

Since $C$ is a closed curve and $\mathbf{F}$ is a conservative vector field, $\int_C (\mathbf{F} \cdot \mathbf{T})\, ds = 0$.

**5.** Let $M(x, y) = x + y + 2$ and $N(x, y) = x - 3y - 4$.

$\frac{\partial N}{\partial x} = 1 = \frac{\partial M}{\partial y}$, so **F** satisfies the hypotheses of the Gradient Test.

$f(x, y) = \int M(x, y)\,dx = \int (x+y+2)\,dx = \frac{1}{2}x^2 + xy + 2x + C(y)$

Differentiate the above with respect to $y$, so $\frac{\partial f}{\partial y} = x + C'(y) = N(x, y)$.

Thus $C'(y) = -3y - 4$, so $C(y) = -\frac{3}{2}y^2 - 4y + c$.

Hence $f(x, y) = \frac{1}{2}x^2 + xy - \frac{3}{2}y^2 + 2x - 4y$ is a potential function for **F**.

**7.** Let $M(x, y) = y^2 \sec^2(xy)$ and $N(x, y) = xy\sec^2(xy) + \tan(xy)$.

$\frac{\partial N}{\partial x} = y\sec^2(xy) + 2xy^2\sec^2(xy)\tan(xy) + y\sec^2(xy) = 2y\sec^2(xy) + 2xy^2\sec^2(xy)\tan(xy)$

$\frac{\partial M}{\partial y} = 2y\sec^2(xy) + 2xy^2\sec^2(xy)\tan(xy)$

Since $\frac{\partial N}{\partial x} = \frac{\partial M}{\partial y}$, **F** satisfies the hypotheses of the Gradient Test.

$f(x, y) = \int M(x, y)\,dx = y\tan(xy) + C(y)$

Differentiate the above with respect to $y$.

$\frac{\partial f}{\partial y} = \tan(xy) + xy\sec^2(xy) + C'(y) = N(x, y)$

Thus $C'(y) = 0$, so $C(y) = c$.

Hence $f(x, y) = y\tan(xy)$ is a potential function for **F**.

**9.** Let $P(x, y, z) = 0$, $Q(x, y, z) = 2yz - z^2$, and $R(x, y, z) = y^2 - 2yz$.

$\frac{\partial Q}{\partial x} = 0 = \frac{\partial P}{\partial y}$, $\frac{\partial R}{\partial x} = 0 = \frac{\partial P}{\partial z}$, $\frac{\partial R}{\partial y} = 2y - 2z = \frac{\partial Q}{\partial z}$

$F$ satisfies the hypotheses of the Gradient Test.

$f(x, y, z) = \int Q(x, y, z)\,dy = y^2z - yz^2 + C_1(x, z)$

Differentiate the above with respect to $z$.

$\frac{\partial f}{\partial z} = y^2 - 2yz + \frac{\partial C_1}{\partial y}(x, z) = R(x, y, z)$

Thus $\frac{\partial C_1}{\partial y}(x, z) = 0$, so $C_1(x, z) = C_2(x)$ and $f(x, y, z) = y^2z - yz^2 + C_2(x)$.

Differentiate the above with respect to $x$.

$\frac{\partial f}{\partial x} = C_2'(x) = P(x, y, z)$

Thus $C_2'(x) = 0$, so $C_2(x) = c$.

Hence $f(x, y, z) = y^2z - yz^2$ is a potential function.

**11.** Let $P(x, y, z) = e^{yz}$, $Q(x, y, z) = e^{xz}$, $R(x, y, z) = e^{xy}$.

$\frac{\partial Q}{\partial x} = ze^{xz}$; $\frac{\partial P}{\partial y} = ze^{yz}$

Thus $\frac{\partial Q}{\partial x} \neq \frac{\partial P}{\partial y}$ so $F$ does not satisfy all the conditions of the Gradient Test.

**F** does not have a potential function.

**13.** Let $\mathbf{F}(x, y)=\left\langle \frac{x}{\sqrt{x^2+y^2}}, \frac{y}{\sqrt{x^2+y^2}} \right\rangle$. Observe that $f(x, y)=\sqrt{x^2+y^2}$ is a potential function for **F**:

$$\frac{\partial f}{\partial x}=\frac{x}{\sqrt{x^2+y^2}};\ \frac{\partial f}{\partial y}=\frac{y}{\sqrt{x^2+y^2}}$$

Thus, $\int_C \frac{x}{\sqrt{x^2+y^2}}dx+\frac{y}{\sqrt{x^2+y^2}}dy=\int_C (\mathbf{F}\cdot\mathbf{T})ds=f(2, 1)-f(0, 1)=\sqrt{5}-1\approx 1.24.$

**15.** Let $\mathbf{F}(x, y, z) = \langle y, z, x\rangle$. Note that since $\frac{\partial}{\partial x}z=0\neq 1=\frac{\partial}{\partial y}y$, **F** does not have a potential function.

$C$ can be parameterized as $C_1$: $\mathbf{r}_1(t)=\langle t, t, 0\rangle$, $0\le t\le 1$ followed by $C_2$: $\mathbf{r}_2(t)=\langle 1-t, 1-t, t\rangle$, $0\le t\le 1$.

$$\begin{aligned}\int_C y\,dx+z\,dy+x\,dz &= \int_{C_1} y\,dx+z\,dy+x\,dz+\int_{C_2} y\,dx+z\,dy+x\,dz\\ &=\int_0^1 (t+0+0)dt+\int_0^1 [(1-t)(-1)+(t)(-1)+(1-t)(1)]dt\\ &=\int_0^1 t\,dt-\int_0^1 t\,dt\\ &=0\end{aligned}$$

**17.** Let $M(x, y) = y - 2$ and $N(x, y) = x + 1$.

$f(x, y)=\int M(x, y)\,dx=xy-2x+C(y)$

Differentiate the above with respect to $y$, so $\frac{\partial f}{\partial y}=x+C'(y)=N(x, y)$.

Thus $C'(y)=1$, so $C(y) = y + c$ and $f(x, y) = xy - 2x + y + c$. $f(1, 1) = c = 0$, so $c = 0$.

$f(x, y) = xy - 2x + y$

**19.** Let $M(x, y, z) = \sin(yz)$, $N(x, y, z) = xz\cos(yz) - y$, and $P(x, y, z) = xy\cos(yz)$.

$f(x, y, z)=\int \sin(yz)\,dx=x\sin(yz)+C_1(y, z)$

Differentiate the above with respect to $y$.

$\frac{\partial f}{\partial y}=xz\cos(yz)+\frac{\partial C_1}{\partial y}(y, z)=N(x, y, z)$

Thus $\frac{\partial C_1}{\partial y}=(y, z)=-y$, so $C_1(y, z)=-\frac{1}{2}y^2+C_2(z)$.

$f(x, y, z)=x\sin(yz)-\frac{1}{2}y^2+C_2(z)$

Differentiate the above with respect to $z$.

$\frac{\partial f}{\partial z}=xy\cos(yz)+C_2'(z)=P(x, y, z)$

Thus $C_2'(z)=0$, so $C_2=c$.

$f(x, y, z)=x\sin(yz)-\frac{1}{2}y^2+c$

$f\left(2, 1, \frac{\pi}{2}\right)=\frac{3}{2}+c=-3$, so $c=-\frac{9}{2}$.

$f(x, y, z)=x\sin(yz)-\frac{1}{2}y^2-\frac{9}{2}$

**21. a.** Represent $C$ parametrically by $\mathbf{r}(t) = \langle \cos t, \sin t \rangle$, $0 \le t \le 2\pi$.

Then $\mathbf{N}(t) = \langle \cos t, \sin t \rangle$ and $\mathbf{v}(\mathbf{r}(t)) = \langle \cos^3 t - 3\cos t \sin^2 t, 3\cos^2 t \sin t - \sin^3 t \rangle$.

$$\begin{aligned}
\text{flux} = \int_C \mathbf{v}(\mathbf{r}(t)) \cdot \mathbf{N}\, ds &= \int_0^{2\pi} \langle \cos^3 t - 3\cos t \sin^2 t, 3\cos^2 t \sin t - \sin^3 t \rangle \cdot \langle \cos t, \sin t \rangle dt \\
&= \int_0^{2\pi} (\cos^4 t - 3\cos^2 t \sin^2 t + 3\cos^2 t \sin^2 t - \sin^4 t) dt \\
&= \int_0^{2\pi} (\cos^4 t - \sin^4 t) dt \\
&= \int_0^{2\pi} (\cos^2 t - \sin^2 t)(\cos^2 t + \sin^2 t) dt \\
&= \int_0^{2\pi} \cos 2t\, dt \\
&= \left(\frac{1}{2}\sin 2t\right)\Big|_0^{2\pi} \\
&= 0
\end{aligned}$$

**b.** Represent $C$ parametrically by $\mathbf{r}(t) = \langle \cos t, \sin t \rangle$. $\mathbf{r}'(t) = \langle -\sin t, \cos t \rangle$ and $\mathbf{v}(\mathbf{r}(t)) = \langle \cos^3 t - 3\cos t \sin^2 t, 3\cos^2 t \sin t - \sin^3 t \rangle$.

Thus,

$$\begin{aligned}
\text{circulation} = \int_C \mathbf{v} \cdot \mathbf{T}\, ds &= \int_0^{2\pi} \mathbf{v}(\mathbf{r}(t)) \cdot \mathbf{r}'(t)\, dt \\
&= \int_0^{2\pi} \langle \cos^3 t - 3\cos t \sin^2 t, 3\cos^2 t \sin t - \sin^3 t \rangle \cdot \langle -\sin t, \cos t \rangle dt \\
&= \int_0^{2\pi} (-\cos^3 t \sin t + 3\cos t \sin^3 t + 3\cos^3 t \sin t - \cos t \sin^3 t)\, dt \\
&= \int_0^{2\pi} (2\cos^3 t \sin t + 2\cos t \sin^3 t)\, dt \\
&= \left(-\frac{1}{2}\cos^4 t + \frac{1}{2}\sin^4 t\right)\Big|_0^{2\pi} \\
&= 0
\end{aligned}$$

**23.** Observe that $f(x, y, z) = -\dfrac{GM}{\sqrt{x^2 + y^2 + z^2}}$ is a potential function for $\mathbf{F}$ and therefore $\mathbf{F}$ is a conservative vector field. The domain of $f$ is $\{(x, y, z) : x^2 + y^2 + z^2 > R^2\}$ where $R$ is the radius of the Earth.

**25.** Describe $C$ by $\mathbf{r}(t) = \langle t, t, t \rangle$, $1 \le t \le 3$.

$\mathbf{r}'(t) = \langle 1, 1, 1 \rangle$; $\mathbf{F}(\mathbf{r}(t)) = \langle 1, 1, 1 \rangle$

$$\text{work} = \int_C \mathbf{F} \cdot d\mathbf{r} = \int_1^3 \langle 1, 1, 1 \rangle \cdot \langle 1, 1, 1 \rangle dt = \int_1^3 3\, dt = 6$$

**27.** Suppose $\mathbf{G} = \nabla g$, so $P = g_x$, $Q = g_y$, and $R = g_z$.

$Q_x = g_{yx} = g_{xy} = P_y$

$R_x = g_{zx} = g_{xz} = P_z$

$R_y = g_{zy} = g_{yz} = Q_z$

**29.** **a.** Let $M = \dfrac{-x}{(x^2+y^2)^2}$ and $N = \dfrac{-y}{(x^2+y^2)^2}$.

$\dfrac{\partial M}{\partial y} = \dfrac{4xy}{(x^2+y^2)^3}$; $\dfrac{\partial N}{\partial x} = \dfrac{4xy}{(x^2+y^2)^3}$

Observe that if $f(x, y) = \dfrac{1}{2(x^2+y^2)}$, then $\nabla f = \mathbf{F}$ on $D$. Thus $\mathbf{F}$ satisfies the conditions of the Gradient Test.

**b.** The theorem does not imply that if $D$ is not simply connected that a function $\mathbf{F}$ cannot have a potential function when $\dfrac{\partial N}{\partial y} = \dfrac{\partial M}{\partial x}$.

**31.** Let $M = (2xy + y^2 + 2)e^{xy}$ and $N = (2x^2 + xy)e^{xy} + u(x, y)$.

$\dfrac{\partial N}{\partial x} = (4x + y + 2x^2y + xy^2)e^{xy} + \dfrac{\partial u}{\partial x}$

$\dfrac{\partial M}{\partial y} = (4x + 2y + 2x^2y + xy^2)e^{xy}$

If $\dfrac{\partial N}{\partial x} = \dfrac{\partial M}{\partial y}$, then $\dfrac{\partial u}{\partial x} = ye^{xy}$.

Let $u(x, y) = e^{xy}$. If $\mathbf{G}(x, y) = \left\langle (2xy + y^2 + 2)e^{xy}, (2x^2 + xy + 1)e^{xy} \right\rangle$, a potential function for $\mathbf{G}$ is $g(x, y) = (2x + y)e^{xy}$.

$\displaystyle\int_C \mathbf{G}\cdot\mathbf{T}\,ds = g\left(2, \frac{1}{2}\right) - g(1, 1) = \frac{3}{2}e$

$\mathbf{F} - \mathbf{G} = (0, -u)$, so $\displaystyle\int_C (\mathbf{F}-\mathbf{G})\cdot\mathbf{T}\,ds = \int_C -e^{xy}dy.$

With $x = \dfrac{1}{y}$, $\displaystyle\int_C (\mathbf{F}-\mathbf{G})\cdot\mathbf{T}\,ds = \int_1^{1/2} -e\,dy = \frac{1}{2}e$

$\displaystyle\int_C (\mathbf{F}\cdot\mathbf{T})ds = \frac{3}{2}e + \frac{1}{2}e = 2e$

**33.** Let $P(x, y, z) = 2xy + z^2$, $Q(x, y, z) = x^2 + 2yz$, and $R(x, y, z) = y^2 + u(x, y, z)$.

$\dfrac{\partial P}{\partial z} = 2z$; $\dfrac{\partial R}{\partial x} = \dfrac{\partial u}{\partial x}$; $\dfrac{\partial Q}{\partial z} = 2y$; $\dfrac{\partial R}{\partial y} = 2y + \dfrac{\partial u}{\partial y}$

Thus $\dfrac{\partial u}{\partial x} = 2z$ and $\dfrac{\partial u}{\partial y} = 0$.

Let $u(x, y, z) = 2xz$. If $\mathbf{G}(x, y, z) = \left\langle 2xy + z^2, x^2 + 2yz, y^2 + 2xz \right\rangle$, then a potential function for $\mathbf{G}$ is $g(x, y, z) = x^2y + xz^2 + y^2z$.

$\displaystyle\int_C (\mathbf{G}\cdot\mathbf{T})ds = g(1, 0, 4\pi) - g(1, 0, 0) = 16\pi^2$

$\displaystyle\int_C ((\mathbf{F}-\mathbf{G})\cdot\mathbf{T})ds = \int_C (-2xz)dz = \int_0^{2\pi} (-8t\cos t)dt = (-8t\sin t - 8\cos t)\Big|_0^{2\pi} = 0$

Thus $\displaystyle\int_C (\mathbf{F}\cdot\mathbf{T})ds = 16\pi^2$.

**35. a.** $\mathbf{v}(\mathbf{c}_k)\cdot\mathbf{T}_k$ gives the signed length of the velocity component relative to $\mathbf{T}_k$. Then $(\mathbf{v}(\mathbf{c}_k)\cdot\mathbf{T}_k)\Delta s$ gives the real amount over time that moves with or against the direction of the unit tangent. Thus the sum gives an approximation of the "flow tangent to $C$."

**b.** As $n\to\infty$, the sum becomes an integral and flow $=\int_C(\mathbf{v}\cdot\mathbf{T})\,ds$.

**c.** The units are the amount of fluid per time. "Circulation" is a good name since it implies amount of fluid going around the curve over time.

## Section 12.4 Green's Theorem

**1.** Let $C$ be parameterized by $\mathbf{r}(t)=\langle\cos t,\sin t\rangle$, $0\le t\le 2\pi$.

$\mathbf{r}'(t)=\langle-\sin t,\cos t\rangle$

$$\int_C 2y\,dx+3\,dy=\int_0^{2\pi}[2\sin t(-\sin t)+3\cos t]\,dt$$

$$=\int_0^{2\pi}(-2\sin^2 t+3\cos t)\,dt$$

$$=\int_0^{2\pi}(-1+\cos 2t+3\cos t)\,dt$$

$$=\left(-t+\frac{1}{2}\sin 2t+3\sin t\right)\Big]_0^{2\pi}$$

$$=-2\pi$$

Next we use Green's Theorem.

$$\int_C 2y\,dx+3\,dy=\iint_R\left(\frac{\partial}{\partial x}3-\frac{\partial}{\partial y}2y\right)dA=\iint_R -2\,dA$$

Since the region is a circle of radius 1, $\iint_R dA=\pi$. Thus, $\iint_R -2\,dA=-2\pi$.

**3.** Let $C$ be parameterized by $\mathbf{r}(t)=\langle\sqrt{2}\cos t,\sqrt{2}\sin t\rangle$, $0\le t\le 2\pi$.

$\mathbf{r}'(t)=\langle-\sqrt{2}\sin t,\sqrt{2}\cos t\rangle$

$$\int_C(\mathbf{F}\cdot\mathbf{T})\,ds=\int_C\frac{x}{x^2+y^2+2}dx+\frac{y}{x^2+y^2+2}dy$$

$$=\int_0^{2\pi}\left[\frac{\sqrt{2}}{3}\cos t\left(-\sqrt{2}\sin t\right)+\frac{\sqrt{2}}{3}\sin t\left(\sqrt{2}\cos t\right)\right]dt$$

$$=\int_0^{2\pi}0\,dt=0$$

Next we use Green's Theorem.

$$\int_C(\mathbf{F}\cdot\mathbf{T})\,ds=\iint_R\left(\frac{\partial}{\partial x}\left(\frac{y}{x^2+y^2+2}\right)-\frac{\partial}{\partial y}\left(\frac{x}{x^2+y^2+2}\right)\right)dA$$

$$=\iint_R\left(\frac{2xy}{(x^2+y^2+2)^2}-\frac{2xy}{(x^2+y^2+2)^2}\right)dA$$

$$=\iint_R 0\,dA=0$$

**5.** Let $C = L_1 + L_2 + L_3 + L_4$ where each $L_j$ is a side of the square. They are parameterized by

$L_1$: $\mathbf{r}_1(t) = \langle t, 0\rangle$, $0 \le t \le 2$

$L_2$: $\mathbf{r}_2(t) = \langle 2, t\rangle$, $0 \le t \le 2$

$L_3$: $\mathbf{r}_3(t) = \langle 2-t, 2\rangle$, $0 \le t \le 2$

$L_4$: $\mathbf{r}_4(t) = \langle 0, 2-t\rangle$, $0 \le t \le 2$

$$\int_{L_1} (\mathbf{F}\cdot\mathbf{T})ds = \int_{L_1} (x^2y - 2xy)dx + (xy^2 + 3xy)dy = \int_0^2 0\,dt = 0$$

$$\int_{L_2} (\mathbf{F}\cdot\mathbf{T})ds = \int_{L_2} (x^2y - 2xy)dx + (xy^2 + 3xy)\,dy = \int_0^2 (2t^2 + 6t)\,dt = \left(\frac{2}{3}t^3 + 3t^2\right)\Bigg|_0^2 = \frac{52}{3}$$

$$\int_{L_3} (\mathbf{F}\cdot\mathbf{T})ds = \int_{L_3} (x^2y - 2xy)dx + (xy^2 + 3xy)\,dy = \int_0^2 (2(2-t)^2 - 4(2-t))(-1)\,dt = \int_0^2 (-2t^2 + 4t)dt$$

$$= \left(-\frac{2}{3}t^3 + 2t^2\right)\Bigg|_0^2 = \frac{8}{3}$$

$$\int_{L_4} (\mathbf{F}\cdot\mathbf{T})\,ds = \int_{L_4} (x^2y - 2xy)dx + (xy^2 + 3xy)\,dy = \int_0^2 0\,dt = 0$$

Thus, $\int_C (\mathbf{F}\cdot\mathbf{T})\,ds = \int_{L_1} (\mathbf{F}\cdot\mathbf{T})\,ds + \int_{L_2} (\mathbf{F}\cdot\mathbf{T})\,ds + \int_{L_3} (\mathbf{F}\cdot\mathbf{T})\,ds + \int_{L_4} (\mathbf{F}\cdot\mathbf{T})\,ds = 20.$

Next we use Green's Theorem.

$$\int_C (\mathbf{F}\cdot\mathbf{T})\,ds = \iint_R \left(\frac{\partial}{\partial x}(xy^2 + 3xy) - \frac{\partial}{\partial y}(x^2y - 2xy)\right)dA$$

$$= \iint_R ((y^2 + 3y) - (x^2 - 2x))dA$$

$$= \int_0^2\int_0^2 (-x^2 + y^2 + 2x + 3y)dx\,dy$$

$$= \int_0^2 \left(-\frac{1}{3}x^3 + xy^2 + x^2 + 3xy\right)\Bigg|_{x=0}^{x=2} dy$$

$$= \int_0^2 \left(2y^2 + 6y + \frac{4}{3}\right)dy$$

$$= \left(\frac{2}{3}y^3 + 3y^2 + \frac{4}{3}y\right)\Bigg|_0^2$$

$$= 20$$

**7.** Let $C$ be parameterized by $\mathbf{r}(t) = \langle a\cos t, a\sin t\rangle$, $0 \le t \le 2\pi$.

$\mathbf{r}'(t) = \langle -a\sin t, a\cos t\rangle$

$$\text{area} = \frac{1}{2}\int_C -y\,dx + x\,dy = \frac{1}{2}\int_0^{2\pi} a^2 dt = \pi a^2$$

**9.** Let $C$ be parameterized by

$\mathbf{r}(\theta) = \langle(1+\cos\theta)\cos\theta,\ (1+\cos\theta)\sin\theta\rangle = \left\langle \cos\theta + \cos^2\theta,\ \sin\theta + \cos\theta\sin\theta \right\rangle,\ 0 \le \theta \le 2\pi.$

$\mathbf{r}'(\theta) = \left\langle -\sin\theta - 2\cos\theta\sin\theta,\ \cos\theta + \cos^2\theta - \sin^2\theta \right\rangle$

$$\begin{aligned}
\text{area} &= \frac{1}{2}\int_C -y\,dx + x\,dy \\
&= \frac{1}{2}\int_C (-(\sin\theta + \cos\theta\sin\theta)(-\sin\theta - 2\cos\theta\sin\theta) + (\cos\theta + \cos^2\theta)(\cos\theta + \cos^2\theta - \sin^2\theta))d\theta \\
&= \frac{1}{2}\int_0^{2\pi} (\cos^4\theta + 2\cos^3\theta + \cos^2\theta + \sin^2\theta + 2\cos\theta\sin^2\theta + \cos^2\theta\sin^2\theta)d\theta \\
&= \frac{1}{2}\int_0^{2\pi} (\cos^2\theta + 2\cos\theta + 1)d\theta \\
&= \frac{1}{2}\int_0^{2\pi} \left(\frac{1}{2}\cos 2\theta + 2\cos\theta + \frac{3}{2}\right)d\theta \\
&= \frac{1}{2}\left(\frac{1}{4}\sin 2\theta + 2\sin\theta + \frac{3}{2}\theta\right)\Bigg|_0^{2\pi} \\
&= \frac{3\pi}{2}
\end{aligned}$$

**11.** 
$$\begin{aligned}
\text{flux} = \int_C (\mathbf{v}\cdot\mathbf{N})ds = \int_C (\tilde{\mathbf{v}}\cdot\mathbf{T})ds &= \int_C -(x-y)\,dx + (x+y)\,dy \\
&= \iint_R \left(\frac{\partial}{\partial x}(x+y) - \frac{\partial}{\partial y}(-(x-y))\right)dA \\
&= \iint_R 0\,dA \\
&= 0
\end{aligned}$$

**13.** 
$$\begin{aligned}
\text{flux} = \int_C (\mathbf{v}\cdot\mathbf{N})\,ds = \int_C (\tilde{\mathbf{v}}\cdot\mathbf{T})\,ds &= \int_C \cos x\sin y\,dx + \sin x\cos y\,dy \\
&= \iint_R \left(\frac{\partial}{\partial x}(\sin x\cos y) - \frac{\partial}{\partial y}(\cos x\sin y)\right)dA \\
&= \iint_R 0\,dA \\
&= 0
\end{aligned}$$

**15.** $P(x, y) = \dfrac{x}{\sqrt{x^2+y^2}}$; $Q(x, y) = \dfrac{y}{\sqrt{x^2+y^2}}$

$$\frac{\partial Q}{\partial x} = -\frac{xy}{(x^2+y^2)^{3/2}} = \frac{\partial P}{\partial y}$$

Let $C_1$ be the circle with center (0, 0) and radius 1. Parameterize $C_1$ by $\mathbf{r}(t) = \langle \cos t, \sin t \rangle$.

$\mathbf{r}'(t) = \langle -\sin t, \cos t \rangle$

$$\begin{aligned}\int_C \frac{x}{\sqrt{x^2+y^2}}dx + \frac{y}{\sqrt{x^2+y^2}}dy &= \int_{C_1} \frac{x}{\sqrt{x^2+y^2}}dx + \frac{y}{\sqrt{x^2+y^2}}dy \\ &= \int_0^{2\pi} ((\cos t)(-\sin t) + (\sin t)(\cos t))\,dt \\ &= \int_0^{2\pi} 0\,dt \\ &= 0\end{aligned}$$

**17.** $P(x, y) = 4xy - \dfrac{y}{x^2+y^2}$; $Q(x, y) = 2x^2 + \dfrac{x}{x^2+y^2}$

$$\frac{\partial Q}{\partial x} = 4x + \frac{y^2-x^2}{(x^2+y^2)^2} = \frac{\partial P}{\partial y}$$

Observe that $C$ contains the hole at (0, 0). Let $C_1$ be the circle with center (0, 0) and radius 1. Parameterize $C_1$ by $\mathbf{r}(t) = \langle \cos t, \sin t \rangle$.

$\mathbf{r}'(t) = \langle -\sin t, \cos t \rangle$

$$\begin{aligned}\int_C \left(4xy - \frac{y}{x^2+y^2}\right)dx + \left(2x^2 + \frac{x}{x^2+y^2}\right)dy &= \int_{C_1} \left(4xy - \frac{y}{x^2+y^2}\right)dx + \left(2x^2 + \frac{x}{x^2+y^2}\right)dy \\ &= \int_0^{2\pi} ((4\cos t \sin t - \sin t)(-\sin t) + (2\cos^2 t + \cos t)(\cos t))dt \\ &= \int_0^{2\pi} (-4\cos t \sin^2 t + 2\cos^3 t + 1)dt \\ &= \int_0^{2\pi} (-4\cos t \sin^2 t + 2\cos t(1 - \sin^2 t) + 1)dt \\ &= \int_0^{2\pi} (-6\cos t \sin^2 t + 2\cos t + 1)dt \\ &= (-2\sin^3 t + 2\sin t + t)\Big|_0^{2\pi} \\ &= 2\pi\end{aligned}$$

**19. a.** $\displaystyle\int_C (\mathbf{F}\cdot\mathbf{T})ds = \int_C \mathbf{F}\cdot d\mathbf{r} = \int_C P\,dx + Q\,dy$

Then using Green's Theorem, $\displaystyle\int_C (\mathbf{F}\cdot\mathbf{T})ds = \iint_R \left(\frac{\partial Q}{\partial x} - \frac{\partial P}{\partial y}\right)dA$

**b.** Recall that $\tilde{\mathbf{F}} = (-Q, P)$.

$\displaystyle\int_C (\mathbf{F}\cdot\mathbf{N})ds = \int_C (\tilde{\mathbf{F}}\cdot\mathbf{T})ds = \int_C -Q\,dx + P\,dy$

Then using Green's Theorem $\displaystyle\int_C (\mathbf{F}\cdot\mathbf{N})ds = \iint_R \left(\frac{\partial P}{\partial x} + \frac{\partial Q}{\partial y}\right)dA$

**21.** Let $C_1, C_2, \ldots, C_n$ be curves such that if $R_i$ is the region inside of $C_i$, then the union of the $R_i$ is $R$, and such that each curve is a quadrilateral with two sides parallel to the $x$-axis. We allow the case where one of the two sides has length zero so $C_i$ may be a triangle with one side parallel to the $x$-axis. By applying the result of the previous exercise, we get

$$\int_C Q\,dy = \int_{C_1} Q\,dy + \int_{C_2} Q\,dy + \cdots + \int_{C_n} Q\,dy = \iint_{R_1} \frac{\partial Q}{\partial x}\,dA + \iint_{R_2} \frac{\partial Q}{\partial x}\,dA + \cdots + \iint_{R_n} \frac{\partial Q}{\partial x}\,dA = \iint_R \frac{\partial Q}{\partial x}\,dA$$

**23.** By Green's Theorem, $-\int_C y\,dx = -\iint_R -\frac{\partial}{\partial y} y\,dA = \iint_R dA = \text{area}(R)$ and $\int_C x\,dy = \iint_R \frac{\partial}{\partial x} x\,dA = \iint_R dA = \text{area}(R)$.

**25.** Sample answer (other answers are possible. We want $P$ and $Q$ such that $\frac{\partial Q}{\partial x} - \frac{\partial P}{\partial y} = 1$.

For example, let $P = y$ and $Q = 2x$. Thus $\int_C P\,dx + Q\,dy = \iint_R \left(\frac{\partial Q}{\partial x} - \frac{\partial P}{\partial y}\right) dA = \iint_R dA = \text{area}(R)$.

Another possibility is $P = y + \sqrt{x^3 + 1}$, $Q = 2x - e^{\sin y}$.

**27.** Apply Green's Theorem. $\int_C f(x)\,dx + g(y)\,dy = \iint_R \left(\frac{\partial g}{\partial x} - \frac{\partial f}{\partial y}\right) dA = 0$ since $\frac{\partial g}{\partial x} = \frac{\partial f}{\partial y} = 0$.

**29.** Since $P$ and $Q$ satisfy the hypothesis of the theorem, $\frac{\partial Q}{\partial x}(x, y) = \frac{\partial P}{\partial y}(x, y)$. We can find two simply connected subsets of $D$ that contain $\Gamma_1$ and $\Gamma_2$. Hence we can apply Green's Theorem, so

$$\int_{\Gamma_1} P\,dx + Q\,dy = \iint_{R_1} \left(\frac{\partial Q}{\partial x} - \frac{\partial P}{\partial y}\right) dA = \iint_{R_1} 0\,dA = 0 \text{ and } \int_{\Gamma_2} P\,dx + Q\,dy = \iint_{R_2} \left(\frac{\partial Q}{\partial x} - \frac{\partial P}{\partial y}\right) dA = \iint_{R_2} 0\,dA = 0.$$

Let $C_2$, $C_3$, $C_4$, and $C_5$ be paths such that $C_2$ followed by $C_3$ is $C$, $C_4$ followed by $C_5$ is $C_1$, $L_1$ followed by $C_2$ followed by $L_2$ followed by $-C_4$ is $\Gamma_1$, and $-L_1$ followed by $-C_5$ followed by $-L_2$ followed by $C_3$ is $\Gamma_2$.

Thus,

$$\begin{aligned} 0 &= \int_{\Gamma_1} P\,dx + Q\,dy + \int_{\Gamma_2} P\,dx + Q\,dy \\ &= \int_{L_1} P\,dx + Q\,dy + \int_{C_2} P\,dx + Q\,dy + \int_{L_2} P\,dx + Q\,dy + \int_{-C_4} P\,dx + Q\,dy + \int_{-L_1} P\,dx + Q\,dy + \int_{-C_5} P\,dx + Q\,dy \\ &\qquad + \int_{-L_2} P\,dx + Q\,dy + \int_{C_3} P\,dx + Q\,dy. \end{aligned}$$

Since $\int_{-C} \mathbf{F} \cdot \mathbf{T}\,ds = -\int_C \mathbf{F} \cdot \mathbf{T}\,ds$ for any path $C$, we get

$$\begin{aligned} 0 &= \int_{L_1} P\,dx + Q\,dy + \int_{C_2} P\,dx + Q\,dy + \int_{L_2} P\,dx + Q\,dy - \int_{C_4} P\,dx + Q\,dy - \int_{L_1} P\,dx + Q\,dy \\ &\qquad - \int_{C_5} P\,dx + Q\,dy - \int_{L_2} P\,dx + Q\,dy + \int_{C_3} P\,dx + Q\,dy \\ &= \int_C P\,dx + Q\,dy - \int_{C_1} P\,dx + Q\,dy \end{aligned}$$

Therefore, we get $\int_C P\,dx + Q\,dy = \int_{C_1} P\,dx + Q\,dy$.

## Section 12.5 Divergence and Curl

**1.** $\text{div}(\mathbf{F}) = \frac{\partial}{\partial x}(yz) + \frac{\partial}{\partial y}(zx) + \frac{\partial}{\partial z}(xy) = 0$

$$\text{curl}(\mathbf{F}) = \det\begin{pmatrix} \mathbf{i} & \mathbf{j} & \mathbf{k} \\ \frac{\partial}{\partial x} & \frac{\partial}{\partial y} & \frac{\partial}{\partial z} \\ yz & zx & xy \end{pmatrix} = \left\langle \frac{\partial}{\partial y}(xy) - \frac{\partial}{\partial z}(zx), \frac{\partial}{\partial z}(yz) - \frac{\partial}{\partial x}(xy), \frac{\partial}{\partial x}(zx) - \frac{\partial}{\partial y}(yz) \right\rangle = \langle 0, 0, 0 \rangle$$

**3.** $\text{div}(\mathbf{v}) = \frac{\partial}{\partial x}(x^2 - y^2) + \frac{\partial}{\partial y}(2xy) = 4x$

$$\text{curl}(\mathbf{F}) = \det\begin{pmatrix} \mathbf{i} & \mathbf{j} & \mathbf{k} \\ \frac{\partial}{\partial x} & \frac{\partial}{\partial y} & \frac{\partial}{\partial z} \\ x^2 - y^2 & 2xy & 0 \end{pmatrix} = \left\langle 0, 0, \frac{\partial}{\partial x}(2xy) - \frac{\partial}{\partial y}(x^2 - y^2) \right\rangle = \langle 0, 0, 4y \rangle$$

**5.** $\mathbf{H}(x, y, z) = \text{grad}((x + 2y + 3z)^2) = \langle 2(x + 2y + 3z), 4(x + 2y + 3z), 6(x + 2y + 3z) \rangle$

$$\text{div}(\mathbf{H}) = \frac{\partial}{\partial x}2(x + 2y + 3z) + \frac{\partial}{\partial y}4(x + 2y + 3z) + \frac{\partial}{\partial z}6(x + 2y + 3z) = 2 + 8 + 18 = 28$$

$$\text{curl}(\mathbf{H}) = \det\begin{pmatrix} \mathbf{i} & \mathbf{j} & \mathbf{k} \\ \frac{\partial}{\partial x} & \frac{\partial}{\partial y} & \frac{\partial}{\partial z} \\ 2(x + 2y + 3z) & 4(x + 2y + 3z) & 6(x + 2y + 3z) \end{pmatrix} = \langle 0, 0, 0 \rangle$$

**7.** grad $\mathbf{F} = \nabla\mathbf{F}$ is undefined since $\mathbf{F}$ is a vector field.

**9.** curl(div $\mathbf{F}$) = curl($\nabla \cdot \mathbf{F}$) = $\nabla \times (\nabla \cdot \mathbf{F})$ is undefined since $\nabla \cdot \mathbf{F}$ is a scalar function.

**11.** grad(div $f$) = grad($\nabla \cdot f$) = $\nabla(\nabla \cdot f)$ is undefined since $f$ is a scalar function.

**13.** grad(div $\mathbf{F}$) = grad($\nabla \cdot \mathbf{F}$) = $\nabla(\nabla \cdot \mathbf{F})$ is defined.

**15.** div(curl(grad $f$)) = div(curl($\nabla f$)) = div($\nabla \times \nabla f$) = $\nabla \cdot (\nabla \times \nabla f)$ is defined.

**17.** Let $\mathbf{F}(x, y, z) = (M(x, y, z), N(x, y, z), P(x, y, z))$ and $\mathbf{G}(x, y, z) = (R(x, y, z), S(x, y, z), T(x, y, z))$.

$$\begin{aligned} \text{div}(\mathbf{F} + \mathbf{G}) &= \text{div}(M + R, N + S, P + T) \\ &= \frac{\partial}{\partial x}(M + R) + \frac{\partial}{\partial y}(N + S) + \frac{\partial}{\partial z}(P + T) \\ &= \frac{\partial M}{\partial x} + \frac{\partial N}{\partial y} + \frac{\partial P}{\partial z} + \frac{\partial R}{\partial x} + \frac{\partial S}{\partial y} + \frac{\partial T}{\partial z} \\ &= \text{div}\,\mathbf{F} + \text{div}\,\mathbf{G} \end{aligned}$$

Writing this identity with the $\nabla$ symbol, we get $\nabla \cdot (\mathbf{F} + \mathbf{G}) = \nabla \cdot \mathbf{F} + \nabla \cdot \mathbf{G}$.
Thus the vector identity suggested is the distribution of the dot product over the vector sum.

**19.** Let $\mathbf{F}(x, y, z) = \langle P(x, y, z), Q(x, y, z), R(x, y, z) \rangle$.

$$\begin{aligned} \text{div}(f\mathbf{F}) = \text{div}(fP, fQ, fR) &= \frac{\partial}{\partial x}(fP) + \frac{\partial}{\partial y}(fQ) + \frac{\partial}{\partial z}(fR) \\ &= \frac{\partial f}{\partial x}P + f\frac{\partial P}{\partial x} + \frac{\partial f}{\partial y}Q + f\frac{\partial Q}{\partial y} + \frac{\partial f}{\partial z}R + f\frac{\partial R}{\partial z} \\ &= \frac{\partial f}{\partial x}P + \frac{\partial f}{\partial y}Q + \frac{\partial f}{\partial z}R + f\left(\frac{\partial P}{\partial x} + \frac{\partial Q}{\partial y} + \frac{\partial R}{\partial z}\right) \\ &= \left(\frac{\partial f}{\partial x}, \frac{\partial f}{\partial y}, \frac{\partial f}{\partial z}\right) \cdot (P, Q, R) + f(\text{div}(P, Q, Z)) \\ &= \text{grad}\,f \cdot \mathbf{F} + f(\text{div}\,\mathbf{F}) \end{aligned}$$

**21.** $\nabla\|\mathbf{r}\| = \nabla\left(\sqrt{x^2+y^2+z^2}\right)$

$$= \left\langle \frac{x}{\sqrt{x^2+y^2+z^2}}, \frac{y}{\sqrt{x^2+y^2+z^2}}, \frac{z}{\sqrt{x^2+y^2+z^2}} \right\rangle$$

$$= \frac{1}{\sqrt{x^2+y^2+z^2}}\langle x, y, z\rangle$$

$$= \frac{1}{\|\mathbf{r}\|}\mathbf{r}$$

**23.** $\nabla \times \mathbf{r} = \det\begin{pmatrix} \mathbf{i} & \mathbf{j} & \mathbf{k} \\ \frac{\partial}{\partial x} & \frac{\partial}{\partial y} & \frac{\partial}{\partial z} \\ x & y & z \end{pmatrix} = \langle 0, 0, 0\rangle = \mathbf{0}$

**25.** $\nabla\frac{1}{\|\mathbf{r}\|} = \nabla\frac{1}{\sqrt{x^2+y^2+z^2}}$

$$= \left\langle -\frac{x}{(x^2+y^2+z^2)^{3/2}}, -\frac{y}{(x^2+y^2+z^2)^{3/2}}, -\frac{z}{(x^2+y^2+z^2)^{3/2}} \right\rangle$$

$$= -\frac{1}{(x^2+y^2+z^2)^{3/2}}\langle x, y, z\rangle$$

$$= -\frac{1}{\|\mathbf{r}\|^3}\mathbf{r}$$

**27. a.** Since the first partial derivatives are continuous, as $(x, y) \to (x_0, y_0)$, $\frac{\partial P}{\partial x}(x, y) \to \frac{\partial P}{\partial x}(x_0, y_0)$ and $\frac{\partial Q}{\partial y}(x, y) \to \frac{\partial Q}{\partial y}(x_0, y_0)$. Thus $E(x, y) \to 0$ as $(x, y) \to (x_0, y_0)$.

**b.** $\frac{1}{\pi r^2}\iint_{D(A, r)}\left(\frac{\partial P}{\partial x}(x, y)+\frac{\partial Q}{\partial y}(x, y)\right)dA = \frac{1}{\pi r^2}\iint_{D(A, r)}\left(\frac{\partial P}{\partial x}(x_0, y_0)+\frac{\partial Q}{\partial y}(x_0, y_0)+E(x, y)\right)dA$

$$= \frac{1}{\pi r^2}\left(\frac{\partial P}{\partial x}(x_0, y_0)+\frac{\partial Q}{\partial y}(x_0, y_0)\right)\iint_{D(A,r)} dA + \frac{1}{\pi r^2}\iint_{D(A, r)} E(x, y)dA$$

$$= \frac{1}{\pi r^2}\left(\frac{\partial P}{\partial x}(x_0, y_0)+\frac{\partial Q}{\partial y}(x_0, y_0)\right)\pi r^2 + \frac{1}{\pi r^2}\iint_{D(A, r)} E(x, y)dA$$

$$= \frac{\partial P}{\partial x}(x_0, y_0)+\frac{\partial Q}{\partial y}(x_0, y_0)+\frac{1}{\pi r^2}\iint_{D(A, r)} E(x, y)dA$$

**c.** $\lim_{r\to 0^+}\frac{1}{\pi r^2}\iint\left(\frac{\partial P}{\partial x}(x, y)+\frac{\partial Q}{\partial y}(x, y)\right)dA = \lim_{r\to 0^+}\left(\frac{\partial P}{\partial x}(x_0, y_0)+\frac{\partial Q}{\partial y}(x_0, y_0)+\frac{1}{\pi r^2}\iint_{D(A, r)} E(x, y)dA\right)$

$$= \frac{\partial P}{\partial x}(x_0, y_0)+\frac{\partial Q}{\partial y}(x_0, y_0)+\lim_{r\to 0^+}\frac{1}{\pi r^2}\iint_{D(A, r)} E(x, y)dA$$

Since $E(x, y) \to 0$ as $(x, y) \to (x_0, y_0)$, there is $\varepsilon(r)$ such that $|E(x, y)| \le \varepsilon(r)$ for $(x, y)$ in $D(A, r)$, and $\varepsilon(r) \to 0$ as $r \to 0^+$.

Then $\lim_{r\to 0^+} \frac{1}{\pi r^2} \iint_{D(A,\, r)} |E(x,\, y)|dA \le \lim_{r\to 0^+} \frac{1}{\pi r^2} \varepsilon(r)\pi r^2 = \lim_{r\to 0^+} \varepsilon(r) = 0.$

Thus $\lim_{r\to 0^+} \frac{1}{\pi r^2} \iint_{D(A,\, r)} E(x,\, y)dA = 0$, so

$$\lim_{r\to 0^+} \frac{1}{\pi r^2} \iint_{D(A,\, r)} \left(\frac{\partial P}{\partial x}(x,\, y) + \frac{\partial Q}{\partial y}(x,\, y)\right)dA = \frac{\partial P}{\partial x}(x_0,\, y_0) + \frac{\partial Q}{\partial y}(x_0,\, y_0).$$

**29.** $\text{grad} f = \langle f_x, f_y, f_z\rangle$; $\text{grad } g = \langle g_x, g_y, g_z\rangle$

$\text{grad} f \times \text{grad } g = \langle f_y g_z - f_z g_y, f_z g_x - f_x g_z, f_x g_y - f_y g_x\rangle$

$\text{div}(\text{grad} f \times \text{grad } g)$

$= \text{div}(f_y g_z - f_z g_y, f_z g_x - f_x g_z, f_x g_y - f_y g_x)$

$= \frac{\partial}{\partial x}(f_y g_z - f_z g_y) + \frac{\partial}{\partial y}(f_z g_x - f_x g_z) + \frac{\partial}{\partial z}(f_x g_y - f_y g_x)$

$= (f_{xy} g_z + f_y g_{xz} - f_{xz} g_y - f_z g_{xy}) + (f_{yz} g_x + f_z g_{xy} - f_{xy} g_z - f_x g_{yz}) + (f_{xz} g_y + f_x g_{yz} - f_{yz} g_x - f_y g_{xz})$

$= 0$

**31.** If $\mathbf{F}$ is conservative, $\frac{\partial M}{\partial y} = \frac{\partial N}{\partial x}$, so $\frac{\partial N}{\partial x} - \frac{\partial M}{\partial y} = 0.$

$$\text{curl } \mathbf{F} = \begin{pmatrix} \mathbf{i} & \mathbf{j} & \mathbf{k} \\ \frac{\partial}{\partial x} & \frac{\partial}{\partial y} & \frac{\partial}{\partial z} \\ M(x,\, y) & N(x,\, y) & 0 \end{pmatrix} = \left\langle 0,\, 0,\, \frac{\partial N}{\partial x} - \frac{\partial M}{\partial y}\right\rangle = \langle 0,\, 0,\, 0\rangle = \mathbf{0}$$

The analogous result for $R^3$ is as follows:
Suppose $\mathbf{F}(x, y, z) = \langle P(x, y, z), Q(x, y, z), R(x, y, z)\rangle$ is a vector field and assume that $P, Q, R$ have continuous first partial derivatives. If $\mathbf{F}$ is conservative then curl $\mathbf{F} = \mathbf{0}$.

Note that if $\mathbf{F}$ is conservative, $\frac{\partial P}{\partial y} = \frac{\partial Q}{\partial x}$, $\frac{\partial P}{\partial z} = \frac{\partial R}{\partial x}$, and $\frac{\partial Q}{\partial z} = \frac{\partial R}{\partial y}$. Thus $\frac{\partial Q}{\partial x} - \frac{\partial P}{\partial y} = 0$, $\frac{\partial P}{\partial z} - \frac{\partial R}{\partial x} = 0$, and $\frac{\partial R}{\partial y} - \frac{\partial Q}{\partial z} = 0.$

$$\text{curl } \mathbf{F} = \begin{pmatrix} \mathbf{i} & \mathbf{j} & \mathbf{k} \\ \frac{\partial}{\partial x} & \frac{\partial}{\partial y} & \frac{\partial}{\partial z} \\ P(x,\, y,\, z) & Q(x,\, y,\, z) & R(x,\, y,\, z) \end{pmatrix} = \left\langle \frac{\partial R}{\partial y} - \frac{\partial Q}{\partial z}, \frac{\partial P}{\partial z} - \frac{\partial R}{\partial x}, \frac{\partial Q}{\partial x} - \frac{\partial P}{\partial y}\right\rangle = \langle 0,\, 0,\, 0\rangle = \mathbf{0}$$

**33.** $\int_C f(\nabla g)\cdot \mathbf{N}\, ds = \iint_R \text{div} f(\nabla g) dA$ by the "divergence form" of Green's Theorem.

$$\text{div} f(\nabla g) = \text{div}\left(f\frac{\partial g}{\partial x}, f\frac{\partial g}{\partial y}, 0\right) = \frac{\partial f}{\partial x}\frac{\partial g}{\partial x} + f\frac{\partial^2 g}{\partial x^2} + \frac{\partial f}{\partial y}\frac{\partial g}{\partial y} + f\frac{\partial^2 g}{\partial y^2}$$

$$= \left\langle \frac{\partial f}{\partial x}, \frac{\partial f}{\partial y}\right\rangle \cdot \left\langle \frac{\partial g}{\partial x}, \frac{\partial g}{\partial y}\right\rangle + f\left(\frac{\partial^2 g}{\partial x^2} + \frac{\partial^2 g}{\partial x^2}\right) = \nabla f \cdot \nabla g + f\nabla^2 g$$

Thus $\int_C f(\nabla g)\cdot \mathbf{N}\, ds = \iint_R (\nabla f\cdot \nabla g + f\nabla^2 g)\, dA$, so $\iint_R f\nabla^2 g\, dA = \int_C f(\nabla g)\cdot \mathbf{N}\, ds - \iint_R \nabla f \cdot \nabla g\, dA$

**35. a.** From the figure in the text, observe that $\sin\gamma = \frac{R}{\|\mathbf{r}\|}$ $\left(\text{and } \cos\gamma = \frac{z_0}{\|\mathbf{r}\|}\right)$.

Since we are given that $\omega = \frac{v}{R}$, $\omega = \frac{v}{\|\mathbf{r}\|\sin\gamma}$, so $v = \omega\|\mathbf{r}\|\sin\gamma$.

$$\boldsymbol{\omega}\times\mathbf{r} = \left\langle 0, 0, \frac{v}{R}\right\rangle \times \langle x, y, z_0\rangle = \left\langle -\frac{v}{R}y, \frac{v}{R}x, 0\right\rangle = \frac{v}{R}\langle -y, x, 0\rangle$$

Thus $\boldsymbol{\omega}\times\mathbf{r}$ is parallel to and in the same direction as $\mathbf{v}$ since $\frac{v}{R}$ is positive and $\langle -y, x, 0\rangle$ is tangent to the circle at $(x, y, z_0)$.

$\|\boldsymbol{\omega}\times\mathbf{r}\| = \|\boldsymbol{\omega}\|\|\mathbf{r}\|\sin\gamma = \omega\|\mathbf{r}\|\sin\gamma = v$, so $\boldsymbol{\omega}\times\mathbf{r}$ has the same magnitude as $\mathbf{v}$. Thus $\mathbf{v} = \boldsymbol{\omega}\times\mathbf{r}$.

**b.**
$$\begin{aligned}
\text{curl }\mathbf{v} &= \text{curl}(\boldsymbol{\omega}\times\mathbf{r})\\
&= \boldsymbol{\omega}\text{ div }\mathbf{r} - \mathbf{r}\text{ div}\,\boldsymbol{\omega} + (\mathbf{r}\cdot\nabla)\boldsymbol{\omega} - (\boldsymbol{\omega}\cdot\nabla)\mathbf{r}\\
&= \boldsymbol{\omega}(3) - \mathbf{r}(0) + \left(x\frac{\partial}{\partial x} + y\frac{\partial}{\partial y} + z\frac{\partial}{\partial z}\right)\boldsymbol{\omega} - \left(\frac{v}{R}\frac{\partial}{\partial z}\right)\mathbf{r}\\
&= 3\boldsymbol{\omega} - 0 + 0 - \boldsymbol{\omega}\\
&= 2\boldsymbol{\omega}
\end{aligned}$$

## Section 12.6 Surface Integrals

**1.** Describe $S$ parametrically by $\mathbf{r}(u, v) = \langle 2\sin u\cos v, 2\sin u\sin v, 2\cos u\rangle$, $0 \le u \le \pi$ and $0 \le v \le 2\pi$.

$$\frac{\partial\mathbf{r}}{\partial u} = \langle 2\cos u\cos v, 2\cos u\sin v, -2\sin u\rangle$$

$$\frac{\partial\mathbf{r}}{\partial v} = \langle -2\sin u\sin v, 2\sin u\cos v, 0\rangle$$

$$\frac{\partial\mathbf{r}}{\partial u}\times\frac{\partial\mathbf{r}}{\partial v} = \langle 4\sin^2 u\cos v, 4\sin^2 u\sin v, 4\cos u\sin u\rangle$$

$$\left\|\frac{\partial\mathbf{r}}{\partial u}\times\frac{\partial\mathbf{r}}{\partial v}\right\| = \sqrt{16\sin^4 u + 16\cos^2 u\sin^2 u} = 4|\sin u|$$

Since $0 \le u \le \pi$, we can remove the absolute value.

$f(\mathbf{r}(u, v)) = 4\cos^2 u$

$$\begin{aligned}
\iint_S f(x, y, z)\,d\sigma &= \iint_D f(\mathbf{r}(u, v))\left\|\frac{\partial\mathbf{r}}{\partial u}\times\frac{\partial\mathbf{r}}{\partial v}\right\|dA\\
&= \int_0^{2\pi}\int_0^{\pi}(4\cos^2 u)(4\sin u)\,du\,dv\\
&= 16\int_0^{2\pi}\int_0^{\pi}\cos^2 u\sin u\,du\,dv\\
&= 16\int_0^{2\pi}\left(-\frac{1}{3}\cos^3 u\right)\Bigg|_{u=0}^{\pi}dv\\
&= 16\int_0^{2\pi}\frac{2}{3}\,dv\\
&= \frac{64\pi}{3}
\end{aligned}$$

**3.** Describe $S$ parametrically by $\mathbf{r}(u, v) = \langle u\cos v,\ 2\ln u,\ u\sin v\rangle$, $1 \le u \le 2$ and $0 \le v \le 2\pi$.

$$\frac{\partial \mathbf{r}}{\partial u} = \left\langle \cos v, \frac{2}{u}, \sin v \right\rangle$$

$$\frac{\partial \mathbf{r}}{\partial v} = \langle -u\sin v, 0, u\cos v\rangle$$

$$\frac{\partial \mathbf{r}}{\partial u} \times \frac{\partial \mathbf{r}}{\partial v} = \langle 2\cos v, -u, 2\sin v\rangle$$

$$\left\| \frac{\partial \mathbf{r}}{\partial u} \times \frac{\partial \mathbf{r}}{\partial v} \right\| = \sqrt{u^2+4}$$

$$f(\mathbf{r}(u, v)) = \sqrt{u^2+4}$$

$$\iint_S f(x, y, z)d\sigma$$

$$= \iint_D f(\mathbf{r}(u, v)) \left\| \frac{\partial \mathbf{r}}{\partial u} \times \frac{\partial \mathbf{r}}{\partial v} \right\| dA$$

$$= \int_0^{2\pi} \int_1^2 \left(\sqrt{u^2+4}\right)\left(\sqrt{u^2+4}\right) du\, dv$$

$$= \int_0^{2\pi} \int_1^2 (u^2+4) du\, dv$$

$$= \int_0^{2\pi} \frac{19}{3} dv$$

$$= \frac{38\pi}{3}$$

**5.** $\dfrac{\partial \mathbf{r}}{\partial u} = \langle 1, 1, 1\rangle$, $\dfrac{\partial \mathbf{r}}{\partial v} = \langle -1, 1, -2\rangle$

$$\left\| \frac{\partial \mathbf{r}}{\partial u} \times \frac{\partial \mathbf{r}}{\partial v} \right\| = \|\langle -3, 1, 2\rangle\| = \sqrt{14}$$

$$f(\mathbf{r}(u, v)) = (u - v) + (u - 2v + 2) = 2u - 3v + 2$$

$$\iint_S f(x, y, z)d\sigma = \iint_D f(\mathbf{r}(u, v)) \left\| \frac{\partial \mathbf{r}}{\partial u} \times \frac{\partial \mathbf{r}}{\partial v} \right\| dA$$

$$= \int_1^4 \int_0^2 (2u - 3v + 2)\sqrt{14}\, du\, dv$$

$$= \sqrt{14} \int_1^4 (u^2 - 3vu + 2u)\Big|_{u=0}^{u=2} dv$$

$$= \sqrt{14} \int_1^4 (-6v + 8) dv$$

$$= \sqrt{14}(-3v^2 + 8v)\Big|_1^4$$

$$= -21\sqrt{14}$$

**7.** Describe $S$ parametrically by $\mathbf{r}(u, v) = \langle u\cos v,\ u\sin v,\ 2u\rangle$, $0 \le u \le 2$, $0 \le v \le 2\pi$.

$$\frac{\partial \mathbf{r}}{\partial u} = \langle \cos v, \sin v, 2\rangle;\ \frac{\partial \mathbf{r}}{\partial v} = \langle -u\sin v, u\cos v, 0\rangle$$

$$\frac{\partial \mathbf{r}}{\partial u} \times \frac{\partial \mathbf{r}}{\partial v} = \langle -2u\cos v, -2u\sin v, u\rangle$$

$$\left\| \frac{\partial \mathbf{r}}{\partial u} \times \frac{\partial \mathbf{r}}{\partial v} \right\| = \sqrt{5}|u|$$

Since $0 \le u \le 2$, we can remove the absolute value.

$$\delta(\mathbf{r}(u, v)) = 4u^2 \cos v \sin v + 1$$

$$\text{mass} = \iint_S \delta(x, y, z)\, d\sigma$$

$$= \iint_D \delta(\mathbf{r}(u, v)) \left\| \frac{\partial \mathbf{r}}{\partial u} \times \frac{\partial \mathbf{r}}{\partial v} \right\| dA$$

$$= \int_0^{2\pi} \int_0^2 (4u^2 \cos v \sin v + 1)\sqrt{5}u\, du\, dv$$

$$= \sqrt{5} \int_0^{2\pi} \int_0^2 (4u^3 \cos v \sin v + u) du\, dv$$

$$= \sqrt{5} \int_0^{2\pi} (16\cos v \sin v + 2) dv$$

$$= 4\pi\sqrt{5} \text{ grams}$$

**9.** Describe the portion of the cylinder $S_1$ parametrically by $\mathbf{r}_1(u, v) = \langle 3\cos v, 3\sin v, u\rangle$, $0 \le u \le 3$, $0 \le v \le 2\pi$. Describe the top disk $S_2$ by $\mathbf{r}_2(u, v) = \langle u\cos v, u\sin v, 3\rangle$, $0 \le u \le 3$, $0 \le v \le 2\pi$. Describe the bottom disk $S_3$ by $\mathbf{r}_3(u, v) = \langle u\cos v, u\sin v, 0\rangle$, $0 \le u \le 3$, $0 \le v \le 2\pi$.

$$\frac{\partial \mathbf{r}_1}{\partial u} = \langle 0, 0, 1\rangle; \ \frac{\partial \mathbf{r}_1}{\partial v} = \langle -3\sin v, 3\cos v, 0\rangle$$

$$\left\| \frac{\partial \mathbf{r}_1}{\partial u} \times \frac{\partial \mathbf{r}_1}{\partial v} \right\| = \| \langle -3\cos u, -3\sin u, 0\rangle \| = 3$$

$$\delta(\mathbf{r}_1(u, v)) = 9\sin^2 v + u^2 + 9$$

$$\frac{\partial \mathbf{r}_2}{\partial u} = \langle \cos v, \sin v, 0\rangle; \ \frac{\partial \mathbf{r}_2}{\partial v} = \langle -u\sin v, u\cos v, 0\rangle$$

$$\left\| \frac{\partial \mathbf{r}_2}{\partial u} \times \frac{\partial \mathbf{r}_2}{\partial v} \right\| = \| \langle 0, 0, u\rangle \| = u$$

$$\delta(\mathbf{r}_2(u, v)) = u^2\sin^2 v + u^2 + 9$$

$$\frac{\partial \mathbf{r}_3}{\partial u} = \langle \cos v, \sin v, 0\rangle; \ \frac{\partial \mathbf{r}_3}{\partial v} = \langle -u\sin v, u\cos v, 0\rangle$$

$$\left\| \frac{\partial \mathbf{r}_3}{\partial u} \times \frac{\partial \mathbf{r}_3}{\partial v} \right\| = \| \langle 0, 0, u\rangle \| = u$$

$$\delta(\mathbf{r}_3(u, v)) = u^2\sin^2 v + u^2$$

$$\text{mass} = \iint_S \delta(x, y, z)\, d\sigma$$

$$= \iint_{S_1} \delta(x, y, z)\, d\sigma + \iint_{S_2} \delta(x, y, z)\, d\sigma + \iint_{S_3} \delta(x, y, z)\, d\sigma$$

$$= \int_0^{2\pi}\int_0^3 (9\sin^2 v + u^2 + 9)3\, du\, dv + \int_0^{2\pi}\int_0^3 (u^2\sin^2 v + u^2 + 9)u\, du\, dv + \int_0^{2\pi}\int_0^3 (u^2\sin^2 v + u^2)u\, du\, dv$$

$$= \int_0^{2\pi} (27u\sin^2 v + u^3 + 27u)\Big|_{u=0}^{u=3} dv + \int_0^{2\pi} \left(\frac{1}{4}u^4\sin^2 v + \frac{1}{4}u^4 + \frac{9}{2}u^2\right)\Big|_{u=0}^{u=3} dv + \int_0^{2\pi} \left(\frac{1}{4}u^4\sin^2 v + \frac{1}{4}u^4\right)\Big|_{u=0}^{u=3} dv$$

$$= \int_0^{2\pi} (81\sin^2 v + 108)dv + \int_0^{2\pi} \left(\frac{81}{4}\sin^2 v + \frac{243}{4}\right)dv + \int_0^{2\pi} \left(\frac{81}{4}\sin^2 v + \frac{81}{4}\right)dv$$

$$= \int_0^{2\pi} \left(\frac{243}{2}\sin^2 v + 189\right)dv$$

$$= \int_0^{2\pi} \left(\frac{999}{4} - \frac{243}{4}\cos 2v\right)dv$$

$$= \left(\frac{999}{4}v - \frac{243}{8}\sin 2v\right)\Big|_0^{2\pi}$$

$$= \frac{999\pi}{2} \text{ grams}$$

**11.** Parameterize $S$ by $\mathbf{r}(u, v) = \langle u, v, 4-3u-4v\rangle$, $0 \le u \le \frac{4}{3} - \frac{4}{3}v$, $0 \le v \le 1$.

$$\frac{\partial \mathbf{r}}{\partial u} = \langle 1, 0, -3\rangle;\ \frac{\partial \mathbf{r}}{\partial v} = \langle 0, 1, -4\rangle$$

$$\frac{\partial \mathbf{r}}{\partial u} \times \frac{\partial \mathbf{r}}{\partial v} = \langle 3, 4, 1\rangle$$

$$\mathbf{F}(\mathbf{r}(u, v)) = \langle u, v, 4-3u-4v\rangle$$

$$\iint_S \mathbf{F}\cdot\mathbf{N}\,d\sigma = \iint_D \langle u, v, 4-3u-4v\rangle \cdot \langle 3, 4, 1\rangle\,dA = \int_0^1 \int_0^{\frac{4}{3}-\frac{4}{3}v} 4\,du\,dv = \int_0^1 \left(\frac{16}{3} - \frac{16}{3}v\right)dv = \left(\frac{16}{3}v - \frac{8}{3}v^2\right)\Big|_0^1 = \frac{8}{3}$$

**13.** Parameterize $S$ by $\mathbf{r}(u, v) = \langle 2\sin u\cos v, 2\sin u\sin v, 2\cos u\rangle$, $0 \le u \le \pi$, $0 \le v \le \pi$.

$$\frac{\partial \mathbf{r}}{\partial u} = \langle 2\cos u\cos v, 2\cos u\sin v, -2\sin u\rangle;\ \frac{\partial \mathbf{r}}{\partial v} = \langle -2\sin u\sin v, 2\sin u\cos v, 0\rangle$$

$$\frac{\partial \mathbf{r}}{\partial u} \times \frac{\partial \mathbf{r}}{\partial v} = \langle 4\sin^2 u\cos v, 4\sin^2 u\sin v, 4\cos u\sin u\rangle$$

$$\mathbf{F}(\mathbf{r}(u, v)) = \langle 2\sin u\cos v, 4\sin^2 u\sin^2 v, -2\cos u\rangle$$

$$\iint_S \mathbf{F}\cdot\mathbf{N}\,d\sigma = \iint_D \langle 2\sin u\cos v, 4\sin^2 u\sin^2 v, -2\cos u\rangle \cdot \langle 4\sin^2 u\cos v, 4\sin^2 u\sin v, 4\cos u\sin u\rangle dA$$

$$= \int_0^\pi \int_0^\pi (8\sin^3 u\cos^2 v + 16\sin^4 u\sin^3 v - 8\cos^2 u\sin u)\,du\,dv$$

$$= \int_0^\pi \left(\frac{32}{3}\cos^2 v + 6\pi\sin^3 v - \frac{16}{3}\right)dv$$

$$= 8\pi$$

**15.** Parameterize $S$ by $\mathbf{r}(u, v) = \langle 3\cos v, 3\sin v, u\rangle$, $0 \le u \le 3$, $0 \le v \le 2\pi$.

$$\frac{\partial \mathbf{r}}{\partial u} = \langle 0, 0, 1\rangle;\ \frac{\partial \mathbf{r}}{\partial v} = \langle -3\sin v, 3\cos v, 0\rangle$$

$$\frac{\partial \mathbf{r}}{\partial u} \times \frac{\partial \mathbf{r}}{\partial v} = \langle -3\cos v, -3\sin v, 0\rangle$$

$$\mathbf{F}(\mathbf{r}(u, v)) = \langle -u, 2, 3\cos v\rangle$$

$$\iint_S \mathbf{F}\cdot\mathbf{N}\,d\sigma = \iint_D \langle -u, 2, 3\cos v\rangle \cdot \langle -3\cos v, -3\sin v, 0\rangle\,dA$$

$$= \int_0^{2\pi} \int_0^3 (3u\cos v - 6\sin v)\,du\,dv$$

$$= \int_0^{2\pi} \left(\frac{27}{2}\cos v - 18\sin v\right)dv$$

$$= 0$$

**17.** $\frac{\partial \mathbf{r}}{\partial u} = \langle \sin v, \cos v, 1\rangle;\ \frac{\partial \mathbf{r}}{\partial v} = \langle u\cos v, -u\sin v, 0\rangle$

$$\frac{\partial \mathbf{r}}{\partial u} \times \frac{\partial \mathbf{r}}{\partial v} = \langle u\sin v, u\cos v, -u\rangle$$

Since we use $\mathbf{N}$ with a non-negative third component, let $\frac{\partial \mathbf{r}}{\partial u} \times \frac{\partial \mathbf{r}}{\partial v} = \langle -u\sin v, -u\cos v, u\rangle$.

$$\mathbf{F}(\mathbf{r}(u, v)) = \langle u\sin v, u\cos v, u^2\rangle$$

$$\iint_S \mathbf{F}\cdot\mathbf{N}\,d\sigma = \iint_D \langle u\sin v, u\cos v, u^2\rangle \cdot \langle -u\sin v, -u\cos v, u\rangle dA = \int_0^{2\pi} \int_0^1 (u^3 - u^2)\,du\,dv = \int_0^{2\pi} -\frac{1}{12}\,dv = -\frac{\pi}{6}$$

**19.** $\iint_S 1\,d\sigma$ gives the surface area of $S$.

**21.** Parameterize $S$ by $\mathbf{r}(u, v) = \langle u, v, u\arctan v\rangle$.

$$\frac{\partial \mathbf{r}}{\partial u} = \langle 1, 0, \arctan v\rangle; \quad \frac{\partial \mathbf{r}}{\partial v} = \left\langle 0, 1, \frac{u}{1+v^2}\right\rangle$$

$$\frac{\partial \mathbf{r}}{\partial u} \times \frac{\partial \mathbf{r}}{\partial v} = \left\langle -\arctan v, -\frac{u}{1+v^2}, 1\right\rangle$$

$$\mathbf{N} = \frac{1}{\sqrt{(\arctan y)^2 + \left(\frac{x}{1+y^2}\right)^2 + 1}} \left\langle \arctan y, \frac{x}{1+y^2}, -1\right\rangle$$

**23.** Parameterize $S$ by $\mathbf{r}_1(u, v) = \left\langle u\cos v, \sqrt{u^2-1}, u\sin v\right\rangle$, $1 \le u < \infty$, $0 \le v \le 2\pi$ for $y \ge 0$ and by

$\mathbf{r}_2(u, v) = \left\langle u\cos v, -\sqrt{u^2-1}, u\sin v\right\rangle$, $1 \le u < \infty$, $0 \le v \le 2\pi$ for $y < 0$.

$$\frac{\partial \mathbf{r}_1}{\partial u} = \left\langle \cos v, \frac{u}{\sqrt{u^2-1}}, \sin v\right\rangle; \quad \frac{\partial \mathbf{r}_1}{\partial v} = \langle -u\sin v, 0, u\cos v\rangle$$

$$\frac{\partial \mathbf{r}_1}{\partial u} \times \frac{\partial \mathbf{r}_1}{\partial v} = \left\langle \frac{u^2}{\sqrt{u^2-1}}\cos v, -u, \frac{u^2}{\sqrt{u^2-1}}\sin v\right\rangle = \frac{u}{\sqrt{u^2-1}}\left\langle u\cos v, -\sqrt{u^2-1}, u\sin v\right\rangle$$

Hence $\mathbf{N}$ is parallel to $\langle x, -y, z\rangle$ and an orientation on $S$ for $y \ge 0$ is

$$\mathbf{N}(x, y, z) = \frac{1}{\sqrt{x^2+y^2+z^2}}\langle x, -y, z\rangle.$$

With this orientation, $\mathbf{N}(x, y, z)$ points outward from the surface.

$$\frac{\partial \mathbf{r}_2}{\partial u} = \left\langle \cos v, -\frac{u}{\sqrt{u^2-1}}, \sin v\right\rangle; \quad \frac{\partial \mathbf{r}_2}{\partial v} = \langle -u\sin v, 0, u\cos v\rangle$$

$$\frac{\partial \mathbf{r}_2}{\partial u} \times \frac{\partial \mathbf{r}_2}{\partial u} = \left\langle -\frac{u^2}{\sqrt{u^2-1}}\cos v, -u, -\frac{u^2}{\sqrt{u^2+1}}\sin v\right\rangle = \frac{u}{\sqrt{u^2-1}}\left\langle -u\cos v, -\sqrt{u^2-1}, -u\sin v\right\rangle$$

Hence $\mathbf{N}$ is parallel to $\langle -x, y, -z\rangle$, or $\langle x, -y, z\rangle$ so that $\mathbf{N}(x, y, z)$ points outward from the surface.

Thus, an orientation for $S$ is $\mathbf{N}(x, y, z) = \dfrac{1}{\sqrt{x^2+y^2+z^2}}\langle x, -y, z\rangle$.

**25.** Let $S^*$ be the part common to $S_1$ and $S_2$. Let $S_3$ be $S_1 - S^*$ and $S_4$ be $S_2 - S^*$. Therefore the union of $S_3$ and $S_4$ is $S$. Observe that on $S^*$ for $D_1$, $\mathbf{N}_1 = -\mathbf{N}_2$.

$$\iint_{S_1} (\mathbf{F}\cdot\mathbf{N}_1)d\sigma + \iint_{S_2} (\mathbf{F}\cdot\mathbf{N}_2)d\sigma = \iint_{S_3} (\mathbf{F}\cdot\mathbf{N}_1)d\sigma + \iint_{S^*} (\mathbf{F}\cdot\mathbf{N}_1)d\sigma + \iint_{S_4} (\mathbf{F}\cdot\mathbf{N}_2)d\sigma + \iint_{S^*} (\mathbf{F}\cdot\mathbf{N}_2)d\sigma$$
$$= \iint_{S_3} (\mathbf{F}\cdot\mathbf{N}_1)d\sigma - \iint_{S^*} (\mathbf{F}\cdot\mathbf{N}_2)d\sigma + \iint_{S_4} (\mathbf{F}\cdot\mathbf{N}_2)d\sigma + \iint_{S^*} (\mathbf{F}\cdot\mathbf{N}_2)d\sigma$$
$$= \iint_{S_3} (\mathbf{F}\cdot\mathbf{N}_1)d\sigma + \iint_{S_4} (\mathbf{F}\cdot\mathbf{N}_2)d\sigma$$

Also observe that on $S_3$ for $D_1$, $\mathbf{N}_1 = \mathbf{N}$ and that on $S_4$ for $D_2$, $\mathbf{N}_2 = \mathbf{N}$.

Thus $\iint_{S_1} (\mathbf{F}\cdot\mathbf{N})d\sigma + \iint_{S_2} (\mathbf{F}\cdot\mathbf{N}_2)d\sigma = \iint_{S_3} (\mathbf{F}\cdot\mathbf{N})d\sigma + \iint_{S_4} (\mathbf{F}\cdot\mathbf{N})d\sigma = \iint_{S} (\mathbf{F}\cdot\mathbf{N})d\sigma.$

**27.** Describe $S$ parametrically by $\mathbf{r}(u, v) = \langle \cos u \sin v,\ \sin u \sin v,\ \cos v\rangle$, $0 \le u \le 2\pi$, $0 \le v \le \dfrac{\pi}{2}$.

$\mathbf{r}_u = \langle -\sin u \sin v,\ \cos u \sin v,\ 0\rangle$, $\mathbf{r}_v = \langle \cos u \cos v,\ \sin u \cos v,\ -\sin v\rangle$

$\mathbf{r}_u \times \mathbf{r}_v = \langle -\cos u \sin^2 v,\ -\sin u \sin^2 v,\ -\cos v \sin v\rangle$

$\|\mathbf{r}_u \times \mathbf{r}_v\| = |\sin v| = \sin v$ (since $\sin v > 0$ for our parameterization)

$\delta(\mathbf{r}(u, v)) = c$ for some constant $c$.

$$M = \iint_S \delta(x, y, z)d\sigma = \int_0^{\pi/2}\int_0^{2\pi} c \sin v\, du\, dv = \int_0^{\pi/2} 2\pi c \sin v\, dv = 2\pi c$$

Observe that by symmetry, $x_c = 0$, $y_c = 0$.

$$\iint_S z\delta(x, y, z)d\sigma = \int_0^{\pi/2}\int_0^{2\pi} c \cos v \sin v\, du\, dv = \int_0^{\pi/2} 2\pi c \cos v \sin v\, dv = \pi c$$

Thus $z_c = \dfrac{1}{2\pi c}\pi c = \dfrac{1}{2}$.

**29.** Let $S_1$ be the end of the cylinder at $x = 0$, $S_2$ the end of the cylinder at $x = 4$, and $S_3$ be the side of the cylinder. Describe $S_1$ parametrically by $\mathbf{r}_1(u, v) = \langle 0, u\cos v, u \sin v\rangle$, $0 \le u \le 2$, $0 \le v \le 2\pi$, $S_2$ by $\mathbf{r}_2(u, v) = \langle 4, u\cos v, u\sin v\rangle$, $0 \le u \le 2$, $0 \le v \le 2\pi$, and $S_3$ by $\mathbf{r}_3(u, v) = \langle u, 2\cos v, 2\sin v\rangle$, $0 \le u \le 4$, $0 \le v \le 2\pi$.

$\mathbf{r}_{1u} = \langle 0, \cos v, \sin v\rangle$; $\mathbf{r}_{1v} = \langle 0, -u\sin v, u\cos v\rangle$

$\|\mathbf{r}_{1u} \times \mathbf{r}_{1v}\| = \|\langle u, 0, 0\rangle\| = u$

$\mathbf{r}_{2u} = \langle 0, \cos v, \sin v\rangle$; $\mathbf{r}_{2v} = \langle 0, -u\sin v, u\cos v\rangle$

$\|\mathbf{r}_{2u} \times \mathbf{r}_{2v}\| = \|\langle u, 0, 0\rangle\| = u$

$\mathbf{r}_{3u} = \langle 1, 0, 0\rangle$; $\mathbf{r}_{3v} = \langle 0, -2\sin v, 2\cos v\rangle$

$\|\mathbf{r}_{3u} \times \mathbf{r}_{3v}\| = \|\langle 0, -2\cos v, -2\sin v\rangle\| = 2$

$\delta(\mathbf{r}_1(u, v)) = 1$; $\delta(\mathbf{r}_2(u, v)) = 9$; $\delta(\mathbf{r}_3(u, v)) = 2u + 1$

$$M = \iint_S \delta(x, y, z)d\sigma = \iint_{S_1} \delta(x, y, z)d\sigma + \iint_{S_2} \delta(x, y, z)d\sigma + \iint_{S_3} \delta(x, y, z)d\sigma$$
$$= \int_0^{2\pi}\int_0^2 u\, du\, dv + \int_0^{2\pi}\int_0^2 9u\, du\, dv + \int_0^{2\pi}\int_0^4 2(2u+1)\, du\, dv$$
$$= 4\pi + 36\pi + 80\pi$$
$$= 120\pi$$

Observe that by symmetry, $y_c = 0$ and $z_c = 0$.

$$\iint_S x\delta(x, y, z)d\sigma = \iint_{S_1} x\delta(x, y, z)d\sigma + \iint_{S_2} x\delta(x, y, z)d\sigma + \iint_{S_3} x\delta(x, y, z)d\sigma$$
$$= \int_0^{2\pi}\int_0^2 36u\,du\,dv + \int_0^{2\pi}\int_0^4 2(2u^2 + u)du\,dv$$
$$= 144\pi + \frac{608}{3}\pi$$
$$= \frac{1040}{3}\pi$$

Thus $x_c = \left(\frac{1}{120\pi}\right)\left(\frac{1040}{3}\pi\right) = \frac{26}{9}$.

**31.** If you cut the Möbius strip along the centerline, then you get a strip with two twists with the ends attached.
If you cut the Möbius strip along a line one-third of the way, you get two linked loops. One loop is a Möbius strip and the other is a strip with two twists with the ends attached.

## Section 12.7 The Divergence Theorem

**1.** $$\iint_S (\mathbf{F}\cdot\mathbf{N})d\sigma = \iiint_D \operatorname{div}\mathbf{F}\,dV = \iiint_D 3\,dV = 3\left(\frac{4}{3}\pi\right) = 4\pi$$

**3.** $$\iint_S (\mathbf{F}\cdot\mathbf{N})d\sigma = \iiint_D \operatorname{div}\mathbf{F}\,dV = \int_{-1}^{1}\int_{-\sqrt{1-z^2}}^{\sqrt{1-z^2}}\int_0^{1-y^2-z^2}(-1)dx\,dy\,dz$$
$$= \int_{-1}^{1}\int_{-\sqrt{1-z^2}}^{\sqrt{1-z^2}} y^2 + z^2 - 1\,dy\,dz$$
$$= \int_0^{2\pi}\int_0^1 (r^2 - 1)r\,dr\,d\theta$$
$$= \int_0^{2\pi}\left(-\frac{1}{4}\right)d\theta$$
$$= -\frac{\pi}{2}$$

**5.** Parameterize $S$ by $\mathbf{r}(u, v) = \langle \sin u\cos v, \sin u\sin v, \cos u\rangle$, $0 \le u \le \pi$, $0 \le v \le 2\pi$.

$\mathbf{r}_u = \langle \cos u\cos v, \cos u\sin v, -\sin u\rangle$; $\mathbf{r}_v = \langle -\sin u\sin v, \sin u\cos v, 0\rangle$

$\mathbf{r}_u \times \mathbf{r}_v = \langle \sin^2 u\cos v, \sin^2 u\sin v, \cos u\sin u\rangle$

$\mathbf{F}(\mathbf{r}(u, v)) = \langle \cos^2 u\sin u\sin v, \sin^2 u\cos^2 v + 2\sin u\sin v, e^{\sin u(\cos v+\sin v)}\rangle$

$$\iint_S (\mathbf{F}\cdot\mathbf{N})d\sigma$$
$$= \int_0^{2\pi}\int_0^{\pi}(\cos^2 u\sin^3 u\cos v\sin v + \sin^4 u\cos^2 v\sin v + 2\sin^3 u\sin^2 v + \cos u\sin u\, e^{\sin u(\cos v+\sin v)})du\,dv$$

Due to symmetries, the integrals of the first, second, and fourth terms are 0.

Thus, $$\iint_S (\mathbf{F}\cdot\mathbf{N})d\sigma = \int_0^{2\pi}\int_0^{\pi} 2\sin^3 u\sin^2 v\,du\,dv = \int_0^{2\pi}\frac{8}{3}\sin^2 v\,dv = \frac{8\pi}{3}.$$

$$\iiint_D \operatorname{div}\mathbf{F}\,dv = \iiint_D 2\,dV = 2\left(\frac{4}{3}\pi\right) = \frac{8\pi}{3}$$

7. Let $S_1$, $S_2$, and $S_3$ be the side, bottom, and top parts of the cylinder.

Represent $S_1$ by $\mathbf{r}_1 = \langle \sqrt{2}\cos v, \sqrt{2}\sin v, u \rangle$, $1 \le u \le 4$, $0 \le v \le 2\pi$.

$\mathbf{r}_{1u} = \langle 0, 0, 1 \rangle$; $\mathbf{r}_{1v} = \langle -\sqrt{2}\sin v, \sqrt{2}\cos v, 0 \rangle$

$\mathbf{r}_{1u} \times \mathbf{r}_{1v} = \langle -\sqrt{2}\cos v, -\sqrt{2}\sin v, 0 \rangle$

We want $\mathbf{N}$ to point outward, so $\mathbf{N} = \frac{1}{\sqrt{2}}\langle \sqrt{2}\cos v, \sqrt{2}\sin v, 0 \rangle$.

$\mathbf{F}(\mathbf{r}_1(u, v)) = \langle 4\sqrt{2}\cos v \sin^2 v, \sqrt{2}u \sin v, 4u\cos^2 v \rangle$

Represent $S_2$ by $\mathbf{r}_2 = \langle u\cos v, u\sin v, 1 \rangle$, $0 \le u \le \sqrt{2}$, $0 \le v \le 2\pi$.

$\mathbf{r}_{2u} = \langle \cos v, \sin v, 0 \rangle$; $\mathbf{r}_{2v} = \langle -u\sin v, u\cos v, 0 \rangle$

$\mathbf{r}_{2u} \times \mathbf{r}_{2v} = \langle 0, 0, u \rangle$

We want $\mathbf{N}$ to point downward, so $\mathbf{N} = \frac{1}{u}\langle 0, 0, -u \rangle$.

$\mathbf{F}(\mathbf{r}_2(u, v)) = \langle 2u^3\cos v \sin^2 v, u\sin v, 2u^2\cos^2 v \rangle$

Represent $S_3$ by $\mathbf{r}_3 = \langle u\cos v, u\sin v, 4 \rangle$, $0 \le u \le \sqrt{2}$, $0 \le v \le 2\pi$.

$\mathbf{r}_{3u} = \langle \cos v, \sin v, 0 \rangle$; $\mathbf{r}_{3v} = \langle -u\sin v, u\cos v, 0 \rangle$

$\mathbf{r}_{3u} \times \mathbf{r}_{3v} = \langle 0, 0, u \rangle$

We want $\mathbf{N}$ to point upward, so $\mathbf{N} = \frac{1}{u}\langle 0, 0, u \rangle$.

$\mathbf{F}(\mathbf{r}_3(u, v)) = \langle 2u^3\cos v \sin^2 v, 4u\sin v, 8u^2\cos^2 v \rangle$

$$\iint_S (\mathbf{F}\cdot\mathbf{N})\,d\sigma$$

$$= \iint_{S_1} (\mathbf{F}\cdot\mathbf{N})\,d\sigma + \iint_{S_2} (\mathbf{F}\cdot\mathbf{N})\,d\sigma + \iint_{S_3} (\mathbf{F}\cdot\mathbf{N})\,d\sigma$$

$$= \int_0^{2\pi}\int_1^4 (8\cos^2 v\sin^2 v + 2u\sin^2 v)\,du\,dv + \int_0^{2\pi}\int_0^{\sqrt{2}} (-2u^3\cos^2 v)\,du\,dv + \int_0^{2\pi}\int_0^{\sqrt{2}} (8u^3\cos^2 v)\,du\,dv$$

$$= \int_0^{2\pi}\int_1^4 (8\cos^2 v\sin^2 v + 2u\sin^2 v)\,du\,dv + \int_0^{2\pi}\int_0^{\sqrt{2}} (6u^3\cos^2 v)\,du\,dv$$

$$= 21\pi + 6\pi$$

$$= 27\pi$$

Now use the Divergence Theorem.

$$\iiint_D \operatorname{div}\mathbf{F}\,dV = \int_1^4\int_{-\sqrt{2}}^{\sqrt{2}}\int_{-\sqrt{2-y^2}}^{\sqrt{2-y^2}} (2y^2 + z + 2x^2)\,dx\,dy\,dz$$

$$= \int_1^4\int_0^{2\pi}\int_0^{\sqrt{2}} (2r^2 + z)r\,dr\,d\theta\,dz$$

$$= \int_1^4\int_0^{2\pi} (2+z)\,d\theta\,dz$$

$$= 2\pi\int_1^4 (2+z)\,dz$$

$$= 27\pi$$

9. $\text{flux} = \iint_S \delta(\mathbf{v}\cdot\mathbf{N})\,d\sigma = \iint_S \langle 2x, -y, z \rangle\cdot\mathbf{N}\,d\sigma = \iiint_D 2\,dV = \frac{8\pi}{3}$ g/s

**11.** $\text{flux} = \iint_S \delta(\mathbf{v}\cdot\mathbf{N})\,d\sigma = \iint_S \langle 2x^2+2,\ x^3z^2+xz^2,\ x^2ye^x+ye^x\rangle\cdot\mathbf{N}\,d\sigma = \iiint_D 4x\,dV$

$$= \int_0^3\int_0^{2-\frac{2}{3}x}\int_0^{4-\frac{4}{3}x-2y} 4x\,dz\,dy\,dx$$

$$= \int_0^3\int_0^{2-\frac{2}{3}x}\left(16x-\frac{16}{3}x^2-8xy\right)dy\,dx$$

$$= \int_0^3\left(\frac{16}{9}x^3-\frac{32}{3}x^2+16x\right)dx$$

$$= 12\text{ g/s}$$

**13.** **a.** $\text{div}\,\mathbf{v} = 2-1+1 = 2 > 0$ at all points, so the flow has a source at all points $(x, y, z)$.

**b.** Since div **v** is never less than zero, the flow does not have a sink.

**c.** The flow has neither a source nor a sink at (0, 0, 0).

**15.** **a.** $\text{div}\,\mathbf{v} = 3x^2+3y^2+3z^2 > 0$ at all points except (0, 0, 0), so the flow has a source at all points $(x, y, z) \neq (0, 0, 0)$.

**b.** Since div **v** is never less than zero, the flow does not have a sink.

**c.** The flow has neither a source nor a sink at (0, 0, 0).

**17.** heat flow $= -k$ grad $T = -0.408\langle 1, 1, 1\rangle$

heat flux $= -k\iint_S (\text{grad } T)\cdot\mathbf{N}\,d\sigma = -0.408\iiint_D 0\,dV = 0$ cal/s

**19.** heat flow $= -k$ grad $T = -0.005\langle 2x, 2y, 2\rangle$

heat flux $= -k\iint_S (\text{grad } T)\cdot\mathbf{N}\,d\sigma = -0.005\iiint_D 6\,dV = -0.03\left(\frac{4}{3}\pi 2^3\right) = -0.32\pi$ cal/s

**21.** By applying the Divergence Theorem,

$\frac{1}{3}\iint_S \langle x, y, z\rangle\cdot\mathbf{N}\,d\sigma = \frac{1}{3}\iiint_D \text{div}(x, y, z)\,dV = \frac{1}{3}\iiint_D 3\,dV = \iiint_D dV =$ volume of $D$.

Any vector field **F** such that div **F** = 1 can be used. Sample answers (other answers are possible): let **F** = $(x, 0, 0)$ or **F** = $(0, y, 0)$.

**23.** The heat flow is $-k$ grad $T = -k\nabla T$.

heat flux $= -k\iint_S (\nabla T)\cdot\mathbf{N}\,d\sigma = -k\iiint_D \text{div}(\nabla T)\,dV = -k\iiint_D \nabla\cdot\nabla T\,dV = -k\iiint_D \nabla^2 T\,dV$

**25.** **a.** $\iint_{S'} \langle x+yz,\ z^2\sin x,\ 4z\rangle\cdot\mathbf{N}\,d\sigma = \iiint_D 5\,dV = 5\left(\frac{2}{3}\pi\right) = \frac{10}{3}\pi$

**b.** $\iint_{S_d} \langle x+yz,\ z^2\sin x,\ 4z\rangle\cdot\mathbf{N}\,d\sigma = \iint_{S_d} -4z\,d\sigma = 0$ since $z = 0$ on $S_d$.

**c.** $\iint_S \langle x+yz,\ z^2\sin x,\ 4z\rangle\cdot\mathbf{N}\,d\sigma = \iint_{S'} \langle x+yz,\ z^2\sin x,\ 4z\rangle\cdot\mathbf{N}\,d\sigma - \iint_{S_d} \langle x+yz,\ z^2\sin x,\ 4z\rangle\cdot\mathbf{N}\,d\sigma = \frac{10}{3}\pi$

**27.** Let $S_d$ be the disk of center (0, 0, 1) and radius $2\sqrt{2}$ in the plane $z = 1$ and let $S' = S + S_d$ be the surface obtained from joining the surfaces $S'$ and $S_d$.

$$\iint_{S'} \left\langle yze^{yz},\ x^3\ln(2xz+z^3),\ z\right\rangle \cdot \mathbf{N}\,d\sigma = \iiint_D 1\,dV = \int_0^{2\pi}\int_0^{2\sqrt{2}}\int_1^{9-r^2} r\,dz$$
$$= \int_0^{2\pi}\int_0^{2\sqrt{2}} (8r - r^3)\,dr\,d\theta$$
$$= \int_0^{2\pi} 16\,d\theta$$
$$= 32\pi$$

Orient $S_d$ with the downward unit normal $\mathbf{N} = \langle 0, 0, -1\rangle$.

$$\iint_{S_d} \left\langle yze^{yz},\ x^3\ln(2xz+z^3),\ z\right\rangle \cdot \mathbf{N}\,d\sigma = \iint_{S_d} -z\,d\sigma = \iint_{S_d} -1\,d\sigma = -1(8\pi) = -8\pi$$

Thus,

$$\iint_S \left\langle yze^{yz},\ x^3\ln(2xz+z^3),\ z\right\rangle \cdot \mathbf{N}\,d\sigma$$
$$= \iint_{S'} \left\langle yze^{yz},\ x^3\ln(2xz+z^3),\ z\right\rangle \cdot \mathbf{N}\,d\sigma - \iint_{S_d} \left\langle yze^{yz},\ x^3\ln(2xz+z^3),\ z\right\rangle \cdot \mathbf{N}\,d\sigma$$
$$= 32\pi - (-8\pi)$$
$$= 40\pi$$

**29.** $\operatorname{div}(f\nabla g) = f(\operatorname{div}\nabla g) + \operatorname{grad} f \cdot \nabla g$

$= f\nabla^2 g + \nabla f \cdot \nabla g$

Thus, applying the Divergence Theorem,

$$\iint_S (f\nabla g)\cdot\mathbf{N}\,d\sigma = \iiint_D \operatorname{div}(f\nabla g)\,dV = \iiint_B (f\nabla^2 g + \nabla f\cdot\nabla g)\,dV$$

## Section 12.8 Stokes' Theorem

**1.** (i) Since the coordinate functions $r_1(x, y) = x$, $r_2(x, y) = y$, and $r_3(x, y) = 4 - x^2 - y^2$ are continuously differentiable and $\mathbf{r}$ is a one-to-one function, the surface is smooth and simple.

(ii) We can describe the boundary $C^*$ by $\mathbf{h}(t) = \langle 2\cos t,\ 2\sin t\rangle$, $0 \le t \le 2\pi$. Then $\mathbf{r}(\mathbf{h}(t)) = \langle 2\cos t,\ 2\sin t,\ 0\rangle$, $0 \le t \le 2\pi$, describes the boundary $C$. Viewed from above this is the circle of radius 2, being traced out counterclockwise.

(iii) $\mathbf{r}_x = \langle 1, 0, -2x\rangle$; $\mathbf{r}_y = \langle 0, 1, -2y\rangle$

$$\mathbf{r}_x \times \mathbf{r}_y = \langle 1, 0, -2x\rangle \times \langle 0, 1, -2y\rangle$$
$$= \langle 2x, 2y, 1\rangle$$

(iv) $\|\mathbf{r}_x \times \mathbf{r}_y\| = \|\langle 2x, 2y, 1\rangle\| = \sqrt{4x^2 + 4y^2 + 1} = \sqrt{1 + 4(x^2 + y^2)}$

$$\mathbf{N} = \frac{1}{\|\mathbf{r}_x \times \mathbf{r}_y\|}\mathbf{r}_x \times \mathbf{r}_y = \frac{1}{\sqrt{1 + 4(x^2 + y^2)}}\langle 2x, 2y, 1\rangle$$

(v) $\operatorname{curl}\mathbf{F} = \nabla \times \mathbf{F} = \nabla \times \langle -y, x, 1\rangle = \left\langle \frac{\partial}{\partial y}(1) - \frac{\partial}{\partial z}(x),\ \frac{\partial}{\partial z}(-y) - \frac{\partial}{\partial x}(1),\ \frac{\partial}{\partial x}(x) - \frac{\partial}{\partial y}(-y)\right\rangle = \langle 0, 0, 2\rangle$

3. $\iint_S (\nabla\times\mathbf{F})\cdot\mathbf{N}\,d\sigma = \iint_D \langle 0, 0, 2\rangle\cdot\left(\frac{1}{\|\mathbf{r}_x\times\mathbf{r}_y\|}\langle 2x, 2y, 1\rangle\right)\|\mathbf{r}_x\times\mathbf{r}_y\|dx\,dy = \iint_D 2dx\,dy = 2\iint_D dx\,dy$

$\iint_D dx\,dy$ gives the area of $D$ which is the circle of radius 2, so $\iint_S (\nabla\times\mathbf{F})\cdot\mathbf{N}\,d\sigma = 2(\pi 2^2) = 8\pi$

5. Describe $C_1$ parametrically by $\mathbf{r}_1(t) = \langle \sin t, 0, \cos t\rangle$, $0\le t\le\frac{\pi}{2}$.

$\mathbf{r}_1'(t) = \langle \cos t, 0, -\sin t\rangle$; $\mathbf{F}(\mathbf{r}(t)) = \langle 0, \sin t, 0\rangle$.

$\int_{C_1}\mathbf{F}\cdot\mathbf{T}\,ds = \int_0^{\pi/2}\langle 0, \sin t, 0\rangle\cdot\langle \cos t, 0, -\sin t\rangle dt = \int_0^{\pi/2} 0\,dt = 0$

Describe $C_3$ parametrically by $\mathbf{r}_3(t) = \langle 0, \cos t, \sin t\rangle$, $0\le t\le\frac{\pi}{2}$.

$\mathbf{r}_3'(t) = \langle 0, -\sin t, \cos t\rangle$; $\mathbf{F}(\mathbf{r}(t)) = \langle -\cos t, 0, 0\rangle$.

$\int_{C_3}\mathbf{F}\cdot\mathbf{T}\,ds = \int_0^{\pi/2}\langle -\cos t, 0, 0\rangle\cdot\langle 0, -\sin t, \cos t\rangle dt = \int_0^{\pi/2} 0\,dt = 0$

7. $\mathbf{r}(s, t) = s\mathbf{g}(t) + (1-s)\mathbf{g}(\pi - t)$

$$= s\left\langle \cos t, \sin t, 1+\frac{1}{4}\sin 2t\right\rangle + (1-s)\left\langle \cos(\pi - t), \sin(\pi - t), 1+\frac{1}{4}\sin 2(\pi - t)\right\rangle$$

$$= s\left\langle \cos t, \sin t, 1+\frac{1}{4}\sin 2t\right\rangle + (1-s)\left\langle -\cos(t), \sin(t), 1-\frac{1}{4}\sin 2t\right\rangle$$

$$= \left\langle (-1+2s)\cos t, \sin t, 1-\frac{1}{4}\sin 2t + \frac{1}{2}s\sin 2t\right\rangle$$

9. Describe $S$ by $\mathbf{r}(x, y) = \langle x, y, 1-x-y\rangle$, $0\le x\le 1-y$, $0\le y\le 1$.

$\mathbf{N} = \frac{1}{\|\mathbf{r}_x\times\mathbf{r}_y\|}\mathbf{r}_x\times\mathbf{r}_y = \frac{1}{\sqrt{3}}\langle 1, 1, 1\rangle$

$\nabla\times\mathbf{F} = \langle -z, 2, 1\rangle$; $(\nabla\times\mathbf{F})(\mathbf{r}(x, y)) = \langle x+y-1, 2, 1\rangle$

$\iint_S (\nabla\times\mathbf{F})\cdot\mathbf{N}\,d\sigma = \iint_D \langle x+y-1, 2, 1\rangle\cdot\langle 1, 1, 1\rangle dx\,dy = \int_0^1\int_0^{1-y}(x+y+2)dx\,dy = \int_0^1\left(-\frac{1}{2}y^2 - 2y + \frac{5}{2}\right)dy = \frac{4}{3}$

The surface $S$ is the triangle with vertices (0, 0, 1), (1, 0, 0), and (0, 1, 0). Let $C_1$ be the segment from (0, 0, 1) to (1, 0, 0), $C_2$ be the segment from (1, 0, 0) to (0, 1, 0), and $C_3$ be the segment from (0, 1, 0) to (0, 0, 1). Describe $C_1$ by $\mathbf{r}_1(t) = \langle t, 0, 1-t\rangle$, $0\le t\le 1$, $C_2$ by $\mathbf{r}_2(t) = \langle 1-t, t, 0\rangle$, $0\le t\le 1$, and $C_3$ by $\mathbf{r}_3(t) = \langle 0, 1-t, t\rangle$, $0\le t\le 1$.

$\mathbf{r}_1'(t) = \langle 1, 0, -1\rangle$; $\mathbf{r}_2'(t) = \langle -1, 1, 0\rangle$; $\mathbf{r}_3'(t) = \langle 0, -1, 1\rangle$

$\mathbf{F}(\mathbf{r}_1(t)) = \langle 2-2t, t, 0\rangle$; $\mathbf{F}(\mathbf{r}_2(t)) = \langle 0, 1, 0\rangle$; $\mathbf{F}(\mathbf{r}_3(t)) = \langle 2t, 1-t, t^2-t\rangle$

$\int_C \mathbf{F}\cdot\mathbf{T}\,ds = \int_{C_1}\mathbf{F}\cdot\mathbf{T}\,ds + \int_{C_2}\mathbf{F}\cdot\mathbf{T}\,ds + \int_{C_3}\mathbf{F}\cdot\mathbf{T}\,ds = \int_0^1(2-2t)dt + \int_0^1 1\,dt + \int_0^1(t^2-1)dt = 1+1-\frac{2}{3} = \frac{4}{3}$

**11.** $\mathbf{r}_u = \langle 0, 1, 0\rangle, \mathbf{r}_v = \langle 1, 0, 1\rangle$

$$\mathbf{N} = \frac{1}{\|\mathbf{r}_u \times \mathbf{r}_v\|}\mathbf{r}_u \times \mathbf{r}_v = \frac{1}{\sqrt{2}}\langle 1, 0, -1\rangle$$

$\nabla \times \mathbf{F} = \langle -x, 0, z\rangle$; $(\nabla \times \mathbf{F})(\mathbf{r}(u, v)) = \langle -v, 0, v\rangle$

$$\iint_S (\nabla\times\mathbf{F})\cdot\mathbf{N}\,d\sigma = \iint_D \langle -v, 0, v\rangle\cdot\langle 1, 0, -1\rangle du\,dv = \int_0^1\int_0^1(-2v)du\,dv = \int_0^1(-2v)dv = -1$$

The boundary $C$ is composed of the line segments $C_1$ from (0, 0, 0) to (0, 1, 0), $C_2$ from (0, 1, 0) to (1, 1, 1), $C_3$ from (1, 1, 1) to (1, 0, 1), and $C_4$ from (1, 0, 1) to (0, 0, 0). Describe $C_1$ by $\mathbf{r}_1(t) = \langle 0, t, 0\rangle, 0 \le t \le 1$, $C_2$ by $\mathbf{r}_2(t) = \langle t, 1, t\rangle, 0 \le t \le 1$, $C_3$ by $\mathbf{r}_3(t) = \langle 1, 1-t, 1\rangle, 0 \le t \le 1$, and $C_4$ by $\mathbf{r}_4(t) = \langle 1-t, 0, 1-t\rangle, 0 \le t \le 1$. $\mathbf{r}_1'(t) = \langle 0, 1, 0\rangle$; $\mathbf{r}_2'(t) = \langle 1, 0, 1\rangle$; $\mathbf{r}_3'(t) = \langle 0, -1, 0\rangle$; $\mathbf{r}_4'(t) = \langle -1, 0, -1\rangle$

$\mathbf{F}(\mathbf{r}_1(t)) = \langle 0, 0, 0\rangle$; $\mathbf{F}(\mathbf{r}_2(t)) = \langle t, t^2, t\rangle$; $\mathbf{F}(\mathbf{r}_3(t)) = \langle 1, 1, 1\rangle$; $\mathbf{F}(\mathbf{r}_4(t)) = \langle 1-t, (1-t)^2, 1-t\rangle$

$$\begin{aligned}\int_C \mathbf{F}\cdot\mathbf{T}\,ds &= \int_{C_1}\mathbf{F}\cdot\mathbf{T}\,ds + \int_{C_2}\mathbf{F}\cdot\mathbf{T}\,ds + \int_{C_3}\mathbf{F}\cdot\mathbf{T}\,ds + \int_{C_4}\mathbf{F}\cdot\mathbf{T}\,ds \\ &= \int_0^1 0\,dt + \int_0^1 2t\,dt + \int_0^1(-1)dt + \int_0^1 2(t-1)dt \\ &= 0+1-1-1 \\ &= -1\end{aligned}$$

**13.** $$\mathbf{N} = \frac{1}{\|\mathbf{r}_u \times \mathbf{r}_v\|}\mathbf{r}_u \times \mathbf{r}_v = \frac{1}{\sqrt{2}u}\langle u\cos v, u\sin v, u\rangle$$

$\nabla \times \mathbf{F} = \langle 0, 0, -x\rangle$; $(\nabla \times \mathbf{F})(\mathbf{r}(u, v)) = \langle 0, 0, -u\cos v\rangle$

$$\begin{aligned}\iint_S (\nabla\times\mathbf{F})\cdot\mathbf{N}\,d\sigma &= \iint_D \langle 0, 0, -u\cos v\rangle\cdot\langle u\cos v, u\sin v, u\rangle du\,dv \\ &= \int_0^{2\pi}\int_{1/2}^1(-u^2\cos v)du\,dv \\ &= \int_0^{2\pi}\left(-\frac{7}{24}\cos v\right)dv \\ &= 0\end{aligned}$$

The boundary $C$ is composed of two circles $C_1$ traced counterclockwise in the $(x, y)$-plane with radius 1 and $C_2$ traced clockwise in the plane $z = \frac{1}{2}$ with radius $\frac{1}{2}$. Describe $C_1$ by $\mathbf{r}_1(t) = \langle \cos t, \sin t, 0\rangle$, $0 \le t \le 2\pi$, and $C_2$ by $\mathbf{r}_2(t) = \left\langle \frac{1}{2}\cos t, -\frac{1}{2}\sin t, \frac{1}{2}\right\rangle, 0 \le t \le 2\pi$.

$\mathbf{r}_1'(t) = \langle -\sin t, \cos t, 0\rangle$; $\mathbf{r}_2'(t) = \left\langle -\frac{1}{2}\sin t, -\frac{1}{2}\cos t, 0\right\rangle$

$\mathbf{F}(\mathbf{r}_1(t)) = \langle \cos t\sin t, \sin^2 t, 0\rangle$; $\mathbf{F}(\mathbf{r}_2(t)) = \left\langle -\frac{1}{4}\cos t\sin t, \frac{1}{4}\sin^2 t, \frac{1}{4}\right\rangle$

$$\int_C \mathbf{F}\cdot\mathbf{T}\,ds = \int_{C_1}\mathbf{F}\cdot\mathbf{T}\,ds + \int_{C_2}\mathbf{F}\cdot\mathbf{T}\,ds = \int_0^{2\pi}0\,dt + \int_0^{2\pi}0\,dt = 0$$

**15.** $\nabla\times\mathbf{F}=\langle 0,0,2\rangle$

An equation for the unit sphere is $x^2+y^2+z^2=1$. When $y=z$, we get $x^2+2y^2=1$. Let $S$ be the part of the plane within the unit sphere. Describe $S$ by $\mathbf{r}(u,v)=\left\langle u\cos v, \frac{1}{\sqrt{2}}u\sin v, \frac{1}{\sqrt{2}}u\sin v\right\rangle$, $0\le u\le 1$ and
$0\le v\le 2\pi$.

$$\mathbf{N}=\frac{1}{\|\mathbf{r}_u\times\mathbf{r}_v\|}\mathbf{r}_u\times\mathbf{r}_v=\frac{1}{u}\left\langle 0, -\frac{1}{\sqrt{2}}u, \frac{1}{\sqrt{2}}u\right\rangle$$

$$\iint_S(\nabla\times\mathbf{F})\cdot\mathbf{N}\,d\sigma=\iint_D\langle 0,0,2\rangle\cdot\left\langle 0,-\frac{u}{\sqrt{2}},\frac{u}{\sqrt{2}}\right\rangle du\,dv=\int_0^{2\pi}\int_0^1\sqrt{2}u\,du\,dv=\int_0^{2\pi}\frac{\sqrt{2}}{2}dv=\sqrt{2}\pi$$

**17.** $\nabla\times\mathbf{F}=\langle 0,0,2\rangle$

Let $S$ be the part of the plane within the cylinder. Describe $S$ by

$\mathbf{r}(u,v)=\left\langle u\cos v, u\sin v, \frac{1}{3}(1-u\cos v-2u\sin v)\right\rangle$, $0\le u\le 1$ and $0\le v\le 2\pi$.

$$\mathbf{N}=\frac{1}{\|\mathbf{r}_u\times\mathbf{r}_v\|}\mathbf{r}_u\times\mathbf{r}_v=\frac{3}{14u}\left\langle\frac{1}{3}u, \frac{2}{3}u, u\right\rangle$$

$$\iint_S(\nabla\times\mathbf{F})\cdot\mathbf{N}\,d\sigma=\iint_D\langle 0,0,2\rangle\cdot\left\langle\frac{1}{3},\frac{2}{3}u,u\right\rangle du\,dv=\int_0^{2\pi}\int_0^1 2u\,du\,dv=\int_0^{2\pi}dv=2\pi$$

**19.** $\nabla\times\mathbf{v}=\langle 0,0,y\rangle$

Using cylindrical coordinates, an equation for the ellipsoid is $\frac{1}{2}r^2+z^2=1$ and an equation for the cylinder is $r=\sin\theta$. Let $S$ be the part of the upper half of the ellipsoid in the cylinder. Describe $S$ by

$\mathbf{r}(r,\theta)=\left\langle r\cos\theta, r\sin\theta, \sqrt{1-\frac{1}{2}r^2}\right\rangle$, $0\le r\le\sin\theta$, $0\le\theta\le\pi$.

$$\mathbf{N}=\frac{1}{\|\mathbf{r}_r\times\mathbf{r}_\theta\|}\mathbf{r}_r\times\mathbf{r}_\theta=\frac{1}{\|\mathbf{r}_r\times\mathbf{r}_\theta\|}\left\langle\frac{r^2}{2\sqrt{1-\frac{1}{2}r^2}}\cos\theta, \frac{r^2}{2\sqrt{1-\frac{1}{2}r^2}}\sin\theta, r\right\rangle$$

$(\nabla\times\mathbf{v})(\mathbf{r}(r,\theta))=\langle 0,0,r\sin\theta\rangle$

$$\text{circulation}=\int_C\mathbf{v}\cdot\mathbf{T}\,ds=\iint_S(\nabla\times\mathbf{v})\cdot\mathbf{N}\,d\sigma=\int_0^{\pi}\int_0^{\sin\theta}r^2\sin\theta\,dr\,d\theta=\int_0^{\pi}\frac{1}{3}\sin^4\theta\,d\theta=\frac{\pi}{8}$$

**21.** $\mathbf{F}$ is a continuously differentiable vector field defined on $D$ and $D$ is convex. $\nabla\times\mathbf{F}=\langle 0,0,0\rangle$, so $\mathbf{F}$ is a conservative field.

**23.** Let $S^*$ be the disk of radius 2 in the $(x,y)$-plane with center $(0,0,0)$. Describe $S^*$ by $\mathbf{r}(u,v)=\langle u\cos v, u\sin v, 0\rangle$, $0\le u\le 2$, $0\le v\le 2\pi$.

$$\mathbf{N}=\frac{1}{\|\mathbf{r}_u\times\mathbf{r}_v\|}\mathbf{r}_u\times\mathbf{r}_v=\frac{1}{u}\langle 0,0,u\rangle$$

$$\text{circulation}=\int_C\mathbf{F}\cdot\mathbf{T}\,ds=\iint_{S^*}(\nabla\times\mathbf{F})\cdot\mathbf{N}\,d\sigma=\iint_D\langle 0,0,2\rangle\cdot\langle 0,0,u\rangle\,du\,dv=\int_0^{2\pi}\int_0^2 2u\,du\,dv=\int_0^{2\pi}4\,dv=8\pi$$

**25.** Regard $\mathbf{g}(t)$ and $\mathbf{g}(\pi-t)$ as "opposite points" of $C$. Then for fixed $t_0$, $\mathbf{r}(s,t_0)=s\mathbf{g}(t_0)+(1-s)\mathbf{g}(\pi-t_0)$, $0\le s\le 1$, is the line segment from $\mathbf{g}(\pi-t_0)$ to $\mathbf{g}(t_0)$. Thus the surface $S$ is the collection of the line segments for all pairs of "opposite points" of $C$.

**27.** Suppose that $\mathbf{F} = \nabla f = \left\langle \frac{\partial f}{\partial x}, \frac{\partial f}{\partial y}, \frac{\partial f}{\partial z} \right\rangle$.

$$\nabla \times \mathbf{F} = \nabla \times \nabla f = \left\langle \frac{\partial^2 f}{\partial y \partial z} - \frac{\partial^2 f}{\partial z \partial y}, \frac{\partial^2 f}{\partial z \partial x} - \frac{\partial^2 f}{\partial x \partial z}, \frac{\partial^2 f}{\partial x \partial y} - \frac{\partial^2 f}{\partial y \partial x} \right\rangle = \langle 0, 0, 0 \rangle$$

**29. a.**
$$\frac{\partial}{\partial u}\left((F_1 \circ \mathbf{r})\frac{\partial r_1}{\partial v}\right) - \frac{\partial}{\partial v}\left((F_1 \circ \mathbf{r})\frac{\partial r_1}{\partial u}\right) = \frac{\partial}{\partial u}(F_1 \circ \mathbf{r})\frac{\partial r_1}{\partial v} + (F_1 \circ \mathbf{r})\frac{\partial^2 r_1}{\partial u \partial v} - \frac{\partial}{\partial v}(F_1 \circ \mathbf{r})\frac{\partial r_1}{\partial u} - (F_1 \circ \mathbf{r})\frac{\partial^2 r_1}{\partial v \partial u}$$
$$= \frac{\partial}{\partial u}(F_1 \circ \mathbf{r})\frac{\partial r_1}{\partial v} - \frac{\partial}{\partial v}(F_1 \circ \mathbf{r})\frac{\partial r_1}{\partial u}$$

**b.**
$$\frac{\partial}{\partial u}(F_1 \circ \mathbf{r}) = \frac{\partial}{\partial u}(F_1(r_1, r_2, r_3)) = \frac{\partial F_1}{\partial x}\frac{\partial r_1}{\partial u} + \frac{\partial F_1}{\partial y}\frac{\partial r_2}{\partial u} + \frac{\partial F_1}{\partial z}\frac{\partial r_3}{\partial u}$$
$$\frac{\partial}{\partial v}(F_1 \circ \mathbf{r}) = \frac{\partial}{\partial v}(F_1(r_1, r_2, r_3)) = \frac{\partial F_1}{\partial x}\frac{\partial r_1}{\partial v} + \frac{\partial F_1}{\partial y}\frac{\partial r_2}{\partial v} + \frac{\partial F_1}{\partial z}\frac{\partial r_3}{\partial v}$$

Substituting into the integrand from (i) gives

$$\frac{\partial}{\partial u}(F_1 \circ \mathbf{r})\frac{\partial r_1}{\partial v} - \frac{\partial}{\partial v}(F_1 \circ \mathbf{r})\frac{\partial r_1}{\partial u}$$
$$= \left(\frac{\partial F_1}{\partial x}\frac{\partial r_1}{\partial u} + \frac{\partial F_1}{\partial y}\frac{\partial r_2}{\partial u} + \frac{\partial F_1}{\partial z}\frac{\partial r_3}{\partial u}\right)\frac{\partial r_1}{\partial v} - \left(\frac{\partial F_1}{\partial x}\frac{\partial r_1}{\partial v} + \frac{\partial F_1}{\partial y}\frac{\partial r_2}{\partial v} + \frac{\partial F_1}{\partial z}\frac{\partial r_3}{\partial v}\right)\frac{\partial r_1}{\partial u}$$
$$= \frac{\partial F_1}{dy}\frac{\partial r_2}{\partial u}\frac{\partial r_1}{\partial v} + \frac{\partial F_1}{\partial z}\frac{\partial r_3}{\partial u}\frac{\partial r_1}{\partial v} - \frac{\partial F_1}{\partial y}\frac{\partial r_2}{\partial v}\frac{\partial r_1}{\partial u} - \frac{\partial F_1}{\partial z}\frac{\partial r_3}{\partial v}\frac{\partial r_1}{\partial u}$$
$$= \frac{\partial F_1}{\partial z}\left(\frac{\partial r_3}{\partial u}\frac{\partial r_1}{\partial v} - \frac{\partial r_3}{\partial v}\frac{\partial r_1}{\partial u}\right) - \frac{\partial F_1}{\partial y}\left(\frac{\partial r_1}{\partial u}\frac{\partial r_2}{\partial v} - \frac{\partial r_1}{\partial v}\frac{\partial r_2}{\partial u}\right)$$

**c.** $\mathbf{r}_u = \left\langle \frac{\partial r_1}{\partial u}, \frac{\partial r_2}{\partial u}, \frac{\partial r_3}{\partial u} \right\rangle$; $\mathbf{r}_v = \left\langle \frac{\partial r_1}{\partial v}, \frac{\partial r_2}{\partial v}, \frac{\partial r_3}{\partial v} \right\rangle$

$$\mathbf{r}_u \times \mathbf{r}_v = \left\langle \frac{\partial r_2}{\partial u}\frac{\partial r_3}{\partial v} - \frac{\partial r_2}{\partial v}\frac{\partial r_3}{\partial u}, \frac{\partial r_3}{\partial u}\frac{\partial r_1}{\partial v} - \frac{\partial r_3}{\partial v}\frac{\partial r_1}{\partial u}, \frac{\partial r_1}{\partial u}\frac{\partial r_2}{\partial v} - \frac{\partial r_1}{\partial v}\frac{\partial r_2}{\partial u} \right\rangle$$

Since $\mathbf{N} = \frac{1}{\|\mathbf{r}_u \times \mathbf{r}_v\|}\mathbf{r}_u \times \mathbf{r}_v$, $\frac{\partial r_3}{\partial u}\frac{\partial r_1}{\partial v} - \frac{\partial r_3}{\partial v}\frac{\partial r_1}{\partial u} = \|\mathbf{r}_u \times \mathbf{r}_v\| n_2$ and $\frac{\partial r_1}{\partial u}\frac{\partial r_2}{\partial v} - \frac{\partial r_1}{\partial v}\frac{\partial r_2}{\partial u} = \|\mathbf{r}_u \times \mathbf{r}_v\| n_3$.

Hence

$$\iint_D \left[\frac{\partial}{\partial u}\left((F_1 \circ \mathbf{r})\frac{\partial r_1}{\partial v}\right) - \frac{\partial}{\partial v}\left((F_1 \circ \mathbf{r})\frac{\partial r_1}{\partial u}\right)\right] du\, dv = \iint_D \left(\frac{\partial F_1}{\partial z}n_2 - \frac{\partial F_1}{\partial y}n_3\right)\|\mathbf{r}_u \times \mathbf{r}_v\| du\, dv$$
$$= \iint_S \left(\frac{\partial F_1}{\partial z}n_2 - \frac{\partial F_1}{\partial y}n_3\right) d\sigma$$

## Chapter 12 Review Exercises

**1.** $C$ can be described by

$\mathbf{r}(t) = \langle -2 + 3t, 3 - 7t \rangle$, $0 \le t \le 1$.

$\|\mathbf{r}'(t)\| = \|\langle 3, -7 \rangle\| = \sqrt{58}$

$$\int_C xy^2 ds = \int_0^1 (-2 + 3t)(3 - 7t)^2 \sqrt{58}\, dt = \sqrt{58}\int_0^1 (-18 + 111t - 224t^2 + 147t^3) dt = \frac{-5}{12}\sqrt{58} \approx -3.17$$

**3.** Let $\mathbf{F}(x, y, z) = \langle (2+xz)xye^{xz},\ x^2e^{xz},\ x^3ye^{xz} \rangle$.

Then $\mathbf{F} = \operatorname{grad} f$ where $f(x, y, z) = x^2ye^{xz}$.
Hence,

$$\int_C (2+xz)xye^{xz}dx + x^2e^{xz}dy + x^3ye^{xz}dz = \int_C \mathbf{F}\cdot\mathbf{T}ds$$
$$= f(\mathbf{r}(1)) - f(\mathbf{r}(0))$$
$$= f(2,\ e,\ -1) - f(1,\ 0,\ 1)$$
$$= \frac{4}{e}$$
$$\approx 1.47152$$

**5.** We use Green's Theorem to obtain

$$\int_C (\mathbf{F}\cdot\mathbf{N})ds = \int_0^3\int_0^{4-4x/3}(7x+y)\,dy\,dx = \int_0^3\left(\frac{-76}{9}x^2 + \frac{68}{3}x + 8\right)dx = 50.$$

**7.** We use spherical coordinates to rewrite

$$\iint_S (x^2+y^2)d\sigma = \int_0^{2\pi}\int_0^{\pi}\left((\cos\theta\sin\phi)^2 + (\sin\theta\sin\phi)^2\right)|\sin\phi|d\phi\,d\theta$$
$$= \int_0^{2\pi}\int_0^{\pi}\sin^3\phi\,d\phi\,d\theta$$
$$= \int_0^{2\pi}\left(\frac{\cos^3}{3}\phi - \cos\phi\right)\Bigg|_0^{\pi}d\theta$$
$$= \int_0^{2\pi}\frac{4}{3}d\theta$$
$$= \frac{8\pi}{3}$$

**9.** Using Stokes' Theorem we see that $\iint_S (\nabla\times\mathbf{F})\cdot\mathbf{N}\,d\sigma = \int_C (\mathbf{F}\cdot\mathbf{T})ds$ where $C$ is the curve $x^2+y^2=2$ in the $(x, y)$-plane. Now use Green's Theorem to get

$$\int_C (\mathbf{F}\cdot\mathbf{T})ds = \iint_D \frac{\partial}{\partial x}(3x) - \frac{\partial}{\partial y}(2y)dA = \iint_D dA = 2\pi.$$

**11.** Let $\mathbf{F}(x, y, z) = \langle 5,\ 3,\ \sqrt{2}\rangle$. Then $\mathbf{F} = \operatorname{grad} f$ with $f(x, y, z) = 5x + 3y + \sqrt{2}z$ so $\int_C \mathbf{F}\cdot\mathbf{T}\,ds = 0$ since $C$ is a closed curve.

**13.** We use the divergence theorem to rewrite

$$\iint_S \mathbf{F}\cdot\mathbf{N}\,d\sigma = \iiint_B \operatorname{div}\mathbf{F}\,dv = \int_0^3\int_{4\left(1-\frac{x}{3}\right)}^4\int_{-5\left(\frac{x}{3}+\frac{y}{4}-1\right)}^0 2x\,dz\,dy\,dx$$
$$= \int_0^3 2x\int_{4\left(1-\frac{x}{3}\right)}^4\left(-5+\frac{5}{3}x+\frac{5}{4}y\right)dy\,dx$$
$$= \int_0^3\frac{20}{9}x^3dx$$
$$= 45$$

**15.** $\nabla \times \mathbf{F} = \langle 0, 0, -3 \rangle$

$\mathbf{N} = \langle x, y, z \rangle$

Using the parametrization $\mathbf{r}(\theta, \phi) = \langle \cos\theta \sin\phi, \sin\theta \sin\phi, \cos\phi \rangle$ with $0 \le \theta \le \frac{\pi}{2}$, $0 \le \phi \le \frac{\pi}{2}$ we get

$$\iint_S (\nabla \times \mathbf{F}) \cdot \mathbf{N}\, d\sigma = \int_0^{\pi/2} \int_0^{\pi/2} -3\cos\phi \cdot \sin\phi \, d\phi \, d\theta = -\frac{3}{4}\pi.$$

**17.** Let $\mathbf{F}(x, y) = \langle P, Q \rangle$.

$$\frac{\partial Q}{\partial x} = 3x^2 \cos(x^2 y) - 2x^4 y \sin(x^2 y)$$

$$\frac{\partial P}{\partial y} = \sin(x^2 y) + 5x^2 y \cos(x^2 y) - 2x^4 y^2 \sin(x^2 y)$$

$\frac{\partial Q}{\partial x} \ne \frac{\partial P}{\partial y}$, so there is no potential function.

**19.** Let $\mathbf{F} = \langle P, Q \rangle$.

$$\frac{\partial Q}{\partial x} = \frac{-x(x^2 + 2y)^{-1/2}}{x^2 + 2y}$$

$$\frac{\partial P}{\partial y} = \frac{-x(x^2 + 2y)^{-1/2}}{x^2 + 2y}$$

$\frac{\partial Q}{\partial x} = \frac{\partial P}{\partial y}$ and the potential function is $\sqrt{x^2 + 2y} + C$.

**21.** Let $\mathbf{F} = \langle P, Q, R \rangle$.

$\frac{\partial Q}{\partial x} = y^2 e^{x+z}$ $\frac{\partial P}{\partial y} = x^2 e^{y+z}$

$\frac{\partial Q}{\partial x} \ne \frac{\partial P}{\partial y}$ so there is no potential function.

**23.** **a.** Flux across $C = \int_C \delta(x, y)(\mathbf{v} \cdot \mathbf{N}) ds$.

Let $\mathbf{F}(x, y) = \delta(x, y)\mathbf{v}(x, y) = \langle 2x^2 - x^3 + 4xy - 2x^2 y, -6xy - 2y^2 + 3x^2 y + xy^2 \rangle$

Then

$$\int_C \delta(x, y)(\mathbf{v} \cdot \mathbf{N}) ds = \int_C \mathbf{F} \cdot \mathbf{N}\, ds = \iint_D \operatorname{div} \mathbf{F}\, dA$$

$$= \iint_D ((4x - 3x^2 + 4y - 4xy) + (-6x - 4y + 3x^2 + 2xy))\, dA$$

$$= \iint_D (-2x - 2xy)\, dA$$

The symmetry of the region $D$ requires both pieces of the integral to vanish, giving us a flux of 0.

**b.** Describe $C$ by $\mathbf{r}(\theta) = \langle \cos\theta,\ \sin\theta \rangle,\ 0 \le \theta \le 2\pi$.

$$\int_C \mathbf{F}\cdot\mathbf{T}\,ds = \iint_D \left(\frac{\partial Q}{\partial x} - \frac{\partial P}{\partial y}\right) dA$$

$$= \iint_D ((-6y + 6xy + y^2) - (4x - 2x^2))\,dx\,dy$$

$$= \int_{-1}^{1}\int_{-\sqrt{1-y^2}}^{\sqrt{1-y^2}} (2x^2 + y^2 + 6xy - 4x - 6y)\,dx\,dy$$

$$= \frac{3}{4}\pi$$

**25.** Write $B$ as the union of $B_1,\ B_2,\ \ldots B_n$.
Let $R$ be the union of $B$ with $W$, the region bounded by $S_1$. By the divergence theorem we have

$$\iint_S \mathbf{F}\cdot\mathbf{N}\,d\sigma = \iiint_R \operatorname{div}\mathbf{F}\,dV = \left(\sum_{i=1}^{n}\iiint_{B_i} \operatorname{div}\mathbf{F}\,dV\right) + \iiint_W \operatorname{div}\mathbf{F}\,dV = \iiint_B \operatorname{div}\mathbf{F}\,dV + \iint_{S_1} \mathbf{F}\cdot\mathbf{N}\,d\sigma,$$ as required.

**27. a.** $E(\mathbf{r})$ proportional to charge $c$ gives us a factor of $c$. $E(\mathbf{r})$ inversely proportional to square of distance gives us a factor of $\frac{1}{\|\mathbf{r}\|^2}$.

The direction of the force is in unit direction $\frac{\mathbf{r}}{\|\mathbf{r}\|}$ or its opposite.

Thus $\mathbf{E}(\mathbf{r}) = \frac{kc\mathbf{r}}{\|\mathbf{r}\|^3}$ where $c$ is charge and $k$ is a proportionality constant.

**b.**

**c.**

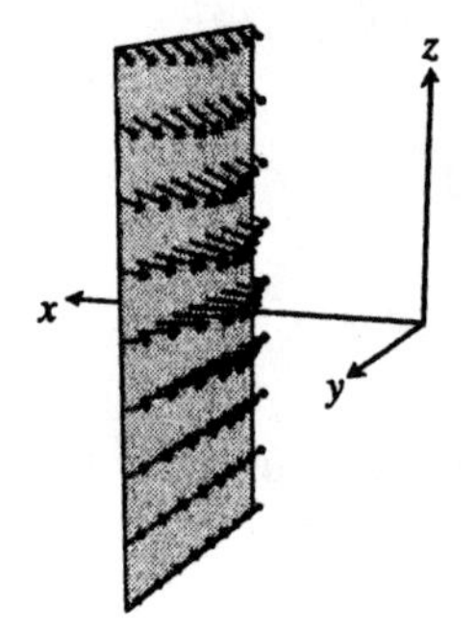

**d.** Flux is $\iint_S \mathbf{E}\cdot\mathbf{N}\,d\sigma$.

Describe $S$ by $\mathbf{r}(\theta,\ \phi)=\langle R\cos\theta\sin\phi,\ R\sin\theta\sin\phi,\ R\cos\phi\rangle$ with $0\le\theta\ne 2\pi,\ 0\le\phi\ne\pi$.

Then

$$\iint_S \mathbf{E}\cdot\mathbf{N}\,d\sigma=\int_0^{\pi}\int_0^{2\pi}\frac{kc\mathbf{r}}{\|\mathbf{r}\|^3}\cdot\frac{\mathbf{r}}{\|\mathbf{r}\|}R^2\sin\phi\,d\theta\,d\phi=\int_0^{\pi}\int_0^{2\pi}kc\sin\phi\,d\theta\,d\phi$$

since $\|\mathbf{r}\|=R$.

This result does not depend on $R$.

**e.** Let $S$ be the boundary surface for $B$.

$$\operatorname{div}\mathbf{F}=kc\frac{\left(x^2+y^2+z^2\right)^{3/2}-x\left(\frac{3}{2}\left(x^2+y^2+z^2\right)^{1/2}\cdot 2x\right)}{\left(x^2+y^2+z^2\right)^3}$$

$$+kc\frac{\left(x^2+y^2+z^2\right)^{3/2}-y\left(\frac{3}{2}\left(x^2+y^2+z^2\right)^{1/2}\cdot 2y\right)}{\left(x^2+y^2+z^2\right)^3}$$

$$+kc\frac{\left(x^2+y^2+z^2\right)^{3/2}-z\left(\frac{3}{2}\left(x^2+y^2+z^2\right)^{1/2}\cdot 2z\right)}{\left(x^2+y^2+z^2\right)^3}$$

$$=kc\frac{3\left(x^2+y^2+z^2\right)^{3/2}-\left(3x^2+3y^2+3z^2\right)\left(x^2+y^2+z^2\right)^{1/2}}{\left(x^2+y^2+z^2\right)^3}$$

$$=0$$

If the origin is not inside $S$ then $\iint_S \mathbf{F}\cdot\mathbf{N}\,d\sigma=\iiint_B \operatorname{div}\mathbf{F}\,dV=0.$

If the origin is inside $S$ then let $S_1$ be the boundary of a sphere around the origin completely inside $S$ and use Exercise 25 to see that flux of $S$ is same as flux through any sphere centered at the origin.

**29.** Applying Stokes' Theorem we get

$$\int_C \mathbf{E}\cdot\mathbf{T}\,ds=\iint_S(\nabla\times\mathbf{E})\cdot\mathbf{N}\,d\sigma$$

Comparing the integral on the right with the integral in the exercise gives

$$\iint_S\left((\nabla\times\mathbf{E})+\frac{\partial\mathbf{B}}{\partial t}\right)\cdot\mathbf{N}\,d\sigma=0.$$

This will hold for any smooth orientable surface bounded by $C$, thus $\left((\nabla\times\mathbf{E})+\frac{\partial B}{\partial t}\right)\cdot\mathbf{N}=0.$

Now consider a variety of smooth orientable surfaces bounded by $C$ and passing through a point $P$. The surface can be changed so that it still passes through $P$ and is bounded by $C$, but where direction of the normal vector $\mathbf{N}$ has changed. Hence, $\left((\nabla\times\mathbf{E})+\frac{\partial B}{\partial t}\right)\cdot\mathbf{N}=0$ for any unit vector $\mathbf{N}$, and thus

$$(\nabla\times\mathbf{E})+\frac{\partial B}{\partial t}=\mathbf{0}.$$